TECTONICS

Edited by **Damien Closson**

INTECHOPEN.COM

Tectonics
Edited by Damien Closson

Published by InTech

Janeza Trdine 9, 51000 Rijeka, Croatia

Edition 2014

Publishing Process Manager Ana Nikolic

Technical Editor Teodora Smiljanic

Cover Designer InTech Design Team

Image Copyright Jarno Gonzalez Zarraonandia, 2010. Used under license from Shutterstock.com

Additional hard copies can be obtained from orders@intechopen.com

Tectonics, Edited by Damien Closson
p. cm.
ISBN 978-953-307-545-7

INTECH

open science | open minds

Contents

Preface

The term tectonics refers to the study dealing with the forces and displacements that have operated to create structures within the lithosphere. The deformations affecting the Earth's crust are the result of the release and the redistribution of energy from Earth's core. Orogenesis and epeirogenesis are two basic types of lithosphere deformations (or tectonism). Orogenesis involves the formation of mountain ranges by means of folding, faulting, and volcanic activity while epeirogenesis takes the form of either uplift or subsidence. The concept of plate tectonics is the chief working principle. Three main categories of tectonics can be distinguished: extensional, thrust, and strike-slip. The subject area is wide and concerns also the development of cratons and terranes, the earthquakes and volcanic belts, igneous processes and metamorphism, erosion patterns in geomorphology etc. Neotectonics is a sub-discipline dedicated to the motions and deformations of the Earth's crust which are current or recent in geologic time. The corresponding time frame is referred to as the neotectonic period and the preceding time as palaeotectonic period. Of course, tectonics has also application to lunar and planetary studies, whether or not those bodies have active tectonic plate systems. Petroleum and mineral prospecting uses this branch of knowledge as guide. An entire library would be necessary to discuss such a broad topic. Indeed, the present book is restricted to the structure and evolution of the terrestrial lithosphere with dominant emphasis on the continents. Thirteen original scientific contributions highlight most recent developments in seven relevant domains. They are:

- "Plates amalgamation and plate destruction, the western Gondwana history" presents a recent model completed with further insights on the Post Pan-African evolution of the belt and inferences to further developments of the western Gondwana.

- Three contributions are dedicated to the tectonics of Europe and the Near East: a) "Lithospheric structure and tectonics of the Eastern Alps – evidence from new seismic data". This article is a contribution to generalized geologic interpretations and tectonic processes inferred on the basis of the imaged structures; b) "Structure and plate tectonic evolution of the northern Outer Carpathians" examines geophysical and geological borehole, surface data, and recently published tectonic work; c) "Tectonic Model of the Sinai Peninsula Based on Geophysical Investigations" provides a kinematic model for Sinai Peninsula based on an integrated analysis of the most recent magnetic, seismic and GPS data.

- "Siberia - From Rodinia to Eurasia" describes about one billion years of the tectonics of Siberia.

- Three studies are devoted to the tectonics of China and its neighbourhood: a) "Proto-Basin Types of North China Craton in Late Triassic and Its Implication for Regional Tectonics of Initial Craton Destruction". The study brings forward new understandings to the proto-basin types, tectonic deformation division and tectonic transition time-limit; b) "Tectonic Implications of Stratigraphy Architecture in Distal Part of Foreland Basin, Southwestern Taiwan" provides a detailed analysis of the subsidence curves to give a tectonic mode of epeirogenic movement; c) "Seismic hazard in Tien Shan: basement structure control over the deformation induced by Indo-Eurasia collision" aims at the reconstruction of the tectonic history of the Central Asia and at the assessment of the role of the inherited basement structure on the recent earthquake events based on the Cenozoic structural position of the Precambrian micro-continents.

- Two chapters discuss on advanced concepts: a)"Lithosphere as a nonlinear system: geodynamic consequences" analyses the results obtained when the lithosphere is considered to be a nonlinear dynamic system with dominant fractal structures formation; b) "Layer-block tectonics, a new concept of plate tectonics—an example from Nansha Micro-plate, southern South China Sea" describes the layer-slip structure of a continental lithosphere plate considered as composed of sub-plates connecting with each other horizontally and overlapping with each other vertically.

- In the frame of neotectonics, two investigation techniques are examined: "The Role of Geoelectrical DC Methods in Determining the Subsurface Tectonics Features: Case Studies from Syria" and "Salt tectonics of the Lisan diapir revealed by Synthetic Aperture Radar images".

- Finally, the relation between tectonics and petroleum researches is illustrated by "Mantle-like trace element composition of Petroleum – contribution from serpentinizing peridotites". In this chapter, chemical analyses have lead to evidence indicating a relationship between mantle and petroleum. This process occurred on an unknown scale.

Damien Closson
Signal and Image Centre,
Belgium

Part 1

Gondwana History

1

Plates Amalgamation and Plate Destruction, the Western Gondwana History

Ngako Vincent[1] and Njonfang Emmanuel[2]
[1]*Mega Uranium Corporation Cameroon PLC (MUCC)*
[2]*Laboratoire de Géologie, Ecole Normale Supérieure, Université de Yaoundé I, Yaoundé, Cameroun*

1. Introduction

Reconstruction of the western Gondwana tectonic setting is a highly challenging topic. Orogenic belts are generally outlined by mountain ranges or the so commonly called "mobile zones" in old shields. They define linear belts following the suture of two plates amalgamated during collision and whose boundaries are usually clearly identifiable. Most often, the original geotectonic settings of the mobile zone are also clearly outlined by sedimentary, magmatic, metamorphic, structural and isotopic zonings that render the reconstructions of the belt relatively easy. Contrastingly, the Pan-African/Brasiliano tectonic setting in western Gondwana corresponds to a broad area that shows no specific orientation and, except at the eastern border of the West African Craton (WAC), plate boundaries, belt polarity and zoning are intensely obliterated by the branching geometry of the regional strain, the intense and widespread deformation, migmatization and granitization, particularly at its extensions towards the northwestern Cameroon and eastern Sahara (Fig.1). These particular features render geotectonic interpretations of the Pan-African mobile zone extremely complex. Thus, the tectonics of western Gondwana appears as a controversial topic and a matter of contention, both in old and recent studies (e.g. Paulsen et al., 2007; Tucker et al., 2007; Goscombe & Gray, 2008; Vaughan & Pankhurst, 2008; Casquet et al., 2008). Indeed, despite the existence of good petrologic and isotopic data and a relatively clear definition of the regional geochrono-stratigraphy in the Pan-African domain of Cameroon and northwestern regions, the lack of detailed structural studies has generated controversial tectonic models. This intermediate domain between the West African Craton (WAC), the São Francisco- Congo Craton (SFCC) and the enigmatic east Saharan craton is currently interpreted either as a Dahomeyan-related basin and range province including meridian directed troughs (Affaton et al., 1991; Castaing et al., 1994), a Neoproterozoic branching belt including numerous micro-plates (Trompette, 1994 and references therein; Toteu et al., 2004; Penaye et al., 2006), or a tectonic indent resulting from the collision between the eastern Sahara rigid prong (ESB) and the northern active margin of the SFCC differentiated into a basin and range province overriding a subduction zone (Ngako et al., 2008). We present in this chapter the more recent model completed with further insights on the Post Pan-African evolution of the belt and inferences to further developments of the western Gondwana.

This study involves modern structural analysis and concepts (phase chronology, strain analysis, deformation and kinematics), petrology (metamorphic parageneses and phase equilibrium, relation with deformation, geothermobarometry), geochemistry (major and trace elements) and geochronology. Data from regional literature including geophysical data are also abundantly used for synthesis and re-interpretation. The most important achievements include:

1. A new tectonic model for the Pan-African/Brasiliano belts; this model includes many original aspects and developments with regards to the collisional setting, tectonics, metamorphic and magmatic processes. In particular, a three-plate model is proposed, and the role of the asthenosphere on crustal melting and reactivation during collision is emphasized.
2. Evidence of a tectonic indent in the central Nigeria-Cameroon domain, inferred from the regional strain field; indentation in central-west Africa strongly suggests that the East Saharan Block (ESB) evolved originally as a rigid prong prior to crustal melting, batholiths emplacement and subsequent dismantling of the prong during the late Pan-African evolution.
3. Clear distinction between Pan-African and Post Pan-African tectonic and magmatic processes which raises the question on the relevance of integrating Phanerozoic lithospheric structures inferred from geophysical data (gravity, seismic tomography, etc.) in the interpretation and reconstruction of the Pan-African tectonic evolution.

Future research would be focused on the Central Cameroon Shear Zone (CCSZ) kinematics and the characterization of vestiges of the dismantled SFCC/ESB cratons and hypothetic pre-tectonic boundaries.

2. Plates amalgamation in western Gondwana

2.1 Orogenic setting and major plate boundaries

2.1.1 The Sao-Francisco-Congo Craton (SFCC) northern boundary

Recent petrologic and isotopic data enable define three Pan-African main geotectonic units at the northern border of the SFCC: the Poli Group (northern Cameroon), the Adamawa domain (central Cameroon) and the Yaoundé Group (southern Cameroon). They likely represent an original basin and range province (Fig.1 & Fig.2).

The Poli Group represents an early Neoproterozoic back-arc basin formed between 830 and 665 Ma (Toteu, 1990; Penaye et al., 2006; Toteu et al., 2006), that includes: detrital and volcaniclastic deposits, metavolcanics (tholeitic basalts and calc-alkaline rhyolites), and pre-, syn- to late-tectonic calk-alkaline intrusions (diorites, granodiorites and granites). These intrusions, emplaced between 660 and 580 Ma (Toteu et al., 2001) form a NNE-SSW corridor referred to as the West Cameroon Domain or WCD (Fig. 2; Toteu et al., 2004; Penaye et al., 2006). They show a north to south potassium increase from LKT to shoshonites that suggests the existence of a southward dipping suture zone to the North (Njonfang et al., 2006). The Massenya-Ounianga (MO) heavy gravity line in Chad (mean values bracketed between 10 and 30 mGal; Poudjom Djomani et al., 1995) likely materializes obducted fragments of an oceanic floor onto the opposite cryptic margin, as suggested by its superimposition upon thick effective elastic thickness curves (Poudjom Djomani et al., 1995; Fig. 2). The concave geometry of these curves to the North suggests a northward extension of a hypothetic paleo-plate in eastern Sahara, represented by inferred cratonic areas located there (Küster & Liegeois, 2001; Fig.1 & Fig.2).

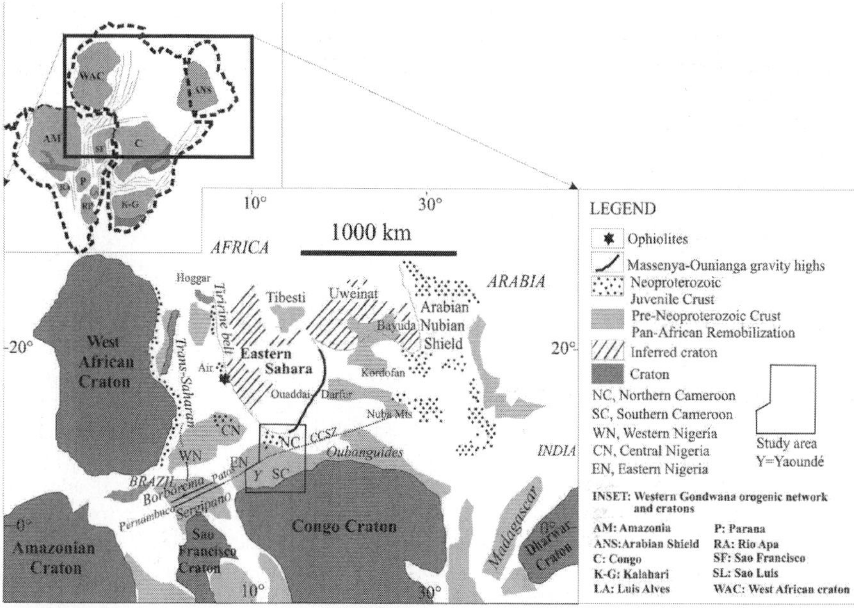

Fig. 1. Sketchy tectonic map of western Gondwana and main geotectonic units (redrawn from Küster & Liegeois, 2001)

The Adamawa domain includes huge Pan-African batholiths and large scale Paleoproterozoic remnants that were emplaced and metamorphosed during the Pan-African tectonic evolution. They represent the basement of the Poli and Yaoundé Groups and likely correspond to the northern and reactivated extension of the Eburnean Nyong series (north-western corner of the Congo craton) within the mobile zone (Fig.2).

The Yaoundé Group represents a syntectonic basin which deposition age is younger than 625 Ma (U-Pb dating of detrital zircons, Toteu et al., 2006). It includes meta-sediments (shales, greywackes, minor quartzites, dolomites and evaporites) and pre- to syn-tectonic alkaline to transitional intrusions represented by pyroxenites, pyroxene-amphibole and pyroxene-plagioclase rocks (Nzenti, 1998); scarce outcrops of serpentinized ultramafic rocks associated with gabbros, diorites and mafic dykes are also found (Seme Mouangue, 1998; Nkoumbou et al., 2006). All these rocks were metamorphosed in the HP-HT granulite facies at ca 620 Ma (U-Pb zircon age by Penaye et al., 1993) prior to exhumation and thrusting over the Congo craton.

2.1.2 The West African Craton (WAC) eastern boundary

The eastern border of the WAC (Fig.1) contains several Archaean to Proterozoic inliers, strongly affected by the Pan-African orogeny. For example, the Tuareg Shield, comprising the Hoggar domains, includes Archaean terranes possibly amalgamated during the Eburnean Orogeny and intruded by 640-580 Ma Pan-African granites. Collision of this Shield with the west African Craton is recorded in the strongly deformed Pharusian Belt which includes ophiolites and arc-related volcanic rocks (Begg et al., 2009 and references

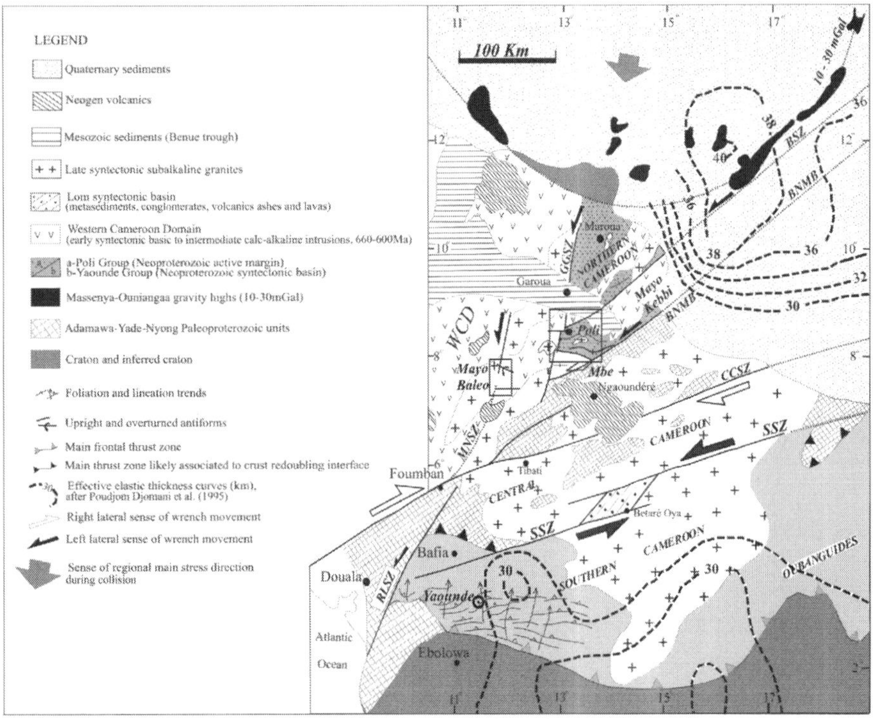

Fig. 2. Structural map of the Cameroon domain (see location in Fig.1)

therein). The WAC boundary is then outlined by a NS-directed linear suture zone revealed by the presence of a high gravity line and the occurrence of ultramafic rocks (ophiolites) straddling parallel to the craton edge in association with ultra-HP metamorphic rocks (eclogites); these features characterize an inter-plates boundary. They mark the transition between the intracratonic Gourma trough and Volta platform deposits and their metamorphic equivalents within internal zones of the Pan-African Trans-Saharan belt.

2.1.3 The Eastern Sahara Block boundary

Except in the Tibesti Archean area (Abdelsalam et al., 2002), the Eastern Sahara Block (ESB, Fig.1) mostly exhibits Pan-African imprints (metamorphism and intrusive) in contrast with the WAC and the SFCC cratons. However, the presence of a stable continental block in eastern Sahara prior to Pan-African events is inferred from the relative position of the Pan-African active margins and back-arc associated volcanics and intrusive rocks in Poli and Central Nigeria (Ferré et al., 2002), the occurrence of an ophiolite complex in SE Air at the southern extension of the Tirririne belt, and the evidence of thick effective elastic thickness characterizing high strength and old tectonothermal ages of the lithospheres across the Massenya-Ounianga high gravity line (Poudjom Djomani et al., 1995). These features suggest the presence of a nearby inter-plates boundary between the Cameroon-Nigeria-Hoggar polycyclic domain and the eastern Sahara (Fig.1 & Fig.2).

2.2 Collision and post-collision evolution

Three main tectonic events related to Pan-African collision and post-collision evolution have been identified in Cameroon: i) crustal thickening; ii) left lateral wrench movements; and iii) right lateral wrench movement, successively.

2.2.1 Crustal thickening (ca 630-620Ma)

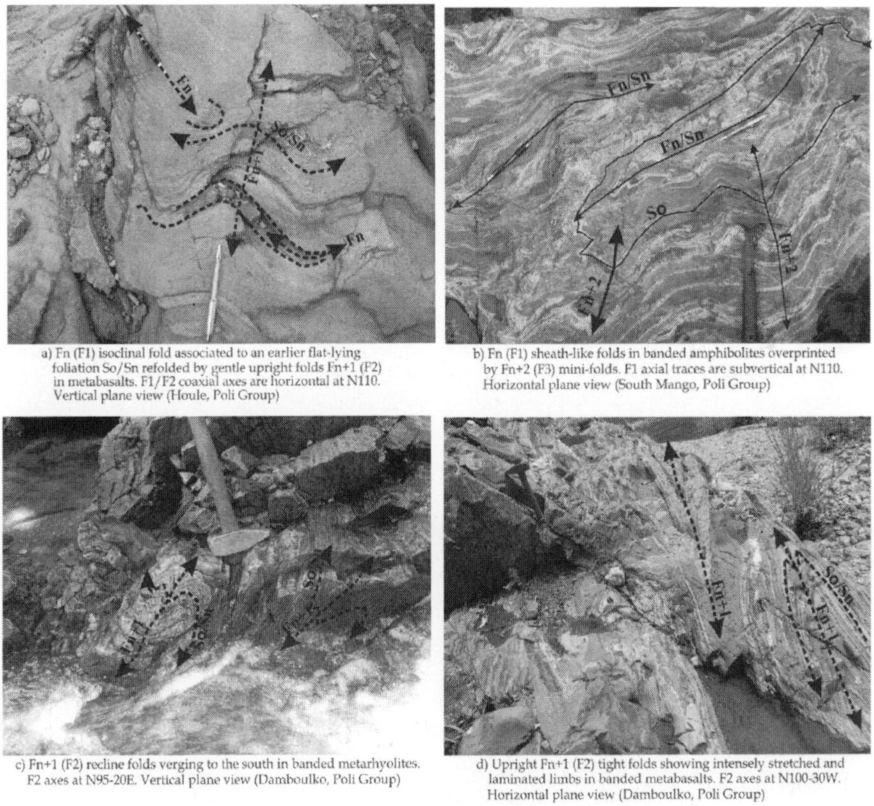

a) Fn (F1) isoclinal fold associated to an earlier flat-lying foliation So/Sn refolded by gentle upright folds Fn+1 (F2) in metabasalts. F1/F2 coaxial axes are horizontal at N110. Vertical plane view (Houle, Poli Group)

b) Fn (F1) sheath-like folds in banded amphibolites overprinted by Fn+2 (F3) mini-folds. F1 axial traces are subvertical at N110. Horizontal plane view (South Mango, Poli Group)

c) Fn+1 (F2) recline folds verging to the south in banded metarhyolites. F2 axes at N95-20E. Vertical plane view (Damboulko, Poli Group)

d) Upright Fn+1 (F2) tight folds showing intensely stretched and laminated limbs in banded metabasalts. F2 axes at N100-30W. Horizontal plane view (Damboulko, Poli Group)

Fig. 3. Fold generations and geometry in Northern Cameroon (Poli Group)

The early Pan-African tectonic evolution includes thrusting and shortening that resulted in crustal thickening. In the Poli region, E-W antiform and synform characterize gentle folding of a regional flat-lying foliation and lineation probably formed during an early thrust evolution (Fig. 3a, c and d). Sheath-like folds coeval with the early foliation materialize intense shear deformation during thrusting (Fig. 3b). The restoration of the folded mineral lineation in primary horizontal position using cyclographic methods enable approximate the kinematic direction to NS during the thrust evolution (Ngako et al., 2008). The HP-HT metamorphic parageneses formed in metapelites, metagranodiorites and basic granulites during thrusting were further retrograded in the amphibolite facies and metamorphic evolution terminated with synkinematic growth of garnet porphyroblasts in the greenschist facies (Ngako et al., 2008; Bouyo et al., 2009).

At the regional scale, subsequent folding of the nappe resulted in the forming of upright to recline folds. In the Poli region, their geometry varies eastwards, from open to tight profiles, correlatively with a progressive change of the axial plane cleavage from crenulation to mylonitic at the contact zone with the Buffle Noir–Mayo Baléo Shear Zone (BNMB). Further to the South, in the Mayo Baléo region, these folds are overturned westwards, indicating large scale thrust movements coeval with E-W stretching lineations (Ngako et al., 2008). Pb-Pb evaporation ages of ca 600±27 Ma on monozircons from syn-thrust intrusions in Mayo Baléo give an age approximation, though imprecise, of this second folding phase (Ngako, 1999).

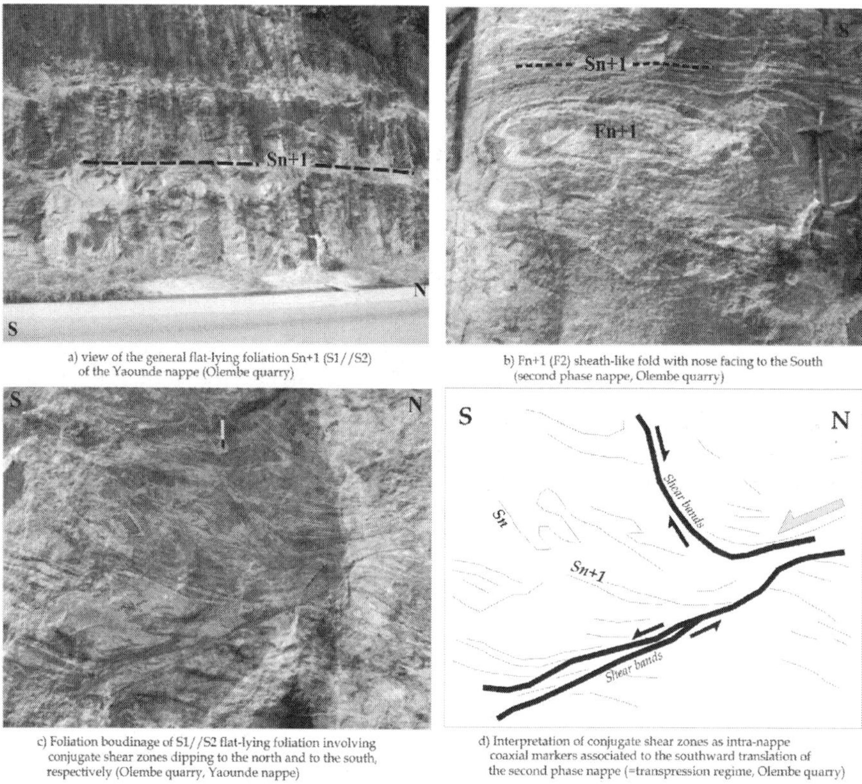

a) view of the general flat-lying foliation Sn+1 (S1//S2) of the Yaounde nappe (Olembe quarry)

b) Fn+1 (F2) sheath-like fold with nose facing to the South (second phase nappe, Olembe quarry)

c) Foliation boudinage of S1//S2 flat-lying foliation involving conjugate shear zones dipping to the north and to the south, respectively (Olembe quarry, Yaounde nappe)

d) Interpretation of conjugate shear zones as intra-nappe coaxial markers associated to the southward translation of the second phase nappe (=transpression regime, Olembe quarry)

Fig. 4. Internal structure of the Yaoundé nappe and inferred kinematics (Olembe quarry)

Further to the South in the Yaoundé Group, the second phase recline folding includes sheath-like folds associated to horizontal stretching lineation at N20 coeval with foliation boudinage and conjugate shear zones (Fig. 4). These structures suggest interaction between pure and simple shear deformations compatible with late transpressional evolution of the Yaoundé granulites. Metamorphic conditions during the granulite evolution culminate at

10-12 kbar/800°-850°C (Barbey et al., 1990) which are the P-T conditions prevailing at the base of the crust. Thus, the early nappe stacking in the Cameroon domain in whole is compatible with crust redoubling and burial of the Yaoundé Group. Concordant ages of ca 630 ± 5 Ma (Toteu et al., 1990) and ca 620±10 Ma (Penaye et al., 1993) on syn-metamorphic zircons in the Poli micaschists and in the Yaoundé granulites give a good age approximation of this early thrust event, suggesting a north to south successive stacking at regional level.
Reversely, the gradual P-T decrease recorded during the second phase folding (Barbey et al., 1990; Mvondo et al., 2003), suggests that the Yaoundé HP-HT granulites were uplifted (exhumation) during regional shortening. Prominent features during this evolution include: (i) a reverse metamorphic gradient marked by northwards increase of metamorphic intensity from the Mbalmayo schist situated at the sole of the nappe, near to the contact zone with the craton towards the Yaoundé migmatites and gneisses; and, (ii) a southwards increase of the deformation intensity, marked by the progressive decrease of the intersection angles between early foliation and thrust plane from the Yaoundé migmatites to the Mbalmayo schists. This angle varies from nearly 45° (partial rotation of foliation into parallelism to thrust plane in the Yaoundé gneisses), to 0° (foliation//thrust plane) in the Mbalmayo schists where C'-type shear zones of Berthé et al. (1979) are widespread. P-T conditions range from 4-9 kbar/675-575°C in the Mbalmayo schists to 7-12 kbar/ 800°-550°C in the Yaoundé migmatites (Mvondo et al., 2003). An alternative interpretation of the exhumation of HP-HT granulites in the Yaoundé Group suggests it is linked with late Pan-African extensional evolution (Mvondo et al. 2003, 2007); this interpretation is neither compatible with the important crust thickness (38-42 km) in the Yaoundé region (Poudjoum Djomani et al., 1995), nor with the Adamawa Tertiary uplift situated far to the north as suggested by the authors.
The Rocher du Loup shear zone (RLSZ, Fig. 2) represents a lateral ramp accommodating late horizontal displacements parallel to the Nyong rigid promontory during thrust movements onto the Congo craton (Jégouzo, 1984). This promontory would have obliterated the reverse limb of associated large scale recline fold and correlative metamorphic isograds as suggested by a recent study of the Boumyebel talcshist (Yonta-Ngoune et al., 2010). U-Pb ages of a syn-shear syenite yield ca 590 Ma (Toteu et al., 1994), that approximately date the wrench and thrust evolution in southern Cameroon.

2.2.2 Wrench evolution

Early Pan-African folds and nappes in Cameroon are cross-cut by a complex shear zone network with complex kinematics.

2.2.2.1 Left lateral wrench movements (613- 585 Ma)

2.2.2.1.1 The northern and central Cameroon shear zones

These shear zones form two main sets represented by major and synthetic shear zones directed N50°E and N-S, respectively (Fig. 2). Major shear zones include the Balché (BSZ) and the Buffle Noir-Mayo Baléo Shear Zones (BNMB), whereas synthetic shear zones are represented by the Godé-Gormaya (GGSZ) and the Mayo Nolti Shear Zones (MNSZ). The Sanaga fault in central Cameroon is a N70-directed SZ marking a major structural boundary between the northern Cameroon characterized by intense shear deformation, and the southern Cameroon where equivalent regional structures are quite absent. An exhaustive study of these shear zones is provided in Ngako et al. (2008), and only the most illustrative per set is presented in this section.

2.2.2.1.1.1 The Buffle Noir – Mayo Baléo (BNMB) Shear Zone

The BNMB Shear Zone is the most representative major sinistral shear zone. This shear zone (Fig. 2) marks the south-eastern contact between the Poli Group and the gneiss units. It cross-cuts the gneisses in the Mbé area where a three-dimension exposure may be observed; the Mbé quarry (Fig. 5) displays alternating metric bands of coeval steep dipping mylonites and upright folds (Fig. 5a). In the YZ plane, various fold profiles (open, tight, sheath-like, crenulated, dissymmetric and sigmoid) can also be observed (Fig. 5 b and c). The fold axes are parallel to mineral stretching lineation and plunge between 20-35° to SW. Most kinematic markers associated with the mylonite bands are foliation sigmoid shapes, σ-type

a) Internal morphology of the N50-directed BNMB shear zone. Note the sharp tectonic contact between the SZ marked by vertical mylonites C/Sn+1/S2 (left) and the refolded Sn (S1) gneiss foliation (right); Fn+1 (F2) plunge 20-30SW. Mbe quarry (southeastern contact of the Poli Group)

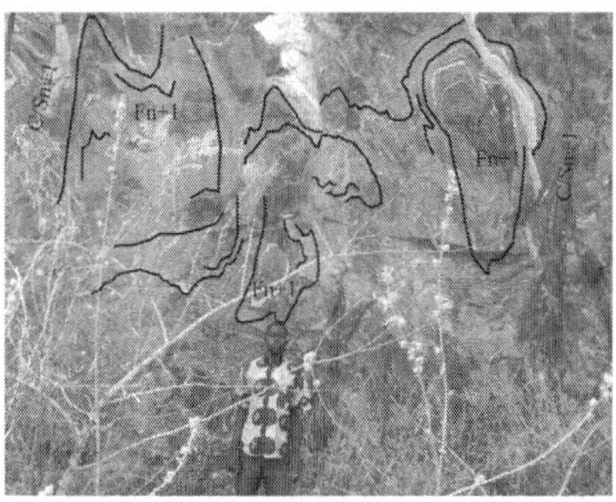

b) Complex association of Fn+1 (F2) sheath-like folds in the BNMB shear zone. Mbe quarry

c) Vertical mylonites exhibiting tight Fn+1 (F2) folds or isolated hinges; also note opened section of a sheath-like fold to the right. Mbe quarry

Fig. 5. Internal structure of the BNMB shear zone (Mbe quarry, Northern Cameroon)

porphyroclasts of feldspar, C'-type fabrics and asymmetric boudins of quartzo-feldspathic veins. All these markers are consistent with a bulk sinistral shear movement associated with a moderate normal component along the shear zone. At the regional scale, the general N50 trend of the foliation shows large scale S-type folds that suggest late clockwise rotation of the mylonite band within potential shear zones striking EW (Ngako et al., 2008). A U-Pb sphene age of 580±11 Ma from a mylonitized hornblende-biotite gneiss in the Mbé area was interpreted as dating the sinistral shear evolution (Toteu et al., 2004).

2.2.2.1.1.2 The Godé-Gormaya Shear Zone (GGSZ) and The Mayo Nolti Shear Zone (MNSZ)

The BSZ and BNMB are cross-cut to the West by the GGSZ and the MNSZ, respectively (Figs. 2). Although differently oriented, these synthetic (secondary) SZ are coeval with the major SZ, as both systems show similar senses of shear movement and similar metamorphic evolution. Secondary SZ are almost parallel to N-S batholiths represented by pre-, syn- and late-tectonic intrusions oblique to the major shear zones. The GGSZ shows the most complex structure among this generation.
It is outlined by an intensely deformed meridian belt which intersects with the Balché mylonite fan to the West (Fig. 2). It shows a right-lateral stepping due to segmentation by the Vallée des Roniers (VRSZ) and the Demsa (DSZ) shear zones. In the Godé segment, the Balché mylonites have been variably transposed into N-S direction parallel to small scale meridian SZ, in the amphibolite and greenschist facies, respectively (Toteu et al., 1991). In the Gormaya segment, an E-W cross-section enables define two metamorphic and structural zones: a western garnet-biotite zone (gneisses and migmatites) and an eastern epidote-chlorite zone overprinting the former. These zones exhibit different directions of mineral lineation, meridian in the garnet-biotite and E-W in the epidote-chlorite zones. Kinematic markers in the gneisses and migmatites are typical Z-type sigmoid foliation, compatible with sinistral shear movement, whereas the epidote-chlorite zone exhibits S$\wedge$C-type markers (Berthé et al., 1979) indicating a down-slip normal-shear movement. Both, S and C planes in the garnet-biotite zone are coeval with the migmatization, suggesting they were formed during a progressive shear evolution, according to the model of Brun and Choukroune (1981). The superimposition of NS bands of epidote-chlorite upon garnet-biotite mylonites in the Gormaya and Godé segments and their association with EW steep plunging and NS sub-horizontal lineations respectively, suggest successive compressional and extensional evolution of these parallel structures and associated metamorphic zones.

2.2.2.1.1.3 The Sanaga Shear zone

The Sanaga Shear Zone is a major crustal ductile fault identified by remote sensing technique (Dumont, 1986). This shear zone strikes N70°E and likely extends into the Central African Republic. In the Lom region, it intersects with N40 sinistral en-echelon shear zones interpreted as coeval with a pull-apart basin (Ngako et al. 2003) and the deposition of the Lom Group. They mark a step-over between the Sanaga shear zone and the Bouzoum-Ndélé fault which likely represents its eastern equivalent reactivated during Cretaceous in Central African Republic. Deposition of the Lom Group bracketed between 600 and 613 Ma (Toteu et al., 2006) probably dates the sinistral wrench movement in the Lom region.

2.2.2.2 Right lateral wrench movement (ca 585-540 Ma).

The Cameroon domain was dextrally rotated during the last Pan-African tectonic event. Rotation was mostly recorded by shear movements in the CCSZ, but also in northern

Cameroon by E-W directed shear zones including: the 'Vallée des Roniers Shear Zone', the 'Demsa Shear Zone (Fig. 2) and potential shear zones defined by S-type foliation trends in the Mbé region.

2.2.2.2.1 The northern Cameroon dextral Shear Zones

In the Poli region (Fig. 2), the 'Vallée des Roniers' (VRSZ) and the 'Demsa' Shear Zones (DSZ) are coeval with down-slip movements parallel to the Godé and Gormaya segments; both E-W and N-S mylonites determine a complex transtensive system formed in the greenschist facies characterized by chlorite-epidote paragenesis. These shear zones seem to be coeval with the Z-shaped rotation of the synform and antiform axis in the Poli region, associated with NS crenulations cleavage Sn+2/S3 (Fig. 6). Rb-Sr ages at 540-538 Ma (Bessoles & Trompette 1980, and the biotite Rb-Sr ages therein quoted) obtained in the Gouna massifs give a rough age approximation of this late tectonic event in the Poli region.

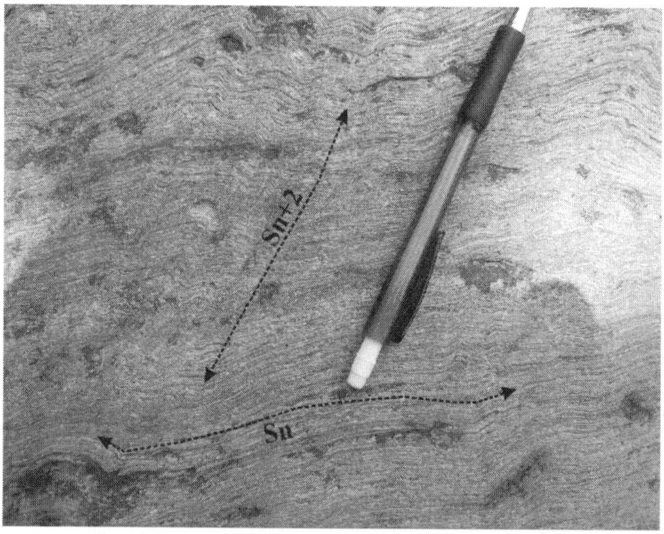

Fig. 6. Crenulation cleavage associated to Late Pan-African folding (Fn+2 or F3) in northern Cameroon (Poli Group)

In the Mbé area (Fig. 2), foliation strikes of the BNMB shear zone show S-type geometry compatible with dextral shear movement (Ngako et al., 2008). These shear zones occurred under amphibolite to greenschist facies metamorphic conditions characterized by biotite ± amphibole ± chlorite ± muscovite ± epidote ± albite.

2.2.2.2.2 The Central Cameroon Shear Zone (CCSZ)

The CCSZ is a N70°E striking major crustal structure parallel to the Sanaga fault and extending from the Soudan region into NE Brazil. It defines a fan-geometry in central Cameroon, due to its interaction with N40°E directed shear zone system (Ngako et al., 2003). Most works in Cameroon and NE Brazil define the CCSZ and correlative extensions as a dextral shear zone (Ngako et al., 1991; Neves & Mariano, 1999; Njanko et al., 2006; Njonfang et al., 2006; Nzenti et al., 2006). However, recent works in Cameroon have also reported the occurrence of sinistral shear sense indicators associated to the same movement direction as that revealed by the dextral shear markers (Ngako et al., 2003; Njonfang et al., 2006), suggesting a more complex kinematic evolution of the CCSZ. Else, superimposition of opposite sense of shear movement at constant direction (Wenberg, 1996) has been debated in the Achankovil shear zone, southern India (Ishii et al., 2006) and reassessed (Rajesh & Chetty, 2006) through an integrated approach including remote sensing data, shaded relief topo-maps and detail field study, in terms of an initial non-coaxial deformation reactivated and superimposed by opposite kinematics. A Similar evolution could be applied to the CCSZ as already proposed by Njonfang et al. (2006). P-T conditions along the CCSZ were estimated at 3.7-5.7 kb and 738-800°C (Njonfang et al., 1998) and are consistent with the absence or scarcity of actinolite and actinolitic hornblende, symptomatic of greenschist facies, within the mylonitic zones (Njonfang et al., 1998, 2006; Njanko et al., 2006). At the regional level, the NW-SE stress direction coeval with the dextral shear evolution of the CCSZ is compatible with the N-S to NNE-SSW direction of large-scale upright open folds mostly reported in the Yaoundé Group where they show horizontal to northward gentle plunging axis (Mvondo et al., 2003) and in the Meiganga region (Ganwa, 2005).

3. Summary of tectonic evolution in the Cameroon domain- Conclusion

The Cameroon domain recorded three major tectonic events during Pan-African evolution (Fig. 7). Crustal thickening generated southward thrust zones that resulted in an initial burial of the Yaoundé Neoproterozoic Group at the base of an overlying crust situated to the North (Fig. 7b); HP-HT granulites coeval with this early evolution were subsequently exhumed and over-thrust onto the Congo craton, during the late stages of the thrust evolution (Fig. 7c). During the post-collisional history, this complex nappe system was differentiated into two main structural units: a northern unit intensely cross-cut and variably rotated by and parallel to deep lithospheric shear zones, and a southern unit exhibiting huge nappes in their original position striking parallel to the tectonic contact with the Congo craton (Fig. 7d). The regional main stress direction during crustal thickening and subsequent sinistral wrench movement was N-S to NNE-SSW across the belt, and almost remained constant between 630 and 585 Ma, but completely changed into a NW-SE direction during the regional dextral shear movement, between 585 and 538 Ma. The resulting regional strain field shows marked NNE-SSW finite trends outlined by the N-S orientation of synthetic SZ and syn- to late-tectonic intrusions emplaced in extensional zones at right angle with both, the EW thrust zones and subsequent antiform/synform more or less overturned during progressive evolution (Fig. 7b, c and d). The CCSZ and the Sanaga fault mark a major tectonic boundary between these units and probably recorded a complex kinematic history.

4. Regional and pre-drift correlations: discussion

4.1 The Trans-Sahara - Nigeria domain

This domain is characterized by a polycyclic basement flanked to the West and to the East by the Pharusian-Dahomeyan and the Tirririne belts, respectively (Fig. 1). These belts belong to an active margin setting and suggest, as in the Poli region, a nearby inter-plates boundary, here marked by the ophiolites in Aïr (Cosson et al., 1987; Black et al., 1994). The Trans-Sahara-Nigeria domain (Fig. 8) includes dextral submeridian and NE-SW synthetic

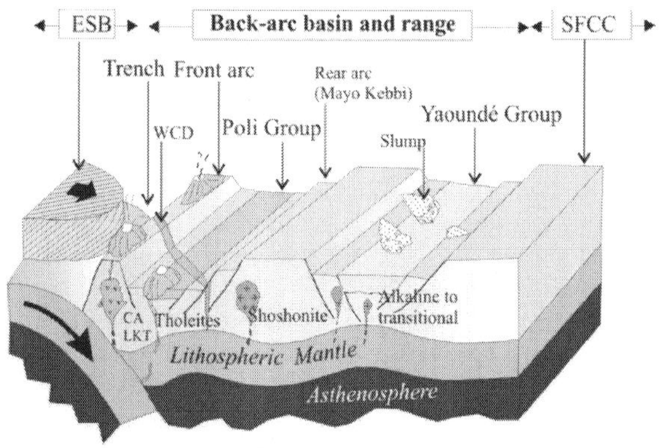

a) Pre-orogenic setting: subduction / back-arc basin and range setting (1000-640 Ma)

Main tectonic, magmatic and sedimentary events:

1) Subduction and back-arc extension
2) Active margin volcanism in Poli (tholeitic basalts and calc-akaline rhyolites at 830 Ma)
3) volcanosediments (Poli group) and sediments (Yaoundé Group: shales, greywackes, minor quartzites, dolomites and evaporites)
4) BIP (660-630 Ma) and dykes varying from calc-alkaline (CA, Poli Group), to alkaline and transitional (Yaoundé Group). **ESB**: Eastern Sahara Block

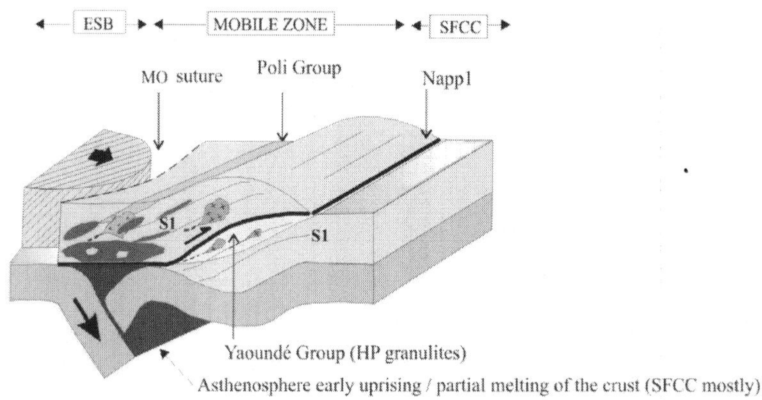

b) PHASE D1: crustal redoubling and thickening-early thrusting/nappe1 (640-630 Ma)

Main Tectonic and metamorphic events:

1) Collision between the ESB prong and the SFCC; crust redoubling / Flat lying S1 foliation
2) Burrial and metamorphism of Yaoundé Group in HP-HT granulite facies (footwall)
3) Retrogression of Eburnean or trench-related granulites in the amphibolite facies (hanging wall)
4) Early stages of delamination of the lithospheric mantle

Fig. 7. (start). Block diagram summarizing the tectonic evolution of the Cameroon domain and the northwestern regions

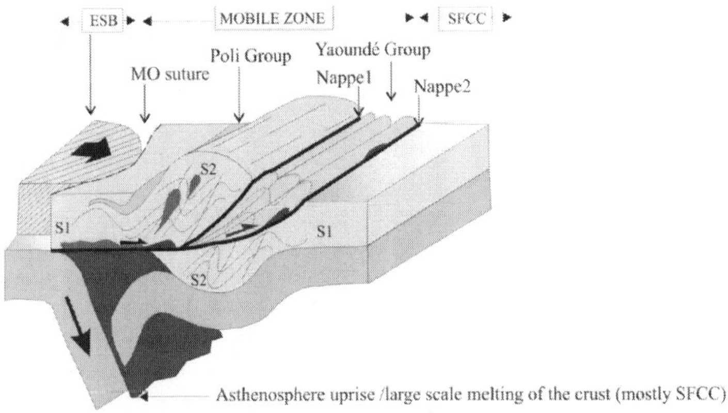

c) PHASE D2: crustal shortening and thickening-upright to recline folds/nappe2 (around 600 Ma)

Main tectonic and metamorphic events:

1) Horizontal shortening (F2 upright to recline), thrusting and exhumation of Yaoundé granulites and Ultra Mafic fragments derived from upper mantle along deep listric faults (Thrusting combines pur and simple shear deformation).
2) Retrogression of the Yaoundé HP-HT granulites and widespread migmatization
3) Increasing delamination of lithospheric mantle and uprise of hot asthenosphere against the crust due to indentation.

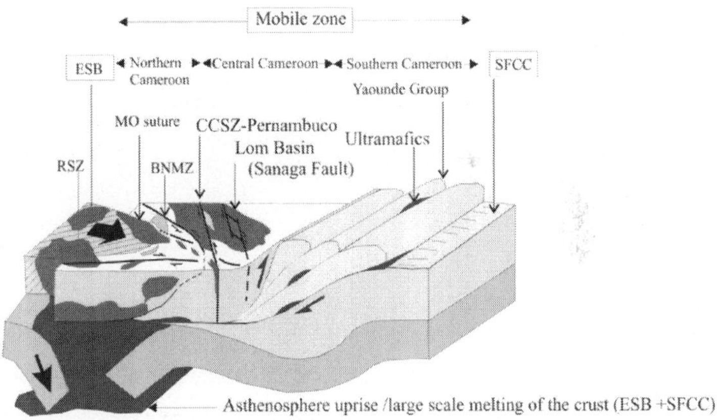

d) PHASE D3 or late-D2: Conjugate wrench movements at the prong front (585-580 Ma)

Main tectonic and metamorphic events:

1) Wrench movements following penetration of the prong / Lom pull-apart basin
2) Advanced delamination of lithospheric mantle due to indentation
3) Widespread melting and granitization of the crust causing dismembering of the ESB, the active margin, reactivated basement and earlier Pan-African structures. D4 (or D3) dextral shear along the major Pernambuco-CCSZ due to prominent active role of the WACis not represented.

Fig. 7 (end). Block diagram summarizing the tectonic evolution of the Cameroon domain and the northwestern regions shear zones. All these shear zones usually cross-cut earlier fold and thrust zones, and both structures are diversely interpreted, either as coeval (Caby, 1989; Caby & Boesse, 2001) or not (Ferré et al., 2002).

The Ifewara-Iwaraja Shear Zone (IISZ, Fig. 8) is coeval with NNE verging thrust zone and its evolution has been correlated with that of wrench and thrust zones in the Hoggar (Caby, 1989; Caby & Boesse, 2001) which age ranges from 614 to 629 Ma (Bertrand et al., 1986). Flat-lying migmatites in the Okene region are folded by NE verging anticline coeval with the thrust zones; they show parallel charnockitic veins sometimes exhibiting stygmatic folds. According to the charnockitization age in western Nigeria (U-Pb zircon ages by Tubosun et al., 1984), the early thrust evolution in the Okene region could be situated between 634 and 620 Ma. Therefore, wrench and thrust evolution in western Nigeria is constrained between 634 and 620 Ma and its evolution was controlled by the NNE-SSW stress direction as in the Cameroon domain.

The Pambegua-Bauchi shear zones (PBSZ, Fig. 8), in Central Nigeria (Ferré et al., 2002) represent NNE-SSW synthetics to the NS system. U-Pb dating of zircons from granites and migmatites associated with the thrust and shear zones yielded average ages of 615 Ma and 585 Ma, which were approximated to the ages of thrust and wrench evolution respectively (Ferré et al., 1998, 2002) as in Cameroon.

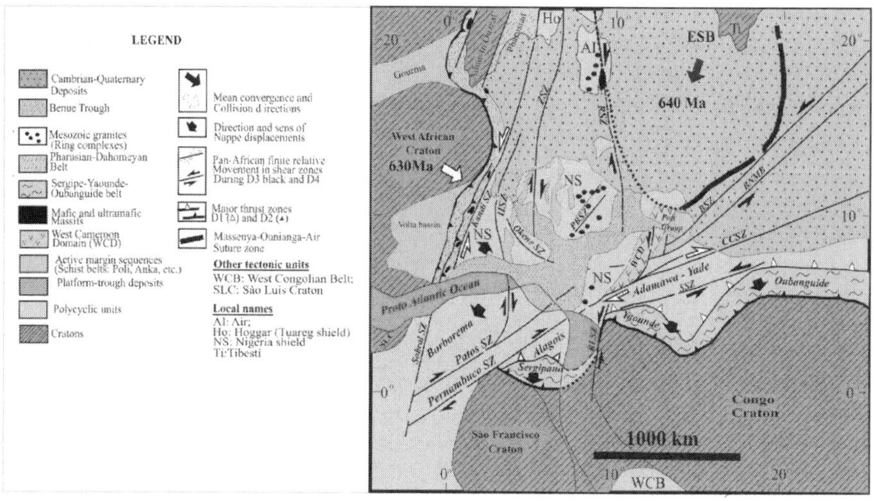

Fig. 8. Pan-African-Brasiliano shear zones and kinematics in pre-drift reconstruction

4.2 The Brasiliano domain

The geology and structure of the Brasiliano domain is summarized in Trompette (1994). Two main geotectonic units are here defined: the Borborema province which represents an old Paleoproterozoic basement reactivated during the Brasiliano orogeny and including metasedimentary belts, and the Canudo-Sergipe (or Sergipano) Neoproterozoic Group overriding the São Francisco Craton to the South (Fig. 8). The Borborema province includes two sets of shear zones: the Cearà-type SZ directed N-S, and the Pernambuco and Patos SZ, E-W-directed. The Patos and N-S branches are interpreted as forming a transpressive dextral

shear zone with duplexes (Corsini et al., 1996), whereas the Pernambuco SZ represents a major structural boundary between the N-S Cearà-type and E-W Sergipano belts.
Tentative correlation of NE-Brazil with its African counterpart (Caby, 1989; Trompette, 1994) shows a good structural fit between the Cearà-type and the Western Nigeria schist belts (Isseyin and Ife), in the one hand, and between the Sergipano and the Yaoundé-Oubanguide Groups, in the other hand; the Patos and Central Cameroon SZ are recognized as structural equivalents (Neves & Mariano, 1999), whereas the Cearà-type SZ are correlated with the trans-Saharan SZ (Caby, 1989). In Brazil, these shear zones are associated to early and late-tectonic transcurrent plutons which ages are bracketed between 600-580 Ma, and 575-565 Ma, respectively (Neves et al., 2004; Souza et al., 2006); both age groups may be correlated with the sinistral and dextral shear evolutions in the Cameroon domain. However, NE-SW sinistral shear zones and respective NS synthetics in northern Cameroon have no equivalent in NE-Brazil. We suggest that the Trans-Sahara- Nigeria - NE Brazil Shear zones in the one hand and the Cameroon-Oubanguides Shear zones in the other hand form a conjugate shear zone system at global scale (Fig.8).

5. The collision and plates amalgamation model-Conclusion

Although based on imprecise data in some cases, correlation of the tectonic events throughout the Pan-African domains and respective kinematics allows determination of a regional strain field compatible with the evolution of a tectonic indent that was progressively formed in northwestern Cameroon between 640 and 580 Ma, during the collision and post-collision evolutions. Indent-related SZ in both western (Trans-Sahara-Nigeria-NE Brazil) and eastern provinces (Cameroon-Oubanguide) were overprinted by right lateral shear movements during a late clockwise rotation of the Cameroon and northwestern domains, indicating a still active collision at the eastern border of the WAC after indentation. The indent related SZ determine a post-collisional branching network system which geometry is different from that of the original belt striking EW. Collision in the Dahomeyan-Pharusian belt is recorded by the early ultra-HP metamorphism between ca 625 Ma (^{40}Ar-^{39}Ar ages on phengite, Jahn et al., 2001) and 633 Ma (^{40}Ar-^{39}Ar ages on muscovite, Attoh et al., 1997), and is almost close to the syn-collisional evolution in northwestern Cameroon, whereas exhumation of the suture zone and nappe stacking at the eastern border of the WAC, dated at ca 590 Ma (^{40}Ar-^{39}Ar ages on muscovite and hornblende, Attoh et al., 1997), coincides with the age of exhumation and over-thrusting of HP-HT granulites onto the SFCC.
In summary, a three-plate collision model involving three major landmasses can be proposed for the Pan-African tectonic evolution of the Trans-Saharan-Central African /CAFB and Brasiliano western Gondwana belts. The model includes: the SFCC, the eastern Saharan block (ESB), and the WAC (Fig. 7 and 8). Collision in the north-south direction involved ESB and the northern margin of the SFCC originally differentiated into a basin and range province (probably during back arc extension related to subduction, Fig. 7a). This collision generated an indent (Fig. 7b and Fig. 8) and intense deformation in the Cameroon and north-western domains, following penetration of the Saharan rigid prong into the SFCC active margin between 640 and 580 Ma. Successive tectonic events recorded in this active margin involved: i) crustal thickening ii) left- and right-lateral conjugate wrench movements (indent), and iii) right-lateral wrench movements in the N70 to EW direction. These structural and kinematic data show that, although collision between the Trans-Sahara active

margin and the WAC started at ca 630 Ma, deformation in north-western Cameroon was mostly controlled by a NNE-SSW stress direction until ca 590 Ma. Further deformation in this domain is recorded by a regional clockwise rotation compatible with a NW-SE stress direction suggesting that the late tectonic event was the only one controlled by the prominent active role of the WAC.

Tectonic indentation requires two major conditions to operate: (1) indenter should be rigid and stronger than indented margin. Considerations to this rheological aspect suggest that active margins should be weaker than passive margins, as particularly submitted to intense seismicity and volcanism during subduction; (2) volume constrains during indentation requires that the denser lithospheric mantle be separated from indenter to enable this one penetrate into the opposite margin. In addition, penetration of indenter would induce differential shortening between continental crust (lighter and deformable) and lithospheric mantle (rigid and denser) in the indented plate causing local delamination of the lithospheric mantle (Fig. 7b, c, d). These fundamental tectonic processes suggest probable ascent of massive volumes of the asthenosphere against the crust, causing large scale melting at the frontal zone of the amalgamating plates. The abundant crustal melting and widespread occurrence of batholiths intrusions that caused intense reactivation and partial dismembering of the colliding plates in north-western Cameroon may be explained by these processes. Increasing metamorphic isograds classically advocated to generate partial melting during thickening would not alone explain these intense and widespread phenomenons, as they would be characterized by a relatively low intensity and apparent regional zoning. The western Gondwana new tectonic model highlights the importance of the boundary geometry of colliding plates, in generating special tectonic processes likely to cause local inversion of the mantle- asthenosphere structure and associated thermal gradients; the absence of apparent belt polarity and zoning in the Gondwana collisional settings would be linked to the presence of a rigid prong. Dismembering of the ESB prong and northern extension of the SFCC during collision may account for these particular processes. Presently, these marginal zones mostly exhibit a marked Pan-African imprint and more detail studies including isotopic and geophysical studies are necessary to enable more precise reconstruction of the respective crustal evolutions.

6. The Post Pan-African history and inferences to plate destruction

The Post Pan-African history in western Gondwana involves sedimentary and magmatic events whose relationships with global tectonics are still not completely understood. This history relates to the Paleozoic, Mesozoic and Tertiary events; it interferes with the breaking of the western Gondwana and subsequent opening of the Atlantic Ocean in the Barremian (Brunet et al., 1988).

6.1 Pan-African molasses

Eroded material identified as Pan-African molasses are found in the belt adjacent foreland basins (e.g. Estancia Group, Dja Group, Proche-Tenere Group, etc.), and within-belt Paleozoic grabens (Trompette, 1994). The Mangbei-type semi-grabens in northern Cameroon (Balche, Nigba, etc.), formed during late Pan-African transtensional movements, are typically located at the intersections between NS secondary and EW shear zones. Main lithologic units in these grabens include: sandstones, conglomerates and continental tholeites represented by acidic, intermediate and basic lavas (Béa et al., 1990).

6.2 The Paleozoic to Quaternary evolution

The Paleozoic to Quaternary alkaline magmatism in Niger, Nigeria and Cameroon, and the associated basins and swells represent one of the most significant and illustrative examples of the post Pan-African evolution in western Gondwana.

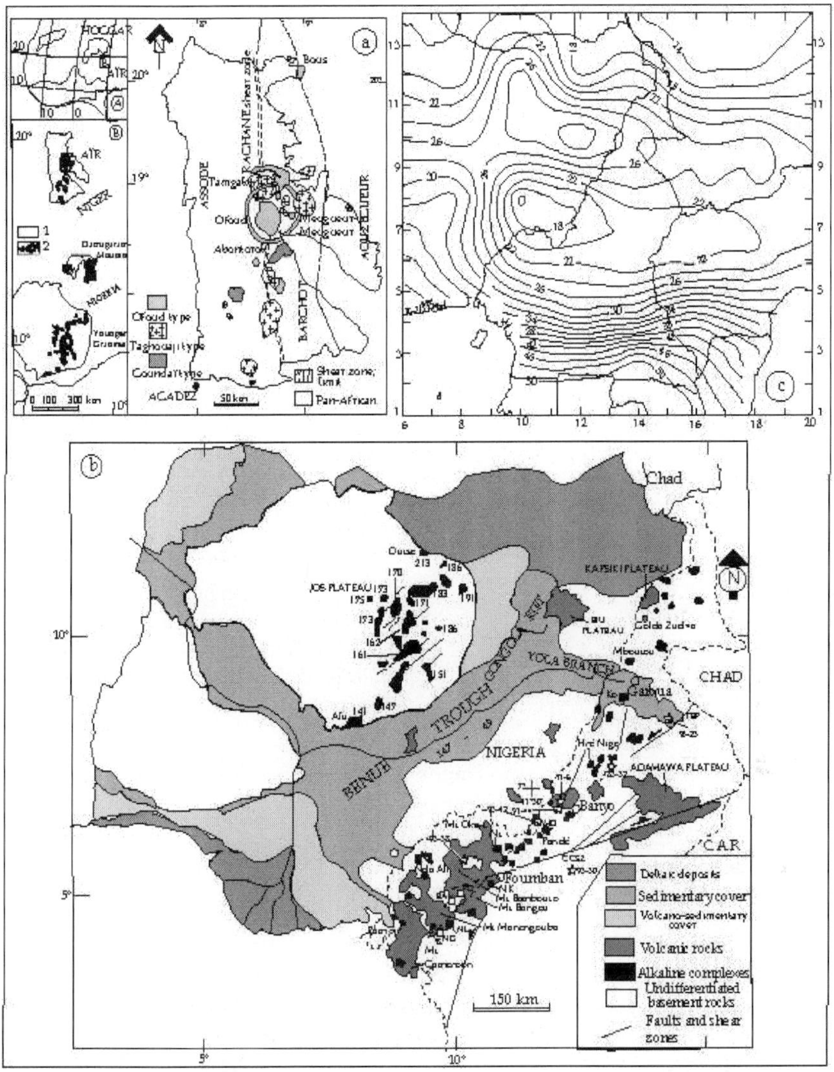

Fig. 9. Phanerozoic major lithospheric structures in Central-West Africa: a) the Air-Damagaram NS-directed magmatic trends; b) The Nigeria (Jos Plateau)-Benue-CL N50-directed magmatic trends, c) Map of crustal thickness in Central Africa based on gravity data (after Poudjom Djomani et al., 1995)

6.2.1 Main magmatic provinces and magmatic activity

A magmatic province is here defined as a broad geographic area (linear or not), characterized by intrusive and/or volcanic bodies, the ages of which approximately correspond to the same geological period, namely Ordovician–Devonian (480-400 Ma), Carboniferous (330-260 Ma), Triassic-Jurassic (215-140 Ma), Cretaceous (147-106 Ma) and Tertiary (73-30 Ma), though age overlapping is reported between some neighboring provinces.

6.2.1.1 The Niger-Nigeria Paleozoic to Mesozoic Super Province

The Niger-Nigeria super province comprises from north to south, the Aïr anorogenic magmatic province (N. Niger), the Damagaram-Mounio province (S. Niger) and the Jos Plateau or 'Younger Granite' province of Nigeria (Fig.9a, inset). A general review of the different types of Paleozoic ring complexes in the Aïr is given by Demaiffe et al. (1991), Moreau et al. (1991, 1994) and Demaiffe & Moreau (1996). The authors recognized 28 complexes among which the largest cone sheet in the world (the Meugueur-Meugueur, 65 km in diameter) and one of the smallest intrusions (Tagueï, 0.8 km in diameter). Most of the intrusions have a circular shape, but some are elliptical or semi-circular. They have been divided into three main types (Fig. 9a) (Moreau et al., 1991; Demaiffe et al., 1991): 1) the Taghouaji type (18 complexes) is composed mainly of alkaline and peralkaline syenite and granite, with or without associated metaluminous granites; 2) the Goundaï type (3 complexes) is composed mainly of acid volcanic rocks (rhyolitic tuffs and ignimbrites) with quartz syenite ring dykes; 3) the Ofoud type (7 complexes) is characterized by a large proportion of basic rocks varying from troctolites and leucogabbros to true anorthosites, intruded by mildly to peralkaline syenites and granites.

In the Younger Granite province of Nigeria, over 50 intrusions forming a series of high level anorogenic complexes have been recognized (Badejoko, 1986; Orajaka, 1986). These circular or elliptical intrusions have an average diameter of 10-25 km. Each of the ring complexes began as a chain of volcanoes. Volcanic rocks commonly occur, especially in the northernmost complexes. Basic intrusive rocks represent less than 1% of the total area of the province (against 40-80% in the Aïr) and include monzonites and monzogabbros (basaltic lavas are more common). Anorthosites occur only as xenoliths in a doleritic dyke near Jos. There are several distinctive granite types (Kinnaird, 1985): i) peralkaline granites and related syenites (with alkali or calcic amphibole); ii) peraluminous biotite alkali feldspar granites and biotite syenogranites; iii) metaluminous fayalite and hornblende-bearing granites and porphyries with amphibole or biotite.

6.2.1.2 The Benue Trough

The magmatic province of the BT constitutes a spatial link between the alkaline to peralkaline province of Nigeria to the north and the CL to the south (Fig.9b). A detail account of its petrology and geochemistry can be found in Baudin (1991) and its main characteristics in Maluski et al. (1995) and Coulon et al. (1996). Two principal magmatic domains are recognized, northern and southern Benue. In the northern domain, magmatism is characterized by transitional alkaline basalts and transitional tholeiitic basalts. Acidic magmatism (rhyolites and granophyres) of peralkaline nature is also present. In the southern Benue, several magmatic districts exhibit alkaline or tholeiitic affinities. Alkaline rocks include basalts, dolerites, rhyolites, trachytes, phonolites, phonotephrites, tephriphonolites, camptonites and nepheline syenites. Only doleritic sills display a clear mineralogical tholeiitic affinity and correspond chemically to quartz tholeiites.

6.2.1.3 The Cameroon Line

The CL, made up of an oceanic (Annobon, São Tomé, Principé and Bioko) and a continental (Mt. Etindé, Mt. Cameroon, Mt. Manengouba, Mt. Bambouto, Mt. Oku, Ngaoundéré plateau, Bui plateau and Mt. Mandara) sectors, has been active since the Cenozoic and is currently defined by an almost SW-NE geological lineament (mean value: N30°E). Its activity includes emplacement of plutonic complexes in the continental sector (more than 60) and volcanic eruptions (Fig. 9b). The plutonic complexes are small in size (mostly 5-10 km in diameter) and except for 4 complexes in north and southwest Cameroon, are mainly constituted of granites or syenites, to which subordinate intermediate and basic rocks are sometimes associated. Volcanic rocks are commonly associated with the plutonic rocks, the whole being referred to as plutonic-volcanic complexes. The Mboutou, Kokoumi and Nigo complexes in north Cameroon comprise basic rocks (Déruelle et al., 1991; Ngako et al., 2006). The Ntumbaw complex in NW Cameroon (Déruelle et al., 1991) contains predominantly intermediate rocks. Syenites are mostly saturated and granites over-saturated, both displaying mineralogical and geochemical characteristics of alkaline to peralkaline rocks. The Kokoumi complex is entirely undersaturated with a gabbro–nepheline monzosyenite - nepheline syenite series. It is also particular in that the plutonic series is cut by lamprophyric dykes (Ngounouno et al., 2001).

Volcanic rocks are more varied mostly undersaturated in the oceanic part and undersaturated to oversaturated in the continental part. At the ocean-continent boundary zone, Bioko and Mt. Cameroon are mainly basaltic. Etindé, located to the intermediate southwestern flank of Mt. Cameroon is made up entirely of undersaturated lavas (Nkoumbou et al., 1995). The other continental volcanoes display either a typical bimodal series (only basic and felsic terms) or a complete series (basic to felsic with few intermediate terms) (Njonfang et al., 2010 in press and references therein). Basic lavas comprise basanites, alkali basalts and hawaiites; intermediate lavas include mugearites and benmoreites and felsic ones comprise trachytes, phonolites and/or rhyolites (Kagou Dongmo et al., 2010 and references therein; Kamgang et al., 2010; Njonfang et al., 2010). Data from the oceanic sector of the CL (Déruelle et al., 1991 and references therein; Lee et al., 1994) indicate basanite to hy-normative basalts, tristanites and trachytes in Annobon; basalts to trachytes and phonolites with no compositional gap in São Tomé; nephelinites, basanites, tristanites, trachyphonolites, phonolites and alkali basalts in Principé and basanites to hy-basalts in Bioko. In Principé, the oldest rocks (31 Ma) are basal hyaloclastite breccias that contain fragments of fresh tholeiite (Fitton & Dunlop, 1985). Aka et al. (2004) noted a geographic control on the distribution of $^{3}He/^{4}He$ ratios along the CL, with high-μ OIB-like values on the ocean-continent boundary zone (Bioko, Mt. Cameroon and Etinde) increasing to mantle (MORB-like) values in its oceanic (towards Annobon) and continental (towards Ngaoundéré) terminals. Even though a review of the CL volcanism by Déruelle et al. (1991) suggests that it is entirely alkaline in nature, some transitional affinities have recently been described in the Mbam, Bangou, Bambouto and Oku volcanic centers in the west and northwest Cameroon (Marzoli et al., 2000; Fosso et al., 2005; Moundi et al., 2007). Some mafic lavas of the CL are rich in ultramafic xenoliths (Lee et al., 1996; Aka et al., 2004). These occur in both the continental sector as in Lake Nyos (Nana et al., 2001; Temdjim et al., 2004), the Kapsiki plateau (Ngounouno et al., 2008) and the Mount Cameroon (Wandji et al., 2009) and in the oceanic sector as in São Tomé (Caldeira & Munhá, 2002).

6.2.2 Spatio-temporal evolution of magmatic activity

On the basis of ages decreasing from the Aïr ring complexes, north of Niger (480-400 Ma) to the Younger Granites of Nigeria in the south (215-140 Ma) through the Damagaram, south of Niger (330-260 Ma), the Niger–Nigeria large magmatic province has been defined as a unique feature in the world of practically continuous within plate anorogenic volcanism and plutonism, with progressive southward shift of centres of magmatic activity (Bowden & Karche, 1984). At the scale of the Aïr province alone, the ages decrease from 487 Ma in the north to 407 Ma in the south described by Bowden & Karche (1984) was not confirmed by new radiometric data (Rb/Sr method) on the same sample powders by Moreau et al. (1994) . Instead the new results show that the emplacement of the Aïr ring complexes took place within a very short time at 407 ± 8 Ma, close to the Silurian-Devonian boundary or the lowest Early Devonian. In Nigeria, the ages of the Younger Granites show that major local migration of magmatic activity was concentrated along two ENE-WSW linear zones at least, from Dutse (213 Ma) in the north, to Afu (141 Ma) in the south (Fig. 9b).

A detailed chronology of emplacement of volcanic rocks of the BT has been established using the Ar/Ar method and the following magmatic evolution is revealed (Maluski et al., 1995). (1) During the Late Jurassic to Albian period (147-106 Ma), magmatism probably occurred in the whole basin. It is particularly expressed in the northern Benue where it is represented by alkaline transitional basalts and associated peralkaline rhyolites (bimodal volcanism) and by tholeiitic transitional basalts. (2) Between 97 and 81 Ma, magmatism was concentrated in the southern Benue, was exclusively alkaline, and predominantly intrusive. (3) During the period 68-49 Ma, the first magmatic products, also restricted to the southern Benue, were alkaline and the last ones tholeiitic. As in the Aïr province (e.g. Moreau et al., 1994), no clear time-space migration of magmatism in the BT is apparent.

Previous radiometric data (Rb/Sr and K/Ar methods) for the anorogenic plutonic complexes of the CL indicate that their emplacement occurred from the Paleocene (*ca* 67 Ma) to the Oligocene (*ca* 30 Ma). The oldest age was obtained on the Golda Zuelva granite in the north, and the youngest on the Mt. Bana granite in the west (Lasserre, 1978). However, the Nkogam and Mboutou complexes in the west and north gave 66 and 60 Ma respectively (Lasserre, 1978; Caen-Vachette et al., 1987), pointing to the lack of time-space migration of plutonism. This conclusion is corroborated by the ages recently obtained for the Kokoumi complex southwest of Mboutou (39 Ma) and for the Hossere Nigo west of Adamawa plateau (65 Ma) (Ngounouno et al., 2001; Kamdem et al., 2002; Montigny et al., 2004).

New K/Ar and Ar/Ar data on volcanic rocks of the CL (Ngako et al., 2006 and reference therein; Moundi et al., 2007; Wandji et al., 2008) show that volcanic activity ranges from Upper Eocene (47 Ma) to the Present and not from Oligocene (30 Ma) as established for the distribution of volcano-capped swells along the line (Burke, 2001). Lee et al. (1994) noted a SW younging of volcanism in the oceanic sector of the CL from Principé (31 Ma) to Pagalu (5 Ma). The oldest volcanic ages (K-Ar) so far obtained (46.7 ± 1.1 Ma and 45.5±1.1 Ma) are respectively in an olivine basalt from the Bamoun plateau (Moundi et al., 2007) and the Mbépit rhyolites in the Noun plain (Wandji et al., 2008), next to the Bamoun plateau. The transitional lavas from the Bamoun plateau give a K-Ar age of 51.8±1.2; Ma (Moundi et al., 2007) and those of Mount Bangou, a K-Ar age of 44.5±1 Ma (Fosso et al., 2005). Therefore, as for anorogenic complexes, it is difficult to consider volcanic activity along the whole CL in terms of a steady time-space migration, symptomatic of a mantle plume reference frame as in the Hawaiian Islands.

From the above synthesis, it is seen that alkaline magmatism in West-Central Africa shows a time-space migration marked by ages decreasing from the Silurian-Devonian boundary (407±8 Ma) in the Aïr province (north Niger) through the Younger Granites of Nigeria (213-141 Ma), the BT (147-49 Ma) to the Paleocene-Present activity on the CL. This seems at first sight to result from a single mantle plume. However, the magmatic period of the BT overlaps the period of plutonic magmatism on the CL and also, the ring complexes of Nigeria display roughly two SW-trending lines of decreasing ages. These observations suggest that magmatism in West-Central Africa may instead be the result of an interaction between mantle plume(s) and other important factors such as lithospheric fractures.

6.2.3 Relationship between tectonics and magmatism

6.2.3.1 The Pan-African structural inheritance

Three main structural units characterize the Pan-African basement in western and central Africa (Fig. 8): shear zones, fold zones and thrust zones. Shear zones are the most prominent and represent lithospheric faults. The Aïr basement that was built during the Pan-African orogeny (Black et al., 1994; Liégeois et al., 1994) resulted from the assembly of three main terranes (Assodé, Barghot and Aouzegueur), separated by N-S trending shear-zones or mega-thrusts (Fig. 9a). These shear and thrust zones control: (1) the variation in thickness and/or deformation of the sedimentary cover; (2) the location of basement topographic highs and (3) the location of the magmatic activity (Moreau et al., 1994). The most important of the shear zones, the Raghane mega-shear zone (Fig. 9a) can be observed over a 400 km distance in Aïr and runs for more than 1000 km, as it corresponds to the southern extension of the 8°30' E lineament of the Hoggar (Demaiffe & Moreau, 1996; Abdelsalam et al., 2002). In general, N-S mega-thrust faults were followed by NW-SE trending sinistral wrench faults and by complementary ENE-WSW and NE-SW trending dextral faults. In Nigeria, the major faults are NE-SW to NNE-SSW and ENE-WSW trending dextral wrench faults (Rahaman et al., 1984 and references therein). The distribution of Triassic-Jurassic anorogenic ring complexes in the Jos Plateau defines a sigmoidal megagash geometry with a ENE-WSW mean direction that is probably linked to shear movement along pre-existing ENE-WSW trending wrench faults in the Pan-African basement (Rahaman et al., 1984; Black et al., 1985). This suggests that the Aïr and Younger Granite provinces were not emplaced under the same stress regime (Demaiffe & Moreau, 1996). The BT is a NE-SW trending extensional sedimentary basin (Benkhelil, 1989) considered to be the failed arm of a Cretaceous triple junction, the two other rift arms of which subsequently developed into the South Atlantic Ocean and the Equatorial mega-shear zone (Burke & Dewey, 1974). Opening of the trough has also been attributed to reactivation during Cretaceous-Early Tertiary times of late Pan-African NE trending dextral shear zones (Guiraud & Maurin, 1992). In Cameroon, the basement is affected by a major translithospheric mega-shear that extends ENE from the Gulf of Guinea, through southern Chad and the Central African Republic into western Sudan. This Central African Shear Zone referred to as the Ngaoundéré or Foumban lineament (Browne & Fairhead, 1983) or the Central Cameroon Shear Zone (Ngako et al., 1991 and reference therein), is a dextral shear that was formed during the Pan-African orogeny and is delineated by broad mylonite zones. The Patos and Pernambuco-Adamawa branches and their N-S relays control the geometry of the BT (Maurin et al., 1986; Guiraud & Maurin, 1992) and the CL respectively (Fig. 8 & 9). Detailed structural studies of some syntectonic Pan-African granitoids in Cameroon show that they are broadly elongated and aligned NE-SW, consistent with their successive emplacement coeval with acting and oblique sinistral shear zones formed during an earlier deformation phase (Ngako et al., 2008). For example, in the Ngondo complex (1000 km^2), the geometry of the internal foliation trajectories and joint orientation suggest that its emplacement was controlled by a N30° sinistral shear zone (Tagne Kamga et al., 1999), a conclusion earlier drawn for the Bandja complex (300 km^2) which outcrops in a N30° elliptic sheet (Nguiessi Chankam et al., 1997). In the Biu Plateau (CL), the distribution of volcanic necks appears to be controlled by a N-S alignment of Tertiary faults. The emplacement of Paleocene and Eocene alkali granitic intrusions in Cameroon appears to be similarly fault controlled (Moreau et al., 1987). Indeed, it is partly superimposed on the Central African Shear Zone (Cornachia & Dars, 1983) which is a pre-existing fracture zone (Fig. 8 and 9). It has been suggested that volcanoes in the offshore part of the CL are located at places where the line crosses fracture zones while inland volcanoes are located along structures related to Riedel-style shear zones (see Meyers et al., 1998 and references therein). Fold zones are characterized by Pan-African domes or diapirs of granitoid (calc-alkaline) rocks (Grant, 1978; Ngako, 1999 and references therein). They determine broad areas bounded by the Nigeria and Cameroon shear zones and are dominantly characterized by the occurrence of Paleozoic to Cenozoic ring complexes. Thrust zones (Ball et al., 1984; Jégouzo, 1984; Nédélec et al., 1993; Penaye et al., 1993) characterize the main structures of the Pan-African belt and overlay the Achaean and Eburnean boundary. These zones reveal no post-Pan-African magmatic activity, neither intrusive, nor effusive.

6.2.3.2 Swell-and-basin structures and magmatic provinces

Topographically, the general structures in western and central Africa suggest a long-lived continental extension history, ongoing at least since Ordovician from one magmatic province to the other. These structures are represented either by large African lithospheric domes or swells such as the Hoggar, Aïr, Adamawa, Tibesti, Darfur and the Great Lakes of East Africa (Ebinger et al., 1989; Burke, 2001), or by grabens, such as the BT (Benkhelil, 1989). Domes correspond to high plateaus and mark the early stages of rift evolution. In Aïr, Jallouli (1989) estimated the thickness of the elastic lithosphere to be 20-30 km which is thin for a continental environment, but comparable to what is commonly observed in rift zones.

Harley et al. (1996) estimated a particularly thin elastic thickness for the lithosphere in the general region of the CL and Poudjom Djomani et al. (1997) found both crustal thickness and elastic thickness of the lithosphere to be unusually thin in the continental part of the CL. For example, their map of crustal thickness (Fig. 9c) has a minimum value of 18 km at around 7°N, 11.5°E, recently referred to by Burke (2001) as the location of a mantle plume (the '711 plume'). The location of these swells coincides with magmatic provinces of different ages as in Aïr and in the CL. In Cameroon, the four islands of Bioko, Principé, São Tomé and Annobon, the two large seamounts (Burke, 2001) and the four central volcanoes of mounts Cameroon, Manengouba, Bambouto and Oku occupy swells of the CL. They define a 1000-km long SW-NE straight line and display a 'swell and basin' geometry which recalls the horst and graben structure of the whole CL (Déruelle et al., 1991). In West and Central Africa, Guiraud & Maurin (1992) recognize two main phases of rifting separated by a major Aptian (*ca* 117 Ma) unconformity. (1) The Neocomian-Early Aptian (*ca* 144-117 Ma) phase that began in the basins of east and northeast Brazil, Gulf of Guinea, south Chad, Sudan, Kenya, north and east Niger, north Egypt and Libya. Late Jurassic magmatic activity appears to precede this rifting phase in northeast Brazil, south Sudan and the BT of Nigeria.

(2) The Middle-Late Aptian-Albian (*ca* 117-98 Ma) phase marked among others by the development of pull-apart basins in an oblique extensional regime from Benue to southern Chad. This is related to strike-slip movements along the Central African Shear Zone (Bosworth, 1992). Magmatic activity in the BT was particularly important during the Cretaceous-Early Tertiary and took place in many phases, contemporaneous with the opening and infilling of the trough (Wilson & Guiraud, 1992). The first phase (Early Cretaceous: 141-106 Ma; Baudin, 1991) is related to the main extensional tectonic regime which affected the trough. During the Tertiary, basaltic magmas were emplaced throughout the entire trough with the greatest concentration in the NE (Wilson & Guiraud, 1992). This Post-Cretaceous extension and volcanism in the Upper Benue corresponds to a period of general stress release after the Santonian or Cretaceous compressional events (Benkhelil, 1989). In general, magmatic activity is predominantly basaltic in the Upper Benue, although rhyolites occur locally. In the Garoua basin, which represents the eastern continuation of the Yola branch of the Upper Benue, the strata are cut by lineaments which may have acted as strike-slip or normal faults. Tertiary-Quaternary magmatism associated with these lineaments occurs as doleritic sills, Paleocene (65-60 Ma) alkaline complexes and Neogene trachy-phonolite necks and dykes (Guiraud et al., 1987). An entirely younger undersaturated anorogenic complex (Kokoumi) belonging to the basin and including a gabbro-nepheline syenite series (39 Ma) and lamprophyre dykes (20.5 Ma) has been studied (Ngounouno et al., 2001).

6.2.4 Continental breaking and possible links with magmatic chains – discussion

From the synthesis above, it appears that (1) There is a time-space link among the magmatic alkaline provinces in West-Central Africa, from 407±8Ma in Aïr (N. Niger) and 330-260Ma in Damagaram-Mounio (S. Niger) to 66 Ma-Present in Cameroon, through 213-141Ma in Jos Plateau (N. Nigeria) and 147-49 Ma in the BT (S. Nigeria). The migration follows an N-S trajectory in the Aïr–Damagaram-Jos provinces, and a NW-SE trajectory in the BT-CL provinces. (2) Only the Aïr province displays a N-S trend of magmatic complexes, the Jos Plateau, BT and CL complexes show three parallel lines rather trending NE–SW. A similar N-S trend followed by a NE-SW one links the N-S East-African Rift System to the NE-SW major igneous lineaments in South Africa (Kinabo et al., 2007, Moore et al., 2008). (3) Each trend is parallel to a shear or fracture zone: N-S Raghane shear zone for the Aïr, NE trending wrench fault for Jos Plateau, NE-trending dextral shear zone for the BT, NE-trending sinistral fault oblique to the Central African Shear Zone or fracture zones directed N70°E for the CL (Fig. 9). (4) Only the ages of the Younger Granites of Nigeria become steadily younger from Dutse (213 Ma) in the NE to Afu (141 Ma) in the SW over a distance of 420 km. (5) There is a great overlap between the third magmatic period in the BT (68-49 Ma) and the ages of anorogenic complexes (66-30 Ma) of the CL, while the beginning of the first magmatic period (147Ma) coincides within analytical error range, with the end of the emplacement of granite complexes of Nigeria (141Ma). An extended discussion on the origin of the magmatic provinces in West Africa, and in the Air-CL line is provided in Ngako et al. (2006) and only summarized in this section.

Following Hieronymus & Bercovici (2000) model, the spreading ridge of the proto Atlantic Ocean during Barremian offers a tentative explanation for the SW-NE Nigeria-BT-CL parallel trends which is susceptible to take into account the whole geometry and structure of the intraplate magmatism in West and Central-Africa. This model predicts island chains aligned with a deviatorically tensile tectonic stress perpendicular to the ridge; the thin elastic lithosphere near the ridge is subjected to strong deviatorically tensile stress field perpendicular to the ridge axis. Under such conditions the model results in parallel lines of volcanoes perpendicular to the spreading ridge. The model explains the development of non-hotspot and seamount chains in terms of the vulnerability of the lithosphere to magma penetration due to lithospheric stresses and the effects of melting of the conduit walls. Their theory is based on the assumption that (1) transport of magma through the brittle part of the lithosphere occurs via fractures and (2) melt is distributed uniformly at the base of the lithosphere underlain by a superswell. It turns out that an initial perturbation is required in all cases to localize volcanic activity. This initial perturbation may be provided by a change in the tectonic stress field due to plate motion reorganization (which is amplified locally by an inhomogeneity in the lithosphere), the formation of a small sublithospheric melting anomaly or a change in convection. Hieronymus & Bercovici (2000) have finally shown that multiple lines of volcanoes can result from interaction of flexural, membrane and tectonic stresses. Indeed: i) due to tensile membrane stresses, several volcanoes typically form in the space between any two volcanoes of the initial chain; ii) membrane stresses perpendicular to the axis of the ridge also interact with the flexural stresses to generate volcanism away from the axis. This off-axis volcanism eventually forms additional lines of volcanoes parallel to the first one. However, at variance to this scheme, the time-space migration of magmatic lines from Jos Plateau (ca 200 Ma) to the CL (ca 60 Ma) suggests an opening of the Gulf of Guinea spreading ridge from NW to the SE.

The origin of intra-plate magmatism on the African continent (and elsewhere in the world, e.g., Smith & Lewis, 1999) is still a topic of great contention. Some workers (e.g. Wilson & Guiraud, 1992; Lee et al., 1994; Ebinger & Sleep, 1998) link it to the role of mantle plumes. However, most African hotspots have been active since the Cretaceous; thus, to explain the extensive Triassic-Early Jurassic magmatism along the West African and North American margins, Wilson (1997) suggested the location of a pre-Pangea continental break-up 'super plume' axis beneath West Africa. Studies of uplift history, topography and seismic tomography of the African plate and underlying mantle (Lithgow-Bertelloni and Silver, 1998; Begg et al., 2009 and references therein) indicate that southern Africa is underlain by a large-scale buoyant, low seismic velocity structure extending from just above the core-mantle boundary to near the base of the African lithosphere and that the ascent of this material could be feeding many hotspots in Africa (Gurnis et al., 1999). However, based on the fact that all plumes born (as traps) in the last 100 Ma (i.e. Ethiopia-Yemen/Afar) are still quite active, whereas those born between 100 and 140 Ma may be failing while those older than 150 Ma do not in general have an active trace (Duncan & Richards, 1991; Courtillot et al., 1999), it is very difficult to explain the whole Air-CL trend by a simple motion of the plate over one stationary hotspot as in the Hawaiian system. The 1650 km distance stretched by the magmatic provinces following this trend is more than three times the postulated diameter of the St. Helena plume tail (O'Connor & Le Roex, 1992; Wilson, 1992). Burke (2001) also demonstrated that a single plume, dubbed the '711' plume, formed the Nigerian granites, generated a topographic dome at *ca* 140 Ma on which the triple-rift system among which the BT developed, and that the plume was also involved in forming the Cameroon granitic complexes: (i) Between 213 and 141 Ma, the 711 plume generated a 400 km-long line of intrusions as the continent moved over it. (ii) From 140 to 66 Ma, the plume moved *ca* 300 km with respect to the overlying continental lithosphere, having been caught up in the evolution of the Benue rift system (Maluski et al., 1995) rather than forming a line of intrusions. (iii) Since 66 Ma, the 711 plume has stayed in the same place, close to lat. 7°N and long. 11.5°E, but has been associated with the development of the CL. Though important differences exist between the tectonic models, an agreement appears on the interaction between preexisting Pan-African faults and one or more superswells in Western Africa.

7. General conclusion

The history of western Gondwana includes complex tectonic processes that led to plate's amalgamation and the formation of this supercontinent, and also to the breaking and dismantling of the amalgamated plates that culminated with the opening of the Atlantic Ocean. Unlike the tectonic evolution in most modern collisional belts, plate's amalgamation involved more than two plates converging in different directions and colliding almost simultaneously. Some of these plates like the ESB were completely dismantled by post-collisional processes and can be presently assessed only through their vestiges.

The Gondwana mobile zones exhibit particular features that render their reconstruction complex. However, the present study has shown that the particularly abundant and widespread partial melting and inferred reactivation that partly caused the dismembering of the colliding SFCC-ESB plates may be explained by the tectonic indentation model. Likely, decoupling of the lithospheric mantle from overlying crust as the result of volume constrains and differential shortening of the lighter and deformable crust during indentation has constituted an efficient mechanism to enhance ascent of the asthenosphere against the crust causing large scale pervasive melting and granitization during the building-up of the Gondwanaland.

Phanerozoic evolution of western Gondwana is mostly characterized by continental extension. The genesis and location of swells and basins as well as the related magmatic provinces was mostly controlled by the interaction of superswells with Pan-African fractures. However, it is likely that the classical hotspot evolution model proposed by previous authors interacted with or was relayed by near axis and superswell chains in Nigeria, BT and CL during progressive opening of the Gulf of Guinea in the Cretaceous (Berriasian to Maestrichian). Obviously, the present prominent lithospheric structures in Africa and Brazil have completely obliterated the Neoproterozoic lithospheric ones and only account for the reconstruction of the Phanerozoic history of the western Gondwana.

8. References

Abdelsalam, M.G., Liégeois, J.P. & Stern, R.J. (2002). The Saharan metacraton, *J. Afr. Earth Sci.*, 34, 119-136.

Affaton, P., Rahaman, M.A., Trompette, R. & Sougy, J. (1991). The Dahomeyide Orogen: Tectonothermal Evolution and relationships with the Volta Basin. In: *The West African Orogens and Circum-Atlantic Correlatives*, Dallmeyer, R.D. & Lécorché, J.P. (Eds), pp. 107-122, Springer-Verlag,.

Aka, F.T., Nagao, K., Kusakabe, M., Sumino, H., Tanyileke, G., Ateba, B. & Hell, J. (2004). Symmetrical Helium isotope distribution on the Cameroon Volcanic Line, West Africa, *Chem. Geol.*, 203, 205-223.

Attoh K., Dallmeyer, R.D. & Affaton, P. (1997). Chronology of nappe assembly in the Pan-African Dahomeyide orogen, West Africa: evidence from $^{40}Ar/^{39}Ar$ mineral ages, *Precambrian Res.*, 82: 153-171.

Badejoko, T.A. (1986). Triclinicity of K-feldspars and trace element content of Mesozoic granites of Central Nigeria, *Chem. Geol.*, 54, 43-51.

Ball, E, Bard, J.P. & Soba, D. (1984). Tectonique tangentielle dans la catazone panafricaine du Cameroun: les gneiss de Yaoundé, *J. Afr. Earth Sci.* 2, 91-95.

Barbey, P., Macaudière, J. & Nzenti, J.P. (1990). High-pressure dehydration melting of metapelites: evidence from the migmatites of Yaoundé (Cameroon), *J. Petrol.*, 31, 401-427.

Baudin, P. (1991). Le magmatisme mésozoïque à cénozoïque du fossé de la Benoué (Nigéria). Géochronologie, pétrogenèse, cadre géodynamique, *Ph.D Thesis*, Univ. Aix-Marseille III. (unpub.)

Béa, A., Cochemé, J.J., Trompette, R., Affaton, P., Soba, D. & Sougy, J. (1990). Grabens d'âge Paléozoïque inférieur et volcanisme tholéiitique associé dans la région de Garoua, Nord Cameroun, *J. Afr. Earth Sci.*, 10, 657-667.

Begg, G.C., Griffin, W.L., Natapov, L.M., O'Reilly, S.Y., Grand, S.P., O'Neill, C.J., Hronsky, J.M.A., Poudjom Djomani, Y., Swain, C.J., Deen, T. & Bowden, P. (2009). The lithospheric architecture of Africa: seimic tomography, mantle petrology and tectonic evolution, *Geosphere*, 5, 23-50.

Benkhelil, J. (1989). The origin and evolution of the Cretaceous Benue Trough (Nigeria), *J. Afr. Earth Sci.*, 8, 251-282.

Berthé, D., Choukroune, P. & Jégouzo, P. (1979). Orthogneiss, Malonites and non coaxial deformation of granites: the example of the South Armorican shear zone, *J. Struct. Geol.*, 1, 31-42.

Bertrand, J.M.L, Michard, A. Boullier, A.M., Dautel, D., 1986. Structure and U-Pb geochronology of central Hoggar (Algeria); A reappraisal of its Pan-African evolution, Tectonophysics, 5, 955-972.

Bessoles, B. & Trompette, R. (1980). Géologie de l'Afrique. La chaîne panafricaine : " zone mobile d'Afrique Centrale (partie sud) et zone mobile soudanaise ", *Mém. B.R.G.M.*, Paris, 92, 396p.

Black, R., Lameyre, J. & Bonin, B. (1985). The structural setting of alkaline complexes, *J. Afr. Earth Sci.*, 3, 5-16.

Black, R., Latouche, L., Liegeois, J.P., Caby, R. & Bertrand, J.M. (1994). Pan-African displaced terranes in the Tuareg shield (Central Sahara), *Geology*, 22, 641-644.

Bosworth, W. (1992). Mesozoic and Early Tertiary rift tectonics in East Africa. In: C.J. Ebinger, H.K. Gupta and I.O. Nyambok (eds.), Seismology and related sciences in Africa, *Tectonophysics*, 209, 115-137.

Bouyo Houketchang, M., Toteu, S.F., Deloule, E., Penaye, J. & Van Schmus, W.R. (2009). U-Pb and Sm-Nd dating of high-pressure granulites from Tchollíré and Banyo regions: evidence for a Pan-African granulite facies metamorphism in north-central Cameroon, *J. Afr. Earth Sci.*, 54, 144-154.

Bowden, P. & Karche, J.P. (1984). Mid-plate A-type magmatism in the Niger-Nigeria province: age variations and implications. In: Klerkx, J., Michot, J. eds. Géologie Africaine, *Musée Royale d'Afrique Centrale*, Tervuren, 167-177.

Browne, S. & Fairhead, J.D. (1983). Gravity study of the Central African Rift System: a model of continental disruption. 1. The Ngaoundere and Abu Gabra rifts, *Tectonophysics*, 94, 187-203.

Brun, J.P. & Choukroune, P. (1981). Déformation progressive et structures crustales, *Rev. Géol. Dyn. et Géog. Phys.*, 23, 177-193.

Brunet, M., Dejax, J., Brillanceau, A., Congleton, J., Downs, W., Duperon-Laudoueneix, M., Eisenmann, V., Flanagan, K., Flynn, L., Heintz, E., Hell, J., Jacobs, L., Jehenne, Y., Ndeng, E., Mouchelin, G. & Pilbeam, D. (1988). Mise en évidence d'une sédimentation précoce d'âge Barrémien dans le fossé de la Bénoué en Afrique occidentale (Bassin du Mayo Oulo Léré, Cameroun), en relation avec l'ouverture de l'Atlantique sud, *C. R. Acad. Sci. Paris,* 306, 1125-1130.

Burke, K. C. & Dewey, J.F. (1974). Two plates in Africa during the Cretaceous? *Nature,* 249, 313-316.

Burke, K. (2001). Origin of the Cameroon Line of volcano-capped swells, *J. Geol.*, 109, 349-362.

Caby, R. (1989). Precambrian terranes of Benin-Nigeria and northeast Brazil and the late Proterozoic south Atlantic fit, *Geol. Soc. Amer. Sp. Paper,* 230, 145-645.

Caby, R.& Boessé, J.M. (2001). Pan-African nappe system in southwest Nigeria: the Ife-Ilesha schist belt, *J. Afr. Earth Sci.*, 33, 211-225.

Caen-Vachette, M., Tempier, P. & Kamgang, P. (1987). Le massif tertiaire du Nkogam (Ouest Cameroun) : caractréristiques principales et géochronologiques Rb/Sr sur roches totales, *J. Afr. Earth Sci.*, 6, 521-524.

Caldeira, R. & Munhá, J.M. (2002). Petrology of ultramafic nodules from São Tomé Island, Cameroon Volcanic Line (oceanic sector), *J. Afr. Earth Sci.*, 34, 231-246.

Casquet, C., Pankhurst, R.J., Rappela, C.W., Galindo, C., Fanning, C.M., Chiaradia, M., Baldo, E., Gonzalez-Casado, J.M. & Dahlquist, J.A. (2008). The Mesoproterozoic Maz terrane in the Western Sierra Pampeanas, Argentina, equivalent to the Arequipa-Antofalla block of southern Peru? Implications for west Gondwana margin evolution, *Gondwana Res.*, 13, 163-175.

Castaing, C., Feybesse, J.L., Thiéblemont, D., Triboulet, C. & Chèvremont, P. (1994). Palaeogeographical reconstructions of the Pan-African/ Brasiliano orogen: closure of oceanic domain or intracontinental convergence between major blocks, *Precambrian Res.*, 69, 327-344.

Cornachia, M. & Dars, R. (1983). Un trait structural majeur du continent africain: les linéaments centrafricains du Cameroun au golfe d'Aden, *Bull. Soc. Géol. Fr.*, 25, 101-109.

Corsini, M. & Vauchez, A., Caby, R. 1996. Ductile duplexing at a bend of a continental scale strike-slip shear zone: example from NE Brazil, *J. Struct. Geol.*, 18, 385-394.

Cosson, Y., Rocci, G. & Boullier, A.M. (1987). Rock deformation at a Pan-African plate boundary outlined by ophiolites in Aïr (Niger), *Terra Cognita,* 7, abstract, p.119.

Coulon, C., Vidal, P., Dupuy, C., Baudin, P., Popoff, M., Maluski, H. & Hermitte, D. (1996). The Mesozoic to Early Cenozoic magmatism of the Benue Trough (Nigeria); geochemical evidence for the involvement of the St. Helena plume, *J. Petrol.*, 37, 1341-1358.

Courtillot, V., Jaupart, C., Manighetti, I., Tapponnier, P. & Besse, J. (1999). On causal links between flood basalts and continental breakup, *Earth Planet Sci. Lett.*, 166, 177-195.

Demaiffe, D., Moreau, C. & Brown, W.L. (1991). Ring-complexes of Ofoud-type in Aïr, Niger, a new anorogenic-type anorthosite association. In: Kampunzu, A.B., Lubala,

R.T. (eds.), *Magmatism in extensional structural settings, The Phanerozoic African plate.*, pp. 359-382, Springer-Verlag, Berlin.

Demaiffe, D. & Moreau, C. (1996). Crustal growth in Aïr (Niger, West Africa): significance of the Palaeozoic anorthosite-bearing anorogenic province. In: Demaiffe, D. (ed.), *Petrology and geochemistry of magmatic suites of rocks in the continental and oceanic crusts*, pp. 35-49, Univ. Libre Bruxelles.

Déruelle, B., Moreau, C., Nkoumbou, C., Kambou, R., Lissom, J., Njonfang, E., Ghogomu, T.R. & Nono, A. (1991). The Cameroon Line: a review. In: Kampunzu, A.B., Lubala, R.T. (eds.), *Magmatism in extensional structural settings, The Phanerozoic African plate,* pp. 274-327, Springer-Verlag, Berlin.

Dumont, J.F. (1986). Identification par télédétection de l'accident de la Sanaga (Cameroun): sa position dans le contexte des grands accidents d'Afrique Centrale et de la limite nord du craton congolais, *Géodynamique*, 1, 13-19

Duncan, R.A. & Richards, M.A. (1991). Hotspots, mantle plumes, flood basalts, and true polar wander, *Rev. Geophys.*, 29, 31-50.

Ebinger, C.J., Bechtel, T.D., Forsyth, D.W. & Bowin, C.O. (1989). Effective elastic plate thickness beneath the East African and Afar plateaus and dynamic compensation of the uplifts, *J. Geophys. Res.*, 94, 2883-2901.

Ebinger, C. J. & Sleep, N. H. (1998). Cenozoic magmatism throughout east Africa resulting from impact of a single plume, *Nature*, 395, 788-791.

Ferré, E.C., Caby, R., Peucat, J.J., Capdevilla, R. & Monié, P. (1998). Pan-African post-collisional ferro-potassic granite and quartz-monzonite plutons of Eastern Nigeria, *Lithos*, 45, 255-279.

Ferré, E., Gleizes, G. & Caby, R. (2002). Obliquely convergent tectonics and granite emplacement in the Trans-Saharan belts of Eastern Nigerian: a synthesis, *Precambrian Res.*, 114, 199-219.

Fitton, J.D. & Dunlop, H.M. (1985). The Cameroon Line, West Africa and its bearing on the origin of oceanic and continental alkali basalts, *Earth Planet. Sci. Lett.*, 72, 23-38.

Fosso J., Ménard, J-J, Bardintgzeff, J-M, Wandji, P., Tchoua, F.M. & Bellon, H. (2005). Les laves du mont Bangou: une première manifestation volcanique éocène, à affinité transitionnelle, de la Ligne du Cameroun, *C.R.GeoScience*, 337, 315-325.

Ganwa, A. A. (2005). Les granitoïdes de Meiganga: étude pétrographique, géochimique structurale et géochronologique. Leur place dans la chaîne panafricaine, *Doct. d'Etat thesis, Univ. Yaoundé I*, Cameroon (unpubl.).

Goscombe, B.D. & Gray, D.R. (2008). Structure and strain variation at mid-crustal levels in a transpressional orogen: a review of Kaoko belt structure and the character of West Gondwana amalgamation and dispersal, *Gondwana Res.*, 13, 48-85.

Grant, N.K. (1978). Structural distinction between a sedimentary cover and an underlying basement in the 600 my. old Pan-African domain of northwestern Nigeria, West Africa, Geol. Soc. Am. Bull., 89, 50-58.

Guiraud, R. & Maurin, J-C. (1992). Early Cretaceous rifts of West and Central Africa: an overview, *Tectonophysics*, 213, 153-168.

Guiraud, R., Bellion, Y., Benkhelil, J. & Moreau, C. (1987). Post-Hercynian tectonics in Northern and Western Africa, *Geol. J.* , 22, 433-466.

Gurnis, M., Mitrovica, J.X., Ritsema, J. & van Heijst, H.J. (1999). Constraining mantle density structure using geological evidence of uplift rates: the case of the African superplume, *Geochem. Geophys. Geosys.*, 1, paper 1999GC000035.

Hartley, R., Watts, A.B. & Fairhead, J.D. (1996). Isostasy of Africa, *Earth Planet. Sci. Lett.*, 137, 1-18.

Hieronymus, C.F. & Bercovici, D. (2000). Non-hotspot formation of volcanic chains: control of tectonic and flexural stresses on magma transport, *Earth Planet. Sci. Lett.*, 181, 539-554.

Ishii, S., Tsunogae, T. & Santosh, M. (2006). Ultrahigh-temperature metamorphism in the Achankovil Zone: implications for the correlation of crustal blocks in southern India, *Gondwana Res.*, 10, 99-114.

Jahn, B., Caby, R. & Monie, P. (2001). The oldest UHP eclogites of the World: age of UHP metamorphism, nature of protoliths and tectonic implications, *Chem. Geol.*, 178, 143-158.

Jallouli (1989). Caractérisation du comportement mécanique de la lithosphère continentale à partir des données gravimétriques. Implications géodynamiques pour la chaîne panafricaine du Sahara Central de l'Afrique (Aïr), *Ph.D thesis, Univ. Paris-sud* (unpubl.).

Jégouzo, P. (1984). Evolution structurale du Sud-Ouest Cameroun durant l'orogène panafricain. Association de tectoniques cisaillante et chevauchante, *Colloque CNRS, Chevauchement et Déformation*, Toulouse, p. 23.

Kagou Dongmo, A., Nkouathio, D.G., Pouclet, A., Bardintzeff, J.M., Wandji, P., Nono, A., & Guillou, H. (2010). The discovery of late Quaternary volcanic basalt on Mount Bambouto: implications for recent widespread volcanic activity in the southern Cameroon Line, *J. Afr. Earth Sci.*, 57, 96-108.

Kamdem, J.B., Kraml, M., Keller, J. & Henjes-Kunst, F. (2002). Cameroon Line magmatism: conventional K/Ar and single-crystal laser $^{40}Ar/^{39}Ar$ ages of rocks and minerals from the Hossere Nigo anorogenic complex, Cameroon, *J. Afr. Earth Sci.*, 35, 99-105.

Kamgang, P., Njonfang, E., Nono, A., Gountie Dedzo, M., Tchoua, F.M., 2010. Petrogenesis of a silicic magma system: geochemical evidence from Bamenda Mountains, NW Cameroon, Cameroon Volcanic Line, J. Afr, Earth Sci., 58, 285-304.

Kinabo, B.D., Atekwana, E.A., Hogan, J.P., Modisi, M.P., Wheaton, D.D. & Kampunzu, A.B. (2007). Early structural development of the African rift zone, NW Botswana, *J. Afr. Earth Sci.*, 48, 125-136.

Kinnaird, J.A. (1985). Hydrothermal alteration and mineralization of the alkaline anorogenic ring-complexes of northern Nigeria, *J. Afr. Earth Sci.*, 3, 185-222.

Küster, D. & Liégeois, J.P. (2001). Sr, Nd isotopes and geochemistry of the Bayuda Desert high-grade metamorphic basement (Sudan): an early Pan-African oceanic convergent margin, not the edge of the East Saharan ghost craton? *Precambrian Res.*, 109, 1-23.

Lasserre, M. (1978). Mise au point sur les granitoïdes dits «ultimes» du Cameroun: Gisements, pétrographie et géochronologie, *Bull. BRGM.*, 2, 143-159.

Lee, D.C., Halliday, A.N., Fitton, J.G. & Poli, G. (1994). Isotopic variation with distance and time in the volcanic islands of the Cameroon Line: evidence for a mantle plume origin, *Earth Planet. Sci. Let. 123, 119-138.*

Lee, D.C., Halliday, A.N., Davies, G. R., Essene, E.J., Fitton, J.G. & Temdjin, R. (1996). Melt enrichment of shallow depleted mantle: a detailed petrological, trace element and isotopic study of mantle derived xenoliths and megacrysts from the Cameroon Line, *J. Petrol.*, 17, 415-441.

Liégeois, J.P., Black, R., Navez, J. & Latouche, L. (1994). Early and late Pan-African orogenies in the Aïr assembly of terranes (Tuareg shield, Niger), *Precambrian Res.*, 67, 59-88.

Lithgow-Bertelloni, C. & Silver, P. (1998). Dynamic tomography, plate driving forces and the African superswell, *Nature*, 395, 269-272.

Maluski, H., Coulon, C., Popoff, M. & Baudin, P. (1995). $^{40}Ar/^{39}Ar$ chronology, petrology and geodynamic setting of Mesozoic to early Cenozoic magmatism from the Benue Trough, Nigeria, *J. Geol. Soc. Lon.*, 152, 311-326.

Marzoli, A., Piccirillo, E.M., Renne, P.R., Bellieni, G., Lacumin, M., Nyobe, J.B. & Aka, F.T. (2000). The Cameroon Volcanic Line revisited: petrogenesis of continental basaltic magmas from lithospheric and asthenospheric mantles sources, *J. Petrol.*, 41, 87-109.

Maurin, J.C., Benkhelil, J., Robineau, B., 1986. Fault rocks of the Kaltungo lineament, NE Nigeria, and their relationship with Benue Trough tectonics, J. Geol. Soc. Lon. 143, 587-599.

Meyers, J.B., Rosendahl, B.R., Harrison, C.G.A. & Ding, Z.-D. (1998). Deep-imaging seismic and gravity results from offshore Cameroon Volcanic Line and speculation of African hot-lines, *Tectonophysics*, 284, 31-63.

Montigny, R., Ngounouno, I. & Déruelle, B. (2004). Âges K-Ar des roches magmatiques du fossé de Garoua (Cameroun) : leur place dans le cadre de la « Ligne du Cameroun », *C.R. Geoscience*, 336, 1463-1471.

Moore, A., Blenkinsop, T., and Cotterill (Woody), F. (2008). Controls on post-Gondwana alkaline volcanism in Southern Africa, *Earth Planet. Sci. Lett.*, 268, 151-164.

Moreau, C., Regnoult, J.M., Déruelle, B. & Robineau, B. (1987). A new tectonic model for the Cameroon Line, Central Africa, *Tectonophysics*, 139, 317-334.

Moreau, C. Brown, W.L., Demaiffe, D., Rocci, G. & Perez, J.B. (1991). Palaeozoic magmatism in the Aïr massif (Niger). In: Kampunzu, A.B., Lubala, R.T. (eds.). *Magmatism in extensional structural settings, The Phanerozoic African plate*, pp. 345-358, Springer-Verlag, Berlin.

Moreau, C., Demaiffe, D., Bellion, Y. & Boullier, A-M. (1994). A tectonic model for the location of Palaeozoic ring complexes in Aïr (Niger, West Africa), *Tectonophysics*, 234, 129-146.

Moundi, A., Wandji, P., Bardintzeff, J.-M., Ménard J.-J., Okomo Atouba, L.C., Reusser, E., Bellon, H. & Tchoua, M.F. (2007). Les basaltes éocènes à affinité transitionnelle du plateau Bamoun, témoins d'un réservoir mantellique enrichi sous la Ligne Volcanique du Cameroun, *C. R. Geoscience*, 339, 396-406.

Mvondo, H., den Brok, S.W.J. & Mvondo Ondoa, J. (2003). Evidence for symmetric extension and exhumation of the Yaoundé nappe (Pan-African fold belt, Cameroon), *J. Afr. Earth Sci*, 36, 215-231.

Mvondo, H., Owona, S., Mvondo Ondoua, J. & Essono, J. (2007). Tectonic evolution of the Yaoundé segment of the Neoproterozoic Central African orogenic belt in southern Cameroon, *Can. J. Earth Sci.*, 44, 433-444.

Nana, R., Dietrich, V.J., Tchoua, F.M. & Reusser, E. (2001). Pétrologie des péridotites en enclaves dans les basaltes récents de la région de Nyos (Ligne du Cameroun), *Geoscience Soc. Cameroon, Spec. Publ.*, Abstract Issue (GSAF12) 1, 83-84.

Nédélec, A., Myniem, D. & Barbey, P. (1993). High-P–high-T anatexis of Archaean tonalitic grey gneisses: the Eseka migmatites, Cameroon, *Precambrian Res.*, 62, 191-205.

Neves, S.P. & Mariano, G. (1999). Assessing the tectonic significance of a large transcurrent shear zone system: the Pernambuco lineament, northeastern Brazil, *J. Struct. Geol.*, 21, 1369-1383.

Neves, S.P., Melo, S.C., Moura, C.A.V., Mariano, G. & Rangel Da Silva, J.M. (2004). Zircon Pb-Pb geochronology of the Caruaru Area, Northeastern Brazil : Temporal Constraints on the Proterozoic Evolution of Borborema Province, *Int. Geol. Rev.*, 46, 52-63.

Ngako, V., Jégouzo, P. & Nzenti, J.P. (1991). Le Cisaillement Centre Camerounais. Rôle structural et géodynamique dans l'orogenèse panafricaine, *C. R. Acad. Sci. Paris*, 313, 457-463.

Ngako, V. (1999). Les Déformations continentales panafricaines en Afrique Centrale. Résultat d'un poinçonnement de type himalayen, *Doct. d'Etat thesis, Univ. Yaoundé I*, Cameroon (unpubl.)

Ngako, V., Affaton, P., Nnange, J.M. & Njanko, Th. (2003). Pan-African tectonic evolution in central and Southern Cameroon: transpression and transtension during sinistral shear movements, *J. Afr. Earth Sci.*, 36, 207-214.

Ngako, V., Njonfang, E., Aka Tongwa, F., Affaton, P. & Nnange Metuk, J. (2006). The North-South Paleozoic to Quaternary trend of alkaline magmatism from Niger-Nigeria to Cameroon : complex interaction between hotspots and precambrian faults, *J. Afr. Earth Sci.*, 45, 241-256.

Ngako, V., Affaton, P. & Njonfang, E. (2008). Pan-African tectonics in northwestern Cameroon: implication for the history of western Gondwana, *Gondwana Res.*, 14, 509-522.

Ngounouno, I., Moreau, C., Déruelle, B., Demaiffe, D. & Montigny, R. (2001). Pétrologie du complexe alcalin sous-saturé de Kokoumi (Cameroun), *Bull. Soc. Géol. Fr.*, 172, 675-686.

Ngounouno, I., Nguihdama, D., Kamgang, P. & Déruelle, B. (2008). Petrology of spinel lherzolite xenoliths in alkali basalts from Liri, South of Kapsiki plateau (northernmost Cameroon Hot Line), *J. Cam. Acad. Sci.*, 8, 31-42.

Nguiessi Tchankam, C., Nzenti, J.P., Nkonguin Nsifa, Tempier, P. & Tchoua F.M. (1997). Les granitoïdes calco-alcalins syn-cisaillement de Bandja dans la chaîne panafricaine nord-équatoriale au Cameroun, *C.R. Acad. Sci. Paris*, 325, 95-101.

Njanko, T., Nédélec, A. & Affaton, P. (2006). Synkinematic high-K calc-alkaline plutons associated with the Pan-African Central Cameroon Shear Zone (W-Tibati area): petrology and geodynamic significance, *J. Afri. Earth Sci.*, 44, 494-310.

Njonfang, E., Moreau, C. & Tchoua, F.M. (1998). La bande mylonitique Foumban-Bankim, Ouest-Cameroun: une zone de cisaillement de haute temperature, *C. R. Acad. Sci. Paris*, 327, 735-741.

Njonfang, E., Ngako, V., Kwekam, M. & Affaton, P. (2006). Les orthogneiss calco-alcalins de Foumban - Bankim: témoins d'une zone interne de marge active panafricaine en cisaillement, *C. R. Geoscience*, 338, 606-616.

Njonfang, E., Nono, A., Kamgang, P., Ngako, V. & Tchoua, M.F, 2010. The Cameroon Line alkaline magmatism (Central Africa) : a reappraisal, *Geol. Soc. Amer., Spec. Publ.* (in press) .

Nkoumbou, C., Déruelle, B. & Velde, D. (1995). Petrology of Mt Etinde nephelinites series, *J. Petrol.*, 36, 373-395.

Nkoumbou, C., Yonta Goune, C., Villiéras, F., Njopwouo, D., Yvon, J., Ekodeck, G.E. & Tchoua, F. (2006). Découverte des roches à affinité ophiolitique dans la chaîne panafricaine au Cameroun: les talcschistes de Ngoung, Lamal Pougue et Bibodi Lamal, *C. R. Geoscience*, 338, 1167-1175.

Nzenti, J.P. (1998). Neoproterozoic alkaline meta-igneous rocks from the Pan-African North Equatorial Fold Belt (Yaoundé, Cameroon): biotitites and magnetite rich pyroxenites, *J. Afr. Earth Sci.*, 26, 37-47.

Nzenti, J.P., Kapajika, B., Wörner, G. & Lubala, T.R. (2006). Synkinematic emplacement of granitoids in a Pan-African shear zone in Central Cameroon, *J. Afr. Earth Sci.*, 45, 74-86.

O'Connor, J.M. & Le Roex, A.P. (1992). South Atlantic hot spot-plume systems: 1. Distribution of volcanism in time and space, *Earth Planet. Sci. Lett.*, 113, 343-364.

Orajaka, J.P. (1986). Geochemistry of Kaffo Valley albite-riebeckite-granite, Liruei granite ring-complex, Northern Nigeria, *Chem. Geol.*, 56, 85-92.

Paulsen, T.S., Encarnacion, J., Grunov, A.M., Layer, P.W. & Watkeys, M. (2007). New age constraints for a short pulse on Ross orogen deformation triggered by East-West Gondwana suturing, *Gondwana Res.*, 12, 417-427.

Penaye, J., Toteu, S.F., Van Schmus, W.R. & Nzenti, J.P. (1993). U-Pb and Sm-Nd preliminary geochronologic data on the Yaoundé series, Cameroon: re-interpretation of the granulitic rocks as the suture of a collision in the "Centrafrican" belt, *C. R. Acad. Sci. Paris*, 317, 789-794.

Penaye, J., Kröner, A., Toteu, S.F, Van Schmus, W.R. & Doumnang, J.C. (2006). Evolution of the Mayo Kebbi region as revealed by zircon dating: An early (ca. 740 Ma) Pan-African magmatic arc in southwestern Chad, *J. Afr. Earth Sci.*, 44, 530-542.

Poudjom Djomani, Y.H., Nnange, J.M., Diament, M., Ebinger, C.J. & Fairhead, J.D. (1995). Effective elastic thickness and crustal thickness variations in west central Africa inferred from gravity data, *J. Geophys. Res.*, 100, 22047-22070.

Poudjom Djomani, Y.H., Diament, M. & Wilson, M. (1997). Lithospheric structure across the Adamawa plateau (Cameroon) from gravity studies, *Tectonophysics*, 273, 317-327.

Rahaman, M.A., Van Breemen, O., Bowden, P. & Bennett, J.N. (1984). Age migration of anorogenic ring-complexes in northern Nigeria, *J. Geol.*, 92, 173-184.

Rajesh, K.G. & Chetty, T.R.K. (2006). Structure and tectonics of the Achankovil Shear Zone, southern India, *Gondwana Res.*, 10, 86-98

Seme Mouangue, A. (1998). Géochimie, métamorphisme et métallogénie des formations ultrabasiques du secteur de Lomié (Sud-Est Cameroun), *Doct. 3ème cycle thesis, Univ. Yaoundé I*, Cameroon (unpubl.).

Smith, A.D. & Lewis, C. (1999). The planet beyond the plume hypothesis, *Earth Sci. Rev.*, 48, 135-182.

Souza, Z.S., Montel, J.-M., Costa Giola, S.M.L., Hollanda, M.H.B.M, Nascimento, M.A.L., Sà, E.F.J., Amaro, V.E., Pimentel, M.M., Lardeaux, J.-M. & Veschambre, M. (2006).

Electron microprobe dating of monazite from high-T shear zones in the São José de Campestre Massif, NE Brazil, *Gondwana Res.*, 9, 441-455.

Tagne-Kamga, G., Mercier, E., Rossy, M. & Nsifa, E.N. (1999). Synkinimatic emplacement of the Pan-African Ngondo complex (west Cameroon, central Africa), *J. Afr. Earth Sci.*, 28, 675-691.

Temdjim, R., Boivin, P., Chazot, G., Robin, C. & Rouleau, E. (2004). L'hétérogénéité du manteau supérieur à l'aplomb du volcan du Nyos (Cameroun) révélée par les enclaves ultrabasiques, *C.R. Geoscience Paris*, 336, 1239-1244.

Toteu, S.F. (1990). Geochemical characterisation of the main petrographical and structural units of Northern Cameroon; implications for the Pan-African evolution, *J. Afri. Earth Sci.*, 10, 615-624.

Toteu, S.F., Macaudière, J., Bertrand, J.M. & Dautel, D. (1990). Metamorphic zircon from North Cameroon; implications for the Pan-African evolution of Central Africa, *Geologische Rundschau*, 79, 777-788.

Toteu, S.F., Bertrand, J.M., Penaye, J., Macaudière, J., Angoua, S. & Barbey, P. 1991. Cameroon: a tectonic keystone in the Pan-African network. In the Early Proterozoic Trans-Hudson orogen of North America, Lewry, J.F. and Stauffer, M.R. Eds, *Geol. Assoc. Can., Spec. Paper*, 37, 483-496.

Toteu, S.F., Van Schmus, W.R., Penaye, J. & Nyobé, J.B. (1994). U-Pb and Sm-Nd evidence for Eburnean and Pan-African high-grade metamorphism in cratonic rocks of Southern Cameroon, *Precambrian Res.*, 67, 321-347.

Toteu, S.F., Van Schmus, Penaye, J. & Michard, A. (2001). New U-Pb and Sm-Nd data from North-Central Cameroon and its bearing on the pre- Pan-African history of Central Africa, *Precambrian Res.*, 108, 45-73.

Toteu, S.F., Penaye, J. & Poudjom Djomani, Y.H., 2004. Geodynamic evolution of the Pan-African belt in Central Africa with special reference to Cameroon, *Can. J. Earth Sci.*, 41, 73-85.

Toteu, S. F., Penaye, J., Deloule, E., Van Schmus, W. R. & Tchameni R. (2006). Diachronous evolution of volcano-sedimentary basins North of the Congo craton: Insights from U–Pb ion microprobe dating of zircons from the Poli, Lom and Yaounde´ Groups (Cameroon), *J. Afri. Earth Sci.*, 44, 428-442.

Trompette, R. (1994). *Geology of Western Gondwana (2000–500 Ma). Pan-African–Brasiliano aggregation of South America and Africa.* Balkema, A.A., Rotterdam, 350p.

Tubosun, A., Lancelot, I.R., Rahaman, M.A. & Ocan, O. (1984). U-Pb Pan-African ages of two charnockite-granite associations from south western Nigeria, *Contrib. Mineral. Petrol.*, 88, 188-195.

Tucker, R.D., Kusky, T.M., Buchwaldt, R. & Handke, M.J. (2007). Neoproterozoic nappes and superposed folding of the Itremo Group, west-central Madagascar, *Gondwana Res.*, 12, 356-379.

Vaughan, A.P.M. & Pankhurst, R.J. (2008). Tectonic overview of the West Gondwana margin, *Gondwana Res.*, 13, 150-162.

Wandji, P.; Tchokona Seuwi, D., Bardintzeff, J.-M., Bellon, H. & Platevoet, B. (2008). Rhyolites of the Mbépit Massif in the Cameroon Volcanic Line: an early extrusive volcanic episode of Eocene age, *Mineral. Petrol.*, 94, 271-286.

Wandji, P., Tsafack, , Bardintzeff, J.-M., Nkouathio, D.G., Kagou Dogmo, A., Bellon, H. & Guillou, H. (2009). Xenoliths of dunites, wehrlites and clinopyroxenites in the basanites from Batoke volcanic cone (Mount Cameroon, Central Africa): petrogenetic implications, *Mineral. Petrol.*, 96, 81-98.

Wennberg, O.P. (1996). Superimposed fabrics due to reversal of shear sense: an example from the Bergen Arc Shear Zone, western Norway, *J. Struct. Geol.*, 18, 871-889.

Wilson, M. (1992). Magmatism and continental rifting during the opening of the South Atlantic Ocean: a consequence of Lower Cretaceous super-plume activity? In: Storey, B.C., Alabaster, T., Pankhurst, R.J. (eds.), Magmatism and the causes of continental Break-up, *Geol. Soc. Spec. Publ.*, 68, 241-255.

Wilson, M. (1997). Thermal evolution of the Central Atlantic passive margins: continental break-up above a Mesozoic super plume, *J. Geol. Soc. Lon.*, 154, 491-495.

Wilson, M. & Guiraud, R. (1992). Magmatism and rifting in West and Central Africa, from Late Jurassic to Recent times, *Tectonophysics*, 213, 203-225.

Yonta-Ngoune, Cl., Nkoumbou, C., Barbey, P., Le Breton, N., Montel, J-Mc. & Villieras, F. (2010). Geological context of the Boumnyebel talcschists (Cameroon) : inferences on the Pan-African belt of Central Africa, *C. R. Geoscience Paris*, 342, 108-115.

Part 2

Tectonics of Europe and the Near East

2

Lithospheric Structure and Tectonics of the Eastern Alps – Evidence from New Seismic Data

Ewald Brückl
Vienna University of Technology
Austria

1. Introduction

During the last two decades, seismic campaigns carried out in the Eastern Alps and the surrounding geological units contributed substantially to our knowledge and understanding of the deep lithospheric structure and tectonics in this area. The main topics addressed by these efforts concern the collision between tectonic plates and fragments of the Adriatic realm with the European platform and the many geodynamic reactions induced by this large-scale process. The individual campaigns yielded valuable geophysical data in the form of cross-sections and 3D models of the lithosphere and upper mantle, as well as geologic and tectonic interpretations. However, the diversity of the geologic structures and tectonic processes in the wider Alpine area still pose many unanswered questions. This chapter provides an overview of the main results achieved by the seismic campaigns and presents a current image of the crust and upper mantle. Generalized geological interpretations will be given and tectonic processes inferred on the basis of the imaged structures. Most of the interpretational models will refer to collision and post-collision times and today's tectonics will be emphasized. Although this study considers information from several disciplines the author is aware that it is a limited projection of the comprehensive geology and tectonics in the area.

2. Geological overview

The study of the lithospheric structure and tectonics of the Eastern Alps calls for the consideration of the surrounding geological units. These units are the Western Alps, the European platform and the Molasse Basin in the North, the Carpathians and the Pannonian basin in the east, and the Dinarides and Adriatic foreland in the south. A geological map, generalized after Schmid et al. (2004), is shown in Fig. 1.

The European platform and its most southern promontory, the Bohemian Massif, comprise crystalline rocks which underwent the Caledonian or Variscian orogeny. The northern rim of the Alps and Carpathians is composed of a Flysch belt of Helvetic and Penninic provenance. The main body of the Eastern Alps consists of the Austro-Alpine units. Ocean-derived (Penninic Ocean, Alpine Tethys) metamorphic units and even European basement is exhumed in tectonic windows (Engadin, Tauern, Rechnitz window) (Oberhauser, 1980). The Southern Alps join the Eastern Alps at the Periadriatic or Insubric line (PAL). The peninsula Istria resembles the main part of the tectonically little-disturbed Adriatic foreland (e.g., Poljak et al., 2000). The Dinarides are located south-east of the Southern Alps (e.g., Doglioni, 1987) and represent a bifurcation of the whole Alpine orogenic system. The Molasse basin in the north and the Adriatic and Venetian plain in the south are foreland basins of the Alpine orogen and are proof of the north-south oriented compression regime. The Styrian basin and especially the Pannonian basin resemble large extensional structures. The Vienna basin is a pull-apart basin within the north-east trending Vienna transfer fault system. The thickness of the Neogene sediments in the basins exceeds 5 km in some places according to the results of hydrocarbon exploration.

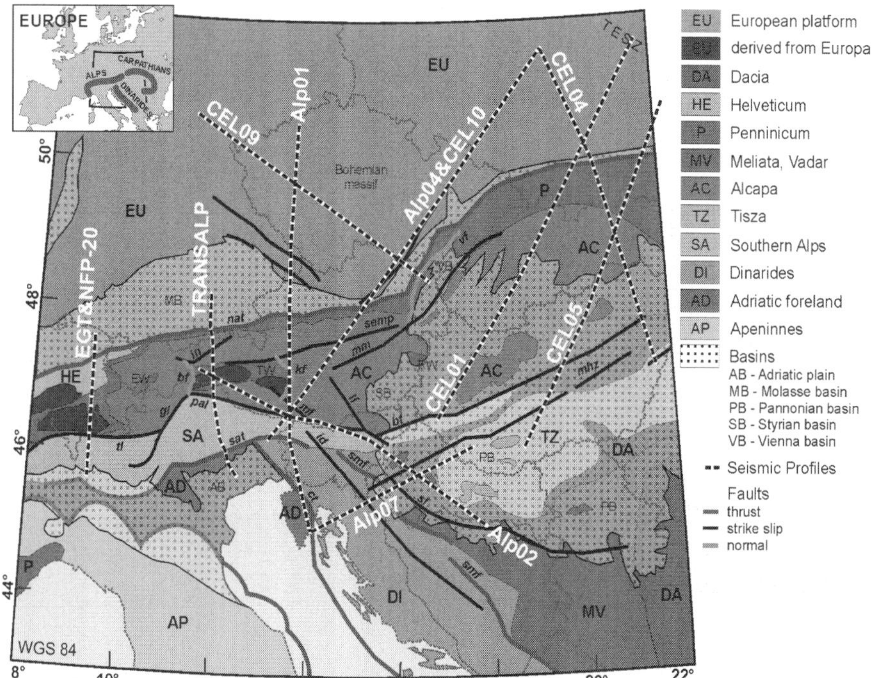

bf - Brenner fault, *bt* - Balaton fault, *ct* - Cicarija fault, *gl* - Giudicarie line, *id* - Idria fault, *in* - Inntal fault, *kf* - Katschberg fault, *lf* - Lavant fault, *mf* - Malta fault, *mhz* - Mid Hungarian fault system, *mm* - Muhr-Mürz fault, *nat* - Northern floor thrust of the Eastern Alps, *pal* - Periadriatic line, *sat* - South alpine thrust fault, *semp* - Salzach-Ennstal-Mariazell-Puchberg fault, *sf* - Sava fault, *smf* - southern marginal fault of the Pannonian basin, *tl* - Tonale line, *vf* - Vienna transfer fault, TW - Tauern window, EW - Engadin window, RW - Rechnitz window, TESZ - Transeuropean suture zone

Fig. 1. Geological map

Main thrust faults are the Northern floor thrust of the Eastern Alps and the South Alpine thrust fault, as well as the Cicarija fault at the south-western front of the Dinarides (e.g., Oberhauser, 1980; Castellarin and Cantelli, 2000; Vrabec and Fodor, 2006). Major sinistral strike-slip faults within the Austro-Alpine units are the Inntal, SEMP (Salzach-Ennstal-Mariazell-Puchberg), and Mur-Mürz faults. Conjugate dextral strike-slip faults are the Möll and Lavant faults. These faults continue to the south-east into the Idria and Sava faults. The Brenner and Katschberg normal faults delimit the Tauern window in the west and east.

Right-lateral movements of about 100 km are reported for the PAL since the Oligocene (Mancktelow et al., 2001; Vrabec and Fodor, 2006). To the east, the PAL transits into the Mid Hungarian fault zone (MHZ), which separates the ALCAPA and TISZA units of the Pannonian basin (e.g., Tomljenović and Csontos, 2001; Horvath, 2006).

The main tectonic processes which built up the Alpine orogenic system started in Early and Mid-Jurassic times (~190 Ma) and are strongly related to the spreading history of the Northern and Central Atlantic. At that time, closure of the Paleotethys by subduction under the Eurasian megaunit was already completed and the Neotethys fully developed. Smaller ocean basins existed in the south-east of ALCAPA (e.g., Stampfli & Kozur, 2006; Le Pichon et al., 1988). In the north-west, rifting initiated the development of the Ligurian and Piemont branches of the Alpine Tethys. The displacement of Africa relative to the European platform led to a complex sequence of rifting, break-up of continental blocks, seafloor spreading, subduction of oceanic and continental lithosphere, and re-amalgamation and suturing of micro-plates. Considering the Eastern Alps and the adjoining Carpathians, Dinarides and Southern Alps, the opening and closure of the Meliata and Vardar oceans in the south and the Alpine Tethys in the north strongly influenced the development of these mountain chains. An early orogenic phase, the Eoalpine orogeny, occurred after the subduction of the Meliata ocean, about 130 - 85 Ma ago. Seafloor spreading of the Alpine Tethys ended about 130 Ma ago and subduction of this ocean initiated ~80 Ma with Europe as the passive margin. At the same time, the Adriatic micro-plate started to move more independently from Africa to the north-west (e.g., Handy et al., 2010).

After subduction of the Alpine Tethys and thrusting of the Penninic Flysch units during the Eocene, ~50 Ma ago (Decker and Peresson, 1996), the collision between Europe and Adria occurred ~35 Ma ago. This process induced far-reaching over-thrusting, north-vergent folding, and stacking of Austro-Alpine nappes to the north, the passive European continental margin. Slab break off (Davies and Blanckenburg, 1995; Wortel and Spakman, 2000) in the central part of the Eastern Alps was initiated in the Oligocene (35 - 30 Ma). Subduction of Adriatic continental mantle to the southwest formed the Apennines. Continued indentation of the Adriatic microplate to the north caused back-thrusting in the Southern Alps and the Dinarides (Castellarin and Cantelli, 2000; Pamić et al., 2002). Post-collision shortening of the orogeny lasted up to the late Miocene (Decker and Peresson, 1996, Ortner et al., 2006). Continuing convergence also left its traces in the central crystalline core including the Tauern Window (Lammerer et al., 2008) and the south-vergent thrust belt of the Southern Alps (Doglioni, 1987; Schönborn, 1999).

During the Miocene, the closure of the remaining ocean in the area of the Carpathian embayment led to the development of the Pannonian Basin and finally the Carpathians (e.g., Royden, 1993). Due to continued indentation of the Adriatic plate in combination with gravitational collapse directed to the unconstrained margin represented by the Pannonian basin, major blocks of the Eastern Alps extruded laterally to the east (Ratschbacher et al., 1991; Linzer et al., 2002). These processes enabled exhumation of ophiolites and even European crystalline basement below oceanic deposits in the Engadin, Tauern, and Rechnitz window (Fügenschuh et al., 1997; Rosenberg et al., 2004). Current tectonic activity due to continuing convergence between Africa and Europe has shifted southward to the subduction zones of the Calabrian and Hellenic arcs. However, the tectonic processes in the Eastern Alps that began in the Miocene are still at work, but diminishing.

3. Seismic profiles through the crust

The exploration of the lithosphere of the Eastern Alps and the surrounding area by WAR/R (wide angle reflection/refraction) experiments started around the Eschenlohe quarry, southern Germany, about 50 years ago (Giese and Prodehl, 1976). The Alpine longitudinal profile (ALP'75) extended along the axis of the Western and Eastern Alps to the Pannonian basin (e.g., Yan and Mechie, 1989; Scarascia and Cassinis, 1997). WAR/R as well as steep angle reflection profiles were also measured in the Bohemian Massif, the Pannonian Basin, and the Dinarides (e.g., Beránek and Zátopek, 1981; Posgay et al., 1981; Joksović and Andrić, 1983). The EGT& NFP-20 traverse provided a detailed crustal image along an approximately north-south oriented profile near the western border of the Eastern Alps. WAR/R data and near-vertical reflection data of the EGT and NFP-20 projects (Valasek et al., 1991; Ye et al., 1995; Pfiffner et al., 1997; Kissling et al., 2006) were combined. TRANSALP is a seismic traverse over the central part of the Eastern Alps from Munich to Venice (e.g., Gebrande et al., 2006 and references therein). The main activity concentrated on steep angle, high resolution reflection seismic profiling. Wide angle reflection and refraction tomography (Bleibinhaus et al., 2006), receiver functions (Kummerov et al., 2004), and teleseismic tomography (Lippitsch et al., 2003) were integrated into the transect. Gravimetric studies supplemented the seismic transect (Ebbing et al., 2006; Zanolla et al., 2006). In 2000 and 2002 the large WAR/R experiments CELEBRATION 2000 and ALP 2002 covered an area from the European Craton in the north-east to the Adriatic foreland in the south-west (Guterch et al., 2003a, 2003b; Brückl et al., 2003). During each deployment, about 1000 receivers were arranged on a network of seismic profiles and each shot was recorded on all profiles, thus realizing a 3D data acquisition. The reflection profiles T2 in the Western Carpathians (Tomek, 1993) and NESTMK (Graßl et al., 2004) near the eastern border of the Eastern Alps yielded valuable data about the crustal structures at the transition of the Bohemian Massif and the Eastern Alps to the Pannonian basin.

For this study 10 profiles were selected from the aforementioned projects and are presented in a generalized style (Fig. 2, for location see Fig. 1). The most important structures are the Moho discontinuity, suture zones within crust and upper mantle, accretion wedges, and decollements. The extent of major tectonic units and the tectonic style of collision, extension and transfer/transform zones are also of major interest.

The cross-section of the EGT&NFP-20 traverse is based on Valasek et al. (1991), Ye et al. (1995), Pfiffner et al. (1997), and Kissling et al. (2006). The WAR/R data reveal the P-wave velocity structure of the crust, Moho depth, and Pn-velocity below the Moho. The near-vertical reflection data allow for the delineation of the Helvetic and Penninic nappes and the basal decollement, a reflective European lower crust and the Moho boundary. Stacking of the wide angle PmP reflections gives further support for the Moho boundary. The geological/tectonic interpretation of Schmid et al. (2004) is adopted herein. Structures of particular interest for this study concern the distinct jump of the Moho boundary from Europe to Adria, the indenting Adriatic lower crust, the Helevetic and Penninic nappes overthrusted to the north, and the seismic signature of back-thrusting in the Southern Alps.

The TRANSALP profile is based on the seismic data of Lüschen et al. (2006), Kummerov et al. (2004), and the interpretations of Castellarin et al. (2006) and Lammerer et al. (2008). The East-Alpine nappes, which overthrusted Flysch and Molasse to the north, are well-imaged down to a depth of 15 km. In the Southern Alps, thrust faults and nappe decollements reach down to 20 km.

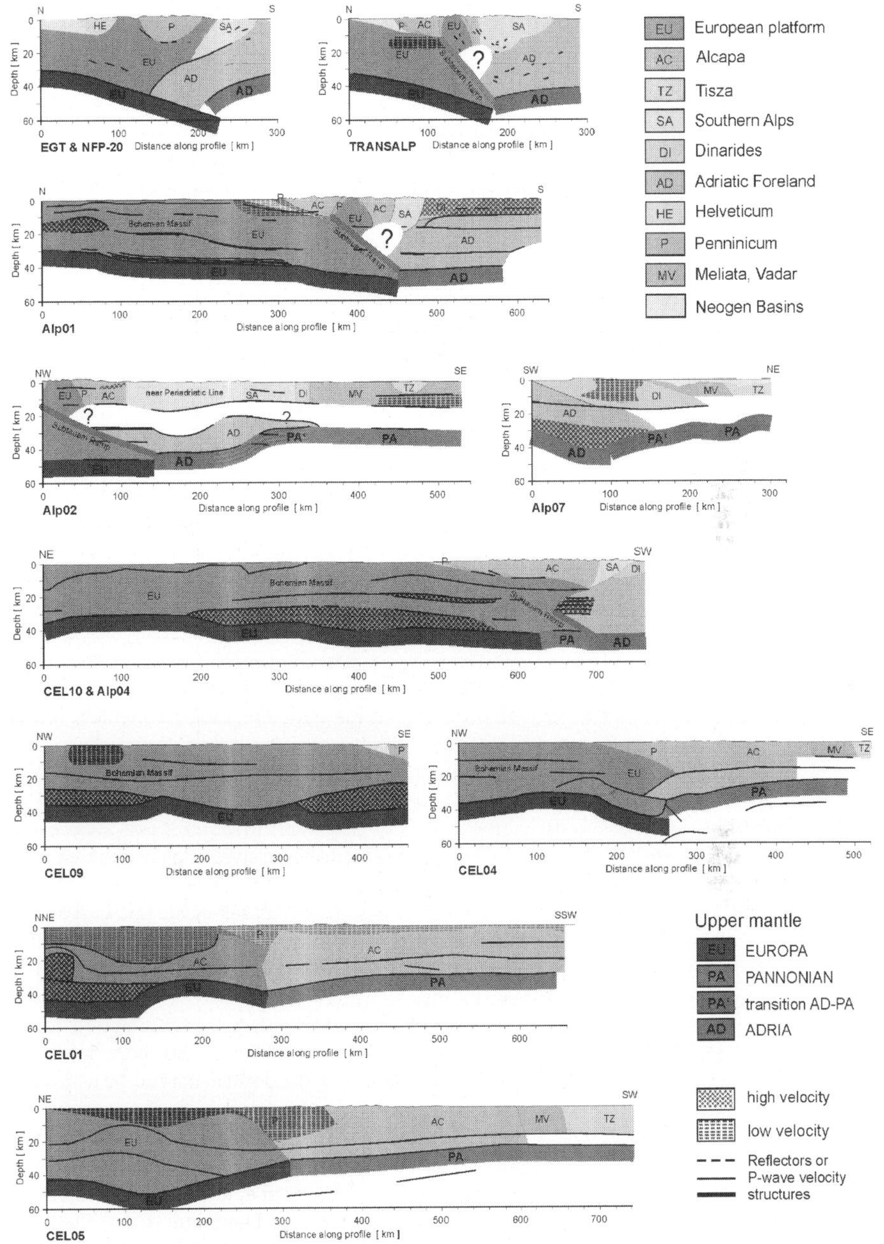

Fig. 2. Seismic profiles

The crystalline European basement below the East-Alpine nappes shows extensional structures which date back to the spreading of the Alpine Tethys. Tomography revealed a low-velocity zone below the Eastern Alps thrust belt (Bleibinhaus and Gebrande, 2006). The Periadriatic line is not uniquely imaged.

The Subtauern ramp is a crustal scale fault over which the Tauern window was exhumed by thrusting and folding. A reflective Adriatic lower crust indicates thickening. The Moho boundary is constrained by steep and wide angle reflections and by receiver functions (Kummerov et al., 2004) which indicate an upward jump from the European to the Adriatic Moho.

Alp01 (Brückl et al., 2007) extends from the Bohemian Massif to the Adriatic foreland (Istria). As a WAR/R profile, Alp01 cannot compete with TRANSALP with regard to resolution. However, the topography of the Moho and the velocity structure of the crust supplied sufficient constraints to transfer the tectonic style of the collision between Europe and Adria (especially the Subtauern ramp) from TRANSALP to Alp01. The shape of the gneiss core of the Tauern window is based on the "lateral extrusion model" proposed for the TRANSALP profile (TRANSALP Working Group, 2002; Castellarin et al., 2006). Clear evidence is also given for mid-, or lower Adriatic crustal thickening and a little-deformed Adriatic foreland.

Alp02 (Brückl et al., 2007) starts at the western end of the Tauern window and extends parallel to the Periadriatic line into the Pannonian basin and further south-east to the TISZA unit. Like Alp01, the interpretation in the area of the Tauern window follows TRANSALP. The Moho is fragmented into Europe (EU), Adria (AD) and the newly interpreted Pannonian fragment (PA), which corresponds to the thin Pannonian crust. Elastic plate modelling, but also the results from Alp07 indicate a transition zone between AD and PA, designated PA'. Most probably, lower Adriatic crust thrusts widely over PA' mantle. There are too few constraints to correlate mid-crustal layers with the geological units. Alp07 (Šumanovac et al., 2009; Šumanovac, 2010) crosses the transition from the Adriatic Sea over the Dinarides to the Pannonian basin at the latitude of Istria. Similar to Alp02, AD, PA', and PA mantle are discerned. Adriatic lower crust, identified by high velocities, thrusts over PA' mantle under the Dinarides. In the upper crust, an extended lower velocity zone exists above the AD-PA' boundary. Again, too few constraints exist for a geological correlation of the mid-crust.

CEL10&Alp04 (Hrubkova & Sroda, 2008; Grad et al., 2009) sample European crust under the Bohemian Massif. High velocity lower crust extends more than 300 km along this profile. This structure was also found on profile CEL09 (Hrubcova et al., 2005). On CEL10&Alp04 the thickness of the Austro-Alpine nappes overthrusting Flysch, Molasse und European basement is constrained by a distinct reflector at ~10 km depth. In the south, the profile just touches the PA fragment at Moho level and finally reaches Adriatic crust and mantle. A significant indenter of Adriatic middle crust into the European crust is marked by a stack of strong reflections.

The profiles CEL01, CEL04 (Sroda et al., 2006), and CEL05 (Grad et al., 2006) cross over the Western Carpathians from the European platform to the Pannonian basin. Near surface low velocity zones mark basins in the Paleozoikum of the European plate, the Flysch trough and the Pannonian basin. CEL04 allows for a clear location of the boundary between EU and PA Moho and upper mantle, approximately below the Pieniny clippen belt at the southern border of the Flysch belt (Outer Carpathians). On CEL01 and CEL05, a wide indentation of PA mantle under EU crust into the European domain was interpreted. In this study the boundary between EU and PA is reinterpreted according to CEL04. The older reflection profile T2 (Tomek, 1993) gives further support for this interpretation. The thinning of EU crust on CEL05 from the EU platform to the contact with the thin PA crust may postdate to the time when EU was the passive margin of the ocean in the Carpathian embayment. ALCAPA lithosphere reached this margin after a considerable extensional phase. One could imagine that the collision was "soft", with little indentation.

4. Moho topography and plate boundaries

Several Moho maps have been generated which cover parts of the Alps (e.g. Waldhauser et al. 1998), continents (e.g., Desez and Ziegler, 2001) or even the whole globe (e.g., Mooney et al., 1998). The interpretation of the Moho structure and topography in a plate tectonic context is a promising task. Destructive plate boundaries can be identified by offsets of an otherwise smooth Moho boundary derived from controlled source seismic data (Waldhauser et al., 1998; Abramovitz and Thybo, 2000; Kissling et al., 2006; Brückl et al., 2007); natural source seismic methods like receiver functions and tomographic inversion also have the potential to clearly resolve destructive Moho boundaries (e.g., Kummerow et al., 2004; Di Stefano et al., 2009). Continental transform or transfer faults may exist without significant offsets of the Moho boundary (Weber et al., 2009; 2005; Thybo et al., 2003; Zhao et al., 2001).

Waldhauser et al., (1998) generated a Moho-map for the Western Alps on the basis of the available seismic data. They approximated the Moho boundary to the data values by smooth surfaces, considering data errors and inferred plate boundaries and allowed for Moho jumps, when otherwise only a rough surface could approximate the depth data. By this procedure, plate boundaries between Europe and Adria and Adria and Liguria were derived. The inferred Moho jumps indicate European upper mantle thrusting under Adriatic upper mantle and Adriatic upper mantle thrusting under Ligurian upper mantle.

Stacking of refracted and reflected P-waves, recorded during the CELEBRATION 2000 and ALP 2002 experiments, in combination with tomographic inversions were applied to generate a Moho-map and a P-wave velocity model of the East Alpine crust (Behm et al., 2007). These data on Moho depth were also integrated into the European Moho-map of Grad et al. (2008). For further considerations, the Moho-map by Grad et al. (2009) is used together with the original data of Behm et al. (2007). Depth and offset of the Moho at the plate boundaries determined by Waldhauser et al. (1998) were integrated. The Moho fragmentation between Adria and Liguria was extended to south east using tomographic data from Di Stefano et al. (2009). In the Eastern Alps, seismic profiles and the Moho-map indicate a fragmentation of the lithosphere into three blocks, the European plate, the Adriatic microplate, and a newly interpreted Pannonian fragment (Brückl et al., 2007; Behm et al., 2007). Brückl et al. (2010) applied 2D elastic plate modelling of the lithosphere to decide if a continuous plate is an appropriate model or if the introduction of a plate boundary yields a better fit. Plate boundaries (or broken plates) were inferred by the conditions that no bending moment is transferred and only vertical force couples act between the two plates at the boundary. The criteria imposed by elastic plate modelling to determine plate boundaries have some similarity to the curvature criteria Waldhauser et al. (1998) introduced. Both methods postulate the smoothness of the Moho discontinuity within one lithopsheric block or plate.

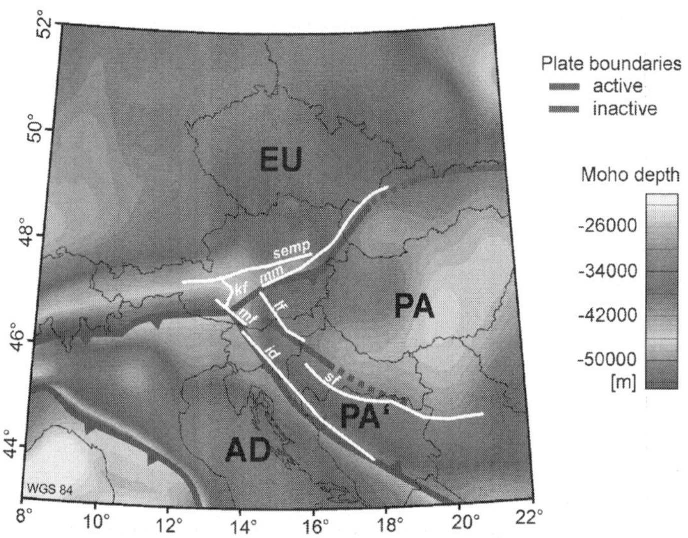

Fig. 3. Moho-map with Moho fragmentation and related fault system

The combined Moho-map of Grad et al. (2009) and Behm et al. (2007), together with the inferred Moho fragmentation is shown in Fig. 3. The Moho is fragmented into EU (European plate), AD (Adriatic microplate), Ligurian plate (LG), and PA (Pannonian fragment). As described before, the profiles Alp02 (Brückl et al., 2007) and Alp07 (Šumanovac, 2009), but also the elastic plate modelling (Brückl et al., 2010) indicate the existence of a transition zone from AD to PA, termed PA′. The general configuration of plates and the polarity of underthrusting almost correspond to the scheme by Doglioni and Carminati (2002). The difference is the plate boundary between EU and PA, which was beyond the scope of their paper. From the Western Alps to the triple junction (AD-EU-PA), a significant upward step from EU to AD is revealed by profiles EGT&NFP-20 and TRANSALP. On Alp01 a step with the same polarity but smaller extent was found. From the triple junction to the Rechnitz window in the east-north-east, a positive Moho step still exists from EU to PA and EU underthrusting PA is probable. The reflection profile NESTMK (Graßl, 2004) gives further support for this interpretation. Strike-slip or transpression may prevail. East of the Rechnitz window, in the area of the Vienna basin, transtension may dominate. The EU-PA boundary in the Western Carpathians does not exhibit a step and has the character of a "soft" collision between ALCAPA and the European platform (see profiles CEL01, CEL04, CEL05). The northern partition of AD-PA′ represents a second order boundary with no significant jump, but a discontinuous change of inclination. In the southern partition, gravity modelling (Šumanovac et al., 2010) reveals a positive jump from AD to PA′. Force couples and the jump in the southern part indicate AD underthrusting PA′. PA′-PA may delimit a transition zone from the Dinaric orogen to the shallow PA crust.

5. 3D seismic models of the crust and implementation of gravity data

The CELEBRATION 2000 and ALP 2002 seismic experiments realized receiver deployments along a net of seismic lines and shots were simultaneously recorded at all receivers.

Therefore about 80% of the data were cross-line traces. Stacking techniques in combination with manually picked P- and S-wave phases and tomographic methods were used to generate 3D P- and S-wave velocity models of the crust and to calculate Poisson's ratio (Behm et al., 2007; Behm, 2009). Data from SUDETES 2003 (Grad et al., 2003) were also implemented for the generation of the S-wave model. These models have an almost complete coverage down to 10 km depth. The maximum penetration depth is about 30 km and is restricted mainly to the Bohemian Massif and the wider area of the Vienna basin.

Gravity studies have a long history within the scope of exploration of the deep crustal structure of the Eastern Alps (e.g., Meurers et al., 1987, Lillie et al., 1994). The seismologic explorations during the TRANSALP project were also accompanied by the compilation of latest gravity data in that region (Zanolla et al., 2006), gravimetric modelling and studies about isostasy (e.g., Ebbing et al., 2006). Studies on the integration of gravity data and the seismic models derived from the CELEBRATION 2000 and ALP 2002 data were carried out by Brückl et al. (2006) and Simeoni and Brückl (2009). Accurate Bouguer gravity data has been provided by Bielik et al. (2006) and Meurers and Ruess (2007) for the Alpine-Carpathian area. Areas not covered by this data have been supplemented by the regional Bouguer gravity data of the West-East Europe Project (http:/www.getech.com). A density model of the upper crust was derived from the seismic model of Behm et al. (2007) using the Vp - density relation of Christensen and Mooney (1995). For the sedimentary basins existing density-depth relations were used (for references see Simeoni and Brückl, 2009). The gravity effects of the upper crust (UC) and the Moho (M) were calculated and subtracted from the Bouguer anomaly (BA).

Depth slices through the 3D Vp-model of the crust (Behm, 2007) and gravity maps are compiled in Fig. 4. Fig. 4a,b,c show the deviation of the Vp-velocities from the average values for orogens according to Christensen and Mooney (1995) at depths of 5 km, 10 km, and 27.5 km. The original Bouguer gravity map (BA) is shown in Fig. 4d, using the same colour code as for the Moho depth map (Fig. 3). Fig. 4e shows BA-M, which is the Bouguer anomaly (BA) reduced by the gravity effect of the Moho. Fig. 4f shows a reduced Bouguer gravity BA-M-UC, where the gravity effect of the Moho (M) as well as the gravity effect of the upper crust (UC) has been subtracted. For Figs. 4e and 4f a similar colour code as for the Vp-depth slices (Fig 4a,b,c) was used in order to visually enhance spatial correlations between Vp velocities and the reduced gravity anomalies BA-M and BA-M-UC.

The Vp velocity at 5 km depth well reflects the tertiary basins (Molasse, Vienna, and western part of the Pannonian basin) as low velocity regions, even the depth of the basins is less than 5 km. The metamorphic eastern part of the Moldanubian, Bohemian Massif, and Adriatic foreland exposed at the peninsula Istria are high velocity zones. A new velocity structure correlates roughly with the west-east striking Greywacke zone, latitude ~ 47°. At their eastern end, these structures bend to the south. The high and low velocity regions at 5 km depth can again be identified at 10 km depth. The low Vp extend in the area of the PA'-PA Moho fragmentation. Vp south of the MHZ shows an increase, which is the only indication of a difference in geophysical data presented in this study between ALCAPA and TISZA. At the depth of 27.5 km the high velocity zone in the Bohemian Massif is still visible. However, while at the higher levels the velocity decreases significantly from the Bohemian Massif to the Vienna basin and further east, at the lower crust levels, the velocity increases toward the Pannonian basin with a maximum around the north-east striking Moho fragmentation between EU and PA. All velocity structures identified in the 3D model can also be seen in the 2D profiles (Fig. 2). An exception is the west-east striking high velocity zone along the Greywacke zone, which is not resolved by the Alp01 and CEL10&Alp04.

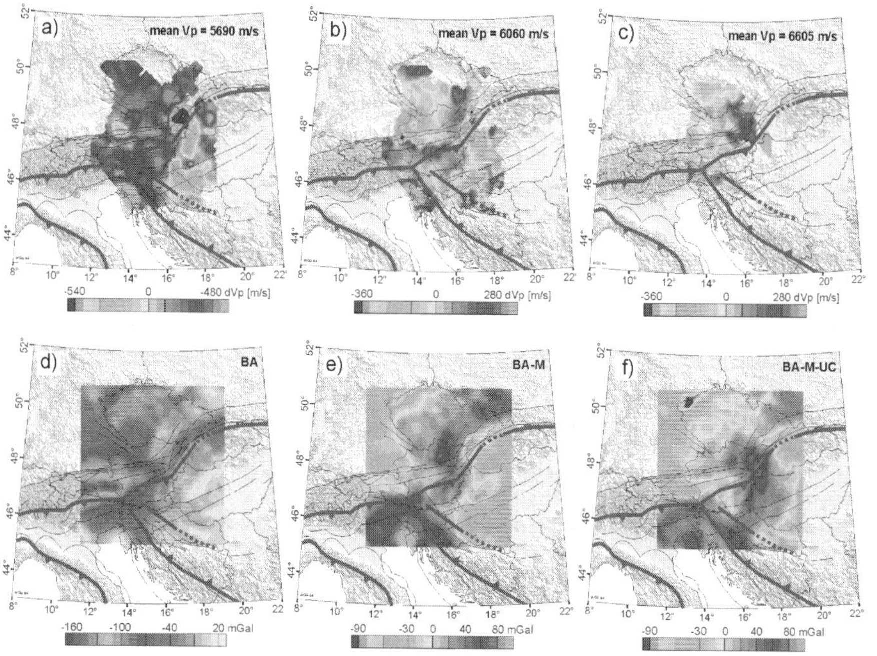

Fig. 4. Slices through the 3D P-wave velocity model of the crust at a) 5 km, b) 10 km, c) 27,5 km depth and gravity data (d, e, f)

The BA (Fig. 4d) reflects mainly the orogenic root and the relatively thin crust in the Pannonian Basin and the Venetian plain. The well-known, narrow gravity minimum of the Tauern window is superimposed on a wider gravity minimum caused by the East Alpine root. A relative gravity high correlates with the metamorphic eastern part of the Moldanubian, Bohemian Massif, and the corresponding high Vp velocity zones at all depth levels. The reduced anomaly BA-M shows again the gravity high in the Moldanubian and enhances the visibility of the gravity minimum at the Tauern window. The Molasse, Vienna, and Pannonian basins are clearly visible. The most prominent structure is an extended gravity high in the south which delineates the extent of the Adriatic indenter. The reduced anomaly BA-M-UC still shows the gravity high in the Moldanubian and the Adriatic indenter. The correlation with the basins is removed and the gravity minimum of the Tauern window is enlarged and now comprises the area around the Greywacke zone, similar to the high Vp zone at depth 5 km and 10 km. A gravity high not recognized on BA or BA-M neighbours the gravity high in the Moldanubian to the SE around the estimated trace of the Moho fragmentation between EU and PA. It coincides with the high velocity zone at 27.5 km depth (Fig. 4c). Most of the structures recognizable in the 3D Vp and gravity data need detailed modelling and analysis of uncertainties for further interpretation. However, there are two significant findings for this study. The first is the extent of the indenting Adriatic crust. It correlates well with the Moho fragmentation, but extends over the AD-EU and AD-PA′ boundaries to the north and north-east. This could be interpreted as an indenting Adriatic crustal wedge beyond the Moho fragmentation into European and Pannonian crust. The profiles TRANSALP, Alp01, CEL10&Alp04, and Alp07 support this interpretation. The second structure essential in the context of regional tectonics is the high velocity and gravity surplus in the lower crust below and slightly south-east of the Vienna basin. In case we assume the Moho fragmentation EU-PA between the triple junction and the Western Carpathians as transform or transfer, the section crossing the high velocity and gravity surplus zone would be trans-tensional. Magmatic intrusions into the lower crust could be an explanation for this anomaly. Additional information on the lithology of the crust comes from the 3D Vs and Poisson's ratio models generated by Behm (2009). This data indicates mafic intrusions in the upper crust of the Adriatic foreland. However, Poisson's ratios of the lower crustal high Vp region below and south-east of the Vienna basin are ~0.25. This value would be an indication that magmatic intrusions are not mafic.

6. Upper mantle tomography

Imaging the lithosphere and upper mantle by seismic tomography can substantially contribute to a better understanding of past and recent plate tectonic processes. Bodies of relatively high seismic velocity (mostly P-wave velocity) can represent down-going slabs of low temperature oceanic crust and lithospheric mantle. Low velocity areas may be related to areas of thinning lithosphere and upwelling astenosphere, mantle plumes or local mantle convections. Global and large-scale tomographic studies are mainly based on ISC travel time data from local, regional, and teleseismic earthquakes (e.g. Piromallo and Morelli, 2003; Wortel and Spakman, 2000; Bijwaard and Spakman, 1998; Koulakov et al. 2009). They may include crustal corrections and take advantage of traveltime reciprocity with respect to the locations of sources and receivers. Because these studies use absolute traveltimes, focal coordinates are relocated and deviations from a symmetric earth are considered for the whole ray paths from hypocentre to receiver. High resolution teleseismic investigations (e.g. Lippitsch et al., 2003; Rawlinson and Kennett, 2008; 2010; Dando, 2010; Mitterbauer et al. 2010) include data from temporary networks, implement crustal corrections based on the best available crustal models, and consider the velocity model by an accurate 3D ray tracing. Only traveltime differences are considered and the disturbance of teleseismic wave fronts by far distance inhomogeneities is neglected in the teleseismic range (~30° - 100°). Results from all these tomographic studies are presented as deviations of Vp or Vs velocities from a reference model, mostly a standard model like ak135.

A complete coverage of the wider Eastern Alps with relatively high resolution is provided by the P- and S-wave models of Koulakov et al. (2009). This P-wave model will be used and discussed together with the high-resolution teleseismic models of Lippitsch et al. (2003) and the latest local teleseismic model of Mitterbauer et al. (2010) derived from ALPASS data (http://info.tuwien.ac.at/geophysik/alpass.htm). The maximum depth extent of these models is 710 km, 400 km, and 600 km. Fig. 5 shows depth slices at 150, 180, 350, and 400 km through the models abbreviated K (Koulakov et al., 2009), L (Lippitsch et al., 2003), and M (Mitterbauer et al., 2010).

The Moho fragmentation is superimposed on the maps of P-wave velocity deviations from the reference values at the particular depths. Seismic stations, from where data were collected for the tomography, are also shown for the M- and the L-model within the frame of the map. The areas of no or low resolution are blank.

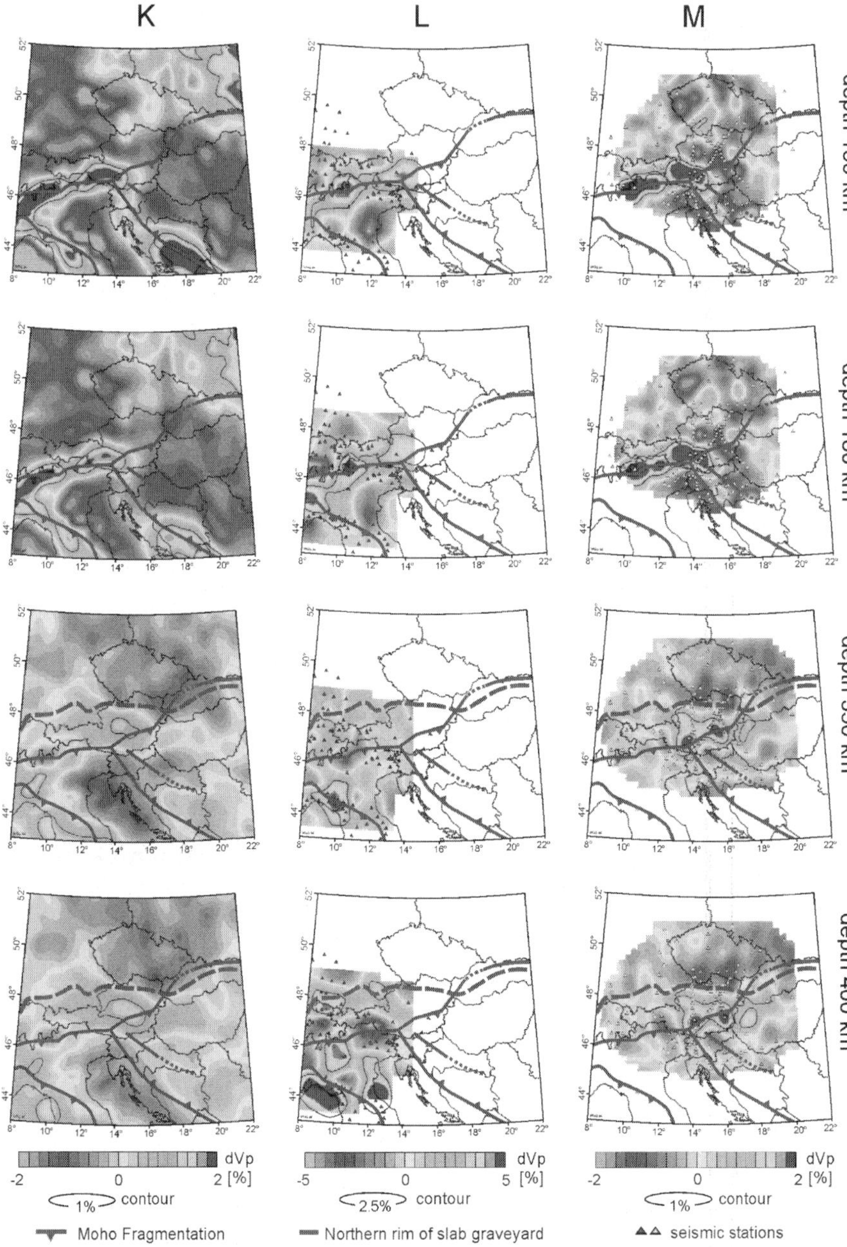

Fig. 5. Depth slices through the K-, L-, and M-models of the upper mantle

High velocity bodies below the Alps, Apennines, and Dinarides at 150 km and 180 km depth correspond to cold subducting slabs. Of main interest is the slab under the Eastern Alps and it will be referred to as "shallow slab". Contours of the shallow slab are superimposed on the depth slices. The axis of the shallow slab agrees well between all three models from 10°E to 14°E and approximately follows the Moho fragmentation. The assumption that the shallow slab was generated by a unique geodynamic process, in particular by subduction of (mainly) lithospheric mantle at the EU-AD boundary may be justified. Along the EU-PA Moho boundary, the K- and M-models do not show any structure which could be interpreted as a shallow slab. This finding is also in agreement with the global or large-scale models of Bijwaard et al. (1998) or Piromallo and Morelli (2003). The L-model does not extend so far to the east.

At 350 km and 400 km depth, the K- and M-models show an E-W striking high velocity body which terminates in the west between 12°E and 13°E. The L-model does not image this high velocity body. We refer to the high velocity body imaged by the K- and M-models as "deep slab". The teleseismic models derived from CBP-data (Dando, 2010) give further support for the existence and extent of this deep slab. Superimposed on these depth slices is also a contour which marks the northern rim of the so-called "slab graveyard" between 410 km and 670 km (e.g., Handy et al., 2010). The deep slab is located above the slab graveyard. The interpretation of the deep slab as subducted oceanic lithosphere of the Alpine Tethys on its way down to the slab graveyard is possible.

Fig. 6 shows 5 vertical profiles (P1 – P5) through the K- and M-teleseismic models down to a depth of 500 km. The corresponding profiles through the L-model are shown for the profiles P1, P2, P3, and P5 down to 400 km. Profiles 1 and 5 correspond to profile B and C after Lippitsch et al. (2003). Profiles 2 and 3 are along the TRANSALP transect and Alp01. Profile 4 crosses the Eastern Alps east of the Tauern window and the extent of the shallow slab. It is not covered by the L-model and the location map of the profiles is put in its place. The topography (enlarged in vertical direction) and the Moho according to the Moho-map (Fig. 3) are superimposed. The shallow slab is clearly imaged by all three models on profiles 1, 2, 3, and 5. The dip, as imaged by the K- and M-models, changes systematically from steeply south or vertical at the profiles P1 (near the boundary between Western and Eastern Alps) and P2 (central Tauern window) to steeply north at profiles P3 and P5 (at and beyond the eastern end of the Tauern window). P4 is located east of the shallow slab. The deep slab can clearly be identified on the K- and M-models at P3, P4 and P5. The high velocity zone widens with greater depth, indicating a continuous transition from the deep slab to the slab graveyard. This transition is better imaged by the K-model than by the M-model.

The L-model confirms the images of the shallow slab, supplied by the K- and M-models at P1, P2, and P3. On P5, the dip of the shallow slab resolved by the L-model deviates significantly from nearly vertical to ~ 60° north. In correspondence with the depth slices at 350 km and 400 km the deep slab cannot be identified on the profiles by the L-model.

As pointed out by Lippitsch et al. (2003), deep velocity anomalies may leak to higher levels in case these deep structures are not covered by the depth extent of the model, or considered by the embedding of the smaller regional model into a larger or global model. This may be the case for the L-model east of 14°-15°E. Therefore, from this latitude further to the east, the interpretations are based on the K- and M-models only.

The image of the L-model at P5 gave rise to the introduction of a subduction polarity change along the Alpine arc near 12°E, the longitude of the TRANSALP profile (e.g. Lippitsch et al.,

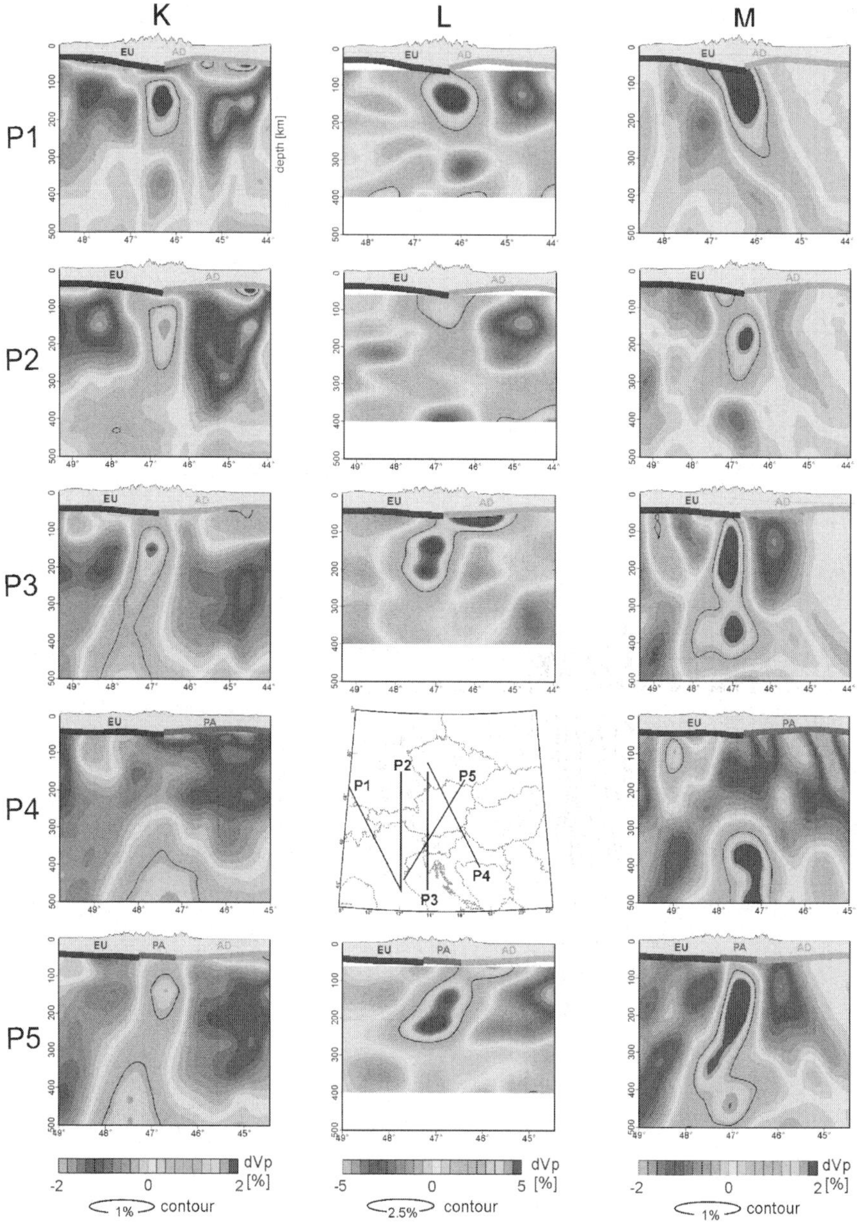

Fig. 6. Profiles through the K-, L-, and M-models of the upper mantle

2003; Schmid et al., 2004; Kissling et al., 2006). The idea is that parts of the Vardar or Meliata ocean, originally situated southeast of the Austro-Alpine, subducted to N and forced Adriatic continental lower lithosphere to subduct north-east beneath the Austro-Alpine. However, based on the arguments for a single shallow slab between 10°E and 14°E and the strong and unquestioned evidence for EU subducting AD at P1 (EGT&NFP-20), the whole shallow slab is interpreted as subducted EU lower continental lithosphere.

The K-model also provides well resolved images of slabs below the Apennines and Dinarides at 150 km and 180 km depth. Both slabs are subducted Adriatic continental lower lithosphere according to the plate tectonic scheme introduced by Doglioni and Carminati (2002). The axes of both slabs correlate well with the Moho fragmentations LG-AD and AD-PA(PA'). The Apennines slab dips to the SW. The Dinarides dip nearly vertically. It cannot be identified near the triple junction, but develops to a large body south of 45°N. Along the EU-PA Moho fragmentation, neither the K- nor the M-model shows any velocity structure comparable to the shallow slabs below the EU-AD, LG-AD, and AD-PA(PA') fragmentations.

7. Crustal thickening, post-collision convergence, lateral extrusion

Crustal thickening is closely related to post-collision convergence of an orogen and quantitative information can be derived from the balancing of these processes. Along the Alpine arc, the indentation of the Adriatic microplate has caused crustal thickening. Near the transition from the Western to the Eastern Alps at EGT&NFP-20, crustal thickening is the effect of a wide indentation of Adriatic lower crust, compression deformation of EU upper crust and Penninic thrusts, and backthrusting of AD upper crust according to seismic data (Valasek et al., 1991; Ye et al., 1995; Pfiffner et al., 1997) and their geologic interpretation (Schmid et al., 2004). In the area of the Tauern Window post-collision convergence and crustal thickening was considered along a cross section from the Molasse basin in the N to the Southern Alps and the Venetian plain (Brückl et al., 2010). Balancing was based on the assumption that no crustal material was subducted and three different regions were considered. North of the Subtauern ramp crustal thickening is mainly achieved by the overthrusting accretion wedge. The crystalline European crust north of the Subtauern ramp formed the Jurassic passive margin of the Penninic Ocean (Alpine Tethys) and experienced extension and crustal thinning before collision (Roeder, 1977). Geological data and reflection seismic profiles in the Molasse show that no or negligible thickening of the European passive margin occurred during collision (Wessely, 1987; Reiter et al., 2005). Ductile thickening of European basement below the Subtauern ramp cannot be excluded. However, an upper bound for crustal shortening of 20 km may be a reasonable assumption.

In the central part of the East Alpine orogen, crustal shortening is connected to the thrusting and folding of the Tauern window up the Subtauern ramp. Fügenschuh et al. (1997) derived a maximum exhumation of ~25 km for the central part of the window. As the southward dip of the Subtauern ramp is about 30° in the western part of the Tauern window (Lüschen et al, 2006) a vertical movement of 25 km corresponds to a horizontal northward directed displacement of about 55 km. This figure can be taken as a measure for the northward movement of the surrounding Adriatic crust over European crust.

S-directed thrusting of the Southern Alps is a relatively young process, most active during the Miocene (Doglioni, 1987, Schönborn, 1999; Schmid et al., 1996, Castellarin and Cantelli, 2000). The average thickness of the AD crust south of the Periadriatic line is 37 km. The major part of the thickening took place either by thrusting in the crystalline upper crust and thrusting of AD lower crust over the Subtauern ramp (Castellarin et al., 2006), or by stacking of lower crust. Thickening of Adriatic crust south of the Periadriatic line corresponds to a crustal shortening of ~55 km assuming a thickness of 30 km for the pre-collision crust. This estimate of crustal shortening is very sensitive to the assumption of the thickness of the pre-collision crust. The estimates of the total north-south convergence since Oligocene derived from crustal thickening range from ~95 km to ~130 km.

Geological investigations and palinspastic reconstructions arrive at significantly larger figures (165 - 195 km) (e.g., Fügenschuh et al., 1997; Ustaszewski et al., 2008). The difference between crustal shortening derived by cross sectional balancing of the whole crust and by palinspastic reconstructions of the upper crust is not equally distributed over the whole profile. There is a fair agreement of N-S convergence derived by both methods in the Southern Alps (Schönborn, 1999; Castellarin and Cantelli, 2000; Castellarin et al., 2006). For the Eastern Alps the situation is significantly different. Frisch et al. (2000) derived 140 km crustal shortening at the western boundary of the Tauern window by a palinspastic reconstruction. The estimates of crustal shortening derived from crustal thickening and thrusting of the Tauern window over the Subtauern ramp range from 55 - 75 km. This discrepancy can be (at least partly) resolved by the assumption of eastward directed lateral extrusion of the central Eastern Alps and the Tauern window between the SEMP fault and the Periadriatic line (Ratschbacher et al. 1991). Linzer et al. (2002) estimated ~120 km of E-W extension between the Brenner and Katschberg normal faults. These faults form the western and eastern boundary of the Tauern window. Accounting for this significant E-directed movement of Tauern crust out of the collision zone by 3D balancing would significantly increase the amount of N-S-shortening across the Tauern window. Shortening is therefore underestimated by the restoration process applied above, which solely considers the amount of exhumation and the dip of the Subtauern ramp.

8. Active tectonics

Stable triple junctions impose constraints on the relative velocities between the plates constituting it. An analysis of relative plate velocities at triple junction between oceanic lithospheric blocks is well established, because the plate boundaries are narrow and clearly defined by topography, magnetic anomalies, and earthquake hypocentres (e.g., Moores & Twiss, 1995). Continental plate boundary processes at collision zones and transform or transfer faults find their expression in whole orogens or systems of strike-slip faults together with transtensional and transpressional structures, covering zones which may be several hundred kilometres wide (e.g., Gordon, 1995). A look at the geological map (Fig. 1) shows that this is also true for our investigation area and also the triple junction AD-EU-PA. One may hypothesize that continental active tectonics and kinematics can be more clearly described at the Moho level, comparable to boundaries between oceanic plates. On the basis of this idea, Brückl et al. (2010) analyzed the kinematics at the triple junction between AD - EU - PA, using the geodetic data of Grenerzcy and Kenyeres (2006), and Weber et al. (2006) to calibrate their model, and made predictions about the seismicity at the EU-PA and AD-PA(PA') plate boundaries.

Fig. 7a shows the relative plate velocities, the extent of shallow slabs, and the velocity triangle at the AD-EU-PA triple junction according to the kinematic model of Brückl et al. (2010). To simplify the analysis the subdivision of the Pannonian fragment into PA and PA'

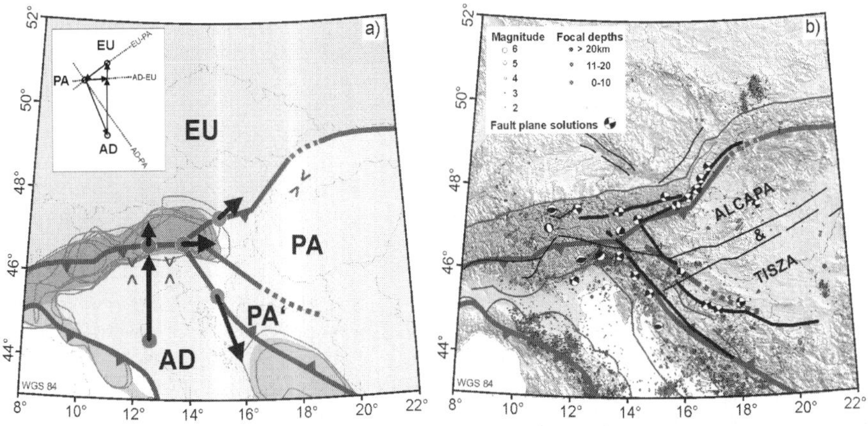

Fig. 7. Triple junction AD-EU-PA: a) kinematics and shallow slabs; b) faults and seismicity

was disregarded and only one Pannonian fragment (PA) considered. The following basic assumptions were made: stable triple junction and pure strike-slip at the EU-PA boundary near the triple junction. The geodetic constraints inferred were: 3 mm/yr N-S oriented convergence between AD and EU including the deformation zone in the Southern Alps and 1 mm/yr eastward escape of PA relative to AD. These assumptions and the geodetic constraints define the relative velocities at the triple junction. The total N-S convergence is partitioned into 0.6 mm/yr AD overthrusting EU and 2.4 mm/yr compression deformation of the Southern Alps. At the triple junction there is eastward spreading of PA relative to AD of 1 mm/yr. This extension is assumed to be compensated by the ongoing lateral extrusion of the Tauern window area as proposed by Ratschbacher et al. (1991) and derived by palinspastic reconstructions for Miocene times (e.g. Linzer et al., 2002). The mass deficit in this area derived by crustal balancing, which follows from lateral extrusion to the east, gives further support for this explanation. South of the compression area in the Southern Alps, the total AD-PA relative velocity is 2.5 mm/yr. The direction of the relative velocity indicates transpression and AD underthrusting PA. The EU-PA boundary near the triple junction has an azimuth of 55°. Further to the east this azimuth increases slightly at the Mur-Mürz fault and a transpressional regime prevails, similar to the situation at AD-PA south of the compression deformation zone in the SA. However, the absolute value of the relative velocity is only 50% at EU-PA in comparison with AD-PA. Seismicity should reflect this ratio. From the Rechnitz window, over the Vienna basin the assumed azimuth of the EU-PA plate boundary is about 35°, generating transtension.

The amount of 0.6 mm/yr AD overthrusting EU may lead to a continued folding and thrusting up the Subtauern ramp. The slope of this ramp is about 30° and a horizontal displacement of 0.6 mm/yr along this ramp would produce about 0.35 mm/yr uplift rate. According to geodetic data (GPS and levelling) the Tauern window still shows an uplift rate of ~1 mm/yr (Höggerl, 2001; Haslinger et al., 2007), which is significantly more than the amount estimated from convergence at the EU-AD plate boundary and the continued folding and thrusting of the Tauern window up the Subtauern ramp. It is a matter of debate how much of the uplift and subsidence rates is related to tectonic activity, especially the ongoing exhumation of the Tauern window, and what may be explained by post glacial rebound. The ice streams in the central Eastern Alps advanced widely into the EU and AD forelands during the last ice age (van Husen, 1987). To the east the maximum extent of glaciation reached into the Styrian basin. But also present day glacier shrinkage should be considered as a possible cause of uplift (Barletta et al., 2006).

Together with geodetically observed crustal deformations, the seismicity proves active tectonic processes. Fig. 7b shows the terrain model, major faults, seismicity (NEIC 1973-2010), and representative focal plane solutions (Reinecker and Lenhardt, 1999; further references in Brückl et al. 2010). The pattern of the Moho fragmentation, also superimposed on this map, is strongly related to seismicity. The area of highest seismicity in the wider EA area are the Southern Alps south of the AD-EU boundary, the Friuli district. The upper crust hypocenter locations and focal mechanisms correspond to back-thrusting of the Southern Alps and strike slip movements. Another area of relatively high seismicity is the Mur-Mürz fault and the southern Vienna basin along the EU-PA Moho fragmentation. Focal plane solutions indicate mostly SW-NE oriented left-lateral strike slip with little dip slip components. Stripes of high seismicity follow the Dinaric thrust belt, southwest of the AD-PA' and PA'-PA Moho fragmentations. Irregular orientations of the principal stress axis at the transition from the Southern Alps to the northern Dinarides are an expression of the complex stress system in this area. Further to the south-east, especially in the costal part of Croatia, reverse or strike-slip faulting with sub-horizontal P-axis in south-west to south directions dominate the active tectonics along these faults. The Periadriatic line shows relatively little seismicity with the exception at its transition to the Mid-Hungarian fault zone. Gutdeutsch and Aric (1987) derived tectonic block models from seismic data alone for the wider Pannonian area. In addition to the European and Adriatic plate they introduced a Dinaric and a Pannonian fragment. The latter correspond approximately to PA' and to PA.

The pattern of the Moho fragmentation correlates well with major active faults (Fig. 7b; see also Fig. 3). The wider area of the newly interpreted Pannonian fragment is bounded by the Katschberg fault in the west, the SEMP fault in the north, the Malta and Idria faults, and probably the Dinaric thrust faults in the southwest. The Mur-Mürz, Lavant and Sava faults are other strike-slip faults, which are related to the Moho fragmentation and which enclose the thin Pannonian crust. Along the Periadriatic line and its prolongation to the Mid-Hungarian fault zone, no significant corresponding features in the deeper crustal structure can be identified. The northern part the Pannonian fragment is composed of ALCAPA units, and to the south, it extends over the Mid-Hungarian fault zone into the TISZA block. The kinematics at the triple junction and the lack of seismic data indicating significantly a boundary between ALCAPA and TISZA give rise to the hypothesis that ALCAPA east of the Tauern Window and TISZA merged to one "soft" plate. The low seismic activity at the EU-PA border in the Western Carpathians may be regarded as a further indication for a "soft" collision between ALCAPA and Europe in this area.

9. Conclusion

Wide angle reflection and refraction profiling supplied information about the seismic velocity structure (mainly Vp) of the crust and resolved the Moho boundary. Steep angle reflection profiles additionally provided detailed images of intra-crustal structures. The most comprehensive information were gained by integrated transects like TRANSALP or EGT&NPF-20. On the basis of this data, suture zones between plates or fragments were delineated and structures, mainly related to compression, were resolved. The topography of the Moho discontinuity was determined and significant steps indicating plate boundaries were recognized. Moho-maps were compiled on the basis of 2D and 3D data. In this study a Moho-map based on the European Moho-map of Grad et al. (2009) and the higher resolution map of Behm et al. (2007) was used. On the basis of this data and additional criteria derived from curvature (Waldhauser et al., 1998) and elastic plate modelling (Brückl et al., 2010), a fragmentation of the Moho and upper mantle into EU, AD, PA (including PA') and the Ligurian plate (LG) was introduced (Fig. 3). The polarity of the Moho steps, the changes of Moho dip, and force couples derived by the elastic plate modelling were considered to identify plate boundary processes. Following this interpretation, the EU underthrusts AD and PA. AD underthrusts LG and PA (PA'). This scheme was also proposed earlier by Doglioni and Carminati, (2002).

3D Vp- and Vs-velocity models of the crust (Behm et al., 2007; Behm, 2009) and gravity data (e.g., Simeoni and Brückl, 2009) support interpolation of velocity structures between profiles and extrapolation to areas not covered by profiles. Important contributions of these models to the general understanding of the geodynamic and tectonic setting were, for instance, outlining the northern and north-eastern boundary of the Adriatic indenter at mid- and lower crustal level and the identification of a high velocity and high density lower crust in the transtensional area in and southeast of the Vienna basin.

Global and large-scale tomographic models of the upper mantle show a wide zone of high Vp-velocity bodies between the 410 km and 670 km discontinuity in the Alpine-Mediterranean region (e.g., Bijwaard and Spakman, 1998; Piromallo and Morelli, 2003). These structures are known as "slab graveyard" and generally interpreted as subducted oceanic lithosphere and continental lower lithosphere. In this study the large scale model of Koulakov at al. (2009) (K-model), the teleseismic model of Lippitsch et al. (2003) (L-model) and the new model of Mitterbauer et al. (2010) (M-model) are compared and used for tectonic interpretation. At the depth levels 350 km and 400 km, the K- and M-model image a west-east striking high velocity body east of 12°E and between 47° - 48°, the so-called "deep slab". It is not imaged by the L-model because it is mostly east of the high resolution area of the L-model. Another point is that the L-model is restricted in depth to 400 km. The deep slab is well imaged along vertical profiles (Fig. 6) by the K- and M-models. Its existence has also been confirmed by a teleseismic model based on data from the CBP-project (Dando, 2010). The deep slab is interpreted as oceanic lithosphere of the Alpine Tethys, which detached from the Alpine orogenic root after collision about 35 Ma ago (Davis and Blanckenburg, 1995) and did not arrive completely at the slab graveyard. A matter of ongoing discussion is the mechanism, how oceanic lithosphere in the Carpathian embayment found its way down to the slab graveyard. The classical model is roll-back as proposed by Royden (1993). However, this process is difficult to explain because of the shape of the deep slab and its relation to the European continental margin. Other mechanisms like delamination (Bird, 1979) or a Rayleigh-Taylor instability (Houseman and Molnar, 1997) should also be taken into consideration.

A "shallow slab" is well resolved on depth slices at 150 km and 180 km by all three models from ~10°E to about 14°-15°E (Fig. 7a-f). The axis of the shallow slab approximately follows the Moho fragmentation EU-AD and the eastern end of the shallow slab coincides roughly with the triple junction between AD-EU-PA. The profiles which cover the Eastern Alps from their western border to ~14°E show the shallow slab down to depths of ~180 km to ~250 km. The dip changes continuously from steeply south in the west to steeply north in the east. The ~60° north dipping slab, imaged by the L-model along a NE-SW oriented profile east of the Tauern window could not be confirmed by the K- and M-models. The EGT&NFP-20 profile (e.g. Valasek et al., 1998), the TRANSALP transect (Lüschen et al., 2006) and the analysis of Moho steps by Waldhauser et al., (1998) and Brückl et al. (2010) support the interpretation that EU continental lower lithosphere (exclusively or mainly upper mantle) was subducted below AD along the whole EU-AD boundary. EU has been the passive margin during the subduction of the Alpine Tethys before collision. After collision, lower continental lithosphere of EU origin was detached from the crust and transported to greater depth forming the shallow slab. This mechanism supports ongoing convergence in the region of the orogenic core from the Western Alps to the eastern border of the Tauern window.

Before collision with EU, ALCAPA underwent an orogenic phase, the Eo-Alpine orogeny. Compression tectonics revived west of the triple junction after collision. East of the triple junction, tectonic escape to the north-east and gravitational collapse were the dominating processes. From the triple junction to the Western Carpathians, the movement of ALCAPA relative to EU has been mainly transfer, with change from pure strike slip to transpression (Mur-Mürz fault near the Rechnitz window) and transtension (Vienna basin, high velocity and density area in the lower crust). During tectonic escape ALCAPA extended significantly (e.g., Ustaszewski et al. 2008). The collision with the EU margin at the Western Carpathians was "soft", because of the weakening of the Pannonian lithosphere during extension into the Carpathian embayment. The crustal cross-sections through the Western Carpathians (Fig. 2) show no clear root below this orogen, but rather a continuous transition from the shallow Pannonian crust to the thick Paleozoic, and further to the north-east, Precambrian crust of the East European craton. This suggests that no shallow slab by subduction of continental lithosphere was generated along EU-PA from the triple junction to and along the Western Carpathians.

The interpretation of profile Alp07 and elastic plate modelling suggest that Adriatic mantle thrusts under the Dinaric crust. However, no clear indication for a subducting slab can be found from the triple junction to the south-east at about 45°N along the AD-PA', or PA'-PA Moho fragmentation. The kinematic analysis of current relative velocities at the triple junction (Fig. 7) revealed extension between AD and PA(PA') at the triple junction and transpression in the SE. In case this plate boundary process was active in roughly the same way since the beginning of lateral extrusion, a major slab must not necessarily have developed by subduction of AD lower lithosphere. This situation gradually changes to the south-east and a pronounced shallow slab is imaged by the K-model below the AD-PA(PA') Moho fragmentation in this area. This is plausible because the Euler pole of the relative movement of AD versus EU is located at 46-47° N and 8-10° E and the relative movement of AD versus PA(PA') changes from transpression to compression south of ~45°N.

In the area of the Tauern window and west of it, the ALCAPA units overthrusted European crust and lack their own lithospheric mantle. East of the Tauern window and the Katschberg normal fault at surface and the triple junction AD-EU-PA at Moho level, the ALCAPA units rest on their own lithospheric mantle. The mid-Hungarian fault zone (MHZ) has been a zone of repeated tectonic inversions (Csontos and Nagymarosy, 1998) since Eocene times. Periods of thrust faulting were followed by phases of extension and transtension. However, no significant geophysical boundary could be found between the ALCAPA and TISZA units by the data presented in this study. Seismicity at the Periadriatic line east of the triple junction and at the MHZ is relatively low. The kinematic model derived from geodetic data and the Moho fragmentation does not consider current deformation between ALCAPA and TISZA. One may speculate that ALCAPA and TISZA merged to one plate.

10. Acknowledgements

The author had the opportunity to participate in the projects CELEBRATION 2000, ALP 2002 (http://www.alp2002.info/) and ALPASS (http://info.tuwien.ac.at/geophysik/alpass.htm) and expresses his thanks to all members of the working groups for their excellent cooperation. The Austrian Academy of Sciences and the Austrian Science Fund (FWF) funded the Austrian contribution to these projects. Johanna Brückl compiled the geological data for this study, prepared all figures, and edited the text. The author gratefully acknowledges this substantial support.

11. References

Abramovitz, T. & Thybo, H. (2000). Seismic images of Caledonian, lithosphere scale collision structures in the southeastern North Sea along Mona Lisa profile 2, *Tectonophysics*, 317(1-2), 27-54, doi:10.1016/S0040-1951(99)00266-8.

Barletta, V. R. (2006). Glacier shrinkage and modeled uplift of the Alps, *Geophys. Res. Lett.*, 33, L14307, doi:10.1029/2006GL026490.

Beránek, B. & Zátopek, A. (1981). Earth`s crust structure in Czechoslovakia and central Europe by methods of explosion seismology, In: *Geophysical Synthesis in Czechoslovakia*, A. Zátopek (Ed.), Veda, Bratislava, Slovakia, 253-264.

Behm, M. et al. (2007). Application of stacking and inversion techniques to 3D wide-angle reflection and refraction seismic data of the Eastern Alps, *Geophys. J. Int.*, 170, 1, 275-298.

Behm, M. (2009). 3-D modelling of the crustal S-wave velocity structure from active source data: application to the Eastern Alps and the Bohemian Massif, *Geophys. J. Int.*, 179, 1, 265-278, doi:10.1111/j.1365-246X.2009.04259.x.

Bielik, M. et al. (2006). Gravity anomaly map of the CELEBRATION 2000 region, *Geol. Carpathica*, 57(3), 145-156.

Bijwaard, H.; Spakman, W. & Engdahl, E.R. (1998). Closing the gap between regional and global travel time tomography, *J. geophys. Res.*, 103, 30 055–30 078.

Bird, P. (1979). Continental delamination and the Colorado Plateau, *JGR*, 84, 7561-7571.

Bleibinhaus, F. & Gebrande, H. (2006). Crustal structure of the Eastern Alps along the TRANSALP profile from wide-angle seismic tomography, *Tectonophysics*, 414 (1-4), 51-69, doi:10.1016/j.tecto.2005.10.028.

Brückl, E. et al. (2003). ALP2002 Seismic Experiment, *Studia geophysica geodaetica*, 47. ISSN 0039-3169, 671-679.

Brückl, E. et al. (2006). Studies on Crustal Structure and Gravity in the Eastern Alps, In: *Geodetic Deformation Monitoring: From Geophysical to Engineering Roles. IAG Symposium 131*, F. Sanso and A.J. Gil (Ed.), Springer, ISBN 3-540-38595-9, 181-192.

Brückl, E. et al. (2007). Crustal structure due to collisional and escape tectonics in the Eastern Alps region based on profiles Alp01 and Alp02 from the ALP 2002 seismic experiment, *J. Geophys. Res*, 112, B06308, doi:10.1029/2006JB004687, 1-25.

Brückl, E. et al. (2010). Crustal structure and active tectonics in the Eastern Alps, *Tectonics*, 29, TC2011, doi:10.1029/2009TC002491.

Castellarin, A. & Cantelli, L. (2000). Neo-Alpine evolution of the Sothern Alps, *Journal of Geodynamics*, 30, Pergamon Press, 251-274.

Castellarin, A. et al. (2006). Structure of the lithosphere beneath the Eastern Alps (southern sector of the TRANSALP transect), *Tectonophysics*, 414, 259-282.

Christensen, N.I. & Mooney, W.D. (1995). Seismic velocity structure and composition of the continental crust: a global view, *J. Geophys. Res.*, 100, 9761-9788.

Csontos, L. & Nagymarosy, A. (1998). The Mid-Hungarian line: a zone of repeated tectonic inversions, *Tectonophysics*, 297, 51-71.

Dando, B. (2010). Seismological structure of the Carpathian -Pannonian region in central Europe, *Ph.D. Thesis*, School of Earth and Environment, University of Leeds.

Davies, J. & von Brandenburg, F. (1995). Slab breakoff: A model of lithosphere detachment and its test in the magmatism and deformation of collisional orogens, *Earth and Planetary Science Letters*, 129, 85-102.

Decker, K. & Peresson, H. (1996). Tertiary kinematics in the Alpine-Carpathian-Pannonian system: links between thrusting, transform faulting and crustal extension, In: *Oil and Gas in Alpidic Thrustbelts and Basins of Central and Eastern Europe*, G. Wessely and W. Liebl (Ed.), EAGE, Spec. Publ., 5, 69-77.

Dezes, P. & Ziegler, P.A. (2001). European Map of the Mohorovicic discontinuity. 2nd EUCOR-URGENT Workshop (Upper Rhine Graben Evolution and Neotectonics), Mt. St. Odile, France.

Di Stefano, R. et al. (2009). Shallow subduction beneath Italy: Three-dimensional images of the Adriatic-European-Tyrrhenian lithosphere system based on high-quality P wave arrival times, *J. Geophys. Res*, 114, B05305, doi:10.1029/2008JB005641.

Doglioni, C. (1987). Tectonics of the Dolomites (Southern Alps Northern Dolomites), *Struct. Geology*, 9, 181-193.

Doglioni, C. & Carminati, E. (2002). The effects of four subductions in NE Italy, *Mem. Sci. Geol.*, 54, 1-4.

Ebbing, J; Braitenberg, C. & Götze, H.-J. (2006). The lithospheric density structure of the Eastern Alps, *Tectonophysics*, 414, 145-155.

Frisch, W.; Dunkl, I. & Kuhlemann, J. (2000). Postcollisional orogen parallel large scale extension in the Eastern Alps, *Tectonophysics*, 327(3-4), 239- 265, doi:10.1016/S0040-1951(00)00204-3.

Fügenschuh, B.; Seward, D. & Mancktelow, N. (1997). Exhumation in a convergent orogen: the western Tauern window, *Terra Nova*, 9, 213-217.

Gebrande, H. et al. (2006). TRANSALP - A transect through a young collisional orogen: Introduction, *Tectonophys*, 414(1-4), 1-7, doi:10.1016/j.tecto.2005.10.030.

Giese, P. & Prodehl, C. (1976). Main features of crustal structure in the Alps, In: *Explosion seismology in Central Europe*, P. Giese and C. Prodehl (Ed.), Springer, 347-375.

Gordon, R.G. (1995). *Present plate motions and plate boundaries, in Global Earth Physics*, A Handbook of Physical Constants, T. J. Ahrens (Ed.), AGU Ref. Shelf, vol. 1, 66-87, AGU, Washington, D. C.

Grad, M. et al. (2003). SUDETES 2003 Seismic Experiment, *Stud. Geophys. Geod*, 47, 681-689.

Grad, M. et al. (2006). Lithospheric structure beneath trans-Carpathian transect from Precambrian platform to Pannonian basin—CELEBRATION 2000 seismic profile CEL05, *J. Geophys. Res*, 111, B03301, doi: 10.1029/2005JB003647.

Grad, M.; Tiira, T. & Group, E.W. (2009). The moho depth map of the european plate, *Geophys. J. Int.*, 176, 279-292.

Graßl, H. et al. (2004). Seismic image of the deep crust at the eastern margin of the Alps (Austria): indications for crustal extension in a convergent origin, *Tectonophysics*, 380, 105-122.

Grenerczy, G. & Kenyeres, A. (2006). Crustal deformation between Adria and the European platform from space geodesy, In: *The Adria Microplate: GPS Geodesy, Tectonics and Hazards*, N. Pinter et al (Ed.), NATO Sci. Ser. IV, vol. 61, 321-334, Springer, Dordrecht, Netherlands, doi:10.1007/1-4020-4235-3_22.

Gutdeutsch, R. & Aric, K. (1987). Tectonic block models based on the seismicity in the East Alpine Carpathian and Pannonian area, In: *Geodynamics of the Eastern Alps*, H. W. Flügl and P. Faupl (Ed.), F. Deuticke, Vienna, 309-324.

Guterch, A. et al. (2003a). An overview of recent seismic refraction experiments in Central Europe, *Stud. Geophys. Geod.*, 47, 651- 657.

Guterch, A. et al. (2003b). CELEBRATION 2000 Seismic Experiment, *Stud. Geophys. Geod.*, 47, 659-669.

Handy, M. et al. (2010). Reconciling plate-tectonic reconstructions of Alpine Tethys with the geological-geophysical record of spreading and subduction in the Alps, *Science Reviews*, doi: 10.1016/j.earscirev.2010.06.002.

Haslinger, C.; Kraus, S. & Stangl, G. (2006). The intraplate velocities of GPS permanent stations of the Eastern Alps, *Vermessung Geoinf.*, 95(2), 66-72.

Höggerl, N. (1989). Rezente Höhenänderungen in Österreich abgeleitet aus Präzisionsnivellement-Messungen, *Österr. Beiträge zu Met. und Geophys.*, 2, 161- 173.

Horváth, F. et al. (2006). Formation and deformation of the Pannonian basin: constraints from observational data, In: *European Lithosphere Dynamics*, D.G. Gee and R.A. Stephenson (Ed.), Geological Society London Memoir, 32, 129-145.

Hrubcova, P. et al. (2005). Crustal and uppermost mantle structure of the Bohemian Massif based on CELEBRATION 2000 data, *J. Geophys. Res*, 110(B11305), doi: 10.1029/2004JB003080.

Hrubcova, P. et al. (2008). Crustal structure at the easternmost termination of the Variscan belt based on CELEBRATION 2000 and ALP 2002 data, *Tectonophysics*, 460, 55-75.

Houseman, G.A. & Molnar, P. (1997). Gravitational (Rayleigh-Taylor) instability of a layer with non-linear viscosity and convective thinning of continental lithosphere, *Geophys. J. Int.*, 128, 125-150.

Joksović, P. & Andrić, B. (1983). Report on examination of earth crust composition by deep seismic sounding on the profile of Pula - Maribor, Expert Document Fund, *Geofizika*, Zagreb.

Kissling, E. et al. (2006). Lithosphere structure and tectonic evolution of the Alpine arc: new evidence from high-resolution teleseismic tomography, In: *European Lithosphere Dynamics*, D.G. Gee and R. A. Stephenson, Mem. Geol. Soc., 32, 129-145.

Koulakov, I. et al. (2009). P- and S-velocity anomalies in the upper mantle beneath Europe from tomographic inversion of ISC data, *Geophys. J. Int.*, 179, 345-366.

Kummerow, J. et al. (2004). A natural and controlled source seismic profile through the Eastern Alps: TRANSALP, *Earth and Planetary Science Letters*, 225, 115-129.

Lammerer, B. et al. (2008). A crustal-scale cross section through the Tauern Window (eastern Alps) from geophysical and geological data, In: *Tectonic Aspects of the Alpine-Dinaride-Carpathian System*, Siegesmund, S. et al. (Ed.), Geological Society, London, Special Publications 298, pp. 219-229.

Le Pichon, X.; Bergerat, F. & Roulet, M.-J. (1988). Plate kinematics and tectonics leading to the Alpine belt formation: A new analysis, In: *Processes in Continental Lithospheric Deformation,* (S.P. Clark, B.C. Burchfiel, J. Suppe, (Ed.), GSA, Special Paper, 111-131.

Linzer, H.-G. et al. (2002). Balancing lateral orogenic float of the Eastern Alps, *Tectonophysics,* 354, 211-237.

Lillie, R.J. et al. (1994). Gravity modelling of the lithosphere in the Eastern Alpine–Western Carpathian–Pannonian basin region, *Tectonophysics,* 231(4), 215-235, doi:10.1016/0040-1951(94)90036-1.

Lippitsch, R.; Kissling, E. & Ansorge, J. (2003). Upper mantle structure beneath the Alpine orogen from high-resolution teleseismic tomography, *J. Geophys. Res,* 108 (B8), 2376, doi:10.1029/2002JB002016.

Lüschen, E. et. al. (2006). TRANSALP-deep crustal Vibroseis and explosive seismic profiling in the Eastern Alps, *Tectonophysics,* 414 (1-4), 9-38.

Mancktelow, N. et al. (2001). The DAV line and Periadriatic fault system in the Eastern Alps south of the Tauern window, *Int. J. Earth Sci.* , 90, 593- 622.

Meurers, B.; Ruess, D. & Steinhauser, P. (1987). The Gravimetric Alpine Traverse, In: *Geodynamics of the Eastern Alps,* H.W. Flügel & P. Faupl (Ed.), Verlag Deuticke, Wien, 334-344.

Meurers, B. & Ruess, D. (2007). Compilation of a new Bouguer gravity data base in Austria., *Österr. Zeitschr. f. Vermessung & Geoinformation,* 95, Heft 2.

Mitterbauer, U. et al. (2010). ALPASS- Neue Erkenntnisse über die seismische Struktur des oberen Erdmantels in den Ostalpen, *Pangeo 2010,* 15.9.-19.9.2010, Leoben.

Mooney, W.D.; Laske, G. & Masters, G. (1998). CRUST5. I: A global crustal model at 5' x 5', *J. Geophys. Res,* 103, 727-747.

Moores, E.M. & Twiss, R.J. (1995). *Tectonics,* W. H. Freeman (Ed.), New York, pp. 415.

Oberhauser, R. (1980). *Der Geologische Aufbau Österreichs,* Springer, New York. pp. 699.

Ortner, H.; Reiter, F. & Brandner, R. (2006). Kinematics of the Inntal shear zone-sub-Tauern ramp fault system section, Eastern Alps, Austria, *Tectonophysics,* 414, 241-258.

Pamić, J. et al. (2002). Geodynamic and petrogenetic evolution of Alpine ophiolites from the central and NW Dinarides: an overview, *Lithos,* 65, 113-142.

Peresson, H. & Decker, K. (1997). Far-filed effect of Late Miocene subduction in the Eastern Carpathians: E-W compression and inversion of structures in the Alpine-Carpathian-Pannonian region, *Tectonics,* 16, 38-56.

Pfiffner, O.A. et al. (1997). *Deep Structure of the Swiss Alps: Results of NRP 20,* Birkhäuser, Basel, pp. 380.

Piromallo, C. & Morelli, A. (2003). P wave tomography of the mantle under the Alpine-Mediterranean area, *J. geophys. Res.,* 108(B2), doi:10.1029/2002JB001757.

Poljak, M. et al. (2000). The seismotectonic characteristics of Slovenia, *Pure Appl. Geophys.,* 157, 37-55, doi:10.1007/PL00001099.

Posgay, K. et al. (1981). Character of the Earth's crust and upper mantle on the basis of seismic reflection measurements in Hungary, *Earth Evolut. Sci.,* 1 3/4, 272-279.

Ratschbacher, L. et al. (1991). Lateral extrusion in the Eastern Alps, Part 2.: Structural analysis, *Tectonics,* 10, 2, 257-271.

Rawlinson, N. & Kennett, B.L.N. (2008). Teleseismic tomography of the upper mantle beneath the southern Lachlan Orogen, Australia, *Physics of the Earth and Planetary Interiors,* 167 (2008) 84-97.

Reinecker, J. & Lenhardt, W. (1999). Present - day stress field and deformation in eastern Austria, *Int. J. Earth Sci.*, 88, 532-550, doi:10.1007/s005310050283.

Reiter, F.; Lenhardt, W. & Brandner, R. (2005). Indications for activity of the Brenner normal fault zone (Tyrol, Austria) from seismological and GPS data, *AJES*, 97, 16-23.

Roeder, D. (1977). Continental convergence in the Alps, *Tectonophysics*, 40(3-4), 339-350, doi:10.1016/0040-1951(77)90072-5.

Rosenberg, C.L. et al. (2004). Indentation model of the Eastern Alps and the origin of the Tauern Window, *Geology*, 32, 997-1000.

Royden, L.H. (1993). Evolution of retreating subduction boundaries formed during continental collision, *Tectonics*, 12 (3), 629-638.

Scarascia, S. & Cassinis R. (1997). Crustal structures in the central-eastern Alpine sector: a revision of available DSS data, *Tectonophysics*, 271, 157-188.

Schmid, S.M. et al. (2004). Tectonic map and overall architecture of the Alpine orogen, Eclogae Geologicae Helvetiae, 97, 93-117.

Schmid, S.M. et al. (1996). Geophysical geological transect and tectonic evolution of the Swiss Italian Alps, *Tectonics*, 15, 1036-1064, doi:10.1029/96TC00433.

Schönborn, G. (1999). Balancing cross sections with kinematic constraints: The Dolomites (northern Italy), *Tectonics*, 18, 527-545.

Simeoni, O. & Brückl, E. (2009). The effect of gravity stripping on the resolution of deep crustal structures in the Eastern Alps and surroundings Regions, *Austrian Journal of Earth Sciences*, 102/2, 157-169.

Sroda, P. (2006). Seismic anisotropy of the upper crust in southeastern Poland—effect of the compressional deformation at the EEC margin: results of CELEBRATION 2000 seismic data inversion, *Geophys. Res. Lett.*, 33, L22302, doi: 10.1029/2006GL027701.

Stampfli, G.M. & Kozur, H.W. (2006). Europe from the Variscan to the Alpine cycles. in European Lithosphere dynamics, In: *European Lithosphere Dynamics*, D.G., Gee, R.A. Stephenson, (Ed.), Geolog. Soc. London Memoirs, 32, London, pp. 129-145.

Šumanovac, F. et al. (2009). Crustal structure at the contact of the Dinarides and Pannonian basin based on 2-D seismic and gravity interpretation of the Alp07 profile in the ALP2002 experiment, *Geophys. J. Int.*, 179, 615-633. doi:10.1111/j.1365-246X.2009.04288.x.

Šumanovac, F. (2010). Lithosphere structure at the contact of the Adriatic microplate and the Pannonian segment based on the gravity modelling, *Tectonophysics*, 485, 94-106.

TRANSALP Working Group (2002). First deep seismic reflection images of the Eastern Alps reveal giant crustal wedges and transcrustal ramps, *Geophys. Res. Let.*, 29, doi:10.1029/2002GL014911.

Thybo, H. et al. (2003). Upper lithospheric seismic velocity structure across the Pripyat Trough and the Ukrainian Shield along the EUROBRIDGE'97 profile, *Tectonophysics*, 371(1-4), 41-79, doi:10.1016/S0040-1951(03)00200-2.

Tomek, C. (1993). Deep crustal structure beneath the Central and Inner West Carpathians, *Tectonophysics*, 226, 417-431.

Tomljenović, B. & Csontos, L. (2001). Neogene-Quaternary structures in the border zone between Alps, Dinarides and Pannonian basin (Hrvatsko Zagorje and Karlovac basins, Croatia), *Int. J. Earth Sciences*, 90, 560-578.

Ustaszewski, K. et al. (2008). A map view restoration of the Alpine Carpathian Dinaridic system for the early Miocene, *Swiss, J. Geosci.*, 101, Suppl. 1, 273-294.

Valasek, P. et al. (1991). Results of NFP 20 seismic reflection profiling along the Alpine section of the European Geotraverse (EGT), *Geophys. J. Int.*, 105, 85-102.

van Husen, D. (1987). *Die Ostalpen in den Eiszeiten*, Geol. Bundesanstalt, Vienna.

Vrabec, M. & Fodor, L. (2006). Late Cenozoic tectonics of Slovenia: Structural styles at the northeastern corner of the Adriatic microplate, In: *The Adria microplates*, N. Pinter. et al. (Ed.), Springer, 151-168.

Waldhauser, F.R. et al. (1998). Three-dimensional interface modelling with two-dimensional seismic data: The Alpine crust - mantle boundary, *Geophys. J. Int.*, 135, 264-278, doi:10.1046/j.1365-246X.1998.00647.x.

Weber, J. et al. (2006). The PIVO-2003 Experiment: A GPS Study of Istria Peninsula and Adria Microplate Motion, and Active Tectonics in Slovenia, In: *The Adria Microplate*, Pinter, N. et al. (Ed.), Springer, 305-320.

Weber, M. et al. (2009). Anatomy of the Dead Sea Transform from lithospheric to microscopic scale, *Rev. Geophys.*, 47, RG2002, doi:10.1029/2008RG000264.

Wessely, G. (1987). Mesozoic and Tertiary evolution of the Alpine Carpathian foreland in eastern Austria, *Tectonophysics*, 137(1-4), 45-59, doi:10.1016/0040-1951(87)90313-1.

Wortel, M.J.R. & Spakman, W. (2000). Subduction and Slab Detachment in the Mediterranean-Carpathian Region, *Science*, 290, no. 5498, 1910 - 1917.

Yan, Q.Z. & Mechie, J. (1989). A fine structural section through the crust and lower lithosphere of the Alps, *Geophys. J.*, 98, 465-488.

Ye, S.; Ansorge, J. & Cloetingh, S. (1995). Crustal structure beneath the Swiss Alps derived from seismic refraction data, *Tectonophysics*, 242, 199-221.

Zanolla, C. (2006). New gravity maps of the Eastern Alps and significance for the crustal structures, *Tectonophysics*, 414, 127-143.

Zhao, W. et al. (2001). Crustal structure of central Tibet as derived from project INDEPTH wide angle seismic data, *Geophys. J. Int.*, 145(2), 486-498, doi:10.1046/j.0956-540x.2001.01402.x.

3

Structure and Plate Tectonic Evolution of the Northern Outer Carpathians

Jan Golonka, Kaja Pietsch and Paweł Marzec
AGH University of Science and Technology; Faculty of Geology, Geophysics and Environmental Protection Mickiewicza 30, 30-059 Kraków, Poland

1. Introduction

The paper published in previous year (Golonka et al., 2009) was dealing with tectonics of the western part of the Outer Carpathians in Poland. Now, present authors expanded the area of investigation into the central and eastern part of the northern Outer Carpathians (Figs. 1, 2). The structural style of this area was recently refined by seismic survey data, especially from the areas southwest and southeast of Krakow (Fig. 2). This survey was performed by Geofizyka Kraków Ltd. and tied to geophysical and geological borehole and surface data. The authors also utilized the recently published tectonic work, integrating geological and geophysical data (Dziadzio, 2006; Kuśmierek, 2010). The structural style is explained by complex plate tectonic evolution of the Carpathian Mountains and adjacent areas.

The Carpathians define an extensive mountain arc, which stretches at a distance of more than 1 300 km, from Vienna area in Austria, to the Iron Gate on the Danube in Romania (Fig. 1). To the west, the Carpathians are linked with the Eastern Alps whereas to the east, they continue into the Balkan mountain chain. Traditionally, the Carpathians are subdivided into their western and eastern parts. The West Carpathians consist of an older, internal orogenic zone known as the Inner or Central Carpathians and the external, younger one, known as the Outer or Flysch Carpathians (Golonka et al., 2005a; Ślączka et al., 2006). The interpreted seismic survey was located in generally northernmost part of the Outer Flysch Carpathians.

2. Previous work

The deep structure of the Polish Outer Carpathians and its basement, that is southern prolongation of the North European Platform, has been recognized previously mainly by extrapolation of outcrops and by deep boreholes supplemented by magnetotelluric, gravimetric, magnetic, geomagnetic, and deep seismic sounding profiles. This work on the stratigraphy of Outer Flysch Carpathians was summarized by Ślączka et al. (2006; see also Oszczypko, 2004). The Carpathian Foredeep as well as relationship between basement, its sedimentary cover and allochthonous flysch nappes was summarized by Oszczypko et al. (2006). The results of these works depict the North European Platform as a great continental plate amalgamated in the Precambrian and Paleozoic. Its basement consists of Proterozoic, Vendian (Cadomian) and Lower Paleozoic (Caledonian) fragments, deformed and metamorphosed. The platform sedimentary cover includes Paleozoic, Mesozoic and Neogene sequences (Oszczypko & Tomaś 1985; Moryc et al., 2005, 2006; Oszczypko et al., 2006; Pietsch et al., 2007; Golonka et al., 2009).

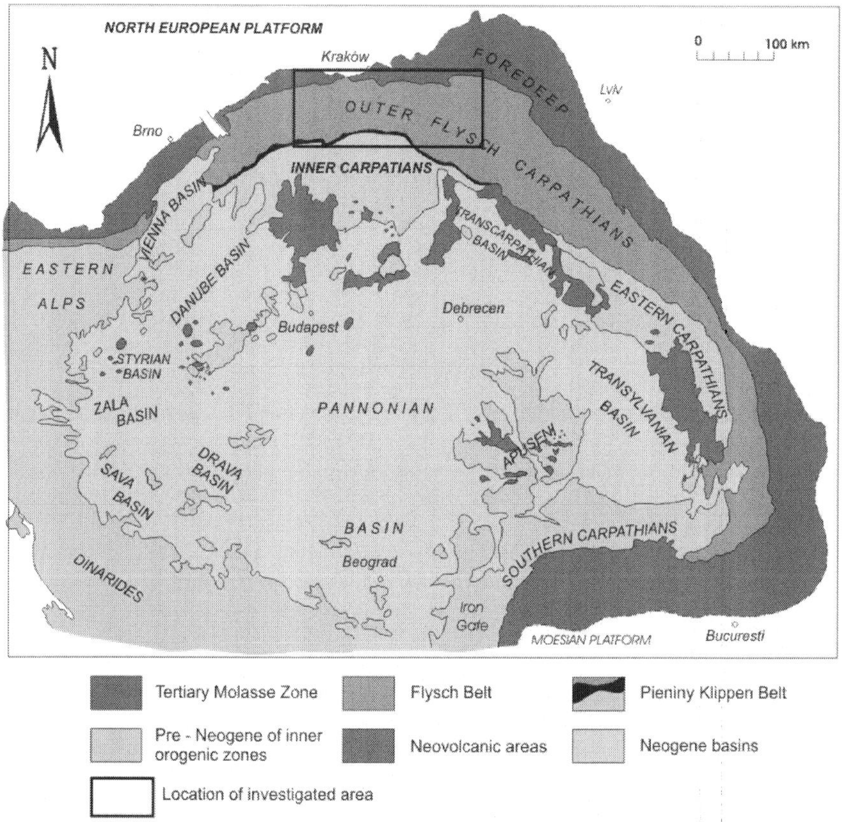

Fig. 1. Geological map of West Carpathians and adjacent areas with location of investigated area. Modified from Kováč et al. (1998); Golonka et al. (2006).

The Polish Outer Carpathians were depicted as a complex structure built up from the Upper Jurassic-Neogene flysch deposits strongly imbricated due to thrusting (Kováč et al., 1998; Oszczypko, 1998, 2004; Oszczypko & Oszczypko-Clowes M., 2003; Golonka et al., 2005a; 2009; Oszczypko et al., 2006, Ślączka et al., 2006). All the Outer Carpathian nappes are thrust over the southern part of the North European platform covered by the autochthonous Miocene deposits of the Carpathian Foredeep on the distance of 70 km, at least. The northern Carpathians nappes became uprooted from the basement during their overthrusting movement and only their basinal parts were preserved. The following Outer Carpathian nappes have been distinguished: Magura Nappe, Fore-Magura group of nappes, Silesian, Subsilesian and Skole nappes. A narrow Zgłobice-Stebnik zone of folded Miocene deposits was developed along the frontal Carpathian thrust (Fig. 2).

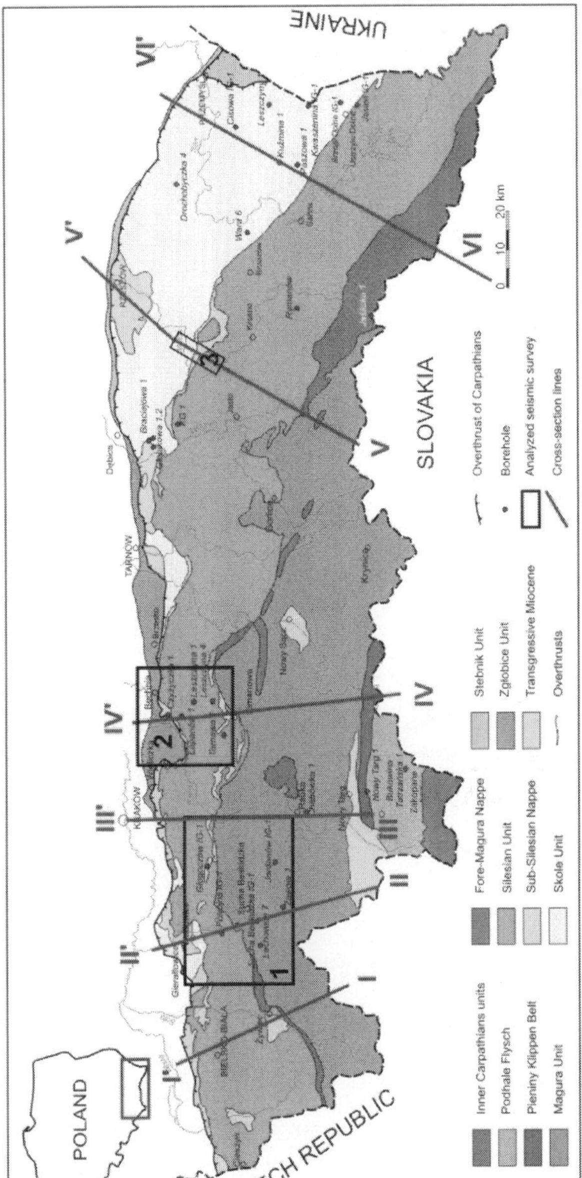

Fig. 2. Map of the northern Outer Carpathians in Poland with locality of cross-sections and seismic survey. Compiled and modified from various sources (Żytko et al., 1989; Pietsch et al., 2007; Golonka et al., 2009).

The seismic survey was conducted in the area southwest of Kraków, between Wadowice and Kraków (see Pietsch et al., 2007; Golonka et al., 2009). The seismic profiles in this area were tied to borehole data from wells: Głogoczów IG1, Potrójna IG1, Sucha Beskidzka IG1, and Zawoja 1 among the others. Similar work was performed and published in the eastern part of the Polish Outer Carpathians (Dziadzio, 2006; Kuśmierek, 2010). The interpretation of seismic survey was integrated with other geophysical research (Grabowska et al., 2007; Stefaniuk et al., 2007 and references therein) as well as with surface mapping and borehole data. The Polish Outer Carpathians were covered by maps in scale at least 1:50 000, performed and published by Polish Geological Institute; the results of this mapping were presented in the compiled maps in scale 1:500 000 (Żytko et al., 1989; Lexa et al, 2000; see also reference therein). The integration of geological and geophysical data enabled reinterpretation of the complex structure of the other Carpathians structure and geodynamic processes leading to the complex tectonics of this area (Golonka et al., 2005a, 2006, 2009; Golonka, 2007; Oszczypko, 2004, 2006; Pietsch et al.,2007; Kuśmierek, 2010).

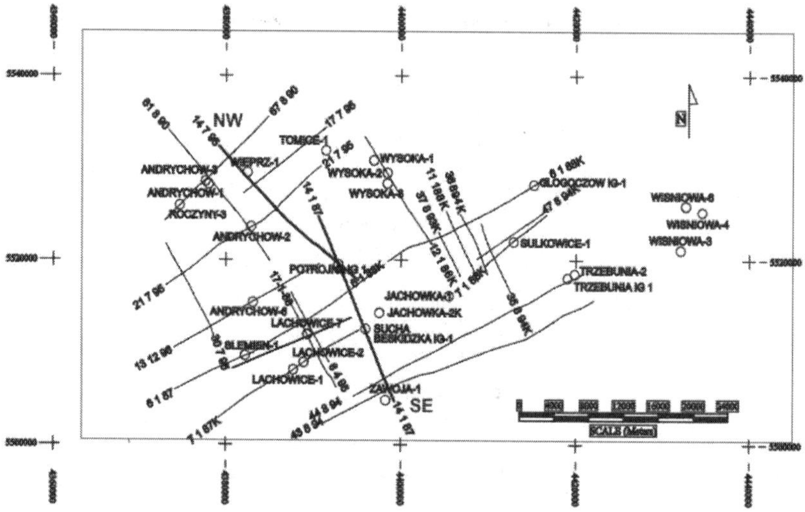

Fig. 3. Seismic survey and well location map of the area southwest from Kraków (location 1 on Fig. 2). Modified from Golonka et al. (2009).

3. Methods

The Carpathian nappes as well as North European platform were investigated by the seismic surveys ordered by Polish Oil and Gas Co. (PGNiG) and Geofizyka Kraków, Ltd. The present authors analyzed these surveys in areas southwest (location 1 on Fig. 2, Figs. 3, 5) and southeast of Kraków (location 2 on Fig. 2, Figs. 4, 6, 7). The seismic boundaries were defined on the basis of interpretation of time reflection seismic sections in version of final sums after migration (Geofizyka Kraków Ltd.) tied to geophysical and geological borehole and surface data. Geological identification of seismic boundaries was made on the basis of synthetic seismograms (seismic modeling 1D), computed with the LogM program in the GeoGraphix (Landmark Graphics Corp.)

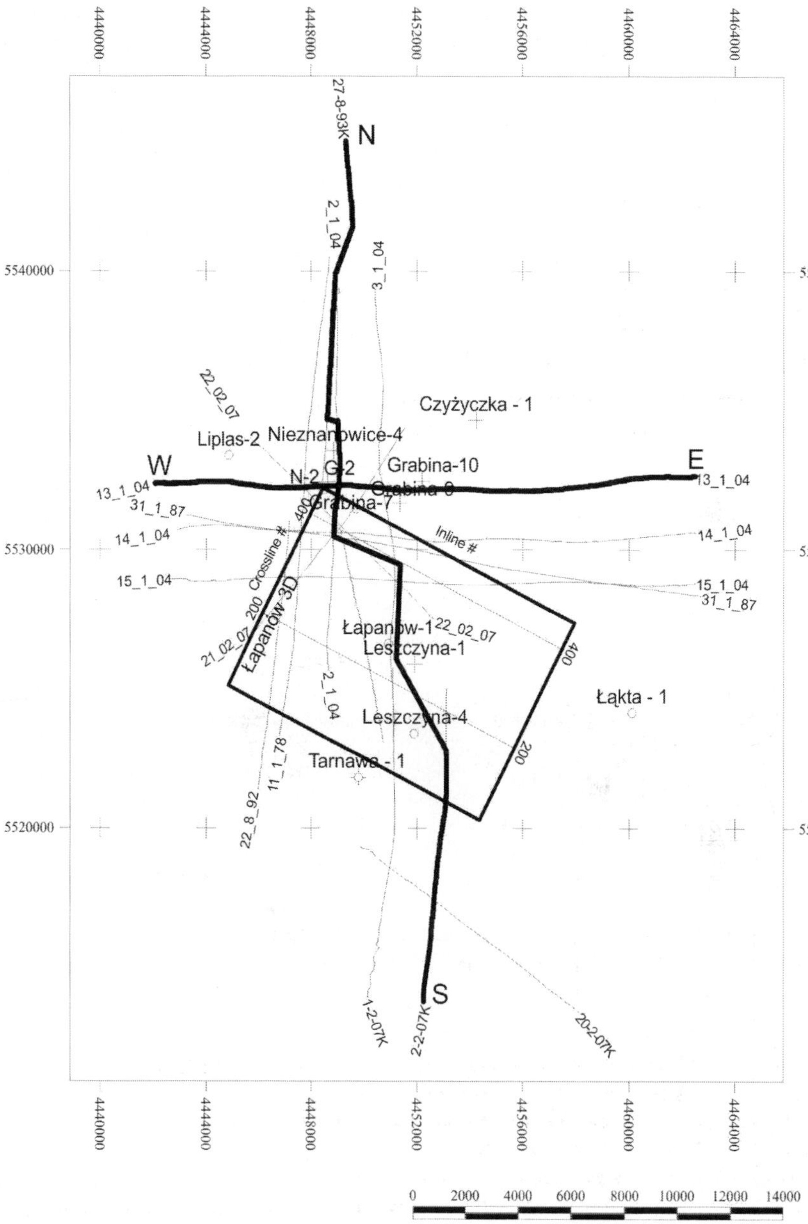

Fig. 4. Seismic survey and well location map of the area southeast from Kraków (location 2 on Fig. 2).

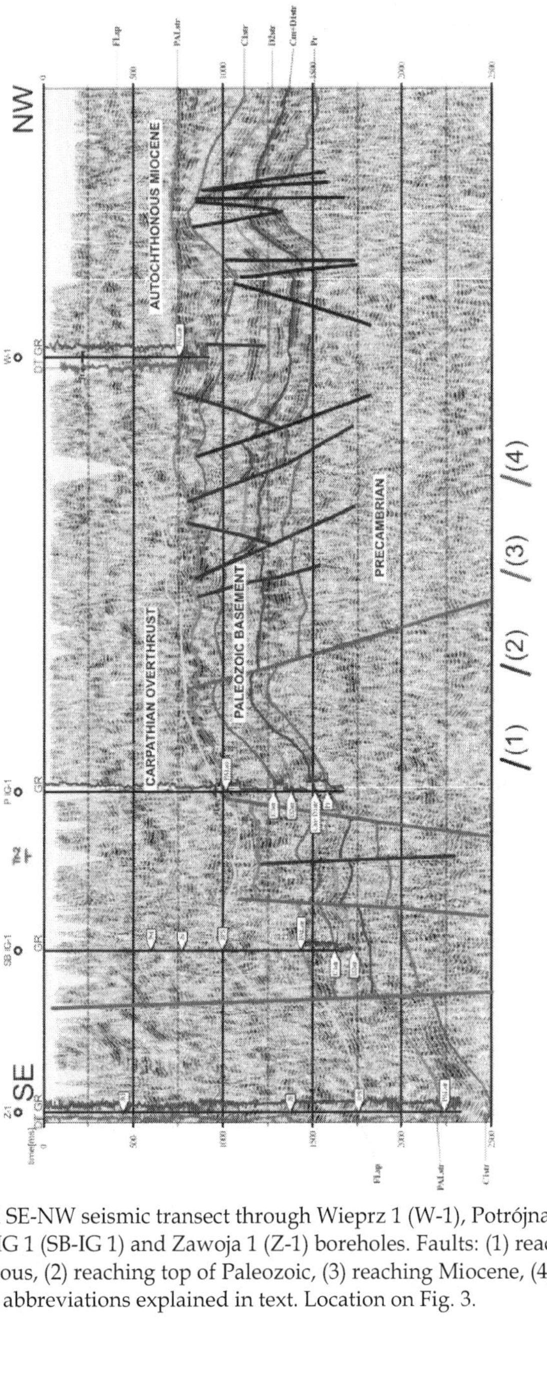

Fig. 5. Interpreted SE-NW seismic transect through Wieprz 1 (W-1), Potrójna IG 1 (P IG 1), Sucha Beskidzka IG 1 (SB-IG 1) and Zawoja 1 (Z-1) boreholes. Faults: (1) reaching top of Lower Carboniferous, (2) reaching top of Paleozoic, (3) reaching Miocene, (4) reaching surface. Horizons abbreviations explained in text. Location on Fig. 3.

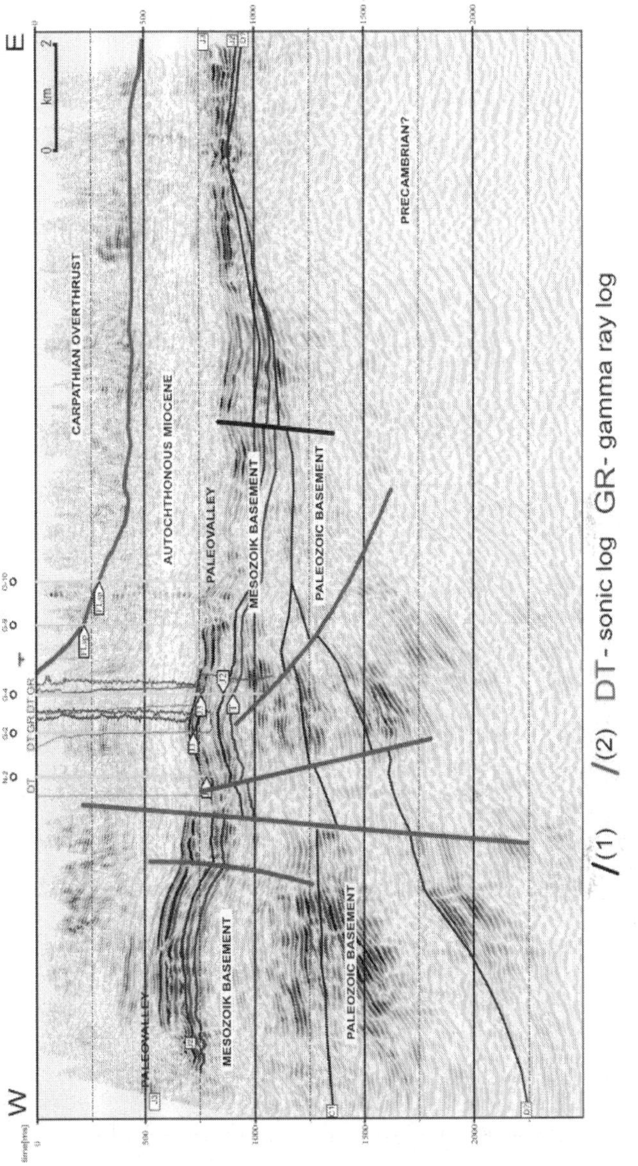

Fig. 6. Interpreted W-E seismic transect through the Nieznanowice-4 (N-2), Grabina-2, 4, 9 and 10 (G-2, G4, G-9, G-10) boreholes. Faults: (1) normal fault (2) thrust fault. Horizons abbreviations explained in text. Location on Fig. 4 (SB-IG 1) and Zawoja 1 (Z-1) boreholes. Faults: (1) reaching top of Lower Carboniferous, (2) reaching top of Paleozoic, (3) reaching Miocene, (4) reaching surface. Horizons abbreviations explained in text. Location on Fig. 3.

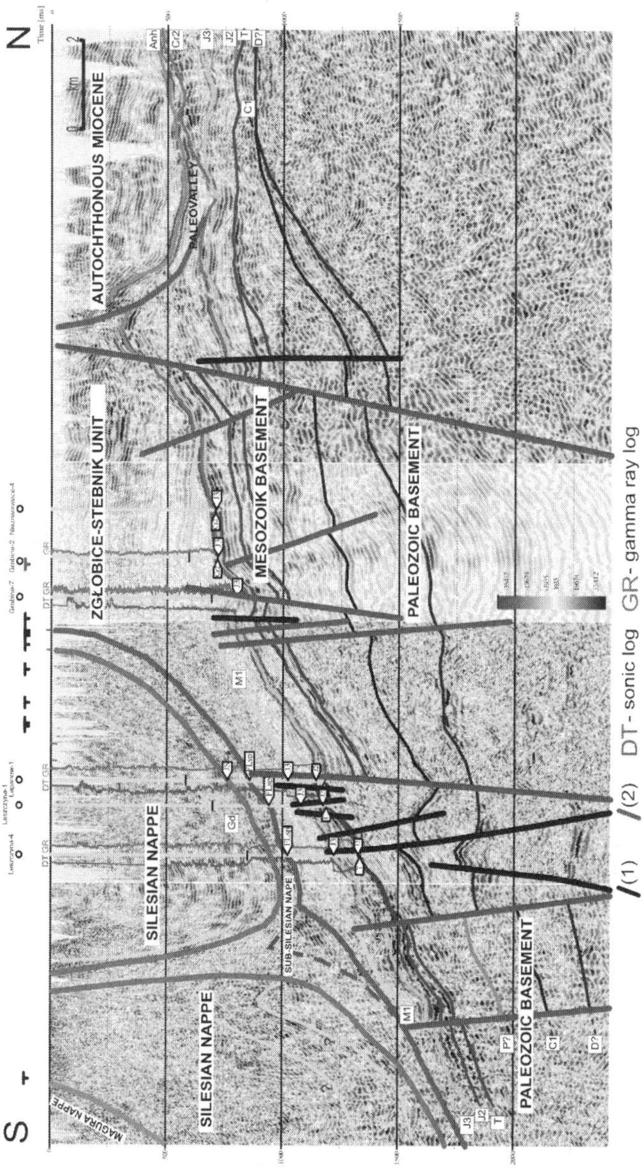

Fig. 7. Interpreted S-N seismic transect through the Nieznanowice-4, Grabina-2, 7, Łapanów-1 and Leszczyna-1, 4 boreholes. Faults: (1) normal fault (2) thrust fault. Horizons abbreviations explained in text. Location on Fig. 4. F1 (SB-IG 1) and Zawoja 1 (Z-1) boreholes. Faults: (1) reaching top of Lower Carboniferous, (2) reaching top of Paleozoic, (3) reaching Miocene, (4) reaching surface. Horizons abbreviations explained in text. Location on Fig. 3.

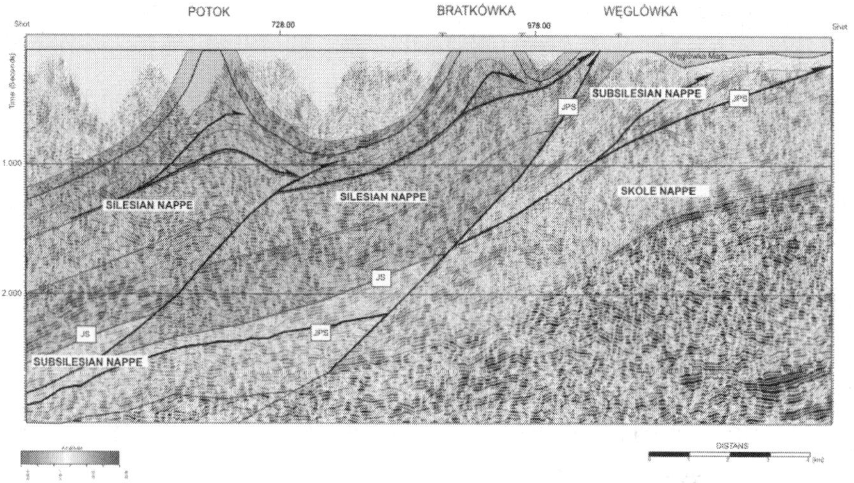

Fig. 8. Seismic transect through the Węglówka area (location on Fig. 2). From Dziadzio (2006), modified.

system for all boreholes where PAP measurements were made (Golonka et al., 2009). The seismic profiles in this area were tied to borehole data from wells (see Pietsch et al.,2007; Golonka et al.,2009): Głogoczów IG1, Potrójna IG1, Sucha Beskidzka IG1, Zawoja 1, Leszczyna 1, 4, and Łapanów 1 among the others (Figs. 5-7). The seismic profile from Węglówka area (Dziadzio, 2006) was also utilized (location 3 on Fig. 2, Fig. 8).

Several seismic boundaries were identified within the North European Platform and allochthonous cover: JMsp - the Magura Nappe base, Gd - top of the Upper Cretaceous Godula Formation (Silesian Nappe), JS - the Silesian Nappe base, JPS - the Subsilesian Nappe, Flsp - the Outer Carpathian flysch base, Anh - top of the Miocene evaporites, M1 - top of the Lower Miocene, Cr2 - top of the autochthonous Upper Cretaceous, J3- top of Upper Jurassic, J2 - top of the Middle Jurassic, T- top of Triassic, P - top of the Permian, PALstr - the top of various Paleozoic formations pinching out to the Sub-Miocene surface, C1str or C1 - the top of the Lower Carboniferous, D2 - top of the undivided Devonian, D2str - the top of carbonate formations of the Middle Devonian, Cm+D1str - the top of shaly-sandstone formations of the Lower Devonian and the Lower Paleozoic (Cambrian) and Pr - the top of consolidated basement, mainly Precambrian. Several normal and thrust faults were also distinguished.

The results of seismic surveys were interpolated, extrapolated as well as integrated with other geophysical data, geological, especially mapping data and other published information (Ślączka, 1976; Książkiewicz, 1977; Oszczypko & Tomaś, 1985; Ryłko & Tomaś, 1995; Żytko et al., 1989; Oszczypko, 1998, 2000, 2004; Paul et al., 1996; Żaba, 1999; Lexa et al., 2000; Poprawa et al., 2001; Zuchiewicz et al., 2002; Czerwiński et al., 2003; Golonka et al., 2005a, 2006, 2009; Moryc, 2005, 2006; Cieszkowski et al., 2006, 2009; Dziadzio, 2006; Guterch & Grad, 2006; Tokarski et al., 2006; Ślączka et al., 2006; Pietsch et al., 2007, 2010; Golonka, 2007; Zuchiewicz & Oszczypko, 2008; Kuśmierek, 2010). The several cross-sections (Figs. 9-14) were constructed on the basis of this integration.

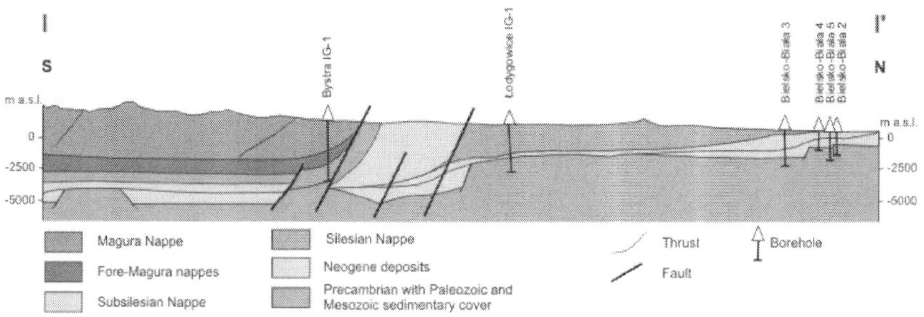

Fig. 9. Cross-section I -I′ through the Outer Carpatians and their foreland. Compiled from various sources (Żytko et al. 1989; Paul et al., 1996; Golonka et al., 2009). Cross-section location on Fig 2.

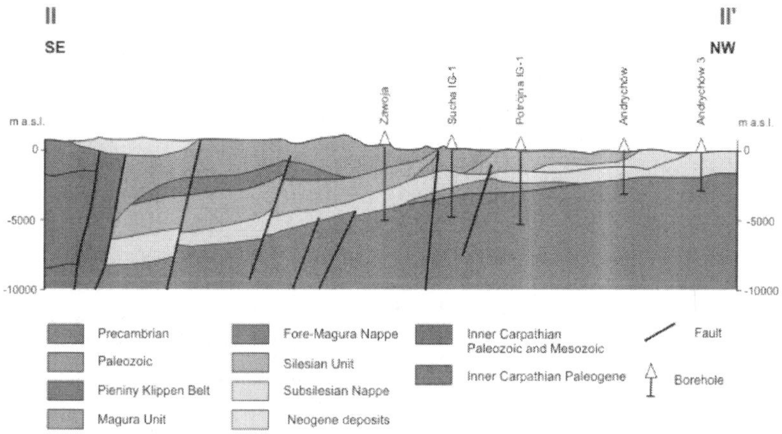

Fig. 10. Cross-section II-II′ through the Outer Carpathians and their foreland. Compiled from various sources (Oszczypko et al., 2006; Pietsch et al., 2007; Golonka et al., 2009). Cross-section location on Fig 2.

4. Structure of the Northern Carpathians

4.1 The North European Platform

The great continental plate, known as North European Platform, forms the basement of the Northern Carpathians. This plate consists of Proterozoic, Vendian (Cadomian) and Lower Paleozoic (Caledonian) fragments, deformed and metamorphosed. The Paleozoic, Mesozoic, Paleogene and Neogene strata cover the crystalline, metamorphosed basement. The crystalline, mainly Precambrian basement is depicted on all cross-sections (Figs. 9-14). It is well constrained in the northern marginal part of the Carpathian and quite speculative in the southern part. The Precambrian metamorphic rocks are well visible in seismic profiles in the western seismic survey area (Figs. 3, 5, 9) and, according to Kuśmierek (2010), in the eastern area (Fig. 14). In the central part, SE from Kraków (Figs. 6, 7) Precambrian is hard to distinguish below thick Paleozoic and Mesozoic sedimentary cover.

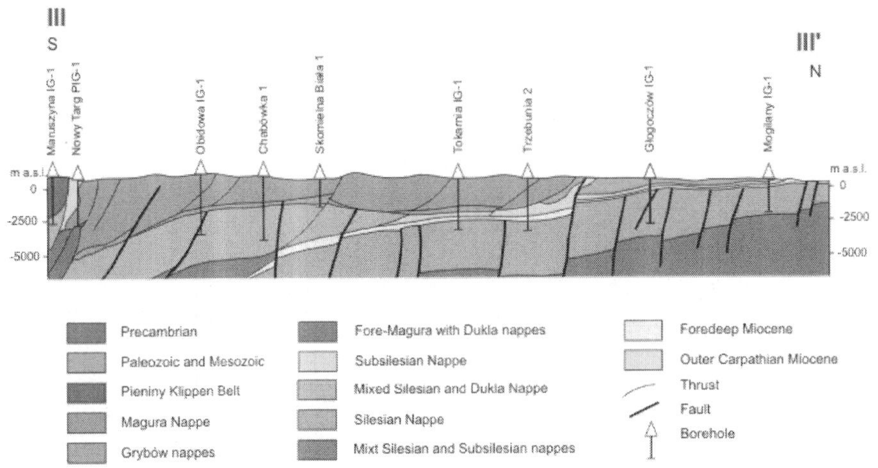

Fig. 11. Cross-section III-III′ through the Outer Carpathians and their foreland. Compiled from various sources (Golonka et al., 2006, 2009). Cross-section location on Fig 2.

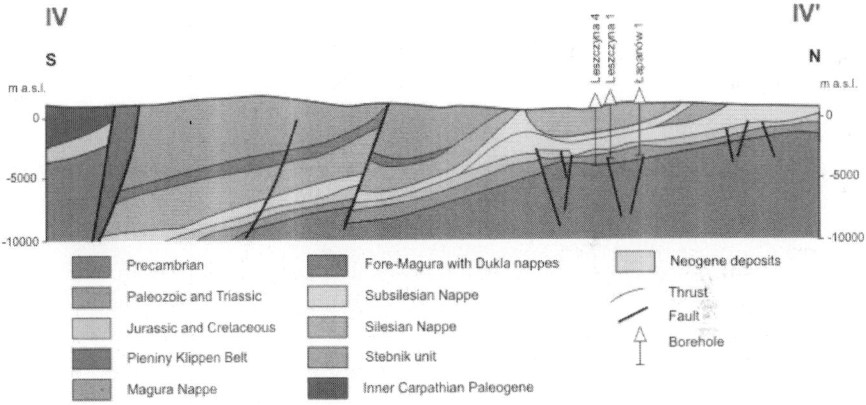

Fig. 12. Cross-section IV-IV′ through the Outer Carpathians and their foreland. Cross-section location on Fig 2.

The alochthonous, mainly flysch rocks were uprooted and thrust over the southern part the North European Platform at least 60-100 km (Ślączka et al., 2006; Golonka et al., 2009). The bottom of the Outer Carpathian flysch nappes is well visible on all seismic profiles from areas southwest and southeast from Kraków (Figs. 5-7). Like the basement, it is well constrained in the northern part and speculative in the southern part of the Polish Outer Carpathians.

The Precambrian basement beneath the Outer West Carpathians is divided into two basement blocks: the Bruno-Vistulicum Block on the west and the Małopolska Block on the east (Dudek, 1980; Buła, 2000). The Krakow-Presov Fault system marks the boundary between these two different tectonic realms within the North European Plate.

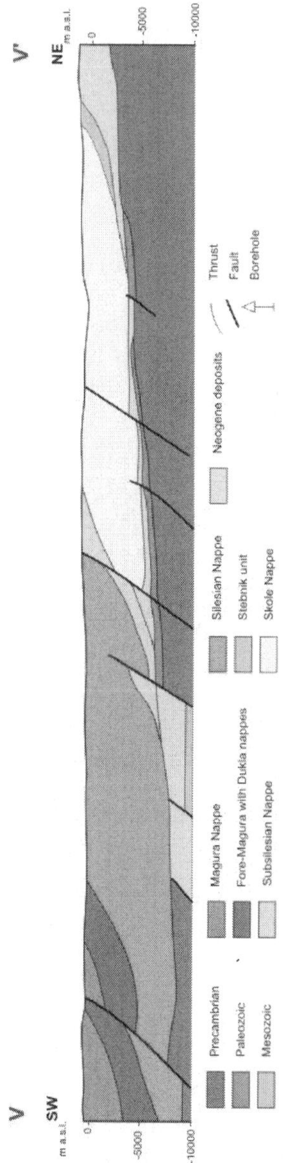

Fig. 13. Cross-section V-V′ through the Outer Carpathians and their foreland. Compiled from various sources (Żytko et al. 1989; Oszczypko, 2004; Ślączka et al., 2006; Dziadzio, 2006). Cross-section location on Fig 2. 1 (SB-IG 1) and Zawoja 1 (Z-1) boreholes. Faults: (1) reaching top of Lower Carboniferous, (2) reaching top of Paleozoic, (3) reaching Miocene, (4) reaching surface. Horizons abbreviations explained in text. Location on Fig. 3.

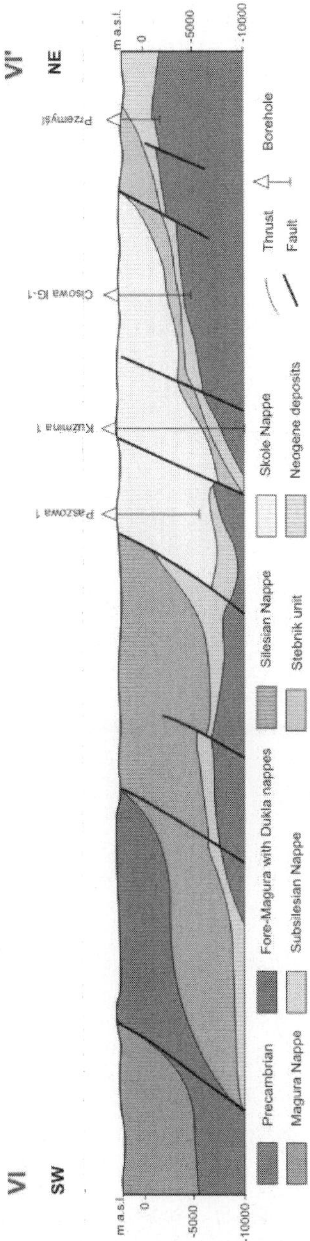

Fig. 14. Cross-section V-V′ through the Outer Carpathians and their foreland. Compiled from various sources (Żytko et al. 1989; Kuśmierek, 2010). Cross-section location on Fig 2.

The Bruno-Vistulicum terrane, consolidated in Vendian, during Cadomian orogeny times and composed of the Precambrian metamorphic rocks, is a main component of the consolidated basement in the western area (Pietsch et al., 2007; Golonka et al., 2006, 2009). The top of Precambrian boundary is cut by several faults into horsts and graben system The Precambrian basement is covered discordantly by Devonian and Upper Paleozoic formations. The Devonian rocks were encountered in numerous wells, their southern extent remains unknown. In some Precambrian horsts, Devonian is missing (Pietsch et al., 2007; Golonka et al., 2009). The Lower and Upper Carboniferous deposits cover the Devonian. The thick Permian-Triassic deposits were encountered in wells SE from Krakow (Poprawa et al, 2001). The Mesozoic sequences are known only from the central part of the northern Outer Carpathians (Figs. 10-11) their presence in the southern part of the Outer Carpathians farther eastward (Fig. 13), postulated by Ślączka et al., (2006) is quite speculative. The clastic Paleogene rocks were encountered locally. The Miocene deposits lay discordantly on the various Paleozoic, Mesozoic and Paleogene rocks.

4.2 The Outer Carpathian Nappes

4.2.1 The Magura Nappe

The Magura Nappe forms the largest tectonic unit of the northern Outer Carpathians (Ślączka et al., 2006, Golonka et al., 2009) The boundary between this nappe and Pieniny Klippen Belt display a flower structure characteristics (Figs. 10, 11, 12), running along the major strike-slip fault. The northern boundary of the Magura Nappe is erosional and runs along the arc from Czech Republic border through Żywiec to Myślenice, turning southeastward south of Kraków. The Magura Nappe has been completely uprooted from its substratum and thrust over the Fore-Magura and Silesian nappes, at least 20, perhaps up to 50 km during the orogenic movements (Figs. 10, 11, 12). The following tectonic units have been distinguished within the Magura Nappe: the Krynica, Bystrzyca, Racza and Siary units. These units display distinctive thrust boundaries in the western part of the northern Outer Carpathians (Paul et al., 2006; Golonka et al, 2009). Especially Bystrzyca Unit is thrust several kilometers over the Racza unit south of Żywiec.

Twenty north-vergent anticlines and associated synclines of W-E orientation were distinguished in the area south of Sucha Beskidzka. The Skawa line system of faults displaces the nappe margin 2 km northward in this area (Cieszkowski et al, 2006; Golonka et al, 2009). The so-called tectonic windows, showing out-of-sequence Fore-Magura nappes basement, are located between Rabka and eastern limit of the Magura Nappe. They are known by names Mszana Dolna, Szczawa, Kłęczany, Ropa, Ujście Gorlickie, Świątkowa and Smilno tectonic windows). Fig. 13 is showing out-of-sequence thrust of Fore-Magura and Dukla nappes in the so-called Świątkowa window, related to the major fault zone, cutting the basemen as well as allochthonous flysch nappes.

4.2.2 The Fore-Magura Group of Nappes

Several units, known as the Fore-Magura Group of nappes (Fig. 2) occur in the northern Outer Carpathians, north from the Magura Nappe. (Ślączka et al., 2006). The Dukla Nappe is the largest tectonic unit belonging to this group. In the East It is stretching from Poland through Slovakia to Ukrainian Carpathians. It consists of several imbricated, thrust-faulted folds with a north-west - south-east strike. The Dukla plunge gradually towards the northwest and disappears below the Magura Nappe. Two subunits were distinguished within the Dukla Nappe. The folds within the internal subunit are generally gently dipping towards the south-west and the unit's overthrust is low dipping, whereas within the external subunit folds are steep and often with a reversed (south-western) vergence (Ślączka et al., 2006).

The narrow zone, known as Fore-Magura sensu stricte nappes runs from Milówka to the area east from Żywiec, where it disappears from the surface (Golonka et al., 2009). It is possible that some elements of these nappes represent olistoliths within Menilite and Krosno formations (Cieszkowski et al., 2009). Scales and scaled north-vergence folds, originated as a result of strong compression between Magura Nappe and sandstone-dominated blocks of Silesian Nappe, dominate Fore-Magura zone structure (Golonka et al., 2009).

In the central part of the northern Outer Carpathians, the Fore-Magura group of nappes occur in the mentioned above so-called tectonic windows from Mszana Dolna tectonic window to Smilno tectonic window in Slovakia and was also encountered in the numerous wells below the Magura Nappe e.g. in the Rabka - Nowy Targ area (Obidowa IG-1, Chabówka 1) and Limanowa - Słopnice area (Słopnice 1, Słopnice 20, Leśniówka 1 and others). Many different names were applied to these units, like Grybów, Słopnice-Obidowa, Jasło, and Zboj. They relationships are unclear and speculative; considering limited data it is perhaps better to leave the traditional local names and to use the broad term "Fore-Magura Group of Nappes" (Golonka et al., 2009). Some of these units probably belonged to a bigger nappe, which was divided into separate units during the Neogene folding, other represent olistoliths within the youngest deposits of the Silesian Nappe (Ślączka et al., 2006;, Jankowski, 2007;, Cieszkowski et al., 2009).

4.2.3 The Silesian Nappe

The Silesian Nappe borders with Fore-Magura Nappes south of Silesian Beskid and with Magura Nappes south of Mały Beskid. The northern border with Sub-Silesian Nappe is erosional. In the western part of the Silesian Nappe, the structures are generally shallow and gently folded, whereas towards the East they pass into long, narrow, steeply dipping, imbricated folds. The southern part of the Silesian Nappe is hidden beneath the Magura Nappe and Fore-Magura nappes (Ślączka et al., 2006). The West of the Soła River, near the western border of Poland, the Silesian Nappe is composed of two subunits. The Cieszyn Subunit is built of a strongly folded Upper Jurassic and Lower Cretaceous Vendryně, Cieszyn Limestone, Hradište and Veřovice formations (Golonka et al., 2009). This unit includes several small anticlines visible between Cieszyn and Soła River. The Godula Unit is built mainly of sandstone-dominated formations. It is divided into two blocks: - Silesian Block and Mały (Lesser) Beskid Block, separated by fault. This blocks dip monoclinally southward, display uplifted northern margin and slightly marked longitudinal folds with Cretaceous rocks. The eastern part of the Silesian Nappe is cut by several transverse faults (Ślączka et al., 2006). The large tectonic window with Subsilesian Nappe is located in the Żywiec area (Fig. 9). Smaller windows are located in Lanckorona-Myślenice area. The border of the Silesian Nappe displays the significant offset along the Skawa dislocation zone. The eastern part of the nappe is located 10 km north of the western part. This fault zone divides the nappe into two segments the western one is characterized by the development of Cretaceous Godula Sandstones, eastern by the occurrence of Krosno Beds. Two separate zones are visible within the Silesian Nappe east of Skawa line. The lower, northern zone, known as Pogórze Lanckorońskie sheet form large syncline with Eocene-Oligocene Menilite and Krosno formations. The upper zone pinches out on the Myślenice area (Golonka et al., 2009).

Farther to the east the Cieszyn Subunit and the Godula Subunit join and the Silesian Nappe is built of several gently folded structures. East of the Dunajec River, these structures pass into imbricated folds. The eastern part of the Silesian Nappe, east of the Wisłok River, is plunging towards the south-east and is represented by a synclinorium (Central Carpathian Synclinorium), which is build mainly of the Oligocene deposits. The Central Carpathian Synclinorium is built of several long, narrow, imbricated, thrust-faulted folds which are often disharmonic. These folds are cut by several transverse faults that divide them into separate blocks. The folds display, along the strike, several axial culminations, where along the northern and southern margins of the Synclinorium the Cretaceous and Eocene strata are exposed.

4.2.4 The Subsilesian Nappe

The Subsilesian Nappe underlies tectonically the Silesian Nappe. In the western sector of the West Carpathians both nappes are thrust over the Miocene molasse of Carpathian Foredeep and in the eastern sector they are thrust over the Skole Nappe (Golonka et al., 2005a; Ślączka et al., 2006). The presence of the Subsilesian Nappe was also established in numerous boreholes beneath the Silesian and the Magura nappes (Figs. 9-12). The Subsilesian Nappe occurs as broken pieces along the northern margin of the Silesian Nappe, as well as in the tectonic windows in the Żywiec and Lanckorona-Myślenice-Żegocina areas within the Silesian Nappe (Golonka et al., 2009). The diapiric-type migration of the less competent formations of Subsilesian nappe along the strike-slip fault forms this so-called tectonic windows or out-of-sequence (see Jankowski 2007) thrust zones (figs. 9, 11, 12). Locally the Silesian Nappe is missing and the Subsilesian Nappe contact directly the Magura Nappe. Several imbricated scale-folds build up the Subsilesian Nappe. In the Andrychów area several blocks known as Andrychów Klippes contain crystalline rocks as well as Jurassic, Cretaceous and Paleogene Limestones. Traditionally they were considered as tectonic slices in linked to the Silesian Nappe front, recently the opinion about their olistostrome origin prevails (Cieszkowski et al., 2009, Golonka et al., 2009). In the frontal part of the Silesian Nappe, north of the town Krosno (Ślączka et al., 2006) the Subsilesian Nappe is exposed in the Węglówka area (Figs. 8,.13), The seismic survey and wells connected with the Węglówka oil field show that it is steeply thrust over the Skole Nappe.

4.2.5 The Skole Nappe

The Skole Nappe forms a large, 40 kilometers wide, portion of the eastern part of the Northern Carpathians and is thrust over the Miocene sediments that cover the North European Platform (Figs. 13, 14). The most distinctive structural feature of this nappe is occurrence of large thrust-folds "skybas" (duplexes) thrust over each other in the north-east direction and traced for several hundreds of kilometers along the stretch of the Carpathian Arc. The width of such "skybas" is from the single kilometers up to 12 km (Ślączka et al., 2006).

Towards the west, the well defined Skole Nappe plunges under the Subsilesian and Silesian nappes near Brzesko (Fig. 2), and its prolongation further towards the West is not clear. According to Żytko et al. (1989) and Paul et al. (1996) the Skole Nappe is thrust over the

Miocene deposits of Carpathian Foredeep in the area north of Wadowice and Andrychów (Golonka et al., 2009) This tectonic element contains the deposits of the basin and slope part of the Skole Basin as wells of Subsilesian Sedimentary Area and is very hard to distinguish from the Subsilesian Nappe. Additionally some of these formations could form huge olistoliths within the Neogene deposits in the frontal part of the Outer Carpathians (Cieszkowski et al, 2009; Golonka et al., 2009).

South-west from the town of Przemyśl the folds create a sigmoidal arc which reflects a similar sigmoidal bend of the Carpathian margin. Near the town of Rzeszów the marginal part of the Skole Nappe is covered by Miocene molasses, which form a piggy-back basin. Farther towards the west, near the town of Pilzno the northern part of the Skole Nappe is probably folded together with the Miocene cover. The inner part of the Skole Nappe is represented by a synclinal area built of several folds with broad synclines composed of the Oligocene/Lower Miocene Krosno Beds (Ślączka et al., 2006). The thrust-plane of the Skole Nappe changes from a very gentle to a very steep.

4.2 The major faults and their origin

The Carpathian nappes are cut by several major faults of different origin. Some of these faults are local, some form the huge systems sometimes over one hundred kilometers long. These systems affected all outer Carpathian nappes, the Pieniny Klippen Belt and the Inner Carpathians. An oblique collision between the North European Plate and the West Carpathians terranes invading it, lead to development of outer accretion prism, formation of a range of flysch nappes and formation of a foredeep (Ślączka et al., 2006). Through the Miocene tectonic movements caused final folding of the basins fill and created several imbricate thrust sheets (nappes) which generally reflect the basin margin configurations after the Cretaceous reorganization and Paleogene development of the Carpathian accretionary prism. The thrust faults dip southward. Most of the older normal faults were covered by allochtonous flysch nappes forming the blind faults. During the last stage of the geodynamic development the Carpathians thrustsheets moved towards their present position. Displacement of the Carpathians northwards is related to development of dextral strike-slip faults of N–S direction. Typical strike-slip fault is limited on the west and on the east by normal faults, configuration of which is similar to an asymmetrical flower structure. The orientation of this strike-slip fault zones zone more or less coincides with the surface position of the major faults (e.g. Skawa river fault zone) perpendicular to the strike of the Outer Carpathian thrust sheets. Some normal faults of E-W orientations were renewed during the final stress period and they controlled formation of morphostructures - horsts and depressions of E-W orientation. A fault of this type is perfectly visible in the seismic profiles (Pietsch et al., 2007). It has the NEE - SWW orientation. This huge fault cuts formations from the Paleozoic basement through the flysch allochthon between the boreholes Zawoja 1, from the south, and Sucha Beskidzka1 and Lachowice 7, from the north (Figs. 5, 10). The displacement of nappes of the Carpathian overthrust and diapiric extrusion of plastic formations of the lower flysch units occurred along this fault, which constitutes fragment of the major Vienna-Krakow fault zone. Another major fault zone is known as Krakow- Presov fault (Żaba 1999). It is an extension of subsequent mutual strike-slip displacements of the two blocks along the Kraków-Lubliniec Fault Zone (Żelaźniewicz in Golonka et al. 2005a). This fault was active through the Phanerozoic times until Quaternary (Zuchiewicz et al. 2002; Tokarski et al. 2006). It extends southeast from Krakow under the Outer Carpathian thrust to Rajbrot south of Bochnia (Kraków-Rajbrot fault, see Moryc 2006) and according to further through Gorlice area to Polish-Slovak border and to Presov area in Slovakia (Żaba 1999; Zuchiewicz et al. 2002).

The thrust faults in the area southeast of Krakow clearly indicate the compressional regime (Figs. 6, 7, 12). Stress data point to present-day compressive reactivation of the Carpathians (Jarosiński in Golonka et al., 2005a) The Alcapa block, advancing towards NNE exerts thin-skinned compression in the flysch nappes of the Outer Carpathians. Alcapa push seems to involve also the autochthonous basement of the Małopolska Massif domain, as analogue SHmax orientation was documented under the front of the accretionary wedge and in the foreland (Jarosiński in Golonka et al. 2005a). The Krakow- Smilno Fault system (Figs. 1, 2) marks the boundary between western, mainly extensional, with strong strike-slip component and eastern compressional regime affecting the basement rocks, their Miocene cover and often also the Carpathian flysch nappes (Pietsch et al., 2010). The thrust system is also related to the triangle zone located in the Miocene sequences north of investigated area (Fig. 7).

The diapiric-type migration of the less competent older formations along the strike-slip fault forms the so-called tectonic windows or out-of-sequence (see Jankowski 2007) thrust zones. The normal thrustsheet (nappe) sequence of the Polish Outer Carpathians is from south to north: Magura, Fore-Magura, Silesian, Subsilesian and Skole. The strike-slip associated migration causing thrusting of Fore-Magura unit over Magura unit and Subsilesian Unit over Silesian Nappe.

The series of deep-rooted transversal faults dislocates the flysch nappes as well and their basement in the eastern part of Poland, close to the Polish-Ukrainian border. These dislocations are known as the so-called Przemyśl Sigmoid, marking the transitional zone between the northern and eastern part of the Carpathian arc.

5. Plate tectonic evolution

The western part of North European platform, known as The Bruno-Vistulicum plate, was consolidated in Late Proterozoic (Vendian) during Cadomian orogeny. It was amalgamated with the eastern part (Małopolska Block) during Paleozoic times.

The southern part of the North European Platform, started to be rifted and new basins were created during the Jurassic times. The Alpine Tethys, which constitutes important paleogeographic elements of the future Outer Carpathians, developed as an oceanic basin during Jurassic as a result if the Pangea break-up (Golonka et al, 2006, 2007). The NE part of the Alpine Tethys between Carpathian- Eastern Alpine terrane and North European Platform is known as Pieniny Magabasin. The NW part of this megabasin is Magura Basin (Fig. 15). This basin is separated by the Czorsztyn Ridge from the Pieniny Klippen Belt Basin. The Silesian Ridge is an uplifted area, originally part of the North European platform separating during Jurassic-Early Cretaceous times the Magura and the Severin-Moldavidic basins. It is known only from exotics and olistoliths occurring within the various allochthonous units of the Outer Carpathians.

The following geodynamic evolution stages could be distinguish in the Outer Carpathians: I -synrift and postrift, formation of passive margin and basin with the attenuated crust, II -- collisional, development of subduction zones partial closing of oceanic basin, development of flysch basin, III - orogenic, perhaps terrane - continent collision with the accompanying convergence of two large continents, IV - postcollisional. These stages correspond with the global sequence stratigraphy, the three supersequences encompassing one stage.

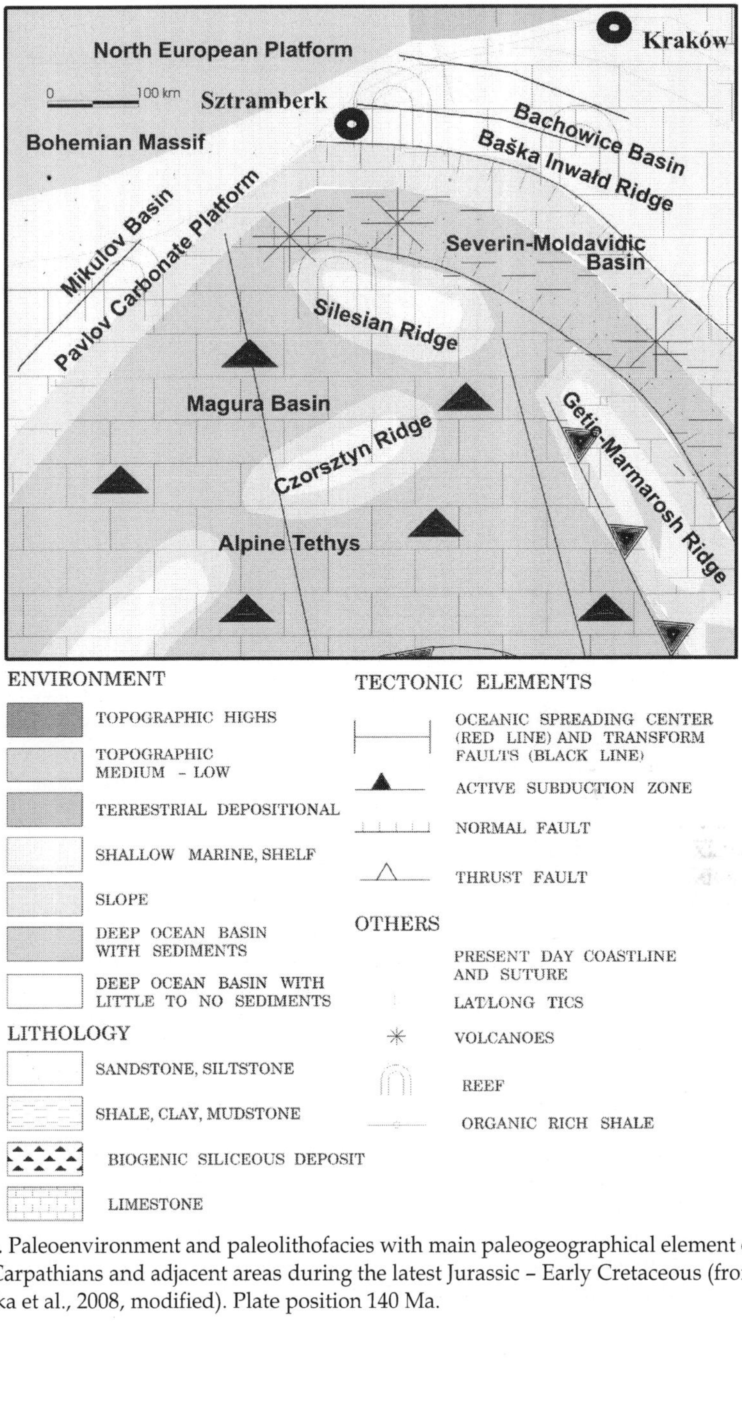

Fig. 15. Paleoenvironment and paleolithofacies with main paleogeographical element of the West Carpathians and adjacent areas during the latest Jurassic – Early Cretaceous (from Golonka et al., 2008, modified). Plate position 140 Ma.

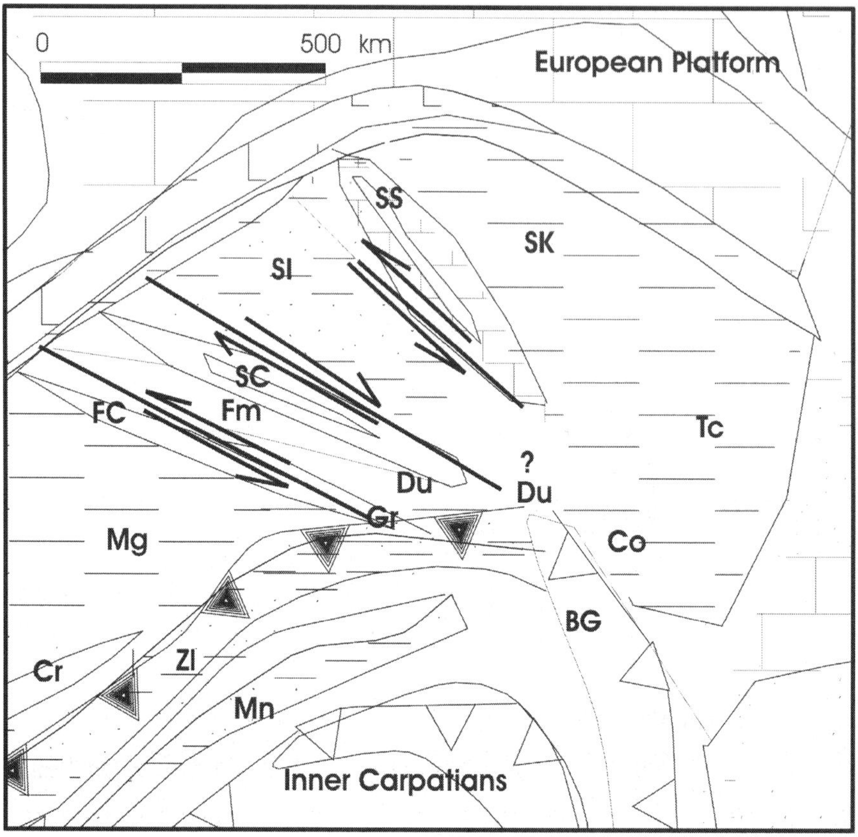

Fig. 16. Paleogeography of the Outer Carpathian basins during Late Cretaceous. Explanations as on Fig. 15. Abbreviations: BG = Bucovinian-Getic, Co = Cornohora, Porkulec, Audia, Teleajen, Cr = Czorsztyn ridge, Du = Dukla, FC = Fore-Magura ridge (cordillera), Fm - Fore-Magura basin, Gr = Grybów, Mg = Magura, Mn = Manin, Si = Silesian basin, SK = Skole, SC = Silesian ridge (cordillera), SS = Sub-Silesian ridge, Tc - Tarcau, Zl = Zlatna. From Golonka et al. (2005b).

The shallow-water marine sedimentation prevailed on the Silesian Ridge during Late Jurassic and earliest Cretaceous times. The carbonate material was transported from the ridge toward the Severin-Moldavidic Basin. Severin-Moldavidic Basin developed within the North European Platform as rift and/or back-arc basin. Its basement is represented by the attenuated crust of the North European plate with perhaps incipient oceanic fragments. The sedimentary cover is represented by several sequences of Late Jurassic - Early Miocene age belonging recently to various tectonic units in Poland and Czech Republic (Golonka et al., 2007).

The rifting process was accompanied by a volcanic activity, which persisted up to the end of Hauterivian. The Late Jurassic - Hauterivian deposition of the Severin-Moldavidic was controlled by syn-rift subsidence and later (Barremian-Cenomanian) by post-rift thermal subsidence, which culminated with the Albian-Cenomanian expansion of deep-water facies. The Cenomanian - late Eocene collisional stage is characterized by formation of subduction zones along the active margin, partial closing of oceanic basin and development of main flysch basins associate with these rifting on the platform (passive margin) with the attenuated crust. Several basins became distinctly separated within the Outer Carpathian realm (Figs. 16-17). Severin-Moldavidic basin was divided by the Subsilesian Ridge into Silesian and Skole basins. Also Fore-Magura basins emerged during these times. From uplifted areas, situated within the Outer Carpathian realm as well as along its northern margin, enormous amount of clastic material was transported by various. Each basin had the specific type of clastic deposits, and sedimentation commenced in different time (Golonka et al., 2005b).

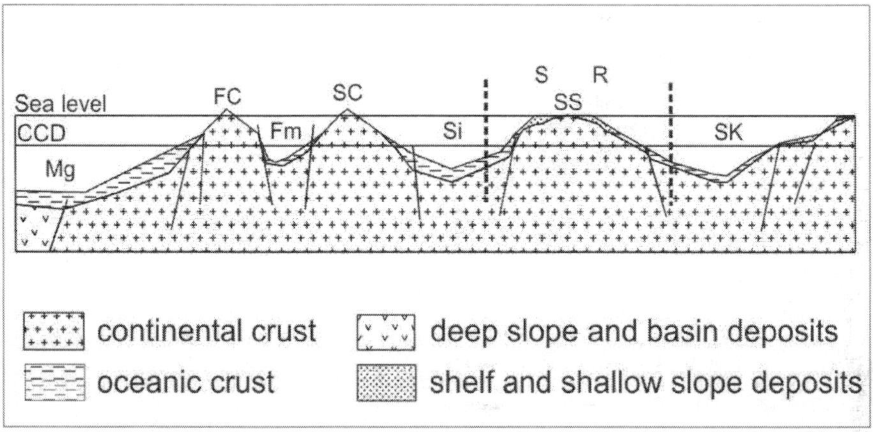

Fig. 17. Palinspastic cross-section showing the Outer Carpathian basins during Paleocene. Abbreviations: FC = Fore-Magura Ridge, Fm = Fore-Magura Basin, Mg = Magura Basin, Si = Silesian Basin, SK = Skole Basin, SC = Silesian Ridge, SS = Sub-Silesian Ridge, SR = Subsilesian sedimentary area. (From Waśkowska et al., 2009).

The Latest Eocene - Burdigalian - orogenic stage is characterized by collision, perhaps terrane - continent, with the accompanying convergence of two large continents. In the circum-Carpathian region, Adria-Alcapa (Inner Carpathians) terranes continued their northward movement during Eocene-Early Miocene times (Golonka et al., 2006). Their oblique collision with the North European plate led to the development of the accretionary wedge of Outer Carpathians. During the Priabonian and Rupelian, a prominent uplift in the Outer Carpathian basin was recorded. The Outer Carpathian remnant oceanic basins turned into foreland basins. Trough the Miocene Africa converged with Eurasia. The direct collision of the supercontinents never happened, but their convergence did not leave much space, leading to the permanent setting of the Alpine-Carpathian system. Tectonic movements caused final folding of the basins infillings and created several imbricated nappes, which generally reflect the original basin configurations. During the overthrusting movements, the marginal part of the advanced nappes has been uplifted, whereas in inner part sedimentation lasted in the remnant basin. Big olistoliths often glided down from uplifted part of the nappes into the adjacent, more outer basins (Cieszkowski et al., 2009). The nappes became uprooted from the basement and the Outer Carpathians allochtonous rocks overthrust northward in the west and eastward in the East onto the North European platform for the distance of 50 km to more than 100 km.

Overthrusting movements migrated along the Carpathians from the West towards the East. In front of the advancing Carpathians nappes the inner part of the platform, in the eastern part also with the marginal part of the flysch basin started to downwarp and tectonic depression formed during the Early Miocene. Thick molasse deposits filled up this depression. At the end of Burdigalian that basin became overthrust by the Carpathians and a new, more external one, developed. Clastic and fine-grained sedimentation of the

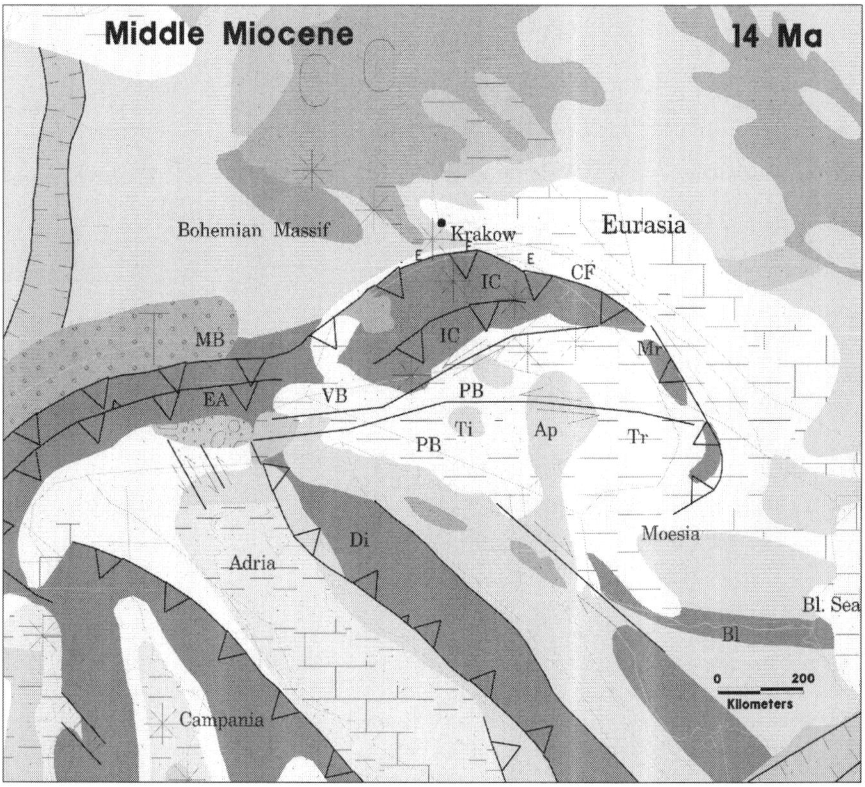

Fig. 18. Paleoenvironment and lithofacies of the circum-Carpathian area during Middle Miocene (Burdigalian- Serravallian); plates position at 14 Ma (Modified from Golonka et al., 2000, 2006). Explanations as on Fig. 15. Abbreviations: Ap - Apusen Mountains, Bl - Balkan foldbelt, CF - Carpathian Foredeep, Di - Dinaric Montains, EA - Eastern Alps, IC - Inner Carpathians, MB - Molasse Basin, Mr - Marmarosh Massif, PB - Pannonian Basin, Ti - Tisa plate, Tl -Teleajen Basin, Tr - Transilvanian Basin, VB - Vienna Basin.

Carpathian and foreland provenance prevailed with a break during the Late Langhian to Early Serravallian, when younger evaporate basin developed. Locally olistostromes were deposited with material derived from the Carpathians and the inner margin of the molasse basin. During Langhian and Serravallian part of the northern Carpathians collapsed and sea invaded the already eroded Carpathians (Fig.18). The Carpathian foreland basin continued its development partly on the top of the thrust front with mainly terrestrial deposits forming the clastic wedge. This clastic wedge along the Carpathians could be comparable with the Lower-freshwater Molasse of the Alpine Foreland Basin. During the Serravallian the marine transgression flooded the foreland basin and adjacent platform (Golonka et al, 2000, 2003, 2005a, 2006).

The foreland basin and its depocenter migrated outwardly and eastward during Langhian and Tortonian times,, contemporary with the advancing Carpathians nappes. As a result the Neogene deposits show diachrony in the foreland area. In the west sedimentation terminated already in Langhian and in the east lasted till Pliocene. These events mark the postcollisional stage in the Outer Carpathian evolution.

The SW-NE direction of the compression and the movement of the Carpatho/Pannonian lithospheric blocks is typical for the Tortonian times. An active front of orogen has been removed far from Western Carpathians area. The NE -SW compression was gradually inverted to extension. The E-W extension was induced in the Carpathian realm due to the westward direction of subducting plate (Royden et al., 1982; Doglioni et al., 1991) and related roll back effect. Numerous new N-S oriented faults developed and many old strike-slip faults were reactivated as normal extensional faults within this youngest tensional event. Normal faults played a dominant role as sedimentation controlling structures. These faults are the most numerous and conspicuous brittle features within the recent architecture of the Western Carpathians (Golonka et al. 2006, 2009.)

6. Conclusions and future work

1. The Late Precambrian to present plate tectonic process contributed to the complex structure of the northern Outer Carpathians and their basement.
2. The newest geological and geophysical research confirmed the nappe structures of the Outer flysch Carpathians and defined relationship between basement and flysch nappes.
3. The out-of-sequence-thrusts within the nappe system reflect the influence of deep rooted faults.
4. The basement structure in the in the western and eastern part of the investigated area, divided by the Krakow-Presov displays different structural styles.
5. The western fault systems developed under mainly extensional regime with strong strike-slip component while the eastern fault systems developed under mainly compressional regime.
6. The deep structure of northern Outer Carpathians is well constrained in the marginal part of the nappe system and speculative in the inner, southern part.
7. The development of Carpatian Froredeep basin is related to the advancing Outer Carpatian nappes.
8. The answer to the remaining scientific questions requires several reflexion seismic transects through the whole Outer Carpathians.
9. The deep drilling in the Orava area has been postulated (Golonka et al., 2005a).

7. Acknowledgements

This research has been partially financially supported by Ministry of Science and Higher Education grants (Działalność Statutowa - Statutory Activities) through the AGH University of Science and Technology in Krakow grants no. 11.11.140.447 (J. Golonka) and 11.11.140.769 (K. Pietsch and P. Marzec).
Authors express their gratitude to Polish Oil and Gas Co. (PGNiG) and Geofizyka Kraków Ltd. for providing the seismic and borehole data, to M. Cieszkowski, N. Oszczypko and M. Stefaniuk for fruitful discussion, and to L. Chudzikiewicz for help in graphics.

8. References

Buła, Z., 2000. Dolny paleozoik Górnego Śląska. (In Polish with English summary), *Prace Państwowego Instytutu Geologicznego*, Vol. 17: 5–83.

Cieszkowski, M. Golonka, J., Waśkowska-Oliwa, A. & Chrustek, M. (2006). Budowa geologiczna rejonu Sucha Beskidzka - Świnna Poręba (polskie Karpaty fliszowe). Geological structure of the Sucha Beskidzka region - Świnna Poręba (Polish Flysch Carpathians) (In Polish with English summary), *Kwartalnik AGH Geologia* Vol. 32, 2: 155-201.

Cieszkowski, M., Golonka, J., Krobicki, M., Slączka, A, Oszczypko, N. Waśkowska, A. & Wendorff, M. (2009). The Northern Carpathians plate tectonic evolutionary stages and origin of olistoliths and olistostromes. *Geodynamica Acta Vol.* 22(1-2): 1-26.

Czerwiński, T., Klityński, W., Florek, R. & Stefaniuk M. (2003). Regionalne badania magnetotelluryczne w polskich Karpatach Zachodnich (In Polish with English summary), *Kwartalnik AGH. Geologia* Vol. 3-4: 169-196.

Doglioni, C., Moretti, I. & Roure, F. (1991). Basal lithospheric detachment, eastward ,mantle flow and Mediterranean Geodynamics: a discussion. *Journal of Geodynamics* Vol 1. 13: 47-65.

Dudek, A., 1980. The crystalline basement block of the Outer Carpathians in Moravia: Bruno-Vistulicum. *Rozpravy Ceskoslovenskey Academii Ved*, 90: 257-264.

Dziadzio, P.S. (2006). The Geology of the Weglowka Oil Field, Subsilesian Unit, Polish Outer Carpathians, *in* Golonka J. & Picha F. (ed.), *The Carpathians and their foreland: Geology and hydrocarbon resources. American Association of Petroleum Geologists Memoir* Vol. 84, 11-46 and CDROM.

Golonka, J. (2007). Tektonika polskich Karpat fliszowych pomiędzy Bielskiem-Białą a Nowym Targiem. Tectonics of the Polish Flysch Carpathians between Bielsko-Biała and Nowy Targ (In Polish with English summary), *Kwartalnik AGH. Geologia* Vol. 33(4/1): 29-38.

Golonka, J., Oszczypko, N. & Ślączka, A. (2000). Late Carboniferous - Neogene geodynamic evolution and paleogeography of the circum-Carpathian region and adjacent areas, *Annales Societatis Geologorum Poloniae*, Vol. 70: 107-136.

Golonka, J., Krobicki, M., Oszczypko, N., Ślączka, A. & Słomka, T. (2003). Geodynamic evolution *in* and palaeogeography of the Polish Carpathians and adjacent areas during Neo-Cimmerian and preceding events (latest Triassic–earliest Cretaceous), McCann, T. & Saintot, A. (ed.), *Tracing Tectonic Deformation Using the*

Sedimentary Record. Geological Society, London, Special Publications, Vol. 208: 138-158.

Golonka, J., Aleksandrowski, P., Aubrecht, M., Chowaniec, J, Chrustek, M., Cieszkowski, M., Florek, R., Gawęda, A., Jarosiński, M., Kępińska, B., Krobicki, M., Lefeld, J., Lewandowski, M., Marko, F., Michalik, M., Oszczypko, N., Picha, F., Potfaj, M., Słaby, E., Ślączka, A., Stefaniuk, M., Uchman, A. & Żelaźniewicz, A. (2005a). Orava Deep Drilling Project and the Post Paleogene tectonics of the Carpathians, *Annales Societatis Geologorum Poloniae* Vol. 75: 211-248.

Golonka, J., Krobicki, M., Matyszkiewicz, J., Olszewska, B., ·Ślączka, A. & Słomka, T. (2005b). Geodynamics of ridges and development of carbonate platform within the Carpathian realm in Poland, *Slovak Geological Magazine* Vol. 11: 5-16.

Golonka J. Gahagan L., Krobicki M., Marko F., Oszczypko N. & Ślączka A. (2006). Plate Tectonic Evolution and Paleogeography of the Circum-Carpathian Region, *in* Golonka J. & Picha F. (ed.), *The Carpathians and their foreland: Geology and hydrocarbon resources. American Association of Petroleum Geologists Memoir* Vol. 84, 11-46.

Golonka, J., Krobicki, M., Waśkowska-Oliwa, A., Vašíček Z. & Skupien P. 2008. Główne elementy paleogeograficzne Zachodnich Karpat zewnętrznych w późnej jurze i wczesnej kredzie. Main paleogeographical elements of the West Outer Carpathians during Late Jurassic and Early Cretaceous times. *in* Krobicki, M. (ed.), *Utwory przełomu jury i kredy w zachodnich Karpatach fliszowych polsko-czeskiego pogranicza.* (In Polish with English summary), *Kwartalnik AGH. Geologia* Vol. 33(4/1): 61-72.

Golonka, J., Pietsch, K., Marzec, P., Stefaniuk, M., Waśkowska, A. & Cieszkowski, M. (2009). Tectonics of the western part of the Polish Outer Carpathians. *Geodynamica Acta* Vol. 22(1-2), 81-97.

Grabowska, T., Bojdys, G., Lemberger, M. & Medoń, Z. (2007). Geophysical-and-geological interpretation of gravity and magnetic anomalis un Polish Western Carpathians, (In Polish with English summary), *Kwartalnik AGH. Geologia* Vol. 33(4/1): 103-126.

Guterch, A. & Grad, M. (2006). Lithospheric structyre of tge TESZ in Poland based on model seismic experiments, *Geological Quarterly* Vol. *50(1): 23-32.*

Jankowski, L. (2007). Chaotic complexes in Gorlice region (Polish outer Carpathians), *Biuletyn Państwowego Instytutu Geologicznego,*Vol. 426: 27-52.

Kováč, M., Nagymarosy, A., Oszczypko, N., Ślączka, A., Csontos, L., Marunteanu, M., Matenco, L. & Márton, M. (1998). Palinspastic reconstruction of the Carpathian-Pannonian region during the Miocene, in Rakús, M. (ed.) *Geodynamic development of the Western Carpathians.* Geological Survey of Slovac Republic, Bratislava, Dionýz Štúr Publishers, pp.189-217.

Książkiewicz, M. (1977). Tectonics of the Carpathians, in Pożaryski, W. (ed.), *Geology of Poland. Vol. IV. Tectonics.* Wydawnictwa Geologiczne, Warszawa, pp. 476-604.

Kuśmierek, J. (2010). Subsurface structure and tectonic style of the NE Outer Carpathians (Poland) on the basis of integrated2D interpretation of geological and geophysical images. *Geologica Carpathica* Vol. 61: 71—85.

Lexa, J., Bzák, V., Elečko, M., Mello, J., Polák, M., Potfaj, M. & Vozár, J., ed. (2000). *Geological map of the Western Carpathians and adjacent areas: Scale 1:500 000*, Geol. Survey of Slovak Republic, Bratislava.

Moryc, W. (2005). Rozwój badań utworów miocenu w Karpatach Zachodnich na obszarze Bielsko - Kraków (Development of research on Western Carpathian Miocene deposits in the Bielsko - Kraków area), *Kwartalnik AGH. Geologia* Vol. 33(4/1): 5-73 (in Polish with English abstract).

Moryc, W. (2006). The geological structure of the Miocene substratum in the Kraków - Pilzno region. Part 1. Precambrian and Paleozoic (without Permian), *Nafta - Gaz* Vol. 62: 197-216 (in Polish with English abstract).

Oszczypko, N. (1998). The Western Carpathian Foredeep - development of the foreland basin in front of the accretionary wedge and its burial history (Poland), *Geologica Carpathica* Vol. 49, 415-431.

Oszczypko, N. (2004). The structural position and tectonosedimentary evolution of the Polish Outer Carpathians, *Przegląd. Geologiczny* Vol. 52: 780-791 (in Polish with English abstract).

Oszczypko N. (2006). Late Jurassic-Miocene evolution of the Outer Carpathian fold-and-thrust belt and its foredeep basin (Western Carpathians, Poland), *Geological Quarterly* Vol. *50* (1): 169–194.

Oszczypko, N. & Tomaś, A. (1985). Tectonic evolution of marginal part of the Polish Flysch Carpathians in the Middle Miocene, *Kwartalnik Geologiczny* Vol. 29: 109-128.

Oszczypko, N. and Oszczypko-Clowes, M. (2003). The Aquitanian marine deposits in the basement of the Polish Outer Carpathians and its paleogeographical and paleotectonic implications. *Acta Geologica Polonica* Vol. 53: 101-122.

Oszczypko, N., Krzywiec, P., Popadiuk, I. & Peryt, T. (2006). Carpathian Foredeep Basin (Poland and Ukraine): Its Sedimentary, Structural, and Geodynamic Evolution, *in* Golonka J. & Picha F. (ed.), *The Carpathians and their foreland: Geology and hydrocarbon resources. American Association of Petroleum Geologists Memoir* Vol. 84: 261-318.

Paul, Z., Ryłko, W. & Tomaś, A. (1996). Geological structure of the western part of the Polish Carpathians. *Geological Quarterly* Vol. 40: 501-521.

Pietsch, K., Golonka, J. & Marzec, P. (2007). Relationship of the basement to the flysch of the Outer Carpathians between Wadowice and Babia Góra in the light of reflection seismic survey. *Kwartalnik AGH. Geologia* Vol. 33(4/1): 143-166 (in Polish with English abstract).

Pietsch, K., Golonka, J. & Marzec, P. (2010). Structural interpretation of seismic data in Polish Outer Carpathians southwest and southeast of Krakow, Proceedings of the 19th Congress of the Carpathian-Balkan Geological Association (CBGA 2010). In press.

Poprawa, P., Jarosiński, M. Pepel, A., Kiersnowski, H. & Jawor, E. (2001). Ewolucja tektoniczna rejonu Liplas-Tarnawa - analiza subsydencji, badania mezostrukturalne oraz analiza danych sejsmicznych i grawimetrycznych. Tectonic evolution of the Liplas-Tarnawa area - analysis of subsidence, mesostructures, seismic and grawimetric data, *in* Matyja, H. (ed.), *Palaeozoic basement of the central segment of the Polish Outer Carpathians. Prace Państwowego Instytutu Geologicznego* Vol. 174: 173-204 (In Polish with English summary).

Royden, L.H., Horvath, F. & Burchfiel, B.C. (1982). Transform faulting, extension and subduction in the Carpathian Pannonian region, *Geological Societ America Bulletin,* Vol. 93: 717-725.

Ryłko W. & Tomaś A. (1995). Morphology of the consolidated basement of the Polish Carpathians in the light of magnetotelluric data. *Kwartalnik Geologiczny* Vol. 39: 1-16.

Stefaniuk, M., Klityński, W., Jarzyna, J. & Golonka, J. (2007). Struktura nasunięcia Karpackiego i jego podłoża w polskich Karpatach Zachodnich w świetle reinterpretacji wybranych profili magnetotellurycznych. The structure of the Carpathian overthrust and its basement in the Polish Western Carpathians in the light of reinterpretation of selected regional magnetotelluric profiles (In Polish with English summary), *Kwartalnik AGH. Geologia* Vol. 33:143-166.

Ślączka, A. (1976). New data on the structure of the basement of the Carpathians south of the Wadowice. *Annales Societatis Geologorum Poloniae* Vol. 46: 337-350.

Ślączka, A., Kruglow, S., Golonka, J. Oszczypko, N. & Popadyuk, I. (2006). The General Geology of the Outer Carpathians, Poland, Slovakia, and Ukraine, *in* Golonka J. & Picha F. (ed.), *The Carpathians and their foreland: Geology and hydrocarbon resources. American Association of Petroleum Geologists Memoir* Vol. 84: 221-258.

Tokarski, A., Świerczewska, A., Zuchiewicz, W., Márton, E., Hurai, V., Anczkiewicz, A., Michalik, M., Szeliga, W. & Rauch-Włodarska M. (2006). Conference Excursion 1: Structural development of the Magura Nappe (Outer Carpathians): from subduction to collapse. In: Proceedings of the 4th Meeting of the Central European Tectonic Studies Group (11th Meeting of the Czech Tectonic Studies Group), 7th Carpathian Tectonic Workshop; Zakopane, Poland, April 19-22, 2006, *Geoline,* Vol. 20: 148-162.

Waśkowska, A., Golonka, J., Strzeboński, P., Krobicki, M., Vašíček Z. & Skupien P. (2009). Utwory wczesnokredowego basenu protośląskiego w polsko-czeskich Karpatach fliszowych. Early Cretaceous deposits of the proto-Silesian Basin in Polish-Czech Flysch Carpathians, *Kwartalnik AGH. Geologia* Vol. 33(2/1): 39-47.

Zuchiewicz, W & Oszczypko, N. (2008). Topography of the Magura floor thrust anf morphotectonics of the Outer West Carpathians in Poland. *Annales Societatis Geologorum Poloniae* Vol. 78: 135-149.

Zuchiewicz, W., Tokarski, A.K., Jarosiński, M. & Márton, E. (2002). Late Miocene to present day structural development of the Polish segment of the Outer Carpathians, *EGU Stephen Mueller Special Publication Series* Vol. 3: 185-2002.

Żaba, J. (1999). The structural evolution of Lower Palaeozoic succession in the Upper Silesia Block and Małopolska Block border zone (southern Poland). *Prace Państwowego Instytutu Geologicznego* Vol. 166: 5-162.

Żytko, K., Gucik, S., Oszczypko, N., Zając, R., Garlicka, I., Nemčok, J., Eliaš, M., Menčik, E, Dworak, J, Stranik, Z, Rakuš, M, & Matejovska, O. (1989) Geological Map of the

Western Outer Carpathians and their foreland without Quaternary formations, *in* Poprawa & Nemčok, J. (ed). *Geological Atlas of the Western Carpathians and their Foreland*. Panstwowy Instytut Geologiczny, Warszawa, 1 sheet.

4

Tectonic Model of the Sinai Peninsula Based on Geophysical Investigations

Taha Rabeh
Faculty of Science, Lisbon University
Portugal

1. Introduction

Sinai Peninsula lies in the northern part of Egypt, between the Gulfs of Suez and Aqaba at the southern end, and the Mediterranean Sea at the northern end. This region is considered to be an active seismic area due to the presence of the triple junction of the Gulfs of Suez and Aqaba and the Red Sea (Khalil, 1998).

Many studies have been undertaken toward understanding the subsurface geological structure of this area. It is considered as a part of a Tertiary cratonic rift between north-eastern Africa and the Arabian Peninsula. The rifting phase essentially ceased during the early-middle Miocene (18-14 Ma), when continental separation became more oblique due to the predominant movements of the left-lateral transform fault that extends north-eastward through the Gulf of Aqaba to the Dead Sea (Patton, et al., 1994; USGS, 1998; and Robert et al., 2006). The dynamics of the Sinai Peninsula based on the geometrical configuration of the basement rocks as revealed by magnetic analysis, the pressure-tension tectonic forces resulting from seismological focal mechanism solutions, as well as the horizontal movements detected by a GPS network (John and Peter, 1969; Mcintyre, 1991; Rabeh and Miranda., 2008).

2. Geological setting

The geology of the Sinai Peninsula ranges from Precambrian basement rocks to Quaternary deposits. According to the surface geologic map (Fig. 1) after Khalil, 1998; McClay et al., 1998, Egyptian Geological Survey (1993), the Quaternary deposits cover the northern part and along the Gulf of Sinai and the Mediterranean Sea coasts. The Mesozoic limestone covers a wide area from the central part of Sinai Peninsula, while the Pre-Cambrian rocks outcrop and covers a wide areas in the southern part of the Peninsula.

The regional stratigraphy of the southern part of the peninsula (Darwish and El Azaby, 1993) shows that the sedimentary sequences overlying the basement comprise rocks from Cambrian to Quaternary. Post-rift sediments include the erosional surface that marks the base of the strata deposited during thermal or post-rift subsidence phases (Purser and Bosence, 1998). The syn-rift strata were deposited in active fault-controlled depocenters of the evolving rift, and the pre-rift strata were deposited prior to rifting according to the paleoenvironment. The upper surface of the pre-rift strata is the syn-rift unconformity or a superimposed post-rift unconformity, according to the geotectonic evolution of the basin.

The Gulf of Aqaba transform apparently lessened extension in the southern part of the gulf, and restricted active rifting to the central area (Steckler et al., 1988; Bosworth et al., 1998). The northern end is comprised of Precambrian basement rocks, Paleozoic sediments of the Carboniferous and Permian, and Mesozoic, Tertiary, and Quaternary deposits. According to Barakat (1982) the geologic sequence can be described from bottom to top as sandstone intercalated with shale and claystone, with dolomitic limestone of the Jurassic occurring at the top. Its maximum thickness reaches about 2200 m. The Cretaceous sediments are divided into Lower and Upper Cretaceous, the former consisting of sandstone with intercalations of clay and limestone, and the latter of thick limestone. This sequence is about 520 m thick. The Tertiary sediments consist of thick limestone with claystone (465 m thick), while the Quaternary is represented by sand and gravels with a maximum thickness of about 100 m.

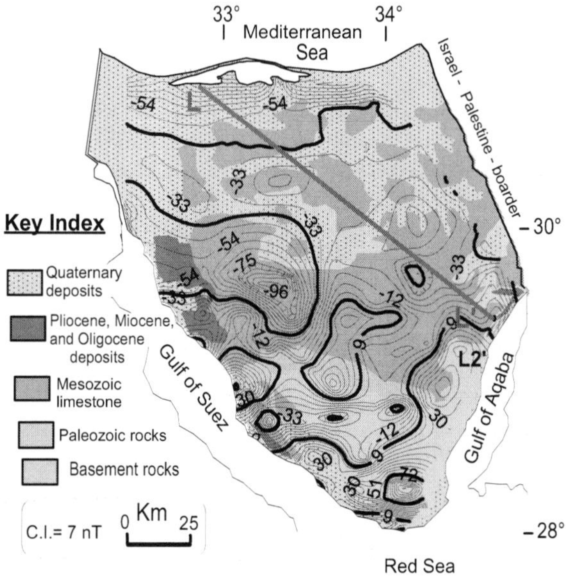

Fig. 1. Compiled geologic map after Khalil, 1998; McClay et al., 1998 and Egyptian Geological Survey (1993), and RTP land magnetic map after Rabeh (2008).

Many studies were performed to understand the subsurface geo-structure of the area. It is considered as a part of a Tertiary cratonic rift between the north-eastern Africa and the Arabian Peninsula. The rifting phase essentially ceased during the early-Middle Miocene (18-14 Ma) when continental separation became more oblique due to the dominant movements on the left-lateral transform fault, that extends through the Gulf of Aqaba north-eastward till the Dead Sea (Patton, et al., 1994, USGS, 1998). The Gulf of the Suez region has long been recognized as one of the best examples of long-axis segmentation with different dip polarities (Colleta et al., 1988; Moustafa, 1993; Bothworth, 1994; Patton et al., 1994; McClay et al., 1994 & 1998). It displays examples of interaction between extensional tectonics and sedimentations (Gawthorpe et al., 1997; Gupta et al., 1999; Sharp et al., 2000). It is remarkably non-volcanic with only a few late pre-rift to early syn-rift basic dykes and isolated basaltic features (Bosworth and McClay, 2001). Four distinct depocenters (sub-basins) separated by complex accommodation zones occur within the Gulf of Suez and Northwestern Red Sea (Bosworth and McClay, 2001). Each sub-basin is a symmetric, bounded on one side by NW trending border fault system with large throws 3-6 Km in general (Gupta et al., 1999). This is providing a good idea about the tectonic position of the Peninsula. The Pre-Cambrian Basement rocks appear in the southern part of the peninsula while the depositional depocenter is dipping towards the northern part (*cf.* Fig. 2). This is due to the compression forces due to Suez rifting that we show it in that Chapter.

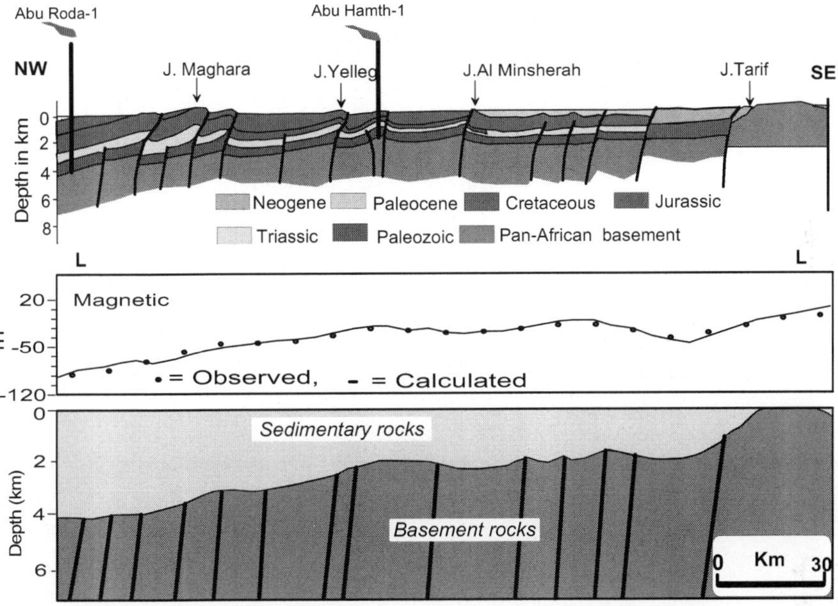

Fig. 2. Schematic geological cross-section along northern Sinai, after Guiraud and Bosworth (1999).

3. Geophysical evaluation

Using the integrated interpretation of seismological, GPS, potential potential-field, geological and well logging data, and several non-outcropping fault zones have been recognized and tentatively mapped in the study area (Rabeh and Miranda, 2008).

Based on Grant & West (1965), Linsser technique (1967) and horizontal gradient method we were able to delineate the subsurface fault trends from the RTP land magnetic map (*cf.* Fig. 3). The Euler deconvolution method, published by Reid et al. (1990) serves to determine source positions and the depths of the geomagnetic inhomogeneities. This method confirmed the existence of the deduced structures whereas the Euler solutions were clustering along these structures. The different directions were then grouped into segments of 10° of azimuth each. These groups are represented according to the tectonic movements/forces prevailed in the studied area by rose diagrams (*cf.* Fig. 3).

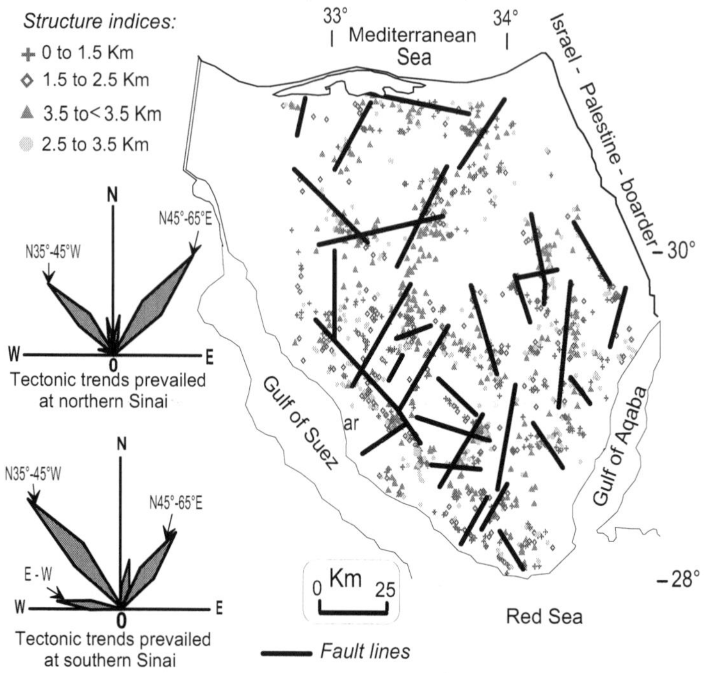

Fig. 3. A deduced structures map shows clustering the Euler solutions along the fault lines and a Rose diagrams show the prevailed tectonic trends at both northern and southern parts.

The results indicate that the N35°-45°W tectonic trend (related to Gulf of Suez and Red Sea tectonics) is more predominant at the southern part of Sinai Peninsula than N35°-65°E tectonic trend (related to Syrian trend) while this arrange is reversed at the northern part. Aqaba trend (N15°-25° E) comes at the third order of predominance is prevailed at the southern part while the E-W trend is predominant at the northern part (related to the Mediterranean tectonics).

The Stress - tension axis prevailed in the studied area derived from the focal mechanism solutions of the events located in the southern Gulf of Suez suggest pure normal faulting mechanism, with a NE-SW trending tension axis. Whereas the mechanism of this event along the Gulf of Aqaba reflects strike slip mechanism with left lateral motion along NW-SE plane (*cf.* Fig. 4). The stress fields based on the deduced focal mechanism of the different seismic zones according to our study have been selected and the average direction of pressure axis (P-axis) and tension axis (T-axis) are calculated for each zone Abu El Enean, (1997). The distribution of the P and T axis along the studied area (*cf.* Fig. 4) shows a dominant T-axis trending N 45° E. The surface faults illustrate that there is an extension stresses act in the region. These results were confirmed from recent analysis of GPS data (Rabeh and Miranda, 200).

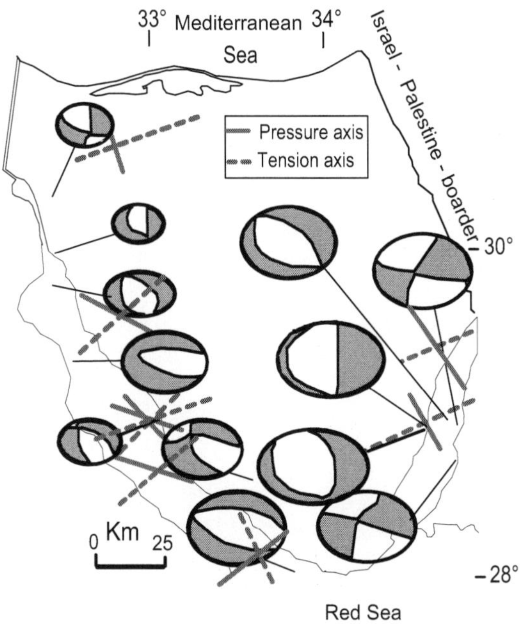

Fig. 4. Map showing focal mechanism solutions and stress - tension axis prevailed in the studied area.

The kinematic model that explains the implications of deformation, stress and tectonic activities in the Sinai Peninsula were interpreted through an integrated study using land magnetic surveys, seismology and geodynamics as well as geological analysis. The most predominant tectonic trends is N35° - 45°W direction. This trend originated due to opening process of the Gulf of Suez and is normal to the NE - SW tension axis (Said 1990). It comes in the first order of predominance while N45° - 65°E (connected to Syrian Arc tectonics) comes in the second order at the southern part of the Peninsula. This order is reversed at the northern part. The N25° - 35°E which is related to Gulf of Aqaba tectonics can be detected at the southern part whereas the E-W (related to the Mediterranean tectonics) tectonic trend is prevailed at the northern part. They are considered as a third order of their predominance.

These forces were confirmed by stress-tension relation derived from focal mechanism solutions of seismological data. Moreover, the results obtained by magnetic and seismological interpretations has been confirmed by GPS data analysis. It indicates that the velocity of Sinai Peninsula ranges from 1.8 to 2.3± 0.5 mm/yr in the NE direction

Finally, the integrated analysis for the magnetic, seismic and GPS interpretations can produce a kinematic model for Sinai Peninsula (*cf.* Fig. 5).

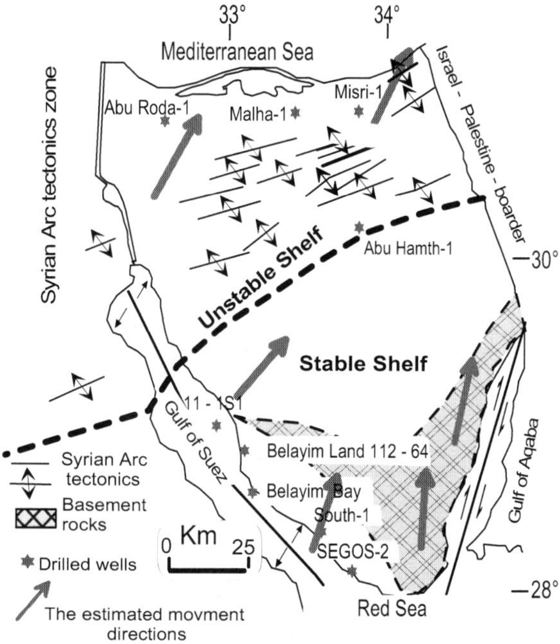

Fig. 5. The deduced kinematic model of the Sinai Peninsula.

4. References

Abu Elenean, K.M., 1997: A study on the seismotectonics of Egypt in relation to the Mediterranean and Red Seas tectonics, Ph.D. Thesis, Faculty of Science, Ain Shams University, Cairo, 187 pp.

Barakat, M. G., 1982: General review of Petroliferous Provinces of Egypt with special emphasis on their geologic setting and oil potentialities. Petro. And Gas Proj., Cairo Univ., M. I. T., P. 86.

Bothworth, W., 1994: A model for the three-dimensional evolution of continental rift basins, north-east Africa. In: Schandelmeier H. & Stern R.J. (eds), Geology of Northeast Africa (Part 2(, Geologisch, Rundschau, 83, p. 671-88.

Bosworth, W. & McClay, K. R. 2001: Structural and stratigraphic evolution of the Gulf of Suez rift, Egypt: A synthesis. in Zeigler, P.A., Cavazza, W., Robertson, A.H.F.R., & Crasquin-Soleau, S. (eds), Peri-Tethys Memoir 6: 'Peritethyan Rift/Wrench Basins and Passive Margins'. Memoires du Museum National d'Historie Naturelle de Paris: v. 186, p. 567 - 606.

Colleta, B. Le Quellec, P. Letouzy, J. & Moretti, I., 1988: Longitudinal evolution of the Suez rift structure (Egypt). Tectonophysics, 153, P. 221 - 233.

Darwish M., and El Azaby, M., 1993: Contributions to Miocene sequences along the western coast of the Gulf of Suez, Egypt. Egyptian Journal of Geology, 37, pp., 21 - 47.

Egyptian Geological Survey and Mining Authority (EGSMA), (1993) Geologic map of Sinai Peninsula, Egypt.

Gawthorpe, R.L., Sharp, I., Underhill, J.R., & Gupta, S., 1997: Linked sequence stratigraphic and structural evolution of propagating normal faults. Geology, 25, p. 795 - 798.

Guiraud, R., Bosworth, W., 1999: Phanerozoic geodynamic evolution of northeastern Africa and the northwestern Arabian platform. Tectonophysics, p 315, 73–108

Grant, F. S., and West, G. F.,1965. Interpretation theory in applied geophysics, MacGraw-Hill.

Gupta, S., Underhill, J.R., Sharp, I.R. & Gawthorpe, R.L., 1999: Role of fault interactions in controlling synrift sediment dispersal patterns: Miocene Abu Alaqa Group, Suez Rift, and Sinai, Egypt. Basin Research, 11, p. 167 - 189.

John C. Behrendt, and Peter Popenoe, 1969: Basement structure contour map of North Park-Middle Park Basin, Colorado. AAPG Bulletin, V. 53; no. 3, p. 678-682.

Khalil, S.M., 1998: Tectonic evolution of the eastern margin of the Gulf of Suez, Egypt. PhD. Thesis, Royal Hollway, University of London, p. 349.

Linsser H. 1967: Investigation of tectonics by gravity detailing, Geophysical Prospecting, V. 15.

McClay, K.R., Nicols, G.J., Khalil, S.M., Darwish, M. & Bosworth, W., 1994: A model for the three-dimensional evolution of continental rift basins, north-east Africa. In: Schandelmeier H. & Stern R.J. (eds), Geology of Northeast Africa (Part 2(, Geologisch, Rundschau, 83, p. 671-88.

McClay, K.R., Nicols, G.J., Khalil, S.M., Darwish, M. & Bosworth, W., 1998: Extensional tectonics and sedimentation, Eastern Gulf of Suez, Egypt. In: Purser B.H. & Bosence D:W:J. (eds), sedimentation and tectonics of Rift Basins, Red Sea-Gulf of Aden, Chapman and Hall, London, P. 223 - 238.

McIntyre J.I., 1991: Northwestern New South Wales regional magnetics and gravity. The Bulletin of the Australian Society of Exploration Geophysicists, V.22, No. 2, P. 261-264.

Moustafa, A.R., 1993: Structural characteristics and tectonic evolution of the east-margin blocks of the Suez rift. Tectonophysics, 223, p. 381 - 399.

Patton T.L., Moustafa, A.R., Nelson, R.A., & Abdine, S.A. 1994: tectonic evaluation and structural setting of the Suez Rift. In: London S.M. (ed), Interior rift basins, American association of Petroleum Geologist, Memoir 59, P 7 - 55.

Purser, B.H. and Bosence, D.W.J. 1998: Sedimentation and tectonics in rift basins. Red Sea . Gulf of Aden, Chapman & Hall.

Rabeh, T.T. and Miranda, J.M., 2008: A tectonic model of the Sinai Peninsula based on magnetic data, Journal of Geophysics and Engineering, V. 5, pp. 469–479.

Reid, J. M., Allsop, H. Granser, 1990, Magnetic interpretation in three dimensions using Euler deconvolution. Geophysics, V., 55, No., 1 , P 80 - 91.

Robert Reilinger, Simon McClusky, Abdullah ArRajehi, Salah Mahmoud, Ali Rayan,

Woldai Ghebreab, Ghebrebrhan Ogubazghi, and A. Al-Aydrus, 2006: Geodetic constraints on Rupturing of Continental Lithosphere along the Red Sea, MARGINS Newsletter No. 17, P. 16.

Said, R., 1990. The geology of Egypt, Elsevier Publishing Co., Amsterdam New York.

Sharp, I.R., Gawthorpe, R.L., Underhill, J.R., & Gupta, S., 2000: Fault-propagation folding in extensional settings: example of structural style and synrift sedimentary response from the Suez rift, Sinai, Egypt. Bulletin of Geological Society of America, v. 112; no. 12; p. 1877-1899.

Steckler, M. S., F. Berthelot, N. Lyberis, and X. Le Pichon, 1988: Subsidence in the Gulf of Suez: implications for rifting and plate kinematics: Tectonophysics, v. 153, p. 249-270.

U. S. Geological Survey, 1998: Open-File Report 99-50-A,U. S. Geological Survey, Denver, Colorado, World Energy Project, October 1998.

Part 3

Tectonics of Siberia

5

Siberia - From Rodinia to Eurasia

Dmitry V. Metelkin, Valery A. Vernikovsky and Alexey Yu. Kazansky
Trofimuk Institute of Petroleum Geology and Geophysics SB RAS
Russia

1. Introduction

The ancient Siberian continent and its Phanerozoic frame occupies the central position in the structure of Northern Asia (Fig. 1). The paleogeography of the Siberian plate attracts considerable interest, as an independent object of research which is throwing light on geographical conditions in the past, features of tectonic development of the region, legitimacies in the distribution of indicators of geological formations and other fundamental questions of geology.

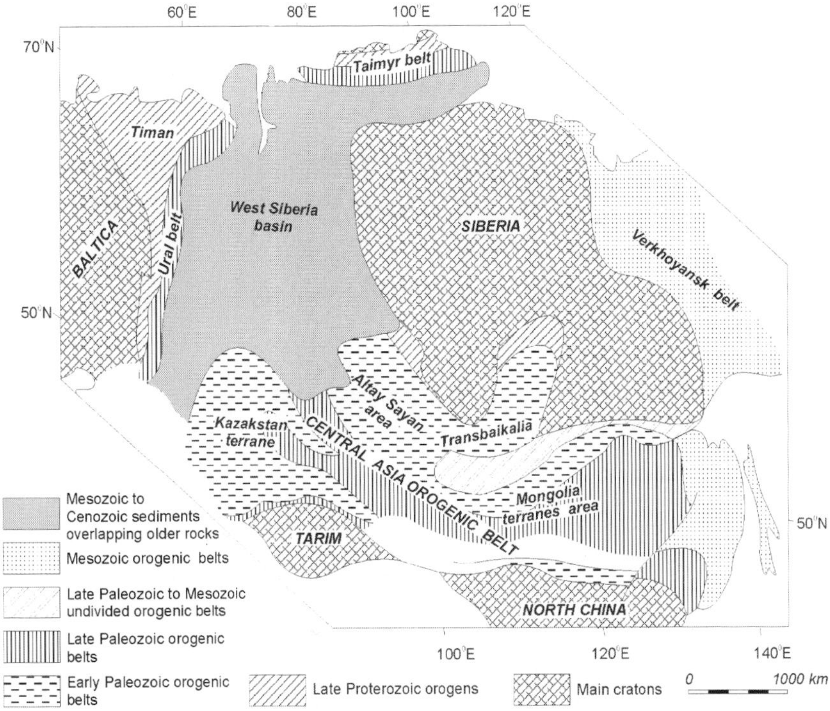

Fig. 1. Main tectonic units of the Northern Asia (adapted from Berzin et al., 1994).

Here we present the results of a short analysis of paleomagnetic data obtained during recent years over the territory of the Siberian Craton and some adjacent terrains. The new data allows us to propose a new version of the apparent polar wander path (APWP), uniting three large intervals of geological history: Neoproterozoic, Paleozoic and Mesozoic. This data allows a reconstruction of step-by-step tectonics of the Siberian continental plate for almost 1.0 Ga. It appears that large-scale long-living strike-slips were playing the major role in a tectonic history of the craton at all stages of development of the crust of its folded frame. Strike-slip motions defined the tectonic style of the evolution of folded systems as in the early stages of formation of oceanic basins, and during active subduction of oceanic crust and, undoubtedly, at the accretional-collisional stage. Intraplate deformations of the newly formed continental crust and accompanying active magmatism were also supervised by strike-slips motion of the fragments of different scale. The present work is an attempt to coordinate most of the paleomagnetic and geological information from different regions of Siberia forming a uniform picture in the context of a strike-slip hypothesis. Here we propose eight paleotectonic reconstructions corresponding to the key moments in the tectonic history of Siberia which describe a change of spatial position of the craton and support the leading role of strike-slip motions in the tectonic evolution of a continental crust of its frame on the base of paleomagnetic estimations. Those reconstructions are partly based on the author's tectonic models published earlier, but the central place here belongs to the Siberian Craton.

2. The tectonics of the Siberian Craton

2.1 The geology of the Siberian Craton

Siberian Craton occupies the central place in the structure of Northern Asia and is located between the largest rivers of Eastern Siberia - the Yenisey and Lena. The southeast boundary of the craton coincides with the Mongol-Okhotsk suture which separates the Early Paleozoic crystal complexes of Stanovoy block from the folded structures of the Mongol-Okhotsk belt developed at the end of Jurassic - Cretaceous (Zorin, 1999; Kravchinsky et al., 2002; Tomurtogoo et al., 2005; Metelkin et al., 2010a). Westward, the fields of Early Cambrian formations of Stanovoy block are "lost" among granitoid batholiths of the Baikal folded area. Here, the Zhuinsk system of faults is accepted as a boundary of the craton crystal complexes. Within the limits of Northern Transbaikalia, the boundary of the Siberian Craton is going inside a well expressed Baikal-Patom paleoisland arc. During Neoproterozoic and Paleozoic this territory was represented by a sedimentary basin on the margin of Siberian continent and was deformed as a result of accretional-collisional events (Parfenov et al., 2003; Khain et al., 2003; Zorin et al., 2009). The Sayan-Yenisey folded-napped structure forms the southwestern margin of the ancient continent. Further to the southwest, a mosaic of terrains of Altay-Sayan fragment of the Central-Asian mobile belt is located. At the western periphery the craton block is overlapped by Mesozoic-Cenozoic cover of the West Siberian plate and the boundary of the craton is conventionally traced over the Yenisey river valley. In the north, the platform deposits of Siberian Craton are buried under sediments of the Yenisey-Khatanga trough, which is considered as a branch of the West Siberia basin, and are limited the by structures of the Taimyr-Severnaya Zemelya folded-napped area. The eastern periphery of the Siberian Craton is formed by the deformed complexes of the Verkhoyansk folded-napped system. Here, predominantly sedimentary complexes deposited within the margin of the Siberian continent during the Paleozoic and Mesozoic, were detached and were broken from the crystal basement and pulled over the craton

(Parfenov et al., 1995; Oxman, 2003). The Verkhoyansk trough, developed at the frontal part of the napes, is accepted as a modern boundary of distribution of the low-deformed cover of the Siberian Craton. Thus, the most ancient crystal complexes of the craton are traced practically everywhere under the mountain ridges of the surrounding folded-napped belts, and the outlined boundaries represent the arbitrary contours used for paleotectonic reconstruction.

The Archean-Paleoproterozoic crystal basement is exposed in the limits of the Aldan-Stanovoy shield in the southeast and in the limits of Anabar-Olenek uplift in the north and also as relatively small missives among folded-napped structures of the cratonic margin in the south-west. Granulite-gneissic and granite-greenschist complexes undoubtedly are prevailing and form a number of terrains developed discretely between 3.3 and 2.5 billion years ago (Rozen et al., 1994; Rosen, 2003; Smelov, Timofeev, 2007). A collision of terrains and a build-up of the craton have taken place about 1.8 billion years ago (Rozen et al., 2005; Smelov, Timofeev, 2007).

The sedimentary cover is formed by Late Proterozoic and Phanerozoic deposits. Mesoproterozoic and Neoproterozoic geological complexes on the Siberian Craton are concentrated over its margins, forming both a sedimentary sequences comparable to conditions of shelf basins (Pisarevsky & Natapov, 2003), and magmatic (volcanic and the volcano-sedimentary) complexes connected with oceanic spreading and subduction processes on the continental margin. The last are included in Neoproterozoic folded belts surrounding the craton: the Central Taimyr, Pre-Yenisey and Baikal-Muya belts. Early Paleozoic sedimentary complexes are widespread and occupy all territory of the plate. Shallow sea and lacustrine terrigenous-carbonate and gypsum-dolomite deposits predominate (Kanygin et al., 2010). A new stage in the development of the plate complex began in the Devonian and has been connected with the continental rift event. Rifting has driven the generation of the Vilyui graben system and the extensive sedimentary basin in the east of the Siberian plate which was developed up to the end of the Mesozoic and resulted in the Vilyui syncline structure which is infilled mainly by terrigenous deposits (Parfenov & Kuzmin, 2001). The Permian-Triassic platobasaltic sequence and underlying Carboniferous- Permian terrigenous and the tuff deposit of the Tungus tectonic province is considered as an independent structural complex of the Siberian platform. The development of the depression here is connected with a stretching and thinning of the continental crust above an extensive hotspot in the mantle, so the thick trapp complex appears to be a direct reflection of the largest plum activity (Dobretsov & Vernikovsky, 2001). Moreover, the Permian-Triassic boundary coincides with rifting in the northwest frame of the Siberian Craton. The giant sedimentary basins of West Siberia, including the Yenisey-Khatanga trough, have occupied the adjusting, lowered margins of the platform. Late Mesozoic collisional processes in the east and the south of the craton have completed the development of the modern structure of the Siberian platform.

2.2 The paleomagnetic record

The apparent polar wander path (APWP) for Siberia is well known only for Paleozoic. Today, not less than four versions of this trend are proposed (Khramov, 1991; Pechersky & Didenko, 1995; Smethurst et al., 1998; Cocks & Torsvik, 2007). Distinctions between the paths are caused by different approaches in data selection, the non-uniform distribution of data over a time scale, and also by a "smoothing" technique during the construction of

APWP. Despite differences in details, the general character of the Paleozoic polar wander is co-coordinated and describes the northward drift of Siberia from the equator to high latitudes of the northern hemisphere with a prevailing clockwise rotation (Pechersky & Didenko, 1995; Cocks & Torsvik, 2007). The maximum drift velocity sometimes exceeded from 5 to 12 cm/year while the amplitude of rotation went up to 1 degree per million years, depending on an APWP version which was used.

We constructed the Neoproterozoic interval of APWP on the basis of a refined summary table of Precambrian poles (Metelkin et al., 2007a) where most of key poles (reliability index (Van der Voo, 1990) more than 3) from Siberia were obtained in resent years (Table 1).

In particular the analysis carried out (Metelkin et al., 2007a), proves nonconventional for the Siberia "eastern" drift (from outside the Indian ocean) of the poles comprising in Neoproterozoic a characteristic loop comparable with the well known "Grenville Loop" of APWP for Laurentia (McElchinny & McFadden, 2000). The similarity of the APWP shapes for Siberia and Laurentia not only quite unequivocally proves a tectonic connection of the cratons within the structure of Neoproterozoic, but also allows a reconstruction of the dynamics of its break-up (Metelkin et al., 2007a, Vernikovsky et al., 2009). The plate kinematics for the first third of the Neoproterozoic can be described by a southward drift with a counter-clockwise rotation from the equatorial to the moderate latitudes of the southern hemisphere. The second third of the Neoproterozoic is characterized by a reversed drift of the plate to the equator with a clockwise rotation. The calculated drift velocity as a rule does not exceed 10 cm/year, and the amplitude of rotation less than 1 degree per million years.

However, the Vendian (Ediacarian - from to 600 million years ago to 540 million years ago) APWP interval connecting the above mentioned Neoproterozoic and Paleozoic APWP trends (Fig. 2, Table 2) still remains ambiguous. For 560 million years we used the mean pole of the group which is concentrated near Madagascar Island (tab. 1). However, we also cannot exclude a more southern pole position for this time - which is near the coast of Antarctica (Shatsillo et al., 2005, 2006). Despite the essential progress in the study of Late Precambrian and a considerable quantity of new paleomagnetic data, the problem of the paleomagnetic pole position for the Vendian time is far from an unequivocal solution. A number of hypotheses were proposed and among them: a non-stationary, non-dipolar state of the geomagnetic field at this time, abnormaly high drift velocities and some others (Kirshvink et al., 1997; Meert et al., 1999; Kravchinsky et al., 2001; Kazansky, 2002; Pavlov et al., 2004; Shatsillo et al., 2005, 2006).

A serious problem whose solution can probably provide the answer to the majority of points of disagreement is the problem of the absolute age of the rocks studied and the age of magnetization preserved in them. Despite the described difficulties, the distribution of Vendian- Early Cambrian poles fits the expected trend between Neoproterozoic and Paleozoic APWP segments (Fig. 2).

Also there is unequivocal substantiation for the Early Mesozoic segment of the Siberian APWP due to the absence of authentic data for Middle and Late Triassic. A combination of Paleozoic and constructed Late Mesozoic segments (Fig. 2, Table 2) assumes the presence of a strongly pronounced casp (an interval with a sharp change in polar wander). The presence of the casp is basically not connected with the tectonic reasons, but is caused by a technique of APWP calculation during the smoothing of selected data over time intervals.

Object	age, Ma	Pole position			Reference
		(°N)	(°E)	A_{95}	
1050 - 640 Ma					
Malga Fm, , Uchur-Maya district	1045±20	25.4	50.4	2.6	Gallet et al., 2000
Lakhanda Group., Uchur-Maya district	1000-1030	13.3	23.2	10.7	Pavlov et al., 2000
Ui Group, including sills, Uchur-Maya district	950-1000	4.9	357.7	4.3	Pavlov et al., 2002
Karagass Group, Pre-Sayan trough	800-740	4.2	292.1	6.2	Metelkin et al., 2010b
Nersa Complex, Pre-Sayan trough	741±4 [1]	22.7	309.8	9.6	Metelkin et al., 2005a
Predivinsk Complex, Yenisei Ridge	637±5.7 [2]	-8.2	7.7	4.7	Metelkin et al., 2004a
600-530 Ma					
Aleshino Fm, , Yenisei Ridge	600-550	-28.3	24.3	7.7	Shatsillo et al., 2006
Carbonates, Igarka district	560-530	-33.4	45.6	12.7	Kazansky, 2002
Carbonates, Lena-Anabar district	560-530	-28.0	66.5	8.2	Kazansky, 2002
Aisin Fm., Pre-Sayan area	600-545	-39.9	75.1	12.1	Shatsillo et al., 2006
Taseevo supergroup Yenisei Ridge	600-545	-32.9	75.1	6.1	Shatsillo et al., 2006
Taseevo supergroup Yenisei Ridge	600-545	-41.0	91.0	15.4	Pavlov & Petrov, 1997
Ushakovka Fm., , Transbaikalia	600-545	-31.6	63.8	9.8	Shatsillo et al., 2005 [3]
Sediments, Pre-Sayan area and Yenisei Ridge	560-530	-29.5	74.1	4.5	Shatsillo et al., 2006 [3]
Kurtun Fm., Transbaikalia	560-530	-25.3	54.5	12.0	Shatsillo et al., 2005 [3]
Irkutsk Fm., Transbaikalia	560-530	-36.1	71.6	3.2	Shatsillo et al., 2005 [3]
Minua Fm., Transbaikalia	600-530	-33.7	37.2	11.2	Kravchinsky et al., 2001
Shaman Fm., Transbaikalia	600-530	-32.0	71.1	9.8	Kravchinsky et al., 2001
MEAN	~ 560	-33.9	62.2	8.9	
200-80 Ma					
Sediments, Lena River	175-245	47.0	129.0	9.0	* Pisarevsky, 1982
Basalts of Tungui depression, Transbaikalia	180-200	43.3	131.4	23.0	Cogné et al., 2005
Sediments, Verkhoynask trough	170-160	59.3	139.2	5.7	Metelkin et al., 2008
Badin Fm., Transbaikalia	150-160	64.4	161.0	7.0	Kravchinsky et al., 2002
Ichechui Fm., Transbaikalia	150-160	63.6	166.8	8.5	Metelkin et al., 2007b
Sediments, Verkhoyansk trough	140-120	67.2	183.8	7.8	Metelkin et al., 2008
Khilok Fm, , Transbaikalia	110-130	72.3	186.4	6.0	Metelkin et al., 2004b
Intrusions, Minusa trough	74-82	82.8	188.5	6.1	Metelkin et al., 2007c

Comment: [1] - age according to (Gladkochub et al., 2006); [2] - age according to (Vernikovsky et al., 1999); 3 - "anomalous" (non-dipolar) field according to the viewpoint of data authors; * Pisarevsky, 1982: pole #4417 from IAGA GPMDB. (http://www.ngu.no/geodynamics/gpmdb/).

Table 1. Selected paleomagnetic poles from Siberia used for calculation of the Neoproterozoic and Mesozoic intervals of the Siberian APWP.

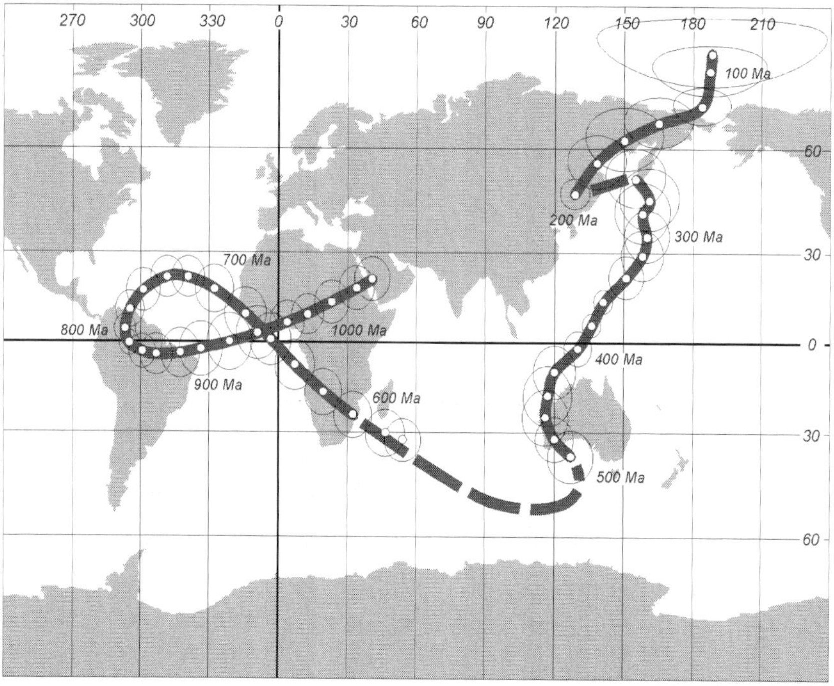

Fig. 2. The apparent polar wander path for Siberia. The pole co-ordinates are listed in the table 2. Dashed lines represent uncertain APWP intervals with poor data, which need verification.

Actually, the Late Mesozoic interval of the APW path is based on the paleomagnetic data obtained for the territory of the Verkhoyansk trough and the southwest periphery of the Siberian platform, generalized in (Metelkin et al., 2010a). We can see that the Late Mesozoic poles for Siberia have demonstrated a regular deviation from the reference poles for Europe (Besse & Courtillot, 2002). The angular distinction in Jurassic positions for Siberia and Europe reaches 45 degrees (Metelkin et al., 2010a) and gradually reduces by the end of Cretaceous. The possible reason of such distinctions appears to be the strike-slip motions between the Siberian and European tectonic domains. The scales of the motions can be estimated firstly as hundreds of kilometers. Under the term "domain", we understand the area with internal heterogeneous structure, but manifesting itself as a tectonicaly rigid block of the Earth's crust. Tectonic rigidity here is understood as the absence of deformations which have led to mutual motions or an essential rotation of blocks, composing the internal structure of the domain. According to the restored paleomagnetic trace, the Siberian domain during the Jurassic was a part of the Eurasian plate and was located in the high latitudes of the northern hemisphere. The whole structure has undergone a general drift in a southern direction (with a maximum velocity 10-12 cm/year) with a gradual clockwise rotation (an amplitude up to 2.5 degrees per million years). Up to the J-K boundary, Siberia has reached the modern co-ordinates and then demonstrates only a clockwise rotation with amplitudes not more than 0.5-1 degrees per million years (Metelkin et al., 2010a).

Mesozoic				Paleozoic				Neoproterozoic			
Age (Ma)	PLat	PLong	A_{95}	Age (Ma)	PLat	PLong	A_{95}	Age (Ma)	PLat	PLong	A_{95}
80	81.3	188.2	6.7	240	52	155	8	560	-32.2	54.3	6.7
100	77.8	187.4	5.2	260	46	161	9	580	-30.0	46.7	7.4
120	70.2	183.9	4.2	280	42	158	9	600	-24.1	32.5	7.5
140	66.3	165.2	6.0	300	35	160	8	620	-16.7	19.6	7.7
160	62.1	150.3	7.8	320	29	158	8	640	-7.6	7.2	8.6
180	56.3	138.3	7.1	340	22	151	7	660	1.0	356.8	8.9
200	47.7	128.8	4.3	360	14	141	4	680	9.7	345.9	8.9
				380	6	136	4	700	18.0	332.4	7.9
				400	-2	130	6	720	22.0	320.9	6.7
				420	-10	120	9	740	21.9	311.7	6.7
				440	-18	117	10	760	17.6	301.2	7.1
				460	-25	116	9	780	11.2	295.3	6.1
				480	-32	120	7	800	4.6	293.2	4.7
				500	-36	129	8	820	-0.4	295.1	4.3
								840	-3.3	300.8	5.8
								860	-4.2	306.9	7.1
								880	-3.7	317.5	8.5
								900	-2.3	326.5	9.4
								920	0.2	339.0	9.8
								940	3.2	351.2	9.1
								960	6.8	3.9	7.4
								980	9.6	12.7	6.8
								1000	13.8	23.2	7.2
								1020	18.4	34.0	7.3
								1040	21.5	40.8	7.2

Comment: The Paleozoic interval is taken from (Pechersky & Didenko, 1995); the Mesozoic and Neoproterozoic intervals are calculated of the basis of poles listed in table. 1. The data set was smoothed using the cubic spline (Torsvik & Smethurst, 1999) and then recalculated using a "sliding window" (window size - 50 Myr, poles through 20 Myr) (Besse & Courtillot, 2002); Plat, Plong - latitude and longitude of paleomagnetic pole; A_{95} - radius of 95% confidence oval.

Table 2. The final APW path for Siberia.

3. The structure of orogenic belts and terrains surrounding the Siberian Craton

3.1 Taimyr orogenic belt

The Late Paleozoic folded-napped structure of the Arctic part of Siberia (Fig. 3) can be divided into three large tectonic elements: the South Taimyr marginal-continental area, the Central Taimyr Neoproterozoic accretonal belt and Kara terrain separated by large thrusts, namely, the Major Taimyr and Pyasino-Faddey (Vernikovsky, 1996; Bogdanov et al., 1998).

3.1.1 South Taimyr area

The southern part of the Taimyr Peninsula is represented by a thick succession of shallow-sea sedimentary sequence with age ranging from Neoproterozoic to Permian. Towards the Siberian platform, the sequence plunges under Mesozoic-Cenozoic deposits of Yenisey-Khatanga trough and overlaps the Archean-Paleoproterozoic crystal basement of the Siberian Craton. The section is basically composed of carbonate and clay deposits and represents a typical sequence of the passive continental margin faced to the north (Ufland et al., 1991; Vernikovsky, 1996; Bogdanov et al., 1998). Development of the oceanic basin in the north of Siberia is assumed to be at the very beginning of the Mesoproterozoic, while in Neoproterozoic the region was developed in a mode of shelf margin of the continent (Pisarevsky & Natapov, 2003). The passive continental margin environment remained up to the Permian. The upper part of the section is sated by Early Triassic volcanic complexes of trapp formation. The formation of the complex has occurred under intraplate conditions under the influence of the North Asian superplum (Vernikovsky et al., 2003) and has probably been connected with the early stage of the opening of the Yenisey-Khatanga rift system and thus corresponds to the model of the forearc trough evolution in the north, in front of the growth of the Hercinian orogen.

3.1.2 Central Taimyr belt

The acrretional structure of the belt was formed during the Neoproterozoic (Vernikovsky & Vernikovskaya, 2001). The structure of the belt basically consists of paleoisland arc and paleoocean terrains which are represented by Neoproterozoic volcano-sedimentary and volcanic successions, alternating with Paleoprtoterosoic cratonic terrains, composed mainly from deeply metamorphosed rock associations (Vernikovsky & Vernikovskaya, 2001, Peace et al., 2001). According to U-Pb dating most of the ancient island arc associations were already developed in the beginning of the Neoproterozoic about 961±3 million years ago (Vernikovsky et al., in progress). Paleomagnetic data obtained for these complexes testifies that the ancient arc was located in the immediate proximity from the South Taimyr margin of Siberia (Vernikovsky et al., in progress). The paleomagnetic pole position (Plat=17.8, Plong=326.8 A95=4.0) is quite close to the one-age pole for the craton (Pavlov et al., 2002): the angular divergence is about 30°, while the latitudal one is less than 9°. Ophiolites and island arcs were developed in the northern margin of Siberia up to the end of the Neoproterozoic: their ages for Cheliuskin and Stanovoy belts are 750- 730 million years ago and 660 million years ago for the Ust-Taimyr belt (Khain et al., 1997; Vernikovsky & Vernikovskaya, 2001; Vernikovsky et al., 2004). The accretion time of the island arcs to Siberia is estimated as Late Neoproterozoic about 600 million years ago (Vernikovsky et al., 1997; 2004) and the overlapping Vendian-Paleozoic sedimentary complex, including characteristic molasse, forms a uniform margin-continental system with the South Taimyr territory (Ufljand et al., 1991; Vernikovsky, 1996; Vernikovsky Vernikovskaya, 2001). The sedimentary complex, along with Vendian coarse-grained molasse, contains siltstone, mudstone and black graptolite shale with layers of limestone and the dolomite, forming the main part of the section from the Lower Cambrian to the Devonian. The presence of graptolite shale proves a more deep-water sedimentary environment, rather than shelf complexes, typical for the South Taimyr area. The axis of a deep-water trough is reconstructed in the frontal part of Pyasino-Faddey thrust (Vernikovsky, 1996).

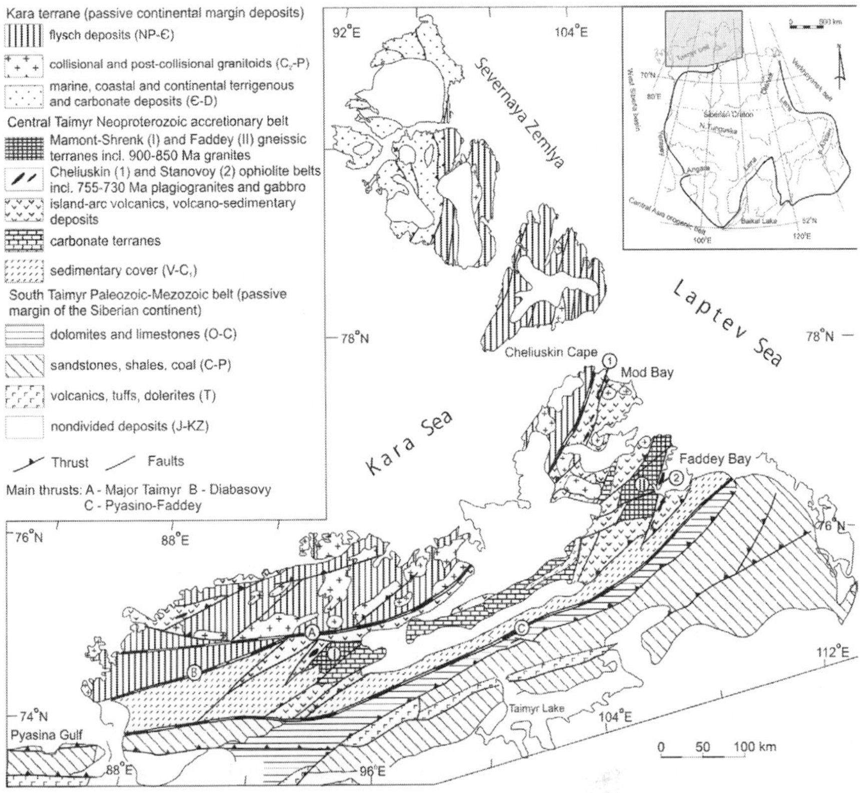

Fig. 3. Tectonic map of the Taimyr - Severnaya Zemlya fold-and-thrust belt (after Vernikovsky & Vernikovskaya, 2001).

3.1.3 Kara terrain

Kara terrain occupies the northern part of the territory, including the coastal part of the Taimyr Peninsula, the islands of the Severnaya Zemlya archipelago and the adjacent floor of the Kara Sea (Fig. 3). Three different rock associations can be distinguished here: i) Paleoproterozoic metamorphic complexes of the crystal basement is represented by palgiogneiss, amphibolite, and granitic gneiss (Ufland et al., 1991; Vernikovsky, 1996); ii) Neoproterozoic-Cambrian deposits of continental slope and slope front has an essential flysch structure and is represented by rhythmically alternating sandstone, siltstone and mudstone, partly metamorphosed from green-schist to amphibolite facies (Bogolepova et al., 2001; Lorenz et al., 2008); iii) relatively low deformed sequences of sedimentary cover with predominant coastal-sea and lagoon facies (limestone, marlstone, sandstone interbeded with gypsum), and also terrigenous continental strata which are often red beds (at the highest levels in the section) (Männik et al., 2002, 2009; Lorenz et al., 2008). The collision of the Kara microcontinent with the Siberian margin in the Carboniferous-Permian is reflected in the development of collisional granitoids and also in synchronous zonal metamorphism

(Vernikovsky, 1996; Vernikovsky et al., 1997). The Most ancient ages (more than 300 million years ago) are obtained for granitoids which are distributed northward and form the Major Taimyr thrust on Bolshevik island (Vernikovsky, 1996; Lorenz et al., 2007; 2008). The paleomagnetic data from the Kara terrain are available only for Early Paleozoic (Metelkin et al., 2005b). Despite the similarity of polar wander and comparable paleolatitudes, the APW trend for the Kara is considerably displaced to the east from the Siberian poles towards APWP for Baltica. Synthesis of the data and analysis of the restored kinematics of the Kara, Siberia and Baltica reveals a terrain history of Kara during the Paleozoic (Metelkin et al., 2005b). From the Ordovician up to the end of the Silurian, the terrain underwent a northern drift from 40°S to 10°N with an average velocity about 5 cm/year and counter-clockwise rotation with 1 degree per million years amplitude. Similar drift, but with outstripping rotation of the plate, is reconstructed for Baltica (Torsvik & Cocks, 2005). The main difference in kinematics of the Siberian plate consists in its clockwise rotation (Cocks & Torsvik, 2007). The opposite rotation of continental masses should provide a development of transform the zone between them and promote a strike-slip motion of the Kara microplate (Metelkin et al., 2005b).

3.2 The Verkhoyansk area

The Folded-napped structures of the Verkhoyansk belt frame the eastern margin of the Siberian platform and are separated by a system of frontal thrusts of the Verkhoyansk trough. Development of the orogen is a result of Late Mesozoic accretion-collision processes (Parfenov et al., 1995; Oxman, 2003). In plan from, the folded area forms a huge loop which in a fan shape extends to the north where it sinks down under the cover of the Arctic shelf. Rock complexes of the Verkhoyansk system are bending around the Olenek uplift and are joined with the South Taimyr, forming a uniform sedimentary paleobasin in the sense of composition, developed on an Early Precambrian crystal basement of the Siberian Craton. Composition and structure of the complex are relatively uniform. It is composed of a monotonous sequence of sandy-siltstone sediments of Carboniferous to Jurassic age which is underlain by more ancient carbonate deposits. The sequence is represented by genetically uniform sedimentary successions deposited under the conditions of passive margin (Parfenov & Kuz'min, 2001; Pisarevsky & Natapov, 2003). Basal horizons of the Mesoproterozoic - Neoproterozoic complex are abundant in the south of the area where they form Kyllakh and Sette-Daban uplifts. They are composed of shallow, lagoon-sea and sometimes continental deposits: dolomite, limestone, marlstone, sandstone, siltstone and shale (Khomentovsky, 2005). The southern part of the Kyllakh uplift (the Judoma-Maya zone) and adjacent Uchur-Maya plate in the west represent the stratotype district for the basic divisions of the Late Mesoproterozoic and Early Neoproterozoic of Siberia (Khomentovsky, 2005). The paleomagnetic data obtained from those strata (Gallet et al., 2000; Pavlov et al., 2000, 2002) give the background for paleotectonic reconstructions for Siberia at the Mesoproterozoic/Neoproterozoic boundary. The most important key poles are yielded from dolerites of subvolcanic intrusions, widely distributed among deposits of Ui group (Pavlov et al., 2002). Development of the intrusions along with the predominantly terrigenous sedimentary environment of the Ui group, distinguish it from the other strata of the Uchur-Maya region and may reflect the riftogenic processes within the craton margins at the boundary of 1 billion years (Rainbird et al., 1998; Khudoley et al., 2001).

3.3 The Baikal-Vitim fold-and-thrust belt

Geological complexes the Baikal-Vitim fold-and-thrust belt occupies an extensive territory to the east from the Baikal Lake up to the Vitim river basin. We consider three tectonic elements in its structure: the Bodajbo-Patom Neoproterozoic marginal-continental area; the Baikal-Muya Neoproterozoic accretional belt; the Barguzin-Vitim Early Caledonian orogenic area (Fig. 4).

3.3.1 Bodaibo-Patom area

The area is characterized mainly by Neoproterozoic sedimentary successions. Sections are usually represented by terrigenous-carbonate deposits but their geodynamic position is interpreted differently (Nemerov & Stanevich, 2001; Pisarevsky & Natapov, 2003; Zorin et al., 2009). The absolute age of the majority of startigraphic units is also under discussion (Stanevich et al., 2007). Nevertheless, the general structural plan does not cause any doubts: the strata represent a thick sedimentary wedge, thickening toward the folded area and overlapping the crystal complexes of the craton margin. The change of shallow-water faces to more deep-water, flysch-like and turbidity deposits, follow the same direction. Such features are inherent in marginal-continental sea basins. Ophiolites and subduction complexes overthrusted on the margin of the craton from the south and comprising the Baikal-Muya belt can be considered as indicators of the Neoproterozoic ocean.

3.3.2 Baikal-Muya belt

Three different units - the Muya cratonic terrain; Kilyan island arc terrain and Param oceanic terrain can be distinguished in the structure of the Neoproterozoic accretion belt (Parfenov et al., 1996; Khain, 2001). The most ancient ophiolite, according to Sm-Nd dates, correspond to the end of the Mesoproterozoic (Rytsk et al., 2001, 2007). The transformation of the continental margin into an active regime possibly occurred in the Middle Neoproterozoic 800-850 million years ago (Khain et al., 2003). Subductional volcano-plutonic associations are represented by the tuff and lava of rhyolite, andesite, basalt, and also gabbro and plagiogranite. The second peak of subduction magmatism corresponds to the interval 750-620 million years (Khain et al., 2003). However, according to some viewpoints the Kilyan island arc system was not connected directly with the margin of the Siberian Craton and the Baikal-Muya belt was a part of another superterrain, and has been developed far from the margin (Kuzmichev, 2005). Paleomagnetic data for this area are absent. It is supposed that just before the Vendian, the initial structure of the superterrain has been disrupted and the Bajkal-Muya fragment moved along the strike-slip and amalgamated to the Siberian Craton (Kuzmichev et al., 2005; Belichenko et al., 2006). The time of this event is supported by stratigraphic data. Acceretional complexes of the Baikal-Muya belt with angular and stratigraphic unconformity are overlapped by Vendian terrigenous deposits and Cambrian, predominantly carbonate deposits which have remained as isolated fragments and are probably the remains of former uniform sedimentary cover with Siberia (Belichenko et al., 2006).

3.3.3 Barguzin-Vitim area

Traditionally it is considered that Late the Precambrian orogenesis resulted from the accretion process between the Siberian Craton and the Barguzin microcontinent. The remains of the microcontinent are expected to be to the south of the Baikal-Muya belt within

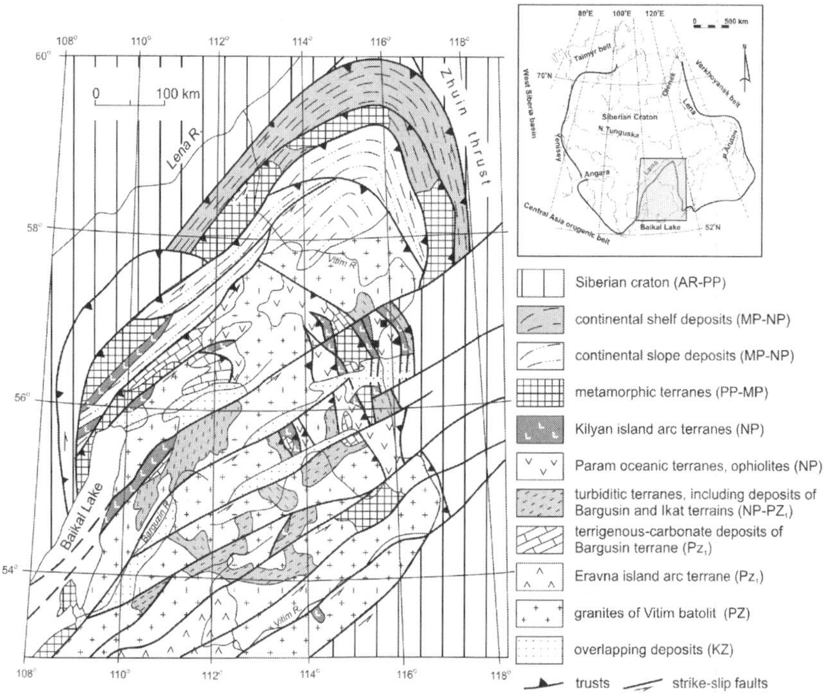

Fig. 4. Schematic geological map of the Baikal-Vitim fold-and-thrust belt (adapted from Parfenov et al., 1996; Vernikovsky et al., 2004)

the Barguzin-Vitim area (Zonenshain et al., 1990). Another point of view dominates now. It is possible to isolate two turbidite terrains within the limits of the area: Barguzin and Ikat terrains (Belichenko et al., 2006; Zorin et al., 2009). The structure of both terrains does not correspond to modern concepts of microcontinents. First of all, both terrains have no Precambrian crystal basement. Deeply metamorphosed deposits of the central part of the Barguzin ridge which were considered as Early Precambrian (Zonenshain et al., 1990) gradually change to weakly metamorphosed Vendian - Early Paleozoic deposits (Belichenko et al., 2006; Zorin et al., 2009), however, the age of metamorphism is not Precambrian but Ordovician-Silurian (Belichenko et al., 2006; Rytsk et al., 2007; Zorin et al., 2009).Cambrian volcanic complexes stretched by a chain along the River Uda in the northeast up to the River Vitim and united as a part of the Eravna terrain which can be considered the structures controlling island arc development (Gordienko, 2006). In the first, paleomagnetic data suggest a westward displaced position of the terrain in relation to its modern arrangement in the structure of the folded area. In the beginning of the Cambrian, the island arc was located about 5-10°N and had a submeridional orientation (Metelkin et al., 2009).

3.4 Sayan-Yenisey fold-and-thrust belt

The folded-napped structure of the Sayan-Yenisey area involves Early Precambrian complexes of southwestern margin of Siberian Craton (Fig. 5). The western boundary is well pronounced in the strike-slip zone - a branch of the Main Sayan deep fault with immediately adjacent Caledonian structures of Altai-Sayan fragment of the Central-Asian orogenic belt which is plunging in the north under Mesozoic-Cenozoic cover of the West Siberian plate. The zone of the deep fault is well traced by geophysical data under the deposits of Western Siberia along the left bank of Yenisey River up to the Turukhansk-Norilsk area (Vernikovsky et al., 2009).

3.4.1 Paleoproterozoic terrains of the craton margins (the Angara belt)

The structures of Sharyzhalgay and Birusa terrains constitute a distinct prominence of the Siberian Craton. They are composed of granulite-gneiss and granite-greenschist type rocks of the Archean-Paleoproterozoic age. Amalgamation of the terrains is associated with the boundary of 1.8 - 1.9 billion years ago which is reflected in the formation of collisional granites (Levitsky et al., 2002; Didenko et al., 2003; Turkina et al., 2007). Formation of similar complexes is also detected in the north (Nozhkin et al., 1999) where Paleoproterozoc structures are built up by granulite and amphibolites of the Angara-Kan terrain. It is supposed that this stage of crustal growth has occurred over the whole western margin of the Siberian Craton, and the resulting structure is united as a part of the Angara belt (Rozen, 2003). Further to the north, within the Yenisey range, the structure of the Angara belt is represented by the East Angara terrain. In the limits of the terrain, Early Precambrian crystal complexes are generally overlapped by low metamorphosed Mesoproterozoic and Neoproterozoic terrigenous-carbonate strata. Among them, to the south, in the limits of Pre-Sayan trough, deposits of the Karagas group are widespread. The group is characterized by a cyclic-constructed succession which was deposited in the coastal-marine environment. (Pisarevsky & Natapov, 2003; Stanevich et al., 2007; Metelkin et al., 2010b). The absolute age of the strata is hotly debated. Considering the available stratigraphic constraints and the indirect geochronological data, the group was deposited between 800 and 740 million years ago (Stanevich et al., 2007). Paleomagnetic data suggest a Middle Neoproterozoic age of the group and a high rate of its deposition (Metelkin et al., 2010b). Deposits of the Karagas group and the underlying Paleoproterozoic metamorphic complexes are saturated by subvolcanic intrusion of gabbro-dolerie, united in the Nersa complex. The age of that complex is estimated as 740 million years, and it has been formed in conditions similar to those of intercontinental riftogenesis (Sklyarov et al., 2003; Gladkochub et al., 2006). The results of paleomagnetic study of the Nersa complex are given in (Metelkin et al., 2005a).

3.4.2 Central Angara belt

The Central-Angara terrain occupies the central part of the Yenisey ridge. The structure of its basement corresponds to Paleoproterozoic metamorphic terrains of the Angara belt. In general, the Early Precambrian basement is overlaid by Mesoproterozoic and Neoproterozoic coastal-marine sedimentary complexes. However, it is separated from the East Angara terrain by the Panimba ophiolite belt accompanied with Ishimba thrust that allows us to consider a tectonic history of the Central-Angara terrain separately from the structures of the Angara belt. Argon-argon age of amphibolites and plagioclases from gabbro-amphibolites of the ophiolite belt is 1050-900 million years (Vernikovsky et al., 2003). Granitoids with 760-720 million years age appear to be indicators of a collisional event (Vernikovsky et al., 2003; Vernikovskaya et al., 2003, 2006).

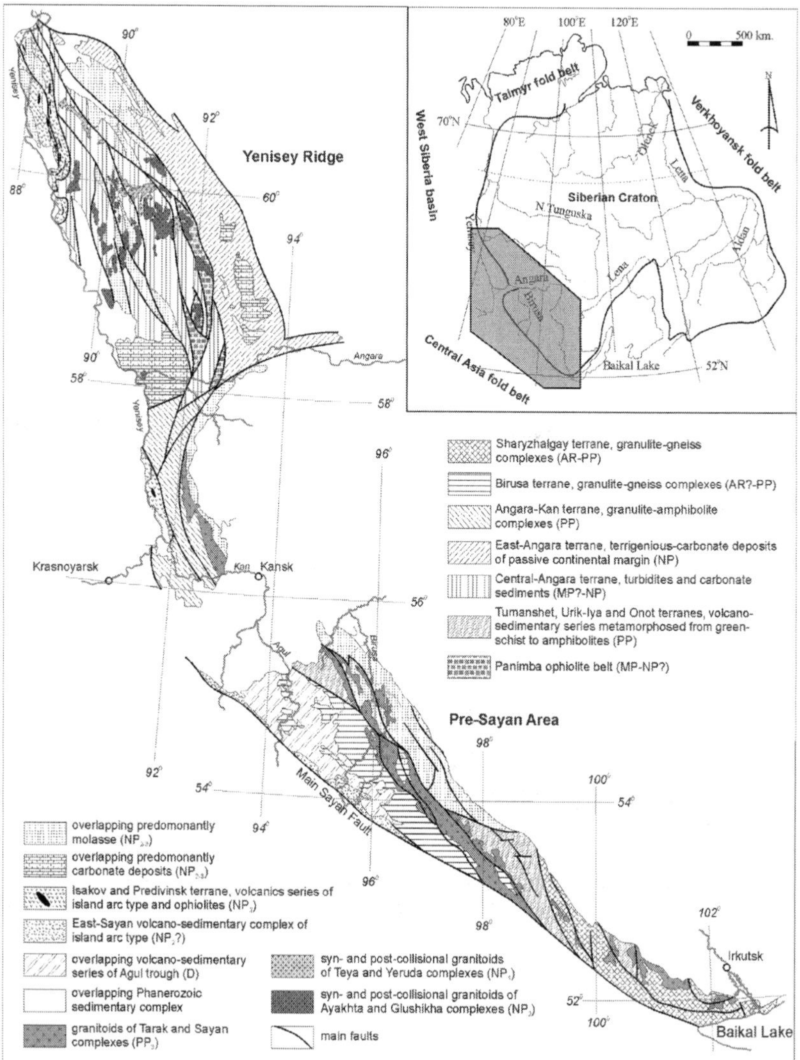

Fig. 5. Geological sketch-map of the Sayan-Yenisey fold-and-thrust area (after Metelkin et al., 2007a)

3.4.3 Pie-Yenisey accretional belt

The western periphery of the Yenisey ridge is occupied by Neoproterozoic accretional structures napped over crystal formations of the Central-Angara and Angara-Kan terrains. Isakov and Predivinsk island arc terrains can be distinguished in the area (Vernikovsky et al., 2003). Structures of the terrains contain subductional volcano-sedimentary complexes including ophiolite fragments, partly metamorphosed in green-schist conditions (Vernikovsky et al., 1999). Results of U-Pb, Ar-Ar, and Rb-Sr isotope study allow an estimate of the evolution time of an active volcanic arc between 700 and 640 million years ago, and the time of accretion as 600 million years ago (Vernikovsky et al., 2003). The time of accretion coordinates well with the Vendian age of overlapping molasse (Sovetov et al., 2000). A paleomagnetic study was carried out on the volcanic complexes of differentiated calc-alkaline series of the Predivinsk terrain with a U-Pb age 637± 5.7 million years ago (Metelkin et al., 2004a). Paleomagnetic pole positions correspond to the general trend of Vendian poles of Siberia (Pisarevsky al., 2000; Kravchinsky et al., 2001; Kazansky, 2002; Shatsillo et al., 2005, 2006). Taking into account that the development of the series has proceeded directly before accretion, the paleomagnetic data allows estimation of a spatial position not only for the arcs, but also as a first approximation for the craton (Metelkin et al., 2004a; Vernikovsky et al., 2009).

3.5 Altai-Sayan area

The Altai-Sayan area occupies the southwest frame of the Siberian Craton and is a fragment of the Central-Asian orogenic belt (Fig. 6). Terrains of island arc genesis compose the basic structure of the Early Caledonian folded area (Dobretsov et al., 2003). Among them are West Sayan, Kuznetsk Alatau, and Gorny Altai which are briefly characterized below. A number of publications describe these terrains as fragments of a formerly uniform island arc system supervising the Vendian-Cambrian subduction zone in the southwest of the Siberian continent (Şengör et al., 1993; Kungurtsev et al., 2001; Buslov et al., 2001; Gordienko, 2006). Accordingly the Early Paleozoic system of the Eravna arc and the Barguzin-Ikat back arc basin, we suggest as a natural continuation of the paleoisland arc system to the east. Accretion of the arcs to the craton started just at the beginning of the Ordovician and is manifested by the occurrence of numerous granite plutons of a collisional type (Dobretsov et al., 2003; Khain et al., 2003). The results of paleomagnetic study of the Altai-Sayan area can be obtained from a generalizing paper (Metelkin et al., 2009) or directly from the publications listed below for concrete regions. Particularly in (Metelkin et al., 2009) it is shown, that paleomagnetic poles for the second half of Cambrian and Ordovician are concentrated within a compact area to the south from Australia, while the Vendian-Early Cambrian poles are distributed along a great circle in the internal area of Africa. The circle center (approximately 61 N, 114° E) is close to the center of the Siberian Craton. On this basis we suggest that the terrains represented some fragments of a uniform island arc, and the deformation of the arc was connected with strike-slips resulting from rotation of those fragments at different angles (or strike-slipped at a different distance). Calculation of the angles leads to the paleoreconstruction which "straightens" the modern mosaic of terrains of the Altai-Transbaikalia area into a linear structure. On the other hand, to reach the coincidence of poles, it is possible to rotate each block around a corresponding sampling point. Actually, we used a series of great circles for the best fit of the pole positions. The average normal direction for all the circles corresponds to the center of the craton. However, a tectonic expression of this fact is an alternate idea and assumes a deformation of the island arc as a result of a rotation of its fragments round their own axes without a displacement from each other. The combination of the mechanism of the general strike-slip motion resulted in a segmentation, and relative displacements of fragments with their local rotations seem to be the most realistic.

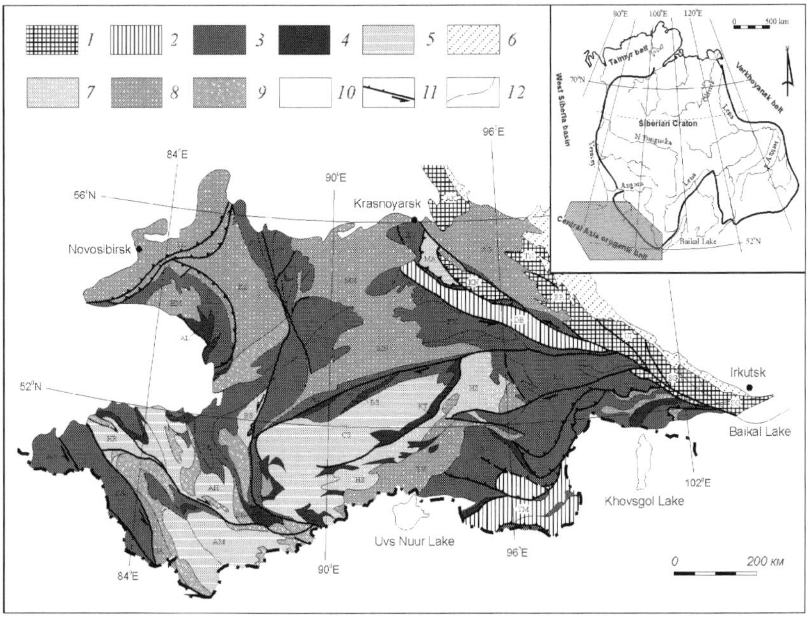

Fig. 6. The main structures of the Altai-Sayan area (modified from Kungurtsev et al., 2001).

Legend: 1-5 - tectonostratigraphic terranes: *1* - Siberian craton and cratonic terranes: BR - Birusa (granulite-gneiss, AR?-PP), SG - Sharyzhalgay (granulite-gneiss, AR-PP), AK - Angara-Kan (granulite-amphibolite, PP), UI - Urik-Iya (greenschist, PP), TN - Tumanshet (amphibolite, PP), KN - Kan (metamorphic, PP); *2* - miogeoclinal terranes (microcontinents): TM - Tuva-Mongolian (passive continental margin, NP), DB - Derba (passive continental margin, NP); *3* - island arc terranes (accretionary wedge, volcanic arc and back-arc basin complexes including): GK - Golden Kitat (accretionary wedge and volcanic arc, V-Є), KI - Kiya (volcanic arc, V-Є), TR - Ters (volcanic arc and back-arc basin, V-Є), BT - Bateni (back-arc basin, V-Є), MR - Mrass (volcanic arc and back-arc basin, V-Є), SL - Salair (island arc undivided, V-Є), NS - North Sayan (accretionary wedge and volcanic arc, V-Є) , GA - Gorny Altay (island arc undivided, V-Є), TK - Terekta (accretionary wedge, V-Є), TN - Tannuola (volcanic arc and back-arc basin, V-Є), KH - Khamsara (volcanic arc, V-Є), KK - Kizir-Kazir (island arc undivided, V-Є), KV - Kuvai (accretionary wedge, NP), RA - Rudny Altay (volcanic arc, D-C), KN - Kalba-Narim (accretionary wedge, D-C); *4* - oceanic terranes (seamounts and ophiolites including as a part of accretionary wedge, basement of island arc and as a product of back-arc spreading): KT - Kurtushiba (ophiolites in accretionary wedge, V-Є), BS - Borus (ophiolite,V- Є), AL - Alambai (ophiolite,V-Є), BR - Baratal (seamount, V-Є); *5* - continental margin turbidite terranes: CS – Central Sayan (Є-S), AH – Anui-Chuya (Є-S), HR – Charysh (Є-S), AM – Altai-Mongolian (NP-S); *6 - 9* – overlap assemblages: *6* – Siberian plate (NP - PZ), *7* – Early Paleozoic basins: HS - Khemchik-Sistigkhem (molasse, O-S), MA – Mana (sedimentary, V-Є), BS – Biya (molasse, O-S); *8* – Late Paleozoic – Early Mesozoic basins: MN – Minusa (volcano-sedimentary, molasse, D-P), KZ - Kuznetsk (molasse including trapps, D-T), AG – Agul (volcano-sedimentary, molasse, D-C), TK – Tom-Kolyvan (back-arc, volcano-sedimentary, D-P), HM – Khmelev (back-arc, volcano-sedimentary, D-C); TV – Tuva (volcano-sedimentary, molasses D-P), SA – South Altai (back-arc, volcano-sedimentary, D-C), *9* – Altai volcano-plutonic belt (D-C), *10* – Mesozoic-Cenozoic sedimentary basins; *11* –faults with specified napped or strike-slip kinematics of post-accretional displacement; *12* -other faults and geological boundaries. The inlay represents the location of the Altai-Sayan area relative to the Siberian Craton.

3.5.1 West Sayan terrain

The West Sayan terrain includes the Vendian-Cambrian complexes of the island arch which are composed of two large fragments. In the frontal part of North Sayan fragment the complexes of the accretionry wedge are represented by terrigenous deposits interbeded with basalt, tuff, marble, and also ophiolite units. The lateral succession to the north is supplemented by turbidities and then by complexes of the volcanic arc. The Vendian-Early Cambrian interval is characterized by tholiite magmatism which was replaced by a differentiated series at the boundary of 520 million years ago. This complex is overlapped by a multicolored terrigenous sequence formed as a result of an intensive washing of the volcanic series. The Kurtushiba fragment corresponds to the frontal part of the arc with widespread oceanic complexes. A paleomagnetic study of the above named complexes proves the preservation of a stable magnetic component which was acquired at the stage of rock formation (Kazansky et al., 1999). Generalization of the data reveals the following tectonic history of the terrain. Both fragments belonged to the uniform arc occupied in the subequatorial position in the Cambrian. The arc has undergone a prevailing submerideonal drift and strike-slip displacement of the Kurtushiba fragment to the southwest, in relation to the North Sayan (Metelkin et al., 2009).

3.5.2 Kuznetsk Alatau terrain

The Kuznetsk Alatau terrain consists of five units which possess the original tectonic style, but undoubtedly are fragments of the uniform paleoisland arc system (Metelkin et al., 2000; Kazansky et al., 2003). We studied four of them: Golden Kitat, Kiya, Ters and Bateni tectonic units, mainly composed of Cambrian subduction complexes with a typical set of rock associations.

The Vendian-Early Cambrian complex is represented by the tholeite series, the Middle-Late Cambrian complex is represented by a differentiated series with an expressed calc-alkaline composition and considerable concentration of pyroclasics. The development of the Late Cambrian-Early Ordovician molasse and alkaline volcano-plutonic complex corresponds to an accretion stage. The ages are proved by the isotope data (Vladimirov et al., 2001). Paleomagnetic data (Metelkin et al., 2000; Kazansky et al., 2003) shows that the structural plan of the region is determined by the strike-slip motions of its fragments. Analysis of APW trends (Kazansky et al., 2003; Metelkin & Koz'min, in progress) allows restoration of a southern drift of the arc with a 5-6 cm/year velocity which was accompanied by a clockwise rotation. The size of displacement of the forearc and back-arc parts were different which caused the strike-slip mechanism of transformation of the initial structure of the arc. The reconstructed position of the arc in the Cambrian corresponds to subequatorial latitudes.

3.5.3 Gorny Altai terrain.

The Gorny Altai terrain stands out because of its repeated alternation of blocks composed of island arc complexes, resulting from strike-slip tectonics which manifested both during Early Caledonian and Hercinian stages of the orogenesis (Buslov et al., 2003). The western part (the Anui-Chuya trough) is represented by terrigenous complexes of forearc basin with characteristic flysch and olistostrome composition. Complexes of accretionary wedge and volcanic arc are predominant in the central part. The basalts are typical representatives of the Kuray zone where the Vendian-Early Cambrian spreading and subduction complexes are widespread. The accretionary wedge is mainly composed of carbonate deposits overlapping pillow-basalts which are considered as seamounts (Safonova et al., 2008). The collision of the seamounts with the island arc occurred 520 million years ago, marking the beginning of the developed stage of subduction magmatism (Buslov et al., 2001). Typical calc-alkaline complexes of Middle-Late Cambrian are distributed in the east in the Ujmen-Lebed zone and volcano-sedimentary sequences of the back-arc basin are also present here. Paleomagnetic data (Pechersky & Didenko, 1995; Kazansky et al., 1998; Kazansky, 2002) testify that the Gorny Altai terrain constitutes a part of the Kuznetsk Alatau island arc and the transformation of its initial structure may be described by the same uniform strike-slip mechanism (Metelkin et al., 2009).

4. The tectonic and paleogeographical reconstructions of Siberia and surrounding terrains

4.1 The Neoproterozoic stage

The history of the Siberian paleocontinent actually begins at the moment of Rodinia break-up. The Neoproterozoic stage of tectonic history corresponds to this event (Li et al., 2008). The set of available geological and paleomagnetic data testifies that at the Mesoproterozoic-Neoproterozoic boundary the Siberian Craton was a part of Rodinia and could represent a large peninsula of the supercontinent in the northeast (Metelkin et al., 2007a; Pisarevsky et al., 2008). In modern co-ordinates, Siberia continued as Laurentia to the north so that the western margin of Siberia was a continuation of the western margin of Laurentia (Fig. 7). The review of geological data on structural position, structure, formation environments and age of the Late Mesoproterozoic and Early Neoproterozoic complexes distributed on the margins of the Siberian Craton shows that this stage of geological history practically in all the periphery of the continent is connected with the dominating conditions of a continental shelf (Pisarevsky & Natapov, 2003; Pisarevsky et al., 2008). The modern northwest margin of Siberia, also as the western and the eastern margins (Semikhatov et al., 2000; Petrov & Semikhatov, 2001), represented a passive continental margin with a typical complex of sedimentary rocks (Pisarevsky & Natapov, 2003). An active tectonic mode, possibly characterizes only the southern margin of Siberia (Rainbird et al., 1998; Pavlov et al., 2002; Yarmolyuk et al., 2005; Metelkin et al., 2007a). Geological complexes in this area can correspond to the mode of intracontinent riftung or an active stage of the development of the ocean. Among them dikes with an age of 950-1000 million years (Rainbird et al., 1998; Pavlov et al., 2002) in the sedimentary cover of Uchur-Maya region in the southeast of the craton, geological complexes of various geodynamic environments of the Baikal-Muya accretion system (Parfenovet al., 1996; Khain et al., 2003; Gordienko, 2006), the youngest at about 750 million years and younger and products of intraplate alkaline magmatism in Pre-Sayan and Baikal regions in the southwest of the craton (Yarmolyuk et al., 2005; Gladkochub et al., 2006) can be distinguished. Thus, the gradual rejuvenation of magmatism due to available isotope data is assumed to be from the east to the west. On the basis of a number of petrologic-geochemical evidences the direct link between the development of intrusive massifs subvolcanic intrusions of the basic composition with rifting processes, and the break-up of the Laurasian part of Rodinia is assumed (Yarmolyuk et al., 2005, Pisarevsky et al., 2008).

Paleomagnetic data suggest that break-up of the continental masses of Siberia and Laurentia during the Neoproterozoic passed gradually along the southern margin of Siberia from the east to the west under the defining role of strike-slips which have set the rotary motion to the Siberian Craton (Metelkin et al., 2007a). Following this model, it was possible to believe

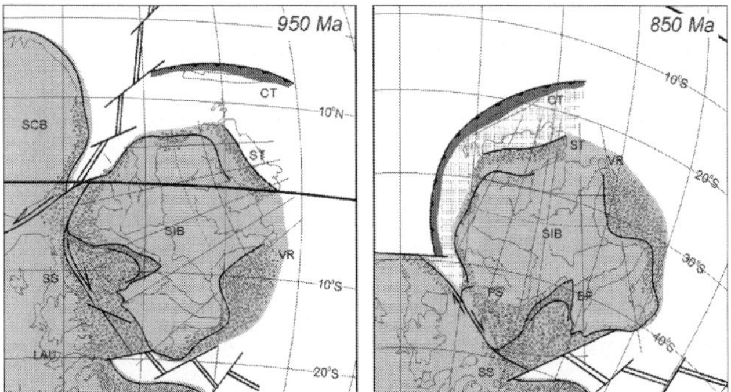

Fig. 7. Paleotectonic evolution of the Siberian Craton and its margins during the Neoproterozoic.

Legend: 1 - continental masses and most important block contours; *2* - accretionary structures, orogenic belts with corresponding age; *3* - subduction systems, including volcanic belts and back arc basins; *4* - marginal seas, shelf basins of passive margins; *5* - suggested spreading zones; *6* - general strike of transform-shear zones with their kinematic style; *7* - schematic area of crust thinning in limits of West-Siberian graben-rift system; *8* - schematic area of the development of the Permian-Triassic Siberian flood basalts; *9* - Mesozoic-Cenozoic cover of the West-Siberian sedimentary basin.

Abbreviations: continental blocks: SIB - Siberia, BAL - Baltica, KAR - Kara, KAZ - Kazakhstan, LAU - Laurentia, NCB - North China Block, TAR - Tarim, SCB - South China Block; basins of passive margins, marginal seas: VR - Verkhoyansk, BP - Bodajbo-Patom, PS - Pre-Sayan, SS - South Siberia (suggested), ST - South Taimyr; orogenic belts: ASB - Altay-Sayan Belt, BMB - Baikal-Muya Belt, VChB - Verkhoyansk-Chukcha Belt, MOB - Mongol-Okhotsk Belt, YEB - Pre-Yenisei belt, CAB - Central-Angara Belt, CAT - Central-Angara terrain, CTB - Central Taimyr Belt; island-arc terranins, fragments of active continental margin and volcano-plutonic belts: BT - Bateni, GA - Gorny Altai, ER - Eravna, GK - Golden Kitat, KI - Kiya, KT - Kurtushiba, NS - North Sayan, TS - Ters, CT - Central Taimyr, OChVB - Okhotsk-Chukcha volcano-plutonic belt; other structures: CPD - Caspian depression, WSB - West Siberian basin.

that at the boundary about 750 million years ago Siberia shifted along the northern margin of Laurentia along a distance of 2,000 km and its southwestern margin was located in the immediate vicinity of the northern margin of Greenland (Fig. 7). At this time, there was a transformation of the passive continental margin environment in the west, in the north (Vernikovsky et al., 2003) and probably in the south of Siberia (Khain et al., 2003) into an active one with the development of a Neoproterozoic system of island arcs. The active belt of the island arc magmatism has been probably separated from the continental margin by an extensive rear basin which provided a predominant quiet shelf sedimentary environment within all western and northwest margins of Siberia (Pisarevky & Natapov, 2003).

The stage of accretion of Neoproterozoic island arcs to the Siberian paleocontinent is manifested in Pre-Vendian - Vendian time (Dobretsov et al., 2003; Vernikovsky et al., 2004). The corresponding age of this accretion event in the west and in the north of Siberia is proved by the combined isotope-geochemical data (Vernikovsky & Vernikovskaya, 2006). Possibly, this stage of development of the active continental margin has resulted in marginal-continental riftogenesis and accompanying magmatism (Yarmolyuk & Kovalenko, 2001; Yarmolyuk et al., 2005; Vernikovsky et al., 2008).

Thus, complete separation of the Siberian continent from structures of Rodinia has occurred more than 100 million years later than the beginning of development of zones of crushing and local riftogenesis on the southern margin (Yarmolyuk & Kovalenko, 2001). The Neoproterozoic transform-strike-slip kinematics of development and transformation of ocean basins around Siberia at the stage of the break-up of Rodinia has predetermined the dynamics of the subsequent accretional-collisional events. From the end of the Neoproterozoic and up to the Mesozoic, Siberia was developed as an independent system of interaction of ocean and continental plates. During this time, the Siberian continental plate underwent a drift of mainly a northern direction from near equatorial latitudes of the southern hemisphere (~ 10°S) in the end of the Precambrian to high latitudes of the northern hemisphere (~60° N) in the end of the Paleozoic (Pechersky & Didenko, 1995; Cocks & Torsvik, 2007). According to paleomagnetic data, during this time the plate gradually rotated clockwise at an angle of about 180° and up to the beginning of the Triassic, the northern margin of Siberia has faced to the west (Fig. 7).

Late Proterozoic formations of the northwest margin of the Siberian Craton are overlapped by the Vendian-Paleozoic plate complex of passive continental margin with a peculiar platform mode of development (Vernikovsky & Vernikovskaya, 2001). The same geodynamic mode characterizes also the western marginal part of the Siberian paleocontinent (Sovetov et al., 2000; Vernikovsky et al., 2009). Against the accumulation of shallow sea carbonate and carbonate-slate deposits, a deep-water basin with distinct features of a linearly extended trough developed along the boundary between Central and South Taimyr. The basin as it was supposed by (Khain, 2001), was connected in the east with a similar basin of the internal areas of the Verkhoyansk system. The axis of this deep-water trough valley was located in the south of a zone of connection of the Central Taimyr accretional block with the continent, in a frontal part of the large Pyasino-Faddey thrust that allows consideration of its development as the foredeep trough (Vernikovsky, 1996).

4.2 The Paleozoic stage

The mode of active continental margin has been renewed, at least, in the southwest Siberian paleocontinent already at the end of the Vendian (Dobretsov et al., 2003). A large number of

Early Paleozoic island arc terrains form a tectonic collage of the Central-Asian belt on the western periphery of the Siberian Craton. Owing to the available paleomagnetic data, the Vendian-Cambrian subduction complexes reconstructed within the limits of the Altai-Sayan orogen represented fragments of a uniform system of island arcs which marked an extended zone of subduction of ocean plate along all the western periphery of the Siberian continent, similar to the modern Pacific margin of Eurasia (fig. 3). Deformation of this system at the stage of accretion to the Siberian Craton in the end of the Cambrian-Ordovician (and later in the Late Paleozoic and Mesozoic), was connected with strike-slip motions which are mainly caused by a clockwise rotation of Siberian continental plate. Such kinematics in the compression environment along the boundary between continental and oceanic plates led to the development of strike-slip zones along the continent periphery, and deformations of the Vendian-Early Cambrian island arc system. Displacements of fragments of this system could occur along strike-slips located both in back, and along zones of oblique subduction (Fig. 8). The rotation of periphery structures resulted in their "lagging behind" the continent forming individual tectonic blocks which interacting with each other have been displaced in a complex way (Kungurtsev et al., 2001; Metelkin et al., 2009).

After the accretion of the island arc system at the end of the Cambrian-Ordovician, a tectonic picture in the west and southwest of Siberia assumed a configuration close to the modern one. Paleomagnetic pole positions for Siberia and for terrains of the Altai-Sayan orogen are very close, though do not coincide completely (Metelkin et al., 2009). Small distinctions in pole positions specify that intensive deformation of paleoisland arc systems and back arc basins under the leading role of the strike-slips began in the Cambrian and proceeded during the whole Paleozoic (Buslov et al., 2003; Van der Voo et al., 2006; Metelkin et al., 2009).

In the north of Siberia, the Vendian to Devonian time interval is characterized by a growth of the Anabar uplift and the development of surrounding large synforms, occupied by epicontinental seas with carbonate sedimentation (Bogdanov et al., 1998). Also a deep-water trough formed at the beginning of Paleozoic along the location of the foredeep trough on the boundary of the South and Central Taimyr zones was still developed. The change of the tectonic mode was manifested in the Carboniferous, when carbonate deposits on the Taimyr shelf were superseded by terrigenous. The principal change in the sedimentary environment was an extremely important event in the tectonic history of the north of Siberia and was connected with the occurrence of a new source of terrigenous material (Zonenshain et al., 1990). Paleotectonic analysis, combined with available paleomagnetic data, shows that this event was caused by the beginning of interaction of the Siberian margin with the Kara microcontinent in the mode of oblique collision under the leading role of strike-slips (Metelkin et al., 2005b). The main role in tectonics of the Kara block belongs to transform zones connecting the Arctic margins of Siberia and Baltica within the limits of the uniform tectonic system which have caused the strike-slip motion of the Kara microplate in a northern direction, from a subtropical zone of the southern hemisphere to subequatorial latitudes of the northern hemisphere with a simultaneous counter-clockwise rotation. Strike-slip tectonics has completely defined the style of deformation of the Paleozoic margin in the north of Siberia during the collisional event in the Late Carboniferous-Permian (Vernikovsky, 1996, Metelkin et al., 2005b) which has also occurred against the contrastive rotation of the continental masses of the Kara and Siberian continents that fits well with the general paleotectonic picture (Fig. 8).

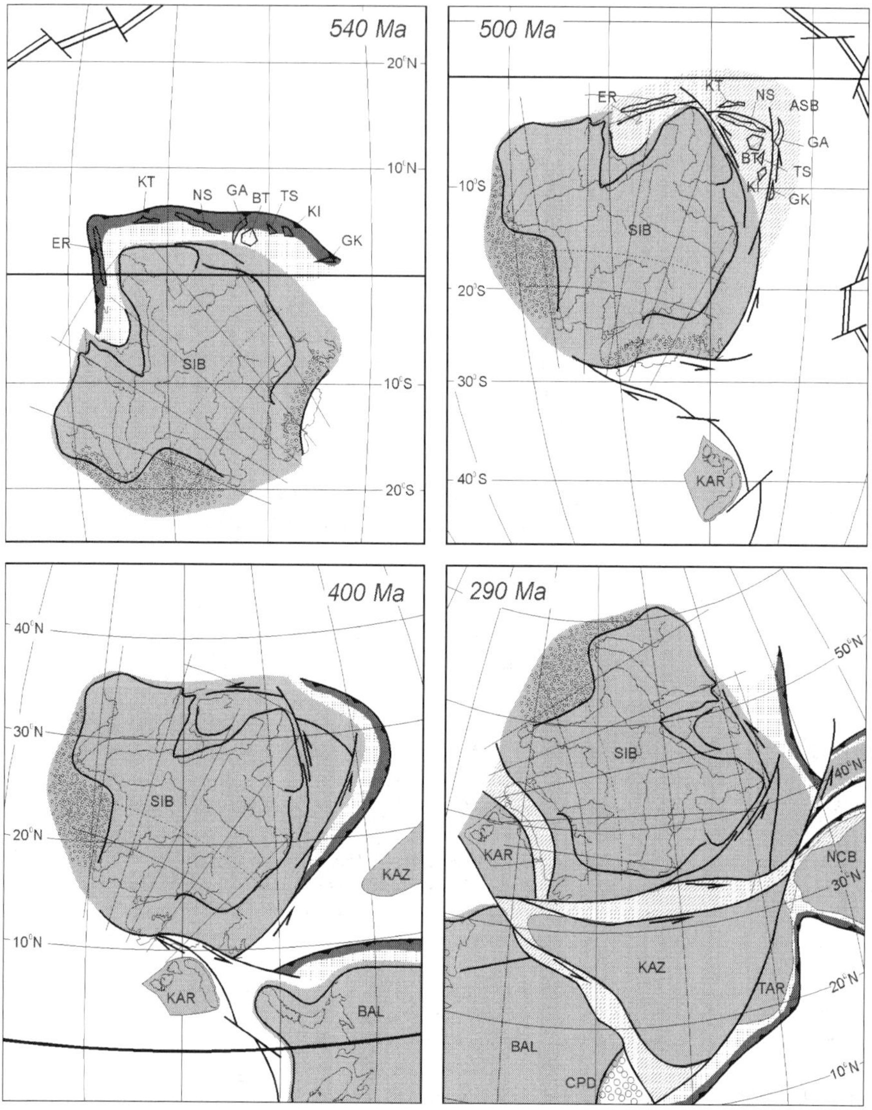

Fig. 8. Paleotectonic evolution of the Siberian Craton and its margins during the Paleozoic. For legend and abbreviations see Fig. 7.

A certain problem is represented by the absence of Paleozoic subduction complexes, which should be located within the Major Taimyr suture, if we assume a space with oceanic crust between Siberia and Kara. The obvious explanation, following from the proposed model, is a soft interaction of sialic masses with the predominating role of strike-slips, under conditions of oblique transform rapprochement and the subsequent collision. The final stages of the orogen the development at the Permian-Triassic boundary have led to development of large extension zones in the frontal part of the Taimyr folded structures and have predetermined the development of a large depression - the Yenisey-Khatanga trough in this segment of the belt.

4.3 The Mesozoic stage

Intracontinental rifting on the Permian-Triassic boundary is well pronounced in Western Siberia. Not only the folded-napped structure of Taimyr-Severnaya Zemlya area was formed up to the beginning of this stage. The closing of the Precambrian-Early Paleozoic oceans resulted in the general structure of the Central-Asian belt sewing the continental masses of the Siberian and East European cratons into a uniform Eurasian plate, which in turn has comprised the basic structure of the Laurasian part of Pangea. This key moment in the tectonic history of Siberia is marked by a giant trapp magmatic event, connected with the activity of the largest mantle plum (Dobretsov, 2005). Within the limits of Siberian platform, flood basalts are concentrated in the Tungus syneclise and continued under the Yenisey-Khatanga trough to South Taimyr. To the west, within the West Siberian plate, trapp basalts were found under the Mesozoic-Cenozoic cover of deposits and traced up to the East Ural trough. As a rule they distributed along rift zones of the Koltogor-Urengoy graben but also exposed in boreholes between rifts. Fields of flood basalts are stretched to the north, covering the bottoms of the Kara and Barents seas (Dobretsov & Vernikovsky, 2001; Dobretsov, 2005). The most southern satellite of Siberian trapps is present in the structure of the Kuznetsk trough (Dobretsov, 2005; Kazansky et al., 2005). Correlations on the basis of available paleomagntic and geochronological data testify that the development of the Siberian trapp province occurred extremely quickly. The duration of intensive magmatism in the different areas is estimated from 1 to 5 million years (Dobretsov, 2005; Kazansky et al., 2005).The magmatism was controlled in the south (the Kuznetsk trough), possibly in the west (West Siberia) and in the north (Yenisey-Khatanga trough), by large scale strike-slips (Fig. 9).

Analysis of paleomagnetic data for the Permian-Triassic boundary gives ground to the assertion that intraplate strike-slip deformation caused by a clockwise rotation of the Siberian tectonic domain of the Eurasian plate, appear to be a possible reason for the generation of submeridional systems of graben structures in the basement of West Siberia which resulted in the development of a large Mesozoic-Cenozoic sedimentary basin (Bazhenov & Mossakovsky, 1986; Voronov, 1997). The east branch of this strike-slip system which caused riftogenesis in West Siberia is an extension connected with the frontal thrust structures of Taimyr. The axial graben of the Yenisey-Khatanga trough and the single age Koltogor-Urengoy graben-rift system (Khain, 2001) form a resemblance to a triple junction that fits the strike-slip model.

Essential deformations in the compression environment, shown in the southwest of Siberia, within the limits of the Altai-Sayan folded area are correlated with the same strike-slip tectonics caused by a rotation of the Siberian domain of the Eurasian plate relative to the European one (Bazhenov & Mossakovsky, 1986; Metelkin et al., 2010a). Strike-slip motions of the described kinematics in the Eurasian continent proceeded up to the end of the Mesozoic (Fig. 9) which is supported by a regular deviation in Mesozoic positions for Siberia and Eastern Europe (Metelkin et al., 2010a). Thus, the crust deformation of the Central Asia in the Mesozoic, against the general clockwise rotation of the Eurasian plate as is shown in reconstruction (Fig. 9), is connected with the motions of separate components of its

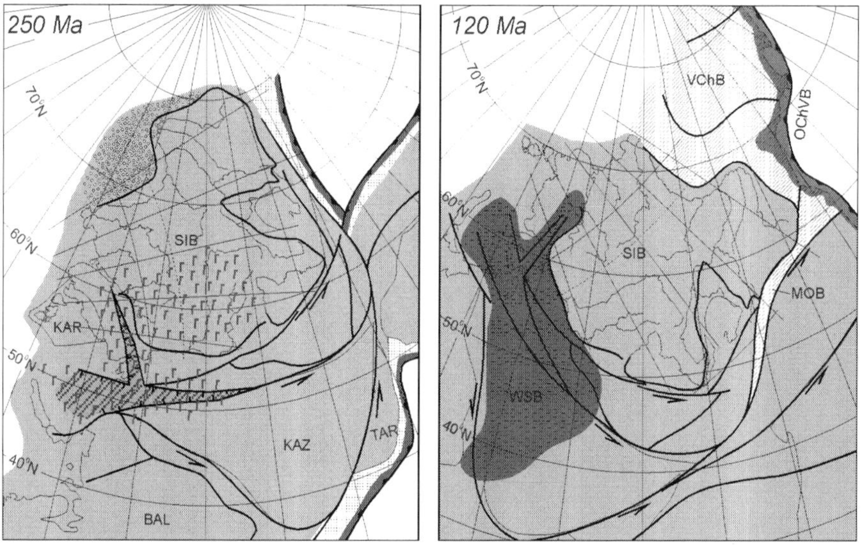

Fig.9. Paleotectonic evolution of the Siberian Craton and its margins during the Mesozoic. For legend and abbreviations see Fig. 7.

composite structure (the Siberian, European and Kazakhstan tectonic domains) along the system of large-scale strike-slip zones of the sinistral (Metelkin et al., 2010a). The deformation of the Mongol-Chinese territory of the plate is also described by a series of strike-slip zones responsible for fragmentation of the Earth's crust against a gradual propagation of the closure of the Mongol-Okhotsk gulf of Paleopacific from the west to the east (in modern co-ordinates), which separated the Siberian margin of Eurasia and the Paleozoic collage of terrains of the territory of Mongolia and China.

The geological consequences of such tectonics are consistent with a viewpoint given in (Bazhenov & Mossakovsky, 1986; Voronov, 1997; Natal'in & Sengör, 2005; Van der Voo et al., 2006). Strike-slip motions of the Siberian domain with a clockwise rotation, due to the configuration of the main structural boundaries, has caused a stable compression environment within the Central-Asian province (the southwestern frame of the Siberian Craton) and in the contrary extension environment within the limits of the north of the West Siberian province. Thus, those motions possibly, had a discrete character that is manifested in the reconstructed multistage character of the main orogenic epochs (De Grave et al., 2007; Buslov et al., 2008) and the correlation of strike-slips and other structural forms disturbing the initial integrity of the Mesozoic sedimentary complex of West Siberia to concrete time boundaries (Belyakov et al., 2000; Koronovsky et al., 2009).

5. Conclusions

The tectonic evolution of Siberia in the Neoproterozoic, Paleozoic and Mesozoic in the global scale can be correlated with the processes of the gathering and disintegrating of two supercontinents: Rodinia and Pangea. The transformation of one tectonic event within the margins of Siberia into another is often defined by the intensity and scale of strike-slip motions. The processes of strike-slip tectonics are presented practically everywhere and in all intervals of geological history of the Siberian plate. They defined the tectonic style of the evolution of the structures of the Siberian region in the early stages of development of the ocean basins, as well as during the active subduction of the oceanic crust and undoubtedly, at the accretion-collision stage. Intraplate deformations of the continental crust, accompanied with active magmatism, were also governed by strike-slip motions of fragments of different sizes. It is an important concept that reconstructed strike-slip zones very largely extended and as a rule correspond to the boundaries of the main tectonic elements that is they are connected with processes of at least regional, and more often of planetary scale.

6. References

Bazhenov, M.L. & Mossakovsky, A.A. (1986). Horizontal movements of the Siberian Platform in the Triassic: paleomagnetic and geological evidence, *Geotektonika*, No.1, 59–69, ISSN 0016-8521, (in Russian)

Belichenko, V.G.; Geletii, N.K. & Barash, I.G. (2006). Barguzin microcontinent (Baikal mountain area): the problem of outlining, *Russ. Geol. Geophys.*, Vol.47, No.10, 1035-1045, ISSN 1068-7971

Belyakov, S.L.; Bondarenko, G.E.; Ivanyuk, V.V. & Smirnov, A.V. (2000). New Data on Late Mesozoic Strike-Slip Deformations in Sedimentary Cover of the Northern Sector of the West Siberian Plate, *Dokl. Earth Sci.*, Vol.372, No.4, 646-649, ISSN 1028-334X

Berzin, N.A.; Coleman, R.G.; Dobretsov, N.L.; Zonenshain, L.P.; Xiao, X.C. & Chang, E.Z. (1994). Geodynamic map of the western Palaeoasian ocean, *Russ. Geol. Geophys.*, Vol.35, No.7–8, 8–28, ISSN 1068-7971

Besse, J. & Courtillot, V. (2002). Apparent and true polar wander and geometry of the geomagnetic field over the last 200 Myr, *J. Geoph. Res.*, Vol.107, No.B11, 1–31, ISSN 0148-0227

Bogdanov, N.A.; Khain, V.E.; Rosen, O.M.; Shipilov, E.V.; Vernikovsky, V.A.; Drachev, S.S. & Kostyuchenko, S.H. (1998). Tectonic Map of the Kara and Laptev Seas and North Siberia (scale 1:2,500,000), *Explanatory notes*, Inst. Lithosphere of Marginal Seas, Moscow

Bogolepova O.K.; Gubanov A.P. & Raevskaya E.G. (2001). The Cambrian of the Severnaya Zemlya archipelago, Russia, *Newsl. Stratigr.*, Vol.39, 73–91, ISSN 0078-0421

Buslov, M.M.; Saphonova, I.Yu.; Watanabe, T.; Obut, O.T.; Fujiwara, Y.; Iwata, K.; Semakov, N.N.; Sugai, Y.; Smirnova, L.V. & Kazansky, A.Yu. (2001). Evolution of the Paleo-Asian Ocean (Altai-Sayan Region, Central Asia) and collision of possible Gondwana-derived terranes with the southern marginal part of the Siberian, *Geosci. J.*, Vol.5, No.3, 203–224, ISSN 1226-4806

Buslov, M.M.; Watanabe, T.; Smirnova, L.V.; Fujiwara, I.; Iwata, K.; De Grave, J.; Semakov, N.N.; Travin, A.V.; Kir'yanova, A.P. & Kokh, D.A. (2003). Role of strike-slip faults in Late Paleozoic-Early Mesozoic tectonics and geodynamics of the Altai-Sayan and East Kazakhstan folded zone, *Russ. Geol. Geophys.*, Vol.44, No.1-2, 49-75, ISSN 1068-7971

Buslov, M.M.; Kokh, D.A. & De Grave, J. (2008). Mesozoic-Cenozoic tectonics and geodynamics of Altai, Tien Shan, and Northern Kazakhstan, from apatite fission-track data, *Russ. Geol. Geophys.*, Vol.49, No.9, 648-654, ISSN 1068-7971

Cocks, L.R.M. & Torsvik, T.H. (2007). Siberia, the wandering northern terrane, and its changing geography through the Palaeozoic, *Earth Sci. Rev.*, Vol.82, 29–74, ISSN 0012-8252

Cogné, J.-P.; Kravchinsky, V.A.; Halim, N. & Hankard, F. (2005). Late Jurassic - Early Cretaceous closure of the Mongol-Okhotsk Ocean demonstrated by new Mesozoic palaeomagnetic results from the Trans-Baikal area (SE Siberia), *Geophys. J. Int.*, Vol.163, 813–832, ISSN 0956-540X

De Grave, J.; Buslov, M.M. & Van den haute, P. (2007). Distant effects of India–Eurasia convergence and Mesozoic intracontinental deformation in Central Asia: Constraints from apatite fission-track thermochronology, *J. Asian Earth Sci.*, Vol.29, 194–213, ISSN 1367-9120

Didenko, A.N.; Kozakov, I.K.; Bibikova, E.V.; Vodovozov, V.Yu.; Khil'tova, V.Ya.; Reznitskii, L.Z.; Ivanov, A.V.; Levitskii, V.I.; Travin, A.V.; Shevchenko, D.O. & Rasskazov S.V. (2003). Paleoproterozoic Granites of the Sharyzhalgai Block, Siberian Craton: Paleomagnetism and Geodynamic Inferences, *Dokl. Earth Sci.*, Vol.390, No.4, 510–515, ISSN 1028-334X

Dobretsov, N.L. (2005). 250 Ma large igneous provinces of Asia: Siberian and Emeishan traps (plateau basalts) and associated granitoids, *Russ. Geol. Geophys.*, Vol.46, No.9, 847-868, ISSN 1068-7971

Dobretsov, N.L. & Vernikovsky, V.A. (2001). Mantle plumes and their geological manifestations, *Int. Geol. Rev.*, Vol.43, 771–787, ISSN 0020-6814

Dobretsov, N.L.; Buslov, M.M. & Vernikovsky, V.A. (2003). Neoproterozoic to Early Ordovician Evolution of the Paleo-Asian Ocean: Implications to the Break-up of Rodinia, *Gondwana Res.*, Vol.6, No.2, 143-159, ISSN 1342-937X

Gallet, Y.; Pavlov, V.E.; Semikhatov, M.A. & Petrov, P.Y. (2000). Late Mesoproterozoic magnetostratigraphic results from Siberia: paleogeographic implications and magnetic field behavior, *J. Geophys. Res.*, Vol.105, 16481-16499, ISSN 0148-0227

Gladkochub, D.P.; Wingate, M.T.D.; Pisarevsky, S.A.; Donskaya, T.V.; Mazukabzov, A.M.; Ponomarchuk, V.A. & Stanevich, A.M. (2006). Mafic intrusions in southwestern Siberia and implications for a Neoproterozoic connection with Laurentia, *Precambrian Res.*, Vol.147, 260–278, ISSN 0301-9268

Gordienko, I.V. (2006). Geodynamic evolution of late Baikalides and Paleozoids in the folded periphery of the Siberian craton, *Russ. Geol. Geophys*, Vol.47, No.1, 51-67, ISSN 1068-7971

Kanygin, A.; Dronov, A.; Timokhin, A. & Gonta, T. (2010). Depositional sequences and palaeoceanographic change in the Ordovician of the Siberian craton, *Palaeogeography, Palaeoclimatology, Palaeoecology*, 2010, doi:10.1016/j.palaeo.2010.02.014, ISSN 0031-0182

Kazansky, A.Yu. (2002). Evolution of structures of western margin of Siberian Platform on paleomagnetic data, *Abstract of ScD thesis*, UIGGM SB RAS, Novosibirsk (in Russian)

Kazansky, A.Yu., Buslov, M.M. & Metelkin, D.V. (1998). Evolution of the Paleozoic accretion-collision structure of Gorny Altai: correlation of paleomagnetic and geological data. *Russ. Geol. Geophys.* Vol.39, No.3, 297-306, ISSN 1068-7971

Kazansky, A.Yu., Metelkin, D.V., Kungurtsev, L.V. & Lavrenchuk, A.V. (1999). New paleomagnetic data on the Early Cambrian of West Sayan, *Russ. Geol. Geophys.*, Vol.40, No(4), 526-545, ISSN 1068-7971

Kazansky, A.Yu., Metelkin, D.V., Kungurtsev, L.V. & Kizub, P.A. (2003). Kinematics of the Martaiga block of the Kuznetsk-Alatau paleoisland arc during the Late Vendian - Early Ordovician as inferred from paleomagnetic data. *Russ. Geol. Geophys. 44(3),* 181-197, ISSN 1068-7971

Kazansky, A.Yu.; Metelkin, D.V.; Bragin, V.Yu. & Kungurtsev, L.V. (2005). Paleomagnetism of the Permian-Triassic traps from the Kuznetsk Basin, southern Siberia, *Russ. Geol. Geophys.*, Vol.46, No.11, 1089–1102, ISSN 1068-7971

Khain, V.E. (2001). *Tectonics of continents and oceans (year 2000)*, Scientific World, ISBN 5-89176-138-6, Moscow, (in Russian)

Khain, V.E.; Gusev, G.S.; Khain, E.V.; Vernikovsky, V.A. & Volobuev, M.I. (1997). Circum-Siberian Neoproterozoic Ophiolite Belt, *Ofioliti*, Vol.22, No.2, 195-200, ISSN 0391-2612

Khain, E.V.; Bibikova, E.V.; Salnikova, E.B.; Kröner, A.; Gibsher, A.S.; Didenko, A.N.; Degtyarev, K.E. & Fedotova, A.A. (2003). The Palaeo-Asian ocean in the Neoproterozoic and Early Palaeozoic: new geochronologic data and palaeotectonic reconstructions, *Precambrian Res.*, Vol.122, 329–358, ISSN 0301-9268

Khomentovsky, V.V. (2005). Topical problems of Neoproterozoic stratigraphy in the Siberian hypostratotype of the Riphean, *Russ. Geol. Geophys.*, Vol.46, No.5, 521–536, ISSN 1068-7971

Khramov, A.N. (1991). Standard series of paleomagnetic poles for the plates of north Eurasia: the connection with the problem of geodynamics of the territory of USSR, In: *Paleomagnetic and Paleogeodynamics of the Territory of USSR*, Khramov, A.N. (Ed.), 154-176, VNIGRI, Leningrad, (in Russian)

Khudoley, A.K.; Rainbird, R.H.; Stern, R.A.; Kropachev, A.P.; Heaman, L.M.; Zanin, A.M.; Podkovyrov, V.N.; Belova, V.N. & Sukhorukov, V.I. (2001). Sedimentary evolution of the Riphean–Vendian basin of southeastern Siberia, *Precambrian. Res.*, Vol.111, 129– 163 ISSN 0301-9268

Kirschvink, J.L.: Ripperdan, R.L. & Evans, D.A. (1997). Evidence for a large-scale reorganization of Early Cambrian continental landmasses by internal interchange true polar wander, *Science*, Vol.277, 541-545, ISSN 0036-8075

Koronovsky, N.V.; Gogonenkov, G.N.; Goncharov, M.A.; Timurziev, A.I. & Frolova, N.S. (2009). Role of Shear along Horizontal Plane in the Formation of Helicoidal Structures, *Geotectonics*, Vol.43, No.5, 379-391, ISSN 0016-8521

Kravchinsky, V.A.; Konstantinov, K.M. & Cogné, J.-P. (2001). Palaeomagnetic study of Vendian and Early Cambrian rocks of South Siberia and Central Mongolia: was the Siberian platform assembled at this time?, *Precambrian Res.*, Vol.110, 61–92, ISSN 0301-9268

Kravchinsky, V.A.; Cogné, J.-P.; Harbert, W.P. & Kuzmin, M.I. (2002). Evolution of the Mongol–Okhotsk Ocean as constrained by new palaeomagnetic data from the Mongol–Okhotsk suture zone, Siberia, *Geophys. J. Int.*, Vol.148, 34–57, ISSN 0956-540X

Kungurtsev, L.V.; Berzin, N.A.; Kazansky, A.Yu. & Metelkin, D.V. (2001). Tectonic evolution of the southwestern framing of the Siberian platform in the Vendian-Cambrian according to paleomagnetic data, *Russ. Geol. Geophys.*, Vol.42, No.7, 1042-1051, ISSN 1068-7971

Kuzmichev, A.; Kröner, A.; Hegner, E.; Dunyi, L. & Yusheng, W. (2005). The Shishkhid ophiolite, northern Mongolia: A key to the reconstruction of a Neoproterozoic island-arc system in central Asia, *Precambrian Res.*, Vol.138, 125–150, ISSN 0301-9268

Levitsky, V.I.; Mel'nikov, A.I.; Reznitsky, L.Z.; Bibikova, E.V.; Kirnozova, T.I.; Kozakov, I.K.; Makarov, V.A. & Plotkina, Yu.V. (2002). Early Proterozoic postcollisional granitoids in southwestern Siberian craton, *Russ. Geol. Geophys.*, Vol.43, No.8, 717-731, ISSN 1068-7971

Li, Z.X.; Bogdanova, S.V.; Collins, A.S.; Davidson, A.; De Waele, B.; Ernst, R.E.; Fitzsimons, I.C.W.; Fuck, R.A.; Gladkochub, D.P.; Jacobs, J.; Karlström, K.E.; Lu, S.; Natapov, L.M.; Pease, V.; Pisarevsky, S.A.; Thrane, K. & Vernikovsky, V. (2008). Assembly, configuration, and break-up history of Rodinia: A synthesis, *Precambrian Res.*, Vol.160, 179–210, ISSN 0301-9268

Lorenz, H.; Gee, D.G. & Whitehouse, M.J. (2007). New geochronological data on Palaeozoic igneous activity and deformation in the Severnaya Zemlya Archipelago, Russia, and implications for the development of the Eurasian Arctic margin, *Geol. Mag.*, Vol.144, No.1, 105–125, ISSN 0016-7568

Lorenz, H.; Männik, P.; Gee, D.G. & Proskurnin, V. (2008). Geology of the Severnaya Zemlya Archipelago and the North Kara Terrane in the Russian high Arctic, *Int. J. Earth Sci.*, Vol.97, 519–547, ISSN 1437-3254

Männik, P.; Menner, V.V.; Matukhin, R.G. & Kurљs V. Silurian and Devonian strata on the Severnaya Zemlya and Sedov archipelagos (Russia). // Geodiversitas, 2002, 24(1), p.99–121. ISSN 1280-9659

Männik, P.; Bogolepova, O.K.; Põldvere, A. & Gubanov, A.P. (2009). New data on Ordovician–Silurian conodonts and stratigraphy from the Severnaya Zemlya Archipelago, Russian Arctic, *Geol. Mag.*, Vol.146, 497-516, ISSN 0016-7568

McElhinny, M.W. & MacFadden, P.L. (2000). *Paleomagnetism: Continents and Oceans*, Academic Press, San Diego, CA.

Meert, J.G. (1999). A paleomagnetic analysis of Cambrian true polar wonder, *Earth Planet. Sci. Lett.*, Vol.168, 131–144, ISSN 0012-821X

Metelkin, D.V.; Kazansky, A.Yu. & Kungurtsev, L.V. (2000). Kuznetsk Alatau structure in Middle - Late Cambrian from paleomagnetic data, *Russ. Geol. Geophys.*, Vol.41, No.6, 772-784, ISSN 1068-7971

Metelkin, D.V.; Vernikovsky, V.A. & Belonosov, I.V. (2004a). Paleomagnetism of volcanogenic complexes from the Predivinsk terrane of the Yenisei Ridge: Geodynamic implications, *Dokl. Earth Sci.*, Vol.399, No.8, 1080–1084, ISSN 1028-334X

Metelkin, D.V.; Gordienko, I.V. & Zhao X. (2004b). Paleomagnetism of Early Cretaceous volcanic rocks from Transbaikalia: argument for Mesozoic strike-slip motions in Central Asian structure, *Russ. Geol. Geophys.*, Vol.45, No.12, 1349–1363, ISSN 1068-7971

Metelkin, D.V.; Belonosov, I.V.; Gladkochub, D.P.; Donskaya, T.V.; Mazukabzov, A.M. & Stanevich, A.M. (2005a). Paleomagnetic directions from Nersa intrusions of the

Biryusa terrane, Siberian craton, as a reflection of tectonic events in the Neoproterozoic, *Russ. Geol. Geophys.*, Vol.46, No.4, 395-410, ISSN 1068-7971

Metelkin, D.V.; Vernikovsky, V.A.; Kazansky, A.Yu.; Bogolepova, O.K. & Gubanov, A.P. (2005b). Paleozoic history of the Kara microcontinent and its relation to Siberia and Baltica: paleomagnetism, paleogeography and tectonics, *Tectonophysics*, Vol.398, 225–243, ISSN 0040-1951

Metelkin, D.V.; Vernikovsky, V.A. & Kazansky, A.Yu. (2007a). Neoproterozoic evolution of Rodinia: constraints from new paleomagnetic data on the western margin of the Siberian craton, *Russ. Geol. Geophys.*, Vol.48, No.1, 32-45, ISSN 1068-7971

Metelkin, D.V.; Gordienko, I.V. & Klimuk, V.S. (2007b). Paleomagnetism of Upper Jurassic basalts from Transbaikalia: new data on the time of closure of the Mongolo-Okhotsk Ocean and Mesozoic intraplate tectonics of Central Asia, *Russ. Geol. Geophys.*, Vol.48, No.10, 825–834, ISSN 1068-7971

Metelkin, D.V.; Kazansky, A.Yu.; Bragin, V.Yu.; Tsel'movich, V.A.; Lavrenchuk, A.V. & Kungurtsev, L.V. (2007c). Paleomagnetism of the Late Cretaceous intrusions from the Minusa trough (southern Siberia), *Russ. Geol. Geophys.*, Vol.48, No.2, 185–198, ISSN 1068-7971

Metelkin, D.V.; Vernikovsky, V.A.; Kazansky, A.Yu.; Kashirtsev, V.A.; Bragin, V.Yu. & Kungurtsev, L.V. (2008). The Mesozoic Apparent Polar Wander Path for the Siberian Domain of the Eurasian Plate, *Dokl. Earth Sci.*, Vol.418, No.1, 62–67, ISSN 1028-334X

Metelkin, D.V.; Kazansky, A.Yu. & Vernikovsky, V.A. (2009) Paleomagnetic Evidence for Siberian Plate Tectonics from Rodinia through Pangaea to Eurasia, In: *Geomorphology and Plate Tectonics*, Ferrari, DM. & Guiseppi, A.R. (Eds.), 159-236, Nova Science Publishers, ISBN 978-1-60741-003-4, NY, USA

Metelkin, D.V.; Vernikovsky, V.A., Kazansky, A.Yu. & Wingate, M.T.D. (2010a). Late Mesozoic tectonics of Central Asia based on paleomagnetic evidence, *Gondwana Res.*, Vol.18, 400-419, ISSN 1342-937X

Metelkin, D.V.; Blagovidov, V.V. & Kazansky, A.Yu. (2010b). The history of the Karagas Supergroup evolution in the Biryusa region: synthesis of paleomagnetic and sedimentological data, *Russ. Geol. Geophys.*, Vol.51, No.8, 868-884, ISSN 1068-7971

Natal'in, B.A. & Sengör, A.M.C. (2005). Late Palaeozoic to Triassic evolution of the Turan and Scythian platforms: the pre-history of the palaeo-Tethyan closure, *Tectonophysics*, Vol.404, 175–202, ISSN 0040-1951

Nemerov, V.K.; & Stanevich, A.M. (2001). Evolution of the Riphean-Vendian biolithogenesis settings in the Baikal Mountainous area, *Russ. Geol. Geophys.*, Vol.42, No.3, 456-470, ISSN 1068-7971

Nozhkin, A.D.; Turkina, O.M.; Bibikova, E.V.; Terleev, A.A. & Khomentovsky, V.V. (1999). Riphean granite gneiss domes of the Yenisei Range: geological structure and U-Pb isotopic age, *Russ. Geol. Geoph.*, Vol.40, No.9, 1284–1292. ISSN 1068-7971

Oxman, V.S. (2003). Tectonic evolution of the Mesozoic Verkhoyansk–Kolyma belt (NE Asia), *Tectonophysics*, Vol.365, 45- 76, ISSN 0040-1951

Parfenov, L.M. & Kuz'min, M.I. (Eds.) (2001). *Tectonics, Geodynamics, and Metallogeny of the Sakha Republic (Yakutia) Territory*, Nauka/Interperiodika, ISBN 5-7846-0046-X, Moscow, (in Russian)

Parfenov, L.M.; Prokopiev, A.V. & Gaiduk, V.V. (1995). Cretaceous frontal trusts of the Verkhoyansk fold belt, Eastern Siberia, *Tectonics*, Vol.4, No.2, 342-358, ISSN 0278-7407

Parfenov, L.M.; Bulgatov, A.N. & Gordienko, I.V. (1996). Terranes and formation of orogens in Transbaikalia, *Tikhookeanskaya Geologiya*, Vol.15, No.4, 3–15, ISSN 0207-4028, (in Russian)

Parfenov, L.M.; Berzin, N.A.; Khanchuk, A.I.; Badarch, G.; Belichenko, V.G.; Bulgatov, A.N.; Dril, S.I.; Kirillova, G.L.; Kuzmin, M.I.; Nokleberg, W.J.; Prokopyev, A.V.; Timofeev, V.F.; Tomurtogoo, O. & Yang, H. (2003). A model for the formation of orogenic belts in Central and Northeast Asia, *Tikhookeanskaya Geologiya*, Vol.22, No.6, 7-41, ISSN 0207-4028, (in Russian)

Pavlov, V.E. & Petrov, P.Yu. (1997). Paleomagnetism of the Riphean Deposits in the Irkineeva Uplift, Yenisei Range: New Evidence for the Integrity of the Siberian Platform in the Middle Riphean, *Izv. Phys. Solid Earth*, Vol.33, No.6, 464-475, ISSN 1069-3513

Pavlov, V.E.; Gallet, Y. & Shatsillo, A.V. (2000). Paleomagnetism of the Upper Riphean Lakhandinskaya Group in the Uchuro-Maiskii Area and the Hypothesis of the Late Proterozoic Supercontinent, *Izv. Phys. Solid Earth*, Vol.36, No.8, 638-648, ISSN 1069-3513

Pavlov, V.E.; Gallet, Y.; Petrov, P.Yu.; Zhuravlev, D.Z. & Shatsillo, A.V. (2002). Uy series and late Riphean sills of the Uchur-Maya area: isotopic and palaeomagnetic data and the problem of the Rodinia supercontinent, *Geotectonics*, Vol.36, 278–292, ISSN 0016-8521

Pavlov, V.E.; Gallet, Y.; Shatsillo, A.V. & Vodovozov, V.Yu. (2004). Paleomagnetism of the Lower Cambrian from the Lower Lena River Valley: Constraints on the Apparent Polar Wander Path from the Siberian Platform and the Anomalous Behavior of the Geomagnetic Field at the Beginning of the Phanerozoic, *Izv. Phys. Solid Earth*, Vol.40, No.2, 114-133 ISSN 1069-3513

Pease, V.; Gee, D.; Vernikovsky, V.; Vernikovskaya, A. & Kireev, S. (2001). The Mamont-Shrenk terrane: a Mesoproterozoic complex in the Neoproterozoic accretionary belt of Central Taimyr, Norhern Siberia, *Terra Nova*, 2001, Vol.13, 270-280, ISSN 0954-4879

Pechersky, D.M. & Didenko, A.N. (1995). *The Paleoasian Ocean: Petromagnetic and Paleomagnetic Information on the Lithosphere*, Schmidt Institute of Physics of the Earth RAS, ISBN 5-201-11898-4, Moscow, (in Russian)

Petrov, P.Y. & Semikhatov, M.A. (2001). Sequence organization and growth patterns of late Mesoproterozoic stromatolite reefs: an example from the Burovaya Formation, Turukhansk Uplift, Siberia, *Precambrian Res.*, Vol.111, 257–281, ISSN 0301-9268

Pisarevsky, S.A. & Natapov, L.M. (2003). Siberia in Rodinia, *Tectonophysics*, Vol.375, 221-245, ISSN 0040-1951

Pisarevsky, S.A.; Komissarova, R.A. & Khramov A.N. (2000). New palaeomagnetic result from Vendian red sediments in Cisbaikalia and the problem of the relationship of Siberia and Laurentia in the Vendian, *Geophys. J. Int.*, Vol.140, 598–610, ISSN 0956-540X

Pisarevsky, S.A.; Murphy, J.B.; Cawood, P.A. & Collins, A.S. (2008). Late Neoproterozoic and Early Cambrian palaeogeography: models and problems, In: *West Gondwana:*

Pre-Cenozoic Correlations Across the South Atlantic Region, Pankhurst, R.J.; Trouw, R.A.J.; Brito Neves B. B. & De Wit M. J. (Eds.), 9-31, Geological Society of London, Spec. Publ., Vol.294, DOI: 10.1144/SP294.2

Rainbird, R.H.; Stern, R.A.; Khudoley, A.K.; Kropachev, A.P.; Heaman, L.M. & Sukhorukov, V.I. (1998). U-Pb geochronology of Riphean sandstone and gabbro from southeast Siberia and its bearing in the Laurentia-Siberia connection, *Earth Planet Sci .Lett.,* Vol.164, 409-420, ISSN 0012-821X

Rosen, O.M. (2003). The Siberian Craton: Tectonic Zonation and Stages of Evolution, *Geotectonics,* Vol.37, No.3, 175-192, ISSN 0016-8521

Rosen, O.M.; Condie K.C.; Natapov, L.M. & Nozhkin, A.D. (1994). Archean and early Proterozoic evolution of the Siberian craton, a preliminary assessment, In: *Archean crustal evolution,* Condie, K.C.(Ed.), 411-459, Elsevier, ISBN: 978-0-444-81621-4 Netherlands

Rosen, O.M.; Manakov, A.V. & Serenko, V.P. (2005). Paleoproterozoic collisional system and diamondiferous lithospheric keel of the Yakutian kimberlite province, *Russ. Geol. Geophys.,* Vol.46, No.12, 1237-1251, ISSN 1068-7971

Rytsk, E.Yu.; Amelin, Yu.V.; Rizvanova, N.G.; Krymsky, R.Sh.; Mitrofanov, G.L.; Mitrofanova, N.N.; Perelyaev, V.I. & Shalaev, B.C. (2001). Age of Rocks in the Baikal-Muya Foldbelt. *Stratigr. Geol. Correl.,* Vol.9, No.4, 315-326, ISSN 0869-5938

Rytsk, E.Yu.; Kovach, V.P.; Kovalenko, V.I. & Yarmolyuk, V.V. (2007). Structure and Evolution of the Continental Crust in the Baikal Fold Region, *Geotectonics,* Vol.41, No.6, 440–464, ISSN 0016-8521

Safonova, I.Yu.; Simonov, V.A.; Buslov, M.M.; Ota, T.; Maruyama, Sh. (2008). Neoproterozoic basalts of the Paleo-Asian Ocean (Kurai accretion zone, Gorny Altai, Russia): geochemistry, petrogenesis, geodynamics. *Russ. Geol. Geoph.,* Vol.49, No.4, 254–271, ISSN 1068-7971

Semikhatov, M.A.; Ovchinnikova, G.V.; Gorokhov, I.M.; Kuznetsov A.B.; Vasil'eva I.M.; Gorokhovskii, B.M. & Podkovyrov, V.N. (2000). Isotope Age of the Middle–Upper Riphean Boundary: Pb–Pb Geochronology of the Lakhanda Group Carbonates, Eastern Siberia, *Dokl. Earth Sci.,* Vol.372, No.4, 625-629, ISSN 1028-334X

Şengör, A.M.C., Natalin B., Burtman V. (1993). Evolution of the Altaide tectonic collage and Paleozoic crustal growth in Eurasia, *Nature,* Vol.34, 299–307, ISSN 0028-0836

Shatsillo, A.V.; Didenko, A.N. & Pavlov, V.E. (2005). Two competing Paleomagnetic directions in the Late Vendian:new data for the SW Region of the Siberian Platform. *Russ. J. Earth Sci.,* Vol.7, No.4, 3–24, ISSN 1681-1208

Shatsillo, A.V.; Pavlov, V.E. & Didenko, A.N. (2006). Paleomagnetism of Vendian rocks in the southwest of the Siberian Platform. *Russ. J. Earth Sci.,* Vol.8, No.2, 1-30, ISSN 1681-1208

Sklyarov, E.V.; Gladkochub, D.P.; Mazukabzov, A.M.; Menshagin, Yu.V.; Watanabe, T. & Pisarevsky, S.A. (2003). Neoproterozoic mafic dike swarms of the Sharyzhalgai metamorphic massif (southern Siberian craton), *Precambrian Res.,* Vol.22, 359–377, ISSN 0301-9268

Smelov, A.P. & Timofeev, V.F. (2007). The age of the North Asian Cratonic basement: An overview, *Gondwana Res.,* Vol.12, p.279–288, ISSN 1342-937X

Smethurst, M.A.; Khramov, A.N. & Torsvik, T.H. (1998). The Neoproterozoic and Palaeozoic palaeomagnetic data for the Siberian Platform: From Rodinia to Pangea, *Earth Sci. Rev.*, Vol.43, No.1-2, 1–24, ISSN 0012-8252

Sovetov, Yu.K.; Blagovidov, V.V.; Luchinina, V.A. & Romashko, A.I. (2000). Vendian Sediments on the Western Enisei Ridge, *Dokl. Earth Sci.*, Vol.372, No.4, 630-632, ISSN 1028-334X

Stanevich, A.M.; Mazukabzov, A.M.; Postnikov, A.A.; Nemerov, V.K.; Pisarevsky, S.A.; Gladkochub, D.P.; Donskaya, T.V. & Kornilova, T.A. (2007). Northern segment of the Paleoasian Ocean: Neoproterozoic deposition history and geodynamics, *Russ. Geol. Geophys.*, Vol.48, No.1, 46-60, ISSN 1068-7971

Tomurtogoo, O.; Windley, B.F.; Kröner, A.; Badarch, G. & Liu, D.Y. (2005). Zircon age and occurrence of the Adaatsag ophiolite and Muron shear zone, central Mongolia: constraints on the evolution of the Mongol–Okhotsk ocean, suture and orogen, *J. Geol Soc. Lond.*, Vol.162, 125–134, ISSN 0016-7649

Torsvik, T.H. & Cocks, L.R.M. (2005). Norway in space and time: A Centennial cavalcade, *Norw. J. Geol.*, Vol.85, 73–86, ISSN 0029-196X

Torsvik, T.H. & Smethurst, M.A. (1999). Plate tectonic modelling: virtual reality with GMAP, *Computers and Geosciences*, Vol.25, 395–402, ISSN 0098-3004

Turkina, O.M.; Nozhkin, A.D.; Bayanova, T.B.; Dmitrieva, N.V. & Travin, A.V. Precambrian terranes in the southwestern framing of the Siberian craton: isotopic provinces, stages of crustal evolution and accretion-collision events, *Russ. Geol. Geophys.*, Vol.48, No.1, 61-70, ISSN 1068-7971

Uflyand, A.K.; Natapov, L.M.; Lopatin, V.M. & Chernov, D.V. (1991). On the tectonic nature of Taimyr, *Geotectonics*, Vol.25, No.6, 76–93, ISSN 0016-8521

Van der Voo, R. (1990). The reliability of paleomagnetic data, *Tectonophysics*, Vol.184, 1–9, ISSN 0040-1951

Van der Voo, R.; Levashova N.M.; Skrinnik, L.I.; Kara, T.V. & Bazhenov, M.L. (2006). Late orogenic, large-scale rotations in the Tien Shan and adjacent mobile belts in Kyrgyzstan and Kazakhstan, *Tectonophysics*, Vol.426, 335–360, ISSN 0040-1951

Vernikovskaya, A.E.; Vernikovsky, V.A.; Sal'nikova, E.B.; Kotov, A.B.; Kovach, V.P.; Travin, A.V.; Palesskii, S.V.; Yakovleva, S.Z.; Yasenev, A.M. & Fedoseenko, A.M. (2003). Neoproterozoic Postcollisional Granitoids of the Glushikha Complex, Yenisei Range, *Petrology*, Vol.11, No1, 48-61, ISSN 0869-5911

Vernikovskaya, A.E.; Vernikovsky, V.A.; Sal'nikova, E.B.; Yasenev, A.M; Kotov, A.B.; Kovach, V.P.; Travin, A.V.; Yakovleva, S.Z. & Fedoseenko, A.M. (2006). Neoproterozoic A-Type Granites of the Garevka Massif, Yenisey Ridge: Age, Sources, and Geodynamic Setting, *Petrology*, Vol.14, No1, 50-61, ISSN 0869-5911

Vernikovsky, V.A. (1996). *Geodynamic Evolution of Taimyr Folded Area*, SB RAS Publicher, ISBN 5-7692-0015-4, Novosibirsk, (in Russian)

Vernikovsky, V.A. & Vernikovskaya, A.E. (2001). Central Taimyr accretionary belt (Arctic Asia): Meso-Neoproterozoic tectonic evolution and Rodinia breakup, *Precambrian Res.*, Vol.110, 127–141, ISSN 0301-9268

Vernikovsky, V.A. & Vernikovskaya, A.E. (2006) Tectonics and evolution of granitoid magmatism in the Yenisei Ridge, *Russ. Geol. Geophys.*, Vol.47, No.1, 32–50, ISSN 1068-7971

Vernikovsky, V.A.; Kotov, A.B.; Ponomarchuk, V.A.; Sal'nikova, E.B., Kovach, V.P., & Travin, A.V. (1997). Late Riphean-Vendian Event in the Northern Taimyr Evolution: Evidence from Sm-Nd, Rb-Sr and K-Ar Dating of the Garnet Amphibolites of the Stanovaya Ophiolite Belt, *Dokl. Earth Sci.*, Vol.352, No.1, 11-14, ISSN 1028-334X

Vernikovsky, V.A.; Vernikovskaya, A.E.; Sal'nikova, E.B.; Kotov, A.B.; Chernykh, A.I.; Kovach, V.P.; Berezhnaya, N.G. & Yakovleva, S.Z. (1999). New U-Pb data on the formation of the Predivinsk paleoisland-arc complex (Yenisei Range), *Russ. Geol. Geophys.*, Vol.40, No.2, 255–259, ISSN 1068-7971

Vernikovsky, V.A.; Vernikovskaya, A.E.; Kotov, A.B.; Sal'nikova, E.B. & Kovach V.P. (2003). Neoproterozoic accretionary and collisional events on the western margin of the Siberian craton: new geological and geochronological evidence from the Yenisey Ridge, *Tectonophysics*, Vol.375, 147–168, ISSN 0040-1951

Vernikovsky, V.A.; Vernikovskaya, A.E.; Pease, V.L. & Gee, D.G. (2004). Neoproterozoic orogeny along the margins of Siberia, In: *The Neoproterozoic Timanide orogen of Eastern Baltica,* Gee, D.G. & Pease, V.L. (Eds.), *Geol. Soc. London Mem.*, Vol.30, 233–247.

Vernikovsky, V.A.; Vernikovskaya, A.E.; Sal'nikova, E.B.; Berezhnaya, N.G.; Larionov, A.N.; Kotov, A.B.; Kovach, V.P.; Vernikovskaya, I.V.; Matushkin, N.Yu. & Yasenev, A.M. (2008). Late Riphean Alkaline Magmatism in the Western Framework of the Siberian Craton: A Result of Continental Rifting or Accretionary Events?, *Dokl. Earth Sci.*, Vol.419, No.2, 226-230, ISSN 1028-334X

Vernikovsky, V.A.; Kazansky, A.Yu.; Matushkin, N.Yu.; Metelkin, D.V. & Sovetov. J.K. (2009). The geodynamic evolution of the folded framing and the western margin of the Siberian craton in the Neoproterozoic: geological, structural, sedimentological, geochronological and paleomagnetic data, *Russ. Geol. Geophys.*, Vol.50, No.4, 372-387, ISSN 1068-7971

Vladimirov, A.G.; Kozlov, M.S.; Shokal'skii, S.P.; Khalilov, V.A.; Rudnev, S.N.; Kruk, N.N.; Vystavnoi, S.A.; Borisov, S.M.; Berezikov, Yu.K.; Metsner, A.N.; Babin, G.A.; Mamlin, A.N.; Murzin, O.M.; Nazarov, G.V. & Makarov, V.A. (2001). Major epochs of intrusive magmatism of Kuznetsk Alatau, Altai, and Kalba (from U-Pb isotopedates), *Russ. Geol. Geophys.*, Vol.42, No.8, 1157-1178, ISSN 1068-7971

Voronov, P.S. (Ed.) (1997). *Role of strike-slip tectonics in the lithosphere structure of the Earth and terrestrial planets.* Nauka, ISBN 5-02-024832-0, St.Petersburg, (in Russian)

Yarmolyuk, V.V. & Kovalenko, V.I. (2001). Late Riphean breakup between Siberia and Laurentia: Evidence from intraplate magmatism, *Dokl. Earth Sci.*, Vol.379, No.5, 525–528, ISSN 1028-334X

Yarmolyuk, V.V.; Kovalenko, V.I.; Sal'nikova, E.B.; Nikiforov, A.V.; Kotov, A.B. & Vladykin, N.V. (2005). Late Riphean Rifting and Breakup of Laurasia: Data on Geochronological Studies of Ultramafic Alkaline Complexes in the Southern Framing of the Siberian Craton, *Dokl. Earth Sci.*, Vol.404, No.7, 1031-1036, ISSN 1028-334X

Zonenshain, L.P., Kuzmin, M.I., Natapov, L.M. (1990). *Geology of the USSR: a plate tectonic synthesis.* Amer. Geoph. Union, ISBN 0-87-590521-8, Washington, USA

Zorin, Yu.A. (1999). Geodynamics of the western part of the Mongolia.Okhotsk collisional belt, Trans-Baikal region (Russia) and Mongolia, *Tectonophysics*, Vol.306, 33–56, ISSN 0040-1951

Zorin, Yu.A.; Sklyarov, E.V.; Belichenko, V.G. & Mazukabzov, A.M. (2009). Island arc–back-arc basin evolution: implications for Late Riphean–Early Paleozoic geodynamic history of the Sayan–Baikal folded area, *Russ. Geol. Geophys.*, Vol.50, No.3, 149-161, ISSN 1068-7971

Part 4

Tectonics of China and its Neighborhood

6

Proto-Basin Types of North China Craton (NCC) in Late Triassic and Its Implication for Regional Tectonics of Initial Craton Destruction[1]

Yang Minghui[2], Liu Chiyang[3], Zeng Peng[1], Bai Hua[1] and Zhou Jin[1]
[1]*College of Geosciences, China University of Petroleum, Beijing 102249;*
[2]*State Key Laboratory of Petroleum Resource and Prospecting, China University of Petroleum, Beijing, 102249;*
[3]*State key Laboratory of Continental Dynamics, Northwest University, Xi'an, 710069*
China

1. Introduction

As a long-term geological process with widespread influence, destruction of the North China Craton (NCC) displays many important geological phenomena and attracts interest of many domestic and overseas researchers engaged in geology, geochemistry and geophysics, etc. The NCC has its own attributes of occurrence and development. Its destruction may have connections with the activities of YYOB and QDOB respectively, distributed in the north and the south, as well as with the subduction between Indian plate and Pacific plate. Even though the inhomogeneity of the destruction process becomes prominent in terms of time, space and destruction intensity, significant disputes still exist on several specific problems. The peak period of NCC destruction was in late Jurassic-early Cretaceous. However, there are various opinions concerning the determination of beginning stage, e. g. 1) late Triassic (Menzies et al., 1998; Gao et al., 2002; Lu Fengxiang et al., 2000; Yang Jinhui et al., 2009); 2) late Mesozoic (Deng Jinfu et al.,1994; Wu Fuyuan et al., 1999; Zhai Mingguo et al., 2003; Lu Fengxiang et al.,2006); 3) Cenozoic (Menzies et al., 1993; Griffin et al., 1998; Xu, 2001) and Meso-Cenozoic (Xu Wenliang et al., 2000); or even in 4) late Paleozoic (Xu Yigang et al., 2009; Li Hongyan et al., 2009). The range of NCC destruction is in the east of the gravity gradient zone of (from Daxing'anling Mountains to) Taihangshan Mountains (Xu Yigang et al., 2006) , according to other opinions, or already had moved westward to nearby of Luliangshan Mountain (Ren Zhanli et al., 2005; Xiao Yuanyuan et al., 2007) around the Shanxi province as transitional or conversional area (Xing Zuoyun et al., 2006). The destruction intensity is getting gradually stronger from west to east, extremely severe in the east (Wei Wenbo et al., 2008), stable in the western Ordos area which remains as typical craton without disturbance in its deep layer by NCC destruction (Qiu Ruizhao et al., 2004; Jia Shixu et al., 2005). NCC destruction, as a geological process, was a kind of gradually

[1] Supported by Key Program of National Natural Science Foundation of China (No: 90814005) and Marine Reservoir Evolution and hydrocarbon enrichment Laboratory Fund.

varying chemical erosion (Xu, 2001; Fan et al., 2001; Zhang et al., 2002; Xu et al., 2003, 2004) or mutational physical delamination (Deng Jinfu et al., 1994, 2006) in deep layer which behaved as lithosphere thinning; however, it acted as drastic crustal deformation (contraction and/or extension) (Yu Fusheng et al., 2002; Duan Qiuliang et al., 2007), or large-scale magmatic activity (Guo et al., 2001; Zhang et al., 2002) and mineralization (Zhai Mingguo et al., 2002; Zhai et al., 2002; Liu Jianming et al., 2003). For the NCC event, there are deficiencies in analytical results about fillings and denudations of the sedimentary basin scale at the moment, although the incomplete or lacking of stratigraphic records can be "windows" of lithosphere evolution in geotectonic research(Liu Shaofeng et al., 1996, 2004; Davis et al., 2001; Yang Geng et al., 2001; Darby et al., 2001; Li Zhong et al., 2007). In recent years, basin research and comparison on NCC destruction become possible, due to 1) expanded knowledge to the deeper stratigraphic framework and structure of the basin obtained from seismic reflection profiles and drilling wells; 2) better restriction to the time series of stratigraphic records of the NCC area based on accurate isotopic dating data. Based on the integration of former data, this paper will focus on the area of comparison and analysis of the NCC basin filling, deposition and tectonic deformation in late Triassic and bring forward new understandings to the proto-basin type, tectonic deformation division and tectonic destruction process of the researching areas.

2. Regional geological background

NCC basin's evolution and its time-space distributing characteristics are closely linked to its regional structural geological background. Since Paleozoic, the tectonic development of China mainland is under the control of three global dynamics systems: paleo-Asian ocean, Tethys ocean and paleo-Pacific dynamics systems (Ren Jishun, 1994). Considering from plate tectonics view, NCC was jointly influenced by Siberian plate, Yangtze plate, west Pacific ocean plate and Indian plate, and collisions, extrusions and subductions among plates directly resulted in the formation of YYOB and QDOB in the north and the south of eastern China mainland (Fig. 1).

YYOB is a collision zone with certain width and numerous fragments of ancient continental crusts resulted from the collision and extrusion between NCC plate and Siberian plate (Li Shuanglin, 1998; Yang Baojun et al., 2003). The patchwork of the north edge of NCC plate and the south edge of Siberian plate was formed in late Paleozoic, and the final suture zone locates in Xilinhot (Hegen Mountains)-Sonidyouqi-Suolunobo in north Inner Mongolia (Wang Quan et al., 1991), which marks the end of influence of the paleo-Asian ocean tectonic domain on basin tectonic evolution in north China, and YYOB has become a limited boundary of Cenozoic tectonic evolution in NCC basin. QDOB lies across the middle of China mainland, as a key to researches on China mainland. Researches result of orogen patterns, characteristics of lithosphere structure, high pressure and ultra-high pressure metamorphism, geodynamic mechanism and other aspects indicate that this orogen was formed in late Triassic due to interplate collision and orogeny caused by subduction of the north edge of Yangtze plate into the lower part of NCC during Yangtze plate's northward drift (Zhang Guowei et al, 2003).

Therefore, the relationship between the orogen formation and the basin evolution shows that the formation of QDOB (resulted from the collision and extrusion of Yangtze plate and NCC) in Indosinian makes the whole NCC in almost south-north extrusion. Because YYOB at the north edge of NCC is a fixed limited boundary, there are a large number of thrusting nappe structures and corresponding interplate foreland basins developed along the south and north edges of NCC. And there are also a great number of EW or NWW striking thrusting nappe structures at Yanshan orogen, southern North China basin and Bohai bay basin. The extrusion effects between Yangtze plate and NCC may be prolonged until the early Yanshan stage (Du Xudong et al, 1999).

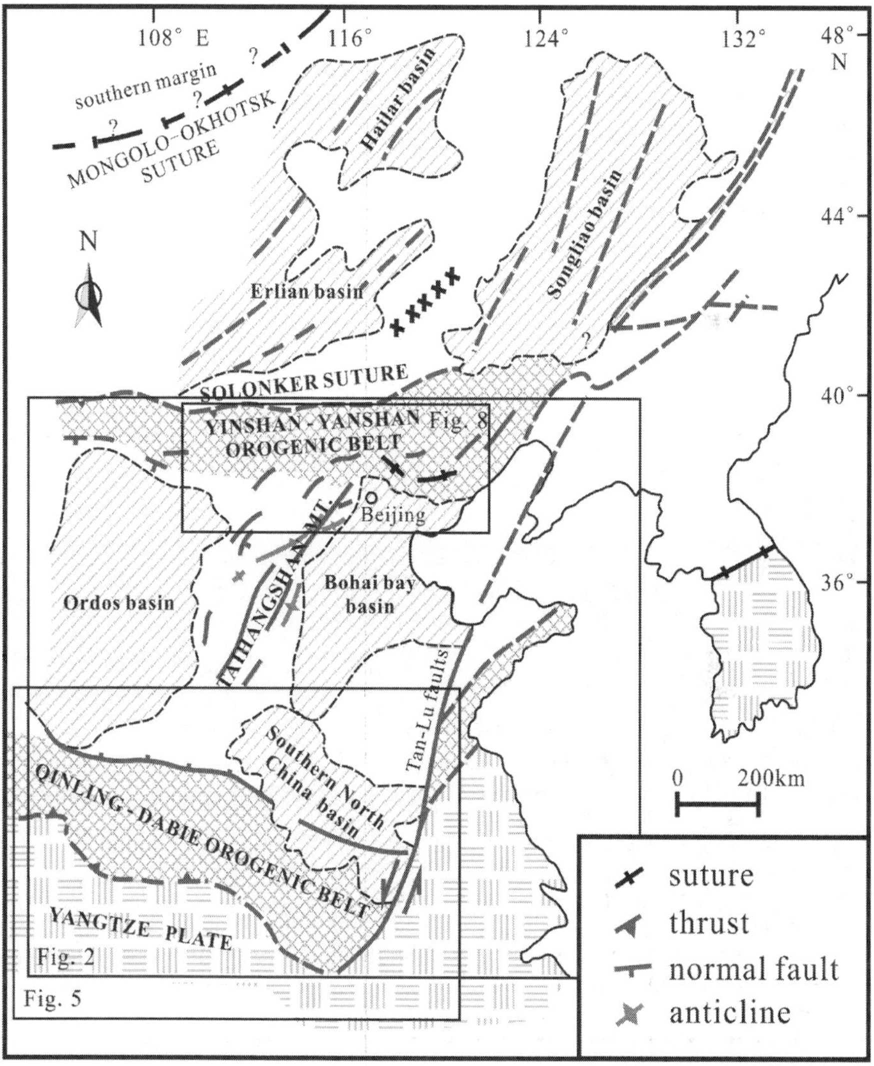

Fig. 1. Regional tectonic patterns of NCC and its adjacent area (simplified from Zheng Yadong et al., 2000).

3. Distribution of contemporaneous residual formations of NCC in late Triassic

The residual formations of late Triassic in NCC was divided into multiple areas: First of all, the sufficiently developed formation of western Ordos in the middle zone and the sedimentary-subsidence center in southwest side, which extend eastward to the middle area of Shanxi province and western Henan province as whole-blocks; secondly, areas in the north zone scattering along western Liaoning-western Beijing-northern Hebei province; finally, areas in the south zone are developed in the north of Mt. Qingling along western Henan-southern Shanxi provinces (Fig. 2). Formations at the north and south edges have been all influenced by the late orogenic deformation, which were mostly developed as fragmentations due to the clamping of thrust fault.

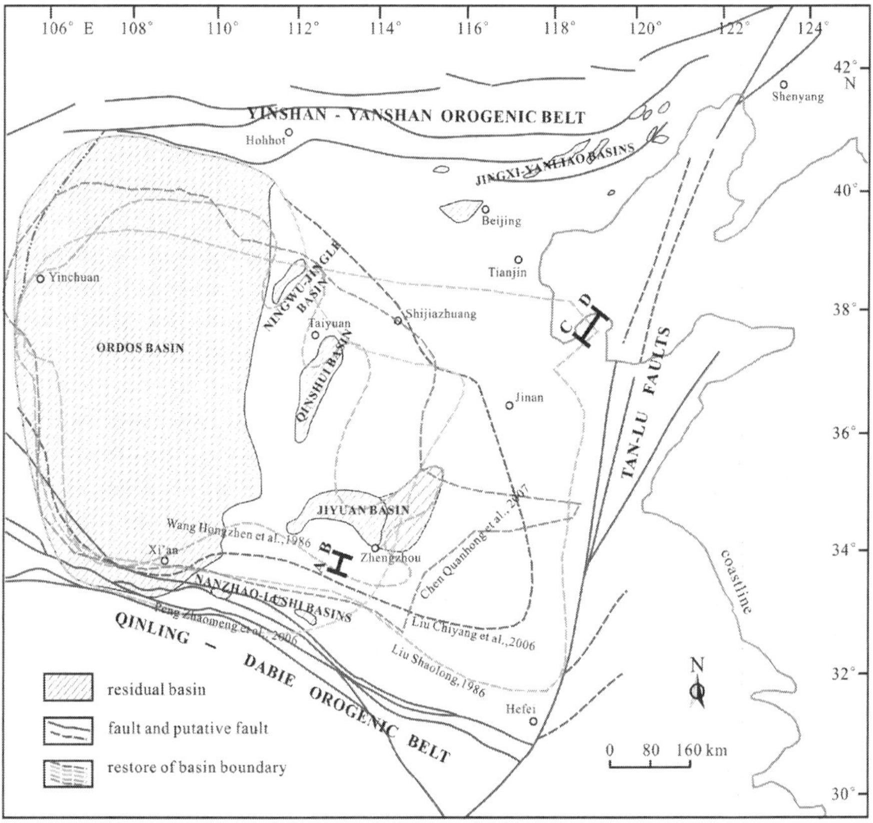

Fig. 2. Distribution map of residual formation of upper Triassic in NCC and its adjacent area.

3.1 Distribution of Late Triassic in Ordos-Middle Shanxi-Western Henan Regions

Upper Triassic in Ordos basin is called Yanchang Formation in general terms (Fig. 3a), which intermittently extends eastward to Qinshui basin of middle Shanxi province (Fig. 3b) and Ningwu-Jingle basin, etc. The approximate thickness of Yanchang Formation is 1300 m, which is a set of terrigenous clastic rocks mainly featured in fluvial and lacustrine deposits: the dominant sedimentary facies of lower part is fluvial and lacustrine with medium-coarse sandstone; the middle part with fluvial-delta-lacustrine; and the upper part with sandstone and mudstone, coals locally developed in the north of Shaanxi province (Fu Jinghua et al., 2005; Deng Xiuqin et al., 2008; Li Wenhou et al., 2009). In the horizontal, the main sedimentary type of Yanchang Formation is fluvial-delta sandstone with significant variations of lithology and lithofacies region. In the north of N38°, the sediments are thinner in thickness and bigger in grain size (100-600 m); while the south part develop in the lacustrine environment with dark mudstone, and sediments are bigger in thickness(1000-1400 m) (Changqing Oilfield, 1992; Yang Minghui et al., 2007; Cao Hongxia et al., 2008). The southwest of Shigouyi-Pingliang-Huanxian region develops fan/fan-deltas with conglomerate or pebbled sandstone; the formation's maximum thickness in area along Shigouyi-Huating is greater than 2,500 m (Liu Shaofeng et al., 1997; Zhao Wenzhi et al., 2006; Zhao Hongge et al., 2007). The outcrops of Yanchang Formation in Ningwu-Jingle basin is about 600 m thick, and is comparatively well developed. The formation in this region develops thick/super-thick layers of gray/green sandstones, and includes thin-layer mudstones or siltstones occasionally. There are parallel unconformities between the formations and purplish-red middle Triassic Ermaying Formation and lower Jurassic Yongdingzhuang Formation.

Upper Triassic in the north western Henan province have been called Chunshuyao Formation and Tanzhuang Formation (Qi Yong'an et al., 1993; Hu Bin et al., 2000; Zhou Xinke et al., 2005), equivalent There are fewer outcrops in the upper Triassic Xingshikou, Laohugou Formations and contemporary strata, but it can be estimated that the late Triassic basin is mainly zonal distribution, that is, the basin is distributed along Shangyi-Pingquan faults on the south side of Luanping and Chengde. To the east, it can extend to Chaoyang region. In the west of Beijing, it is distributed at NEE along the core of basin. The gravel molasse depositing in the two basin groups is the sediment correspondence for the north-side orogen thrust-nappe structure at the same phase, with sediment feature of foreland basin. Similarly, the foreland basin structure is evident in the west segment of south brink of NCC. The foredeep zone is distributed along Shangzhou-Nanzhao in north Qinling, and connects Pingliang Foredeep zone of Ordos basin. to Yanchang Formation in period (Fig. 3c), about 865-1391 m thick, greater than 2500m in Mengxian region southwards. In the areas of Jiyuan and Chengliu, the sedimentary facies of Chunshuyao Formation is fluvial-swamp deposit with yellowish/green medium-fine grained sandstone, siltstone, sandy mudstone and mudstone (Chen Chuanshi et al., 1995), with thickness of 485-567m (see Well Jican 1). Including multilayer coal streak, Tuanzhuang Formation is fluvio-lacustrine deposits with yellowish/grey medium-thick-bedded calcareous siltstone, mudstone, dark mudstone and oil shale, seen in the Well Jican 1 (390 m), Well Deng 5 (Zhang Hongbo et al., 2006). The contemporaneous formations in Luoyang and Yima are called Shifo Formation (Kang Ming, 1988; Yang Shirong, 1994). To the east, the main deposition of upper Triassic in Kaifeng sag is celadon sandstone and mudstone, the top of which develops poor-quality coal seams, coal streak and carbonaceous shale. The fossils combinations are similar with that in the Yanchang Formation. Drilling in Well Tanggu 3 and Well Guangu 2 at Qiuxian depression of Linqing depression (Fig. 3d) indicates that the upper Triassic lithologic combinations is similar to that in the Dongming sag of Kaifeng depression (Wang Congfeng, 1987), and middle-late Triassic fossil assemblages developed in this formation (Liu Shaolong, 1986).

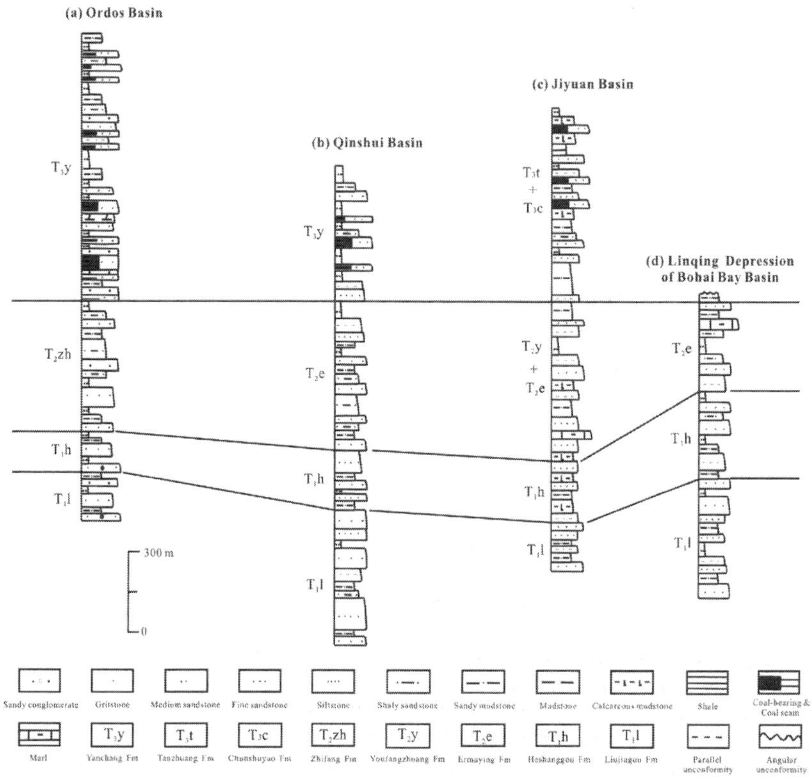

Fig. 3. Sedimentary section of upper Triassic residual formation in intracraton basins of NCC (Section location shown in Fig. 2). a-Ordos basin; b-Qinshui basin; c-Jiyuan basin; d-Linqing depression of Bohai bay basin (Changqing oilfield, 1992; Huabei oilfield, 1988; Bureau of geology and mineral resources of Henan province, 1989; Zhongyuan oilfield, 1993).

3.2 Distribution of late Triassic on the South Edge of NCC

Based mainly on shallow/semi-deep lacustrine or delta-fluvial facies, oil shale inclusions in dark mudstone, the upper Triassic of the southern edge appears along Liuyehe river of Zhouzhi, Mangling of Shangxian, Shuanghuaishu, Tanghe river of Lushi, Mashiping, Yahe river of Nanzhao.

The sedimentary facies of lower upper Triassic in Liuyehe river's west section is grayish/white thick layer of fine quartz-conglomerate, sandy conglomerate, feldspar-quartz sandstone and purple celadon argillaceous-arenaceous slate, 414 m in thickness; the upper part is carbon containing argillites/sandy slate and medium-fine grained quartz sandstone, feldspar-quartz sandstone interbed, and there is a parallel unconformity between the north side of late Triassic and underlying middle carboniferous; the south side is in fault contact with Mesoproterozoic Qinling Group. The lower part of upper Triassic in Mangling includes slate mixed up with feldspar-quartz sandstone, pebbled sandstone, also developed with siderite nodules and partially with inferior coal. The thickness is greater than 370 m; the upper part is slate mixed up with marl, gritstone and siltstone, with a thickness greater than 780 m, unconformable contacting with underlying Paleozoic, whose fossil plants are the important ingredient of Yanchang flora (Bureau of geology and mineral resources of Shaanxi province, 1989; Yin Hongfu, 1996).

As lacustrine-marsh sediments including coal seam, oil shale and carbonaceous mudstone, the lower part of upper Triassic in the east section, Shuanghuaishu-Tanghe river and Yahe river-Mashiping, is grey fine sandstone and dark grey or black mudstone interbed. The upper part is based mainly on grey/brown or yellowish/brown medium-fine grained quartz sandstone and includes fine grained quartz sandstone, black or dark grey mudstone, sandy mudstone. Its sedimentary type is fluvial (Fig. 2). With incomplete outcrops, the formation is gradually thinner from west to east. Its thickness is greater than 1394 m. The thickness of Huangtuling in Mashiping of Nanzhao is greater than 1144 m. eastwards thinning to only 801.5 m in Dashimen of Fangcheng (Jiaozuo Mining Institute, 1982; Bureau of geology and mineral resources of Henan province, 1989; Zhou Xinke et al., 2005).

3.3 Distribution of Late Triassic on the North Edge of NCC

Late Triassic are intermittently distributed throughout Huailai-Zhuolu of northwestern Hebei province, Mt. Xishan of Beijing, Xiabancheng of Chengde-Pingquan in the northeastern Hebei province and Lingyuan-Beipiao in the western Liaoning province (Fig. 4). The formation is called Xingshikou Formation in Huailai-Zhuolu, west Beijing and Chengde; called Dengzhangzi Formation, Laohugou Formation and/or Yangcaogou Formation in Liaoxi; while called Heishanyao Formation in Funings. The lower part is with grayish/brown or grey/purple huge-thick conglomerate and sandy conglomerate. The middle part is with yellow/white or yellow/green pebbled sandstone, sandy shale and thin coal seam; The upper part is with blackish/green silty shale and yellowish/green calcareous fine grained sandstone. And the basal conglomerate has a discordance with middle Triassic in Ermaying Formation or pre-Mesozoic (Yang Nong et al., 1996).

Outcrop of Xingshikou Formation in the west section are distributed in Yaozitou, Chaimulang, Jimingyi, etc. in the northern Hebei province, and is gradually thinning from east to west. Yaozitou is 67.33 m thick while 19.18 m thick in Shuiyaogou and 9.4 m thick in Chaimulang. Further more, the western formation in Jimingyi is only 3.85 m in the thickness, even thinning out. But in the west of Xiahuayuan, this formation outcrops appear again around Xiajiagou and Wujiagou coal-mine of Zhuolu, and is 7-29.3 m thick (Tian Lifu et al., 1996; Wang Lifeng et al., 1998). Becoming thicker upward, Xingshikou Formation is composed of multiple sedimentary cycles consisting of breccia and silty mudstone. With overlying volcanic rocks in early Jurassic of Nandaling Formation, there is an unconformable contact between this formation and underlying Cambrian, which is gradually transiting to Neoproterozoic from east to west.

The middle segment are distributed in west Beijing and Chengde region, with the thickness of 31-610 m, the lithology of which is grey/black or khaki shale, silty shale, grey black

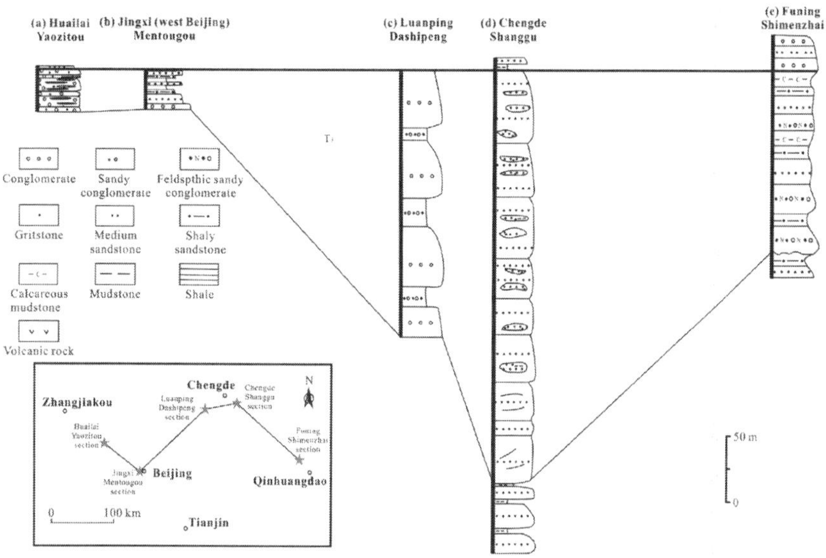

Fig. 4. Profile of late Triassic on the north edge of NCC.

siltstone, sandstone, conglomerate and coal streak, and the sedimentary facies is fluvial (Zhang Jingfang, 2002). The lithology of Dashipeng region in Luanping is yellowish/brown huge-thick conglomerate with small amount of dust shale, carbonaceous shale, silty shale and includes coal streak. At the same time, the lithology of Wangyingzi Formation is celadon/green gravel-bearing tuff and limestone. And the lower part is black/grey tuffaceous polymictic conglomerate. Transiting to medium-fine conglomerate upward, the upper part of Xiabancheng-Weichanggou in Pingquan is greyish/brown or grey/purple huge-thick conglomerate. The middle part is white-yellow or yellowish/green gravel-bearing arkose, silty shale. The upper part is yellowish/green silty shale, including yellowish-brown calcareous fine-grained sandstone (Liu Shaofeng et al., 2004; Xu Gang et al., 2006).

Laohugou Formation of east section Niuyingzi basin of Lingyuan in western Liaoning province is yellowish/green siltstone, shale mixed with yellowish/gray medium-coarse grained arkose, gravel-bearing arkose, sandy-cemented fine conglomerate and silty shale. The main lithology of Dengzhangzi Formation is grey thick-layer limestone and dolomitic conglomerate (Hu Jianming et al., 2005). Both of them are continuous transitional sedimentary formation (Xu Gang et al., 2003). Laohugou Formation of Jinglingsi-Yangshan basin in Beipiao partially outcrops in the southwestern edge. It is composed of purplish/red or variegation sandy conglomerate and a few fluvial and lacustrine sandstones (Yan Yi et al., 2003). The contemporaneous formation of Changheying in Beipiao is called Yangcaogou Formation (Wang Xin et al., 2009).

In addition, late Triassic Heishanyao Formation in Funing area of Qinghuangdao is consisted of four sedimentary cycles formed by yellowish/gray or yellowish/green gravel-bearing gritstone, fine siltstone and black carbonaceous shale. And its sedimentary facies are delta and lacustrine sediment with thickness of 161.1 m (Liu Chengzhi et al., 2006).

4. Regional structural deformation of NCC in late Triassic

Previous studies are carried out on the tectonic deformation in Indosinian of NCC from different points of views (Huang Jiqing et al., 1977; Ren Jishun et al, 1980; 1990; Cui Shengqin et al., 1983). In late Triassic, large-scale extrusion occurred in the south and north of NCC, which agrees with the EW tectonic line develops in interior craton, and generally shows a tectonic pattern of "ramp" (Zheng Yadong et al., 1990; Zhang Guowei et al., 1995). In interior craton area: (1) the western Ordos area is stable, with slight deformation, and relatively complete residual formation; (2) the eastern area of NCC, with intensive deformation and most of upper Triassic; (3) a major concern is that there exists a transitional belt between the upper Triassic remnant area of Ordos and denudation area in the contemporaneous formation of the eastern North China, but it is feasible to conjecture the deformation and existence of Indosinian movement basing on the "lack" of sedimentary formation or the distribution of late Triassic residual formations and denudation areas as well as the tectonic deformation.

4.1 Structural deformation of interior NCC in late Triassic

On the basis of regional tectonic background studies, the N-S collision of late Triassic Yangtze plate and North China plate is in "scissors style", hence NCC is located in a approximate N-S (NNE) compression stress field, the east area began to uplift primarily and extends westward gradually. So the eastern area raised earlier and more intensive while western area raised later and slight. Dabieshan area collided during the end of early-middle Triassic; collision in Sanmenxia area was during the end of middle-late Triassic while the west Qinling collision occurred in the late late-Triassic, therefore Indosinian movement controls the tectonic and sedimentary framework of the south edge of NCC (Xia Bangdong et al., 1996; Wan Tianfeng et al., 2002).

Research on the contact relationship between lower-middle Triassic and overlying formation make it clear that the relationship among Ordos, Qinshui and Jiyuan basins and upper Triassic is conformable or parallel unconformable contact; eastern area is angular unconformity with Jurassic or younger formation, and lower-middle Triassic is eroded in different degree. Jiyang sag lacks Triassic or older formation (Zhu Yanming et al., 2001; Ji Youliang et al., 2006).

Based on paleo-tectonic reconstruction, NNE-trending uplift and western depression appeared in middle Triassic NCC without upper Triassic in most areas of the east, but synchronous deposition began to develop in western Ordos, Qinshui and Jiyuan basins (Liu Chiyang et al., 1987, 2006; Zhao Zhongyuan et al., 1992). The main part of interior craton deformation area in Ordos is Tongchuan sag after uplift (Fig. 5), which was formed after the uplift. Its north and south sedimentary boundary are limited by northern Qinling and Yinshan separately, and the lake-basin is open toward southeast. The late Triassic Yanchang Formation of Qinshui basin is thinner than areas such as Yan'an and Yanchang in the west; marginal facies in upper Triassic of Jiyuan basin does not exist. Yanchang Formation and contemporaneous formations are conjectured to distribute along Datong-Shijiazhuang-Jinan-Zhoukou (Fig. 2), which is beyond the current basin boundary (Liu Shaolong, 1986;

Zhao Zhongyuan et al., 1990, 1992; Liu Chiyang et al., 2005, 2007). Such basin pattern is consistent with the research conclusion for ancient compression stress field in Ordos basin, i.e. the NNE-SSW extrusion in Indosinian (Zhang Hong et al., 1996; Xu Liming et al., 2006).

As the extension part of southeast area of Ordos basin, NNW-trending Yellow river fault in Luoyang basin of western Henan province moved during early late Triassic and formed Daimeizhai anticline (underwater uplift of Carboniferous-Permian; Chen Chuanshi et al., 1995), the area of Mt. Song rise gradually above the water-level while the southeast area is paleo-Funiu uplift, the water between which is deep and connected with the water area of Yichuan-Linru-Dengfeng. In the last late Triassic, the sedimentary area retreated west-northward to Mianchi-Yima-Yiyang-Yichuan and was parallel to paleo-Funiu uplift (Zheng Qiugen et al., 1998). The western uplift of Luoyang basin is lack of upper Triassic; the upper Triassic of eastern sag was eroded, and some of the remains are located in the leading edge of Xiashi-Shimen-Wenquanjie thrust nappe, as well as the footwall of Xiashi-Chenzhaigou thrust fault and some of the foreland areas of Xiashi-Shimen- Wenquanjie, Yichuan basin (Fig. 6a).

The western and north edges are lack of upper Triassic; the thickness of intra-basin thicken from northeast to southwest, lacustrine deposits range mostly between 1000-1400 m thick (Changqing oilfield, 1992). The eroded stratum thickness shows the top of Yanchang

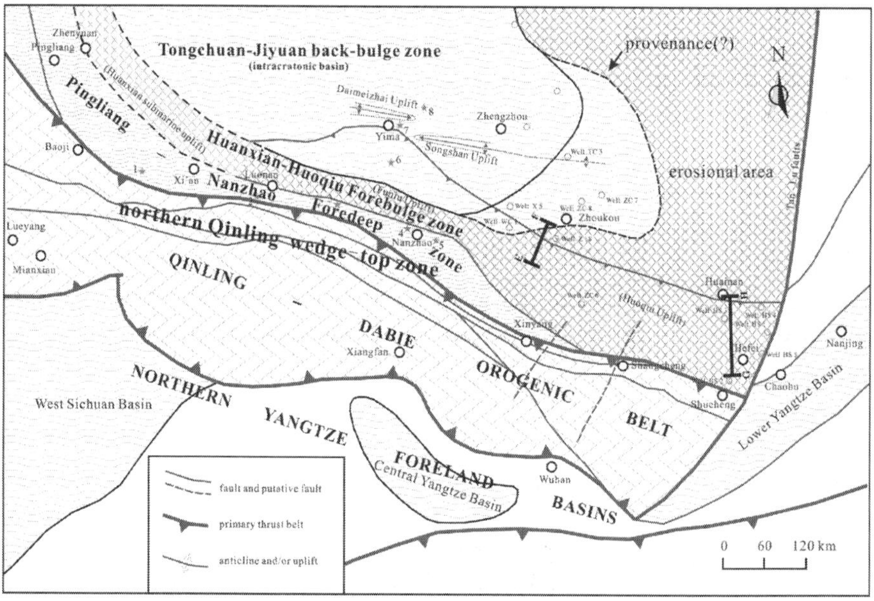

Fig. 5. Regional structure outline map of Indosinian in central-southern part of NCC (Northern Yangtze Plate data is from Xia Bangdong, 1996; Liu Shaofeng et al., 1999; Xu Hanlin et al., 2001; Xu Zhengyu et al., 2004).

Green five-pointed star stands for outcrops: 1-Liuyehe, Zhouzhi; 2-Mangling, Shangxian County; 3-Shuanghuaishu, Tanghe river, Lushi; 4-Mashiping, Nanzhao; 5-Yahe river, Nanzhao; 6-Shimen, Songxian County; 7-Miaoyuan, Yima; 8-Chengliu, Jiyuan; Green circles indicate drilling wells.

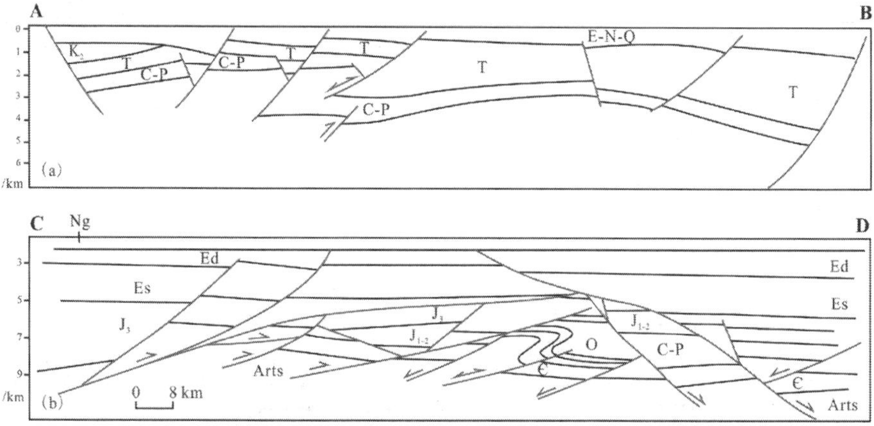

(a) C-P-Carboniferous-Permian; T-Triassic; K_2-Upper Cretaceous; E-N-Q-Paleogene-Neogene-Quaternary
(b) Arts—Archean Taishan Group; €-Cambrian; O-Ordovician; C-P-Carboniferous-Permian;
J_{1-2}-Lower-Middle Jurassic Fangzi Fm; J_3-Upper Jurassic Mengyin Fm; Es-Paleogene Shahejie Fm;
Ed-Paleogene Dongying Fm; Ng-Neogene Guantao Fm

Fig. 6. Profile of overthrust faults in intra-craton basins of NCC (location shown in Fig. 2, 5).

Formation eroded not too much, while Qingyang-Zhenyuan-Huanxian eroded by about 400 m and the east of Jingbian eroded less than 100 m (Chen Ruiyin et al., 2006). Upper Triassic in Qinshui basin is 400-500 m thick; upper Triassic of western Henan province was confined in the area of Jiyuan-Luoyang-Yiyang-Dengfeng and in the north of western Kaifeng depression (Bureau of geology and mineral resources of Henan province, 1989; Xu Hanlin et al., 2004). The thicknesses of Jiyuan and Yima are approximately 1700 m. Upper Triassic of Ordos basin thicken from north to south and depositional grain size become finer, but Upper Triassic thinning from north to south in the western Henan province due to Shigouyi and Huating in the south edge of Ordos is the foredeep of foreland basin, while the latter is depression behind of its uplift (Fig. 5).

NCC lack upper Triassic in the eastern area of Datong-Shijiazhuang-Jinan-Zhoukou at present. Although the tectonic deformation during. Indosinian was reformed intensively by Yanshan and Himalayan movements, but their traces still can be found in the EW-trending fold-thrust belts in areas such as Bohai bay basin, Luxi and Xuhuai area. The axis-track of fold in Bohai bay basin was E-W or NEE-trending, and it changed into NNE-trending at the turning in the west of Liaocheng-Lankao fault. For example, the structure of Huanghua depression in Indosinian changed into NEE-trending in the Yanshan orogenic belt (Jin Chong et al., 2007). Western Jiyang depression, Dawangzhuang, Chengbei and Guxi fault zones, as well as Yihezhuang uplift, Chengdong anticline, eastern part of Chezhen sag appear angular unconformity between the low-middle Jurassic and the Palaeozoic (Zong Guohong et al, 1998); Well Zhuanggu 13 shows overturned syncline of Cambrian-Ordovician in the underlying Jurassic, of which repeated duplication takes place in Yeli-Liangjiashan Formations. The low-middle Jurassic of Huanghua depression formed an approximate EW-NWW-trending fold-thrust belt (Fig. 6b), the Jianhe syncline, Kongdian anticline and Nandagang-Xuyangqiao synclines developed from north to south. Repetition of Carboniferous-Permian can be seen in Well Konggu 5, repetitions of 22.5 m and 54 m in Ordovician are respectively at Well Gang 59 and Well Tai 10. The lower-middle Jurassic unconformity of Well Dengcan 1, Well Donggu 1, Wen'an slope of Jizhong depression and Dacheng uplift covers lower-middle Triassic or older formations (Yang Minghui et al., 2005). Wells Wen 5, 10 and 30 in Wuqing sag show the Carboniferous-Triassic repetition and which are covered by lower-middle Jurassic. The seismic sections cross Well Chengbei 20 in Bohai area and its drillings and Well H8 of Liaoxi low-uplift and Bozhong sag respectively indicate that lower-middle Jurassic appearing as angular unconformity and overlaps of Carboniferous-Permian, Cambrian-Ordovician (Liu Le et al., 2009; Lu Dingyou et al., 2009) and pre-Cambrian metamorphic granite. Indosinian movement appears as lower-middle Jurassic Fangzi Formation with unconformity overlaps of early-middle Triassic or early Paleozoic (Yu Fusheng et al., 2002).

4.2 Structural deformation in the south edge of NCC in late Triassic

The foredeep in the south edge of NCC, develop alluvial-fan, delta and deep lacustrine sediments, distributes in Shangzhou-Nanzhao of north Qinling, and connects with Pingliang foredeep of Ordos basin in the west; fore-bulge depozone is composed of Huanxian-Huoqiu basement uplift, back-bulge depozone distributes inside intra-craton basin, and subsidence center is located in Jiyuan-Tongchuan region, in which deeper lacustrine-delta develops. Therefore, the south edge structure of NCC has relative complete "foreland basin system". From south to north, it develops north Qinling wedge-top zone, Pingliang-Nanzhao foredeep zone, Huanxian-Huoqiu fore-bulge zone and Tongchuan-Jiyuan back-bulge zone which forms after uplift (Liu Shaofeng et al., 1999; Chen Shiyue et al., 2000; Yang Minghui et al., 2007, 2009).

In the late Triassic, east section of south edge in NCC develops fold-thrust belt in the north and northern Qinling orogenic belt in the south, bounded by Luonan-Luanchuan thrusting belt. During the movement of south edge fold-thrust belt in NCC, the foredeep zone and even fore-bulge zone are involved in the thrust deformation. Frontal fault belt includes Luanchuan-Gushi-Feizhong faults and Sanmenxia-Lushan-Fuyang-Huainan faults, and ends at Tanlu faults to the east. The frontal fault belt is about 1000 km long. Along the trending, tectonic segmentation features is stronger in the east segment and weaker in the west segment (Xie Dongning et al., 2006).

According to the research by Guo Xi'nian et al. (1991) and Shi Quanzeng et al. (1990, 2004), the evidences of fold-thrust belt development in western Henan area during the late Triassic can be found including following: (1) fault-belt generally offset Carboniferous-Permian and Triassic, and lateral contemporaneous strata developed a series of miniature NW-trending folds and reverse faults, which are rare in Jurassic; (2) fault plane takes on a group of scratchs which are basically consistent with fault dip; (3) without development of tensile crack; (4) early/middle Yanshanian granite cuts through local sections, and the angular unconformity of lower Jurassic Yima Formation in Jiyuan and Yima region covers on the upper Triassic Tanzhuang Formation. Lower Paleozoic in the Shanxian-Mianchi coalfield override the Carboniferous-Permian strata and is covered by Jurassic unconformity, . The Mesozoic and Paleozoic in the south part of Yichuan basin override the Triassic may be the result of Indosinian movement (Yu Hezhong et al., 2006). Indosinian thrust and uplift-erosion lead southward area of Sanmenxia-Fuyang-Huainan faults lacking of Triassic, only with local upper Triassic confined in the Nanzhao foredeep zone (proto-type of foreland basin system), and its east part lacks late Triassic; the west stratum develops but is subject to late-period formation. Separatrix between them is Xiayi-Woyang-Macheng faults which parallels to Tanlu faults (Yang Minghui et al., 2009).

The thrust strength of Zhoukou depression in the eastern Henan slightly increases and shows in the tectonic pattern of fold-faults (Fig. 7a); thrusting effects get continuously intensive from the eastern part to Huainan region. Archean Wuhe Group in the south edge of Huainan coal basin, Bagongshan Group and Xuhuai Group in Proterozoic, Paleozoic and others thrust and superimposition from south to north to form the imbricate structure (Fig. 7b). In the south, Hefei basin is located in the foreland of north edge in Dabie orogenic belt, and its basement deformation is characterized by a group of south-inclined thrust faults. For example, nearly E-W-trending Shouxian-Dingyuan, Feizhong, Liu'an, Xinyang-Shucheng faults and so on, which were involved into Permian and previous formations. After the levelling, contact relationship between Indosinian movement of Hefei basin and overlying Jurassic is of angular unconformity (Xu Chuanhua et al., 2002). To the south of Liu'an fault, most of the basement is involved into tectonic deformation, while structure of sedimentary cover is dominant to the north (Zhou Jin'gao et al., 1999; Zhao Zongju et al., 2000). In space, the activity is intensive in the south and the slip rate of Liu'an fault is 225 m/Ma, which gradually gets weakened northwards (Xu Shihong et al., 2007). The restructure of balanced section shows that the shortening for crust is about 117.8 km (Zhao Zongju et al., 2001).

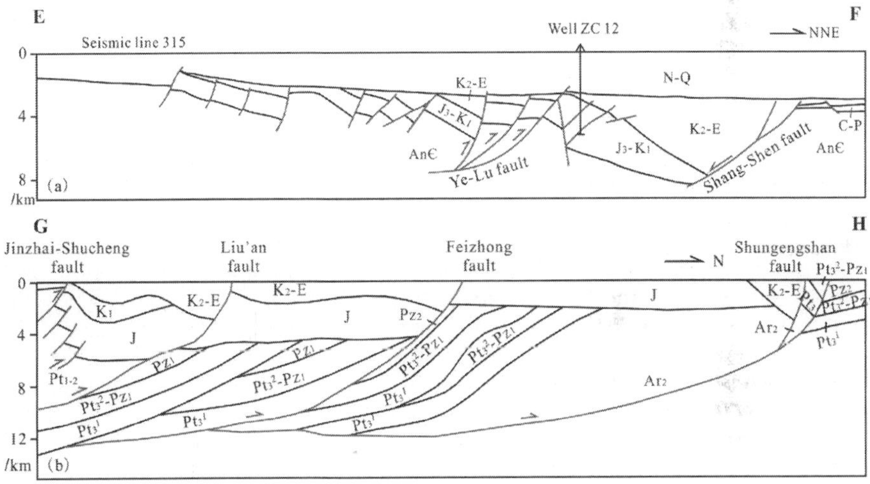

(a) AnЄ-Pre-Cambrian; C-P-Carboniferous-Permian; J_3-K_1-Upper Jurassic-Lower Cretaceous; K_2-E-Upper Cretaceous-Paleogene; N-Q-Neogene-Quaternary;
(b) Ar_2-Mesoarchean; Pt_{1-2}-Paleo-Mesoproterozoic; Pt_3^1-Qingbaikou; Pt_3^2-Pz_1-Zhengdan-Lower Paleozoic; Pz_2-Upper Paleozoic; J-Jurassic; K_1-Lower Cretaceous; K_2-E-Upper Cretaceous-Paleogene

Fig. 7. Profile of overthrusts in South Edge of NCC (location as shown in Fig. 5).

4.3 Structural deformation of north edge of NCC in late Trlassic

The north edge of NCC originates from the Yinshan in the west and passes through Yanshan to the mountainous land of western Liaoning province, with total length of 1400 km from east to west. It is approximately located in a complex tectonic belt at N40°-42° from south to north, with basin-and-range terrain and migrant transformation (Wang Guiliang et al., 1999). Mesozoic Yanshan-western Liaoning province is under intra-continental environment. Its orogenic process during Indosinian last from the middle-late Triassic to

Pre-Jurassic. Pre-Mesozoic NCC put together with the Mongolian block along Soren suture so as to form a typical intra-continental orogenic belt based on craton (Cui Shengqin et al., 2000; Zheng Yadong et al., 2000). In other words, basin deformation in Yanshan and its adjacent regions during Indosinian may show characteristics of intra-continental foreland basin. The late Triassic molasse sediment, distributed on the south side of Luanping-Chengde, develops under Shangyi-Pingquan fault and is subject to control by north-south thrust fault (Zhao Yue et al., 1990; Davis et al., 2001). Basin sediment records the uplift and erosion processes on the north side of this faults in Paleozic, Proterozoic and Archean. The molasse with synclinorium distribution along the west of Beijing also related to strong thrust-uplift of basin edge (Liu Shaofeng et al., 2004). After the E-W-trending tectonic belt was overlaid by NNE tectonic belt, present tectonic framework comes into being.

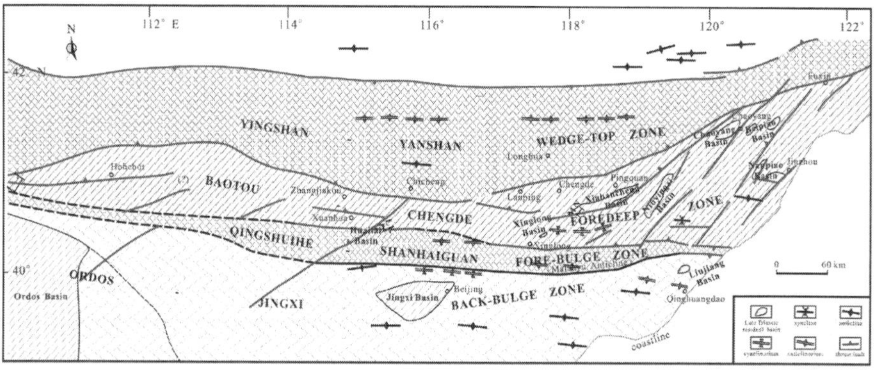

Fig. 8. Tectonic outline map of north edge of NCC in late Triassic.

Indosinian deformation in the north edge of NCC can be traced to Mt. Daqing in Inner Mongolia (in the north of Ordos basin) and China-Mongolia border region in the west, which appears as a S-N compression and forms a series of complex folds and associated overthrusting nappes which are distributed in the E-W direction. Based on the research by Zheng Yadong et al. (1990, 2005), the north nappe in napping tectonics of China-Mongolia border is composed of Neoproterozoic and Mesoproterozoic dolomitic limestone mixed with quartz sandstone. South nappe is involved in tillite of Sinian and black siliceous slate in early Cambrian, with overall napping directs at 180° (Fig. 9a). The wide and gentle synclines are composed of coal series of Wudanggou Fromation in Jurassic under Shiguai basin of Mt. Daqing, and its east part is reformed by Yanshanian nappe structure and shows as rock-slices cover upper Jurassic Daqingshan Formation. Regional survey implies that Baihugou and Sharqin region in the south part of basin develop grayish/white basal conglomerate with middle-thick base-layers in Wudanggou Formation covering lower Triassic-Paleozoic with angular unconformity, as well as thrusts associated with folds. The length of such unconformity is 4 km and inclines northwards. Wudanggou Formation in Baotou city covers lower Triassic Miaogou Formation with angular unconformity of Miaogou region, where occur weathering crust and basal conglomerate. This illustrates that this region underwent late Triassic uplift erosion and subsidence and deposited again in early Jurassic (Liu Zhenghong et al., 2003).

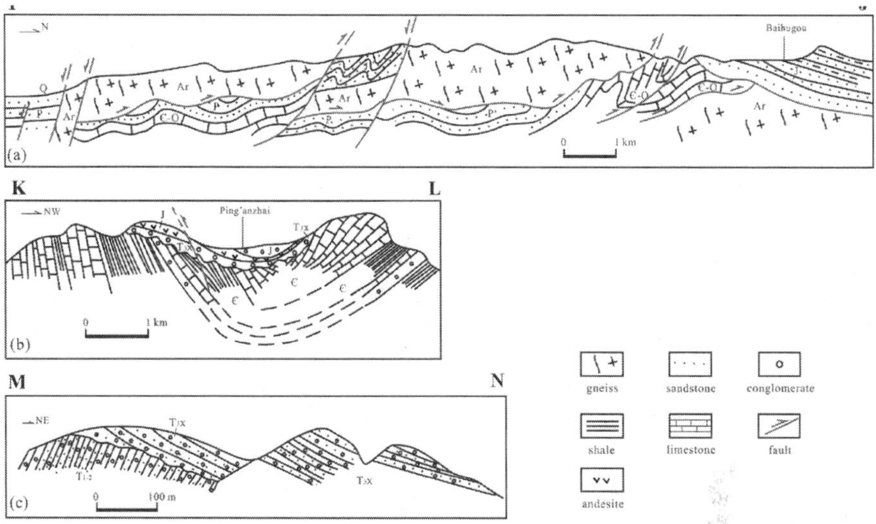

(a) Ar-Archean gneisses; €-O-Cambrian and Ordovician; P-Permian; J_{1-2w}-Wudanggou Fm; Q-Quaternary (Daqingshan area from Liu Zhenghong et al., 2003)
(b) Pt_2-Jixian; Pt_3-Qingbaikou; €-Cambrian; T_3x-late Triassic Xingshikou Fm; J-Jurassic (Huailai basin from Tian Lifu et al., 1996)
(c) T_{1-2}- lower and middle Triassic; T3x-late Triassic Xingshikou Fm (Chengde area from Wang Yu, 1996)

Fig. 9. Profile of thrust-nappe structure in north edge of NCC (location as shown in Fig. 8).

Beijing-Chengde and western Liaoning, located in east part of north edge of NCC, develop upper Triassic Xingshikou Formation, Laohugou Formation and so on. This indicated that there existed the difference of ancient geological environment from east to west, and the tectonic movement of the west part was earlier and stronger than that in the east part. To the west of Beijing, Xingshikou Formation is only distributed on the west side of Shanghuangqi-Wulonggou faults and Huairou-Laishui faults. To the east of Beijing develop Xingshikou Formation and its underlying lower-middle Triassic sediment. In the early Indosinian, Xiahuayuan region developed EW-trending Guchengliang fault (Wang Shide, 1987), to the east of which Xingshikou Formation is distributed in the east of Jimingyi, with thickness increasing westwards, even almost disappear when close to erosion area (Wang Lifeng et al., 1998). The Ping'anzhai syncline core of E-W-trending consist of Xingshikou Formation, early Jurassic Nandaling Formation and Xiahuayuan Formation, while the flanks consist of Jixian, Qingbaikou and Cambrian, with northern flank gentle and southern one steep (Zhang Yong, 2006). According to the studies by Xiao Zongzheng et al. (1995), Indosinian movement in Beijing shows as E-W fold-fault, such as Gujishan anticline and Huiyu syncline, etc. Based on stratigraphic sequence and contact relationship, the huge and thick quartz conglomerate develop on the bottom of Xingshikou Formation (Baozhudong Conglomerate) and overlaps above different strata in different area. Its underlying formation gradually decreases from Badachu, Mt. Xishan in Beijing to Baoershui, with contact relationship is micro-angular unconformity or disconformity. Lithofacies belt, stratigraphic isopach and coal streak in Yaopo Formation of Jiulongshan syncline in north and east Hebei are E-W or nearly E-W-trending, evidently, controlled by Indosinian E-W folds (Chen Ruiqi, 1982).

The unconformity contact on the bottom of Xingshikou Formation was formed during Indosinian in Chengde, where develop later Triassic small basins along fold-thrust belt under uplift background (Fig. 9b). Upper Triassic deposited in the synclines area based on base deformation feature. Sedimentary construction shows the lower part of upper Triassic is alluvial and fluvial sediments, which play a role of basin-filling. The gravels of Xingshikou Formation in Dashipeng region are very well sorted and rounded. They are the products of volcanic activity in the vicinity region (Zhang Jinfang, 2002). Li Zhong et al. (2003) find the provenance comes from orogenic belt on the north side, and paleocurrent direction of conglomerate in Xingshikou Formation is at 168° as the middle part of Yanshan is higher in the north and lower in the south.

To the east, late Triassic small basins develop in west Liaoning, such as Chaoyang, Jianchang and Beipiao basins, etc. Due to strong tectonic uplift and extrusion in the later Triassic, the angular unconformity between lower Jurassic Beipiao Formation and upper Triassic Laohugou Formation occur in Chaoyang basin (Zhang Guoren, 2006). Late Triassic basin in Laohugou region of Lingyuan disappears, sedimentary hiatus appears between lower Jurassic and upper Triassic (Ma Yinsheng, 2001). The NE fault belt with NW incline in Niuyingzi basin in Lingyuan, which controls the sediment during late Triassic Dengzhangzi Formation deposition (Fig. 9c). Some faults are covered by middle Jurassic Guojiadian Formation. The squeezing lenticle and scratches indicate the thrusts from NW to SE (Xu Gang et al., 2003). K-Ar isotopic dating shows the age of consequent layer diabase prophyrite intruding into Mesoproterozoic-Neoproterozoic is 243.4-199.4 Ma, in Indosinian period (Wang Genhou et al., 2001).

5. Proto-basin type of North China basin in late Triassic: Discussion on intra-craton basin edged by foreland basin of South and North Brinks

In early-midle Triassic, NCC still remains as a unified basin. In late Triassic, Yangtze plate collided with North China plate (Xia Bangdong et al., 1996). South edge of NCC was extruded nearly S-N, and ancient Qinling ocean closed. Then QDOB formed. In the north edge, late Paleozoic NCC collided with Siberia plate (Xu Bei et al., 1997). Therefore, between Yangtze plate and Siberia plate, the south edge of late Triassic NCC was actively extruded by Yangtze plate, and the north edge was relatively fixed limited boundary. On the both sides, thrust-nappe structure and frontal intra-continent foreland basin develop. Since late Triassic, the east part of hinterland started uplift. The east part of NCC lacks upper Triassic and suffered from serious erosion, and strata under laid by Jurassic became older and older from west to east. Contemporaneous sedimentary basin retreats southwestwards until turn back into Ordos basin, but still keep stead state as Paleozoic-early Mesozoic intra-craton basin (Fig. 10) (Liu Chiyang et al., 2006).

5.1 Characteristics of intra-continental foreland basin in the north edge of NCC

YYOB extends to the westward region of Baotou from north to west, with the length of is 1100 km. The tectonic line of middle-west segments in this belt strikes dominantly nearly E-W, while the tectonic line of eastern segments in this belt features NNE-NE. And the west part of south edge of YYOB is Mesozoic Ordos basin which is stable, and the east part is covered by late Mesozoic-Cenozoic Bohai bay basin. Mt. Yinshan in the north edge of NCC exists during Indosinian. But the later tectonic movement, contemporaneous foreland basin disappeared gradually because of deformation. Residual intra-continental foreland basin is mainly distributed in west Beijing, Chengde and west Liaoning, called as "west Beijing-Chengde basin" and "Liaoxi basin" (Wang Guidong et al., 1992; Li Zhong et al., 2003; Ma Yinsheng et al., 2003; Liu Shaofeng et al., 2004).

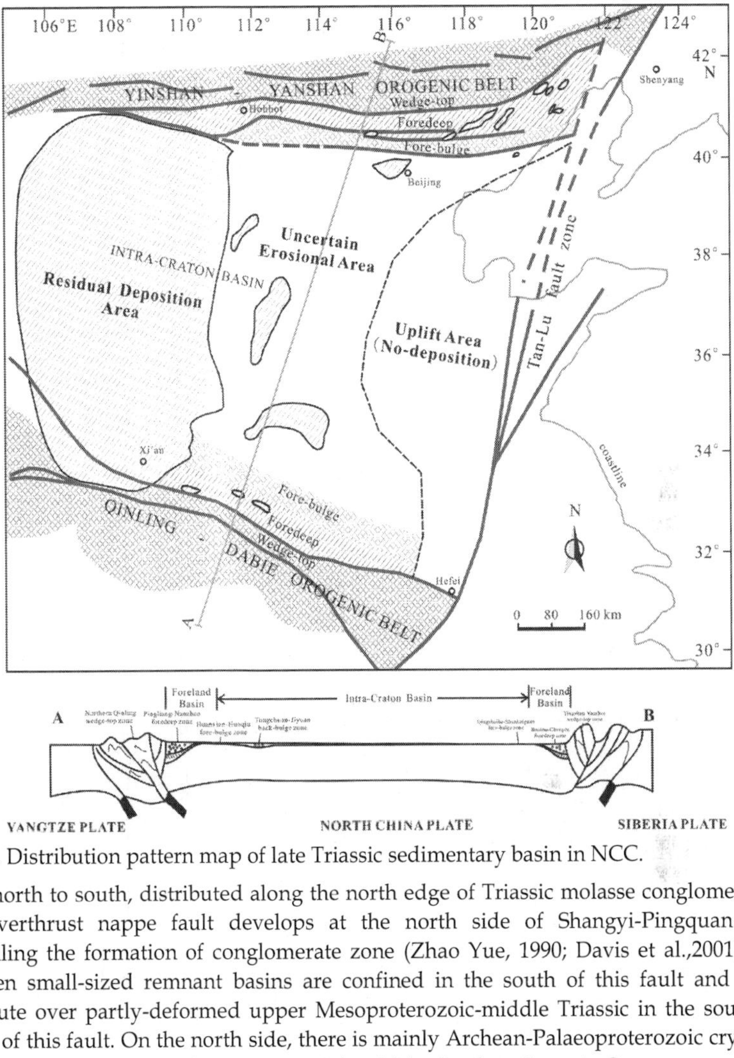

Fig. 10. Distribution pattern map of late Triassic sedimentary basin in NCC.

From north to south, distributed along the north edge of Triassic molasse conglomerate, an N-S overthrust nappe fault develops at the north side of Shangyi-Pingquan faults, controlling the formation of conglomerate zone (Zhao Yue, 1990; Davis et al.,2001). More than ten small-sized remnant basins are confined in the south of this fault and mainly distribute over partly-deformed upper Mesoproterozoic-middle Triassic in the southward region of this fault. On the north side, there is mainly Archean-Palaeoproterozoic crystalline rock. Later, both two sides are covered widely by late Jurassic-Cretaceous volcanic-sedimentary rock. Shangyi-Pingquan faults links with Lingyuan-Donggongyingzi thrust fault in west Liaoning, which is mainly manifested as Archean or Proterozoic thrust above Mesozoic.

Upper Triassic Xingshikou Formation is mainly distributed along the core of west Beijing-Chengde basin and extends westwards to the region around Chaoyang, west Liaoning. On the cross section, basin filling, in wedge shape, shows that subsidence center of the basin is located on north side and sedimentary thickness and coarse clastic contents at north side or northwest side are more than those on the sourth side or southeast side. With slight directional alignment, Xingshikou Formation of Shanggu region in the core of Chengde basin is a set of thick-layer conglomerates. According to the measurement for flat surface of imbricate conglomerates, paleocurrent direction is mainly southward and partly northward. With the sedimentary type of gravelly braided channel, this set of conglomerates is basin margin facies. And Xingshikou Formation conglomerates are slightly rounded and directional in west Beijing basin. Direction of paleocurrent may be from east to west according to few measurements of oblique bedding. Based on the analysis on fragmentary composition and depositional filling polarity characteristics of basin, associated with the formation of nearly EW dispersed linear basin groups, from Mt. Yanshan and Liaoning to north Yinshan region, "Inner Mongolia axis of earth" was ever strong uplifting area, with parallel trend to it. And it becomes the main source of contemporaneous sediments. At the same time, according to the analysis of early-middle Jurassic coal-bearing strata and reverse profile characteristics of late Jurassic detrital composition, there may exist early-middle Jurassic (late Triassic) strata with certain scale before the uplift of late Jurassic on "Inner Mongolia axis of earth", which also was denudated in the strong uplift process of late Jurassic (He Zhengjun et al., 1999). As a whole, Xingshikou Formation is the sedimentary response of strong thrust uplifting in provenance area (Liu Shaofeng et al., 2004).

Most of researchers have accepted that the north part of NCC developed a series of thrust nappe structure during Mesozoic. But the formation time and activity intensity have remained to hang in doubt. As mentioned above, there is coupling relationship in time & space between these thrust nappe structures and intra-continental foreland basin developing E-W in late Triassic. There is obvious genetic relationship between both. According to geological scale, the sedimentary accumulation including lots of coarse clastic constituents have the nature of syngenesis generally. And there are some corresponding relations between conglomerate layers and episodic tectonic phase (Jordan et al., 1988; Burbank et al., 1988; Chen Haihong et al., 1992). In late Triassic, because of ever-increasing thrust loading, the lithosphere in frontal edge of thrust nappe structure presented flexural subsidence, and then formed linear depression basin groups. Therefore, the polarity filling characteristics and huge thick coarse accumulation of late Triassic basin in the north part of NCC responded to the strong nappe-uplift event during the same period.

5.2 The characteristics of intra-continental foreland basin in the south edge of NCC

From the perspective of regional tectonics of east Asia, NCC, Yangtze plate and south Qinling micro-plate are all miniature plates between Euroasian plate and Gondwana Land. In Hercynian, inconsistencies of paleomagnetic polar-wander curve existed among them, but they moved to the north as a whole. During late Triassic-early Jurassic, polar-wander curve tended to be uniform and indicated that they combine together. Most importantly, Yangtze plate and NCC presented large-angle horizontal-rotation that Yangtze plate rotated clockwise and NCC rotated anticlockwise (Wu Hanning et al., 1990; 1992; Liu Yuyan et al., 1993). In terms of the geological evolution, Qinling ocean closed in the late Triassic; NCC spliced with Yangtze plate; Collisional orogeny of Qinling-Dabie comprehensively launched (Zhang Guowei et al., 2001) and ushered in intra-continental orogenic stage (Sengor, 1985; Wang Qingchen, 1989; Li Shuguang et al.,1998). In time & space sequence, the development and evolution of foreland basin in the south edge of NCC and QDOB are very identical. Because of "scissors" collision effect among plates, the foreland basins in south edge of NCC have the characteristic of east-west differentiation.

5.2.1 Foreland basin system in the south edge of Ordos basin

Bridging over the current Weihe garben and north Qinling region, the south edge of Ordos basin can reach the north side of Shangdan suture, with a large sedimentation range. The ancient geographical landscape of Ordos basin was higher in the north and lower in the south, the water-body was shallower in the north and deeper in the south, and deposition was thinner in the north and thicker in the south. All of these asymmetric depression properties show that the development and extension of sedimentary facies belt of south edge of Ordos basin is controlled by the compression and uplift of Qinling orogenic belt. The lacustrine sediment opens westwards and spreads nearly EW-NWW, with the same extension direction of Qinling orogenic belt. The research shows that the foreland basin system in the south edge of Ordos basin is still well preserved nowadays (Liu Shaofeng et al., 1999; Chen Shiyue et al., 2000; Yang Minghui et al., 2007).

The thickness of foredeep zone stratum of south edge foreland basin system can be over 3000 m. In the late Triassic, because of the steep terrain, adjacency to the lake basin, short distance for sediment transport and rapid accumulation, the foreland of overthrust structure of Qinling orogenic belt crossed the foredeep to Mt. Kongtong and other regions, and developed such regions into braided -delta such as Ruishui River-Zhenyuan, the edge of the basin developed into small-scale alluvial fans such as Kongtong Mountain-Mawugou. Conglomerate in Mt. Kongtong comes from Sandaogou in Pingliang, and belongs to marginal facies (Zhao Wenzhi et al., 2006). Braided-delta deposit mainly contains gravelly sand and gritstone, and rock debris content decreases northeastwards. Rock debris which reflects the provenance is from the granite, gneiss and carbonate distribution area in Qinling orogenic belt in the southwest part.

The middle-south part of Ordos basin is a depression belt spreading northwestwards, which is steeper in the south and gentler in the north, and steeper in the south and shallower in the north, i.e., Tongchuan-Jiyuan back-bulge zone. The thickness of Yanchang group is generally more than 1100 m; and the thickness of the south part is more than 1400 m. This back-bulge zone is also the main part of intra-craton basin, with gentle northeast flank, steep southwest flank and opening eastwards. The north edge of basin can reach northwards of Hangjinqi-Dongsheng region, extend eastwards to the west Henan and interior NCC and end in the west Mt. Taihangshan. The isopach map of late Triassic prepared by Zhao Hongge et al. (2007) shows the basin has the characteristics of higher north part and lower southwest and southeast parts on the whole. The thickness of Kongtong-Ruishuihe in foredeep zone of southwest edge can reach up to 3000 m. In the west Huanxian-Tongxin region, there is a low-uplift zone spreading nearly NW, with about 500 m relative thin sedimentary strata on it, which is the contemporaneous fore-bulge zone. The uplifts are covered directly by early-middle Jurassic Yan'an Formation in Mt. Yao, Mt. Tan, Well Yan 11, Huan 26 and other places, which demonstrates that it is the underwater uplift in deposition. The average thickness of intra-craton basin is less than 1300 m, thickest area located in Ansai-Tongchuan region, EW spread and thickening eastwards.

5.2.2 The destroyed foreland basin in south edge of southern North China basin

Because of intensive Indosinian thrust-nappe and uplift erosion, the southward region of Sanmenxia-Lushan-Fuyang-Huainan faults basically lacks Triassic, while thin upper Triassic clastic rocks containing coal only exist in the foredeep zone of Nanzhao region in the western Henan. It indicates that this is a destroyed foreland basin system. Beihuaiyang structural belt in east segment of north edge, Hefei basin and Huainan region develop Indosinian thrust-nappe structure. Although differences in terms of tectonic pattern and strength exists, it forms a unified Indosinian nappe structure and produces nearly EW-trending fold-thrust belt which has the basically consistent direction with spread direction of QDOB, and forms tectonic pattern of interphase uplift-depression (Zhao Zongju et al., 2000). Lushi region of the western Henan in west section and Beihuaiyang structural belt are located in the same tectonic position. Undeformed upper Triassic sandy conglomerate contacts directly with the south-inclined Zhuyangguan-Xiaguan ductile shear zone, the direction of lithofacies spread is controlled by regional tectonics. The southern stratum is thicker than north, and such strata overlap northwards above Erlangping complex with the characteristics of "piggy back" deformation in foreland region.

During Indosinian, Beihuaiyang structural belt was located in the trailing edge of thrust-nappe structure. Based on balanced cross-sections recovery, the crust was shortened by 120 km, which occupied 86% of the whole shortening for Mesozoic-Cenozoic (Cao Gaoshe et al., 2003). According to the combined data of drilling, outcrops, seismic section, interval velocity and MT, a series of nearly EW-trending faults are developed in the south of Hefei basin, such as Gushi-Hefei fault, Shushan fault, Feixi-Hanbaidu fault and Shucheng fault. These faults were shown as slopes in different sizes during Indosinian, and the bedding faults were shown as flat. Upper detachment surface may partly disappear due to later erosion. The northern crystalline basement of Hefei basin is uplift, and overlying Paleozoic cover has been mostly eroded in some areas, with slight the tectonic deformation of remnant sedimentary cover. Therefore, Beihuaiyang structural belt and Hefei basin are involved in deformation of fold-thrust belt, and forward brink reaches into the south side of Huainan region. The core of Huainan syncline includes Carboniferous-Permian and lower Triassic, and the deformation isn't involved into Mesozoic and Cenozoic (Yan Kongde, 1989). It is the Indosinian product. Coalfield exploration data shows that thrust tectonics develop on both side of Huainan syncline. On south side, Shungengshan fault thrusts northwards as the main thrust surface, and it reforms the pre-Mesozoic in varying degrees. On north side, it takes Nanjie fault of Bengbu uplift as the main thrust surface, and the fault thrusts southwards in pre-Mesozoic (Shen Xiuzhi et al., 1995), which forms triangle belt tectonics.

To the west, except southern and northern border fault of NWW, the size of intra-contemporaneous fault is smaller in Xinyang basin. To the west of Xinyang, the thrust-nappe system of Maoji region is involved into the latest stratum, which is middle-lower Triassic containing radiolarian fossils, and the system is also invaded by middle-acid rock-body of middle-late Yanshanian (Du Yuansheng et al., 1997; Guo Hua et al., 2002). The thrust-nappe in Zhoukou depression is covered by the late sediment, but seismic section shows that Cambrian-Ordovician thrust structure makes overthrust above the Carboniferous-Permian and covered by Jurassic. The strength and size of thrust fault in Zhoukou depression are smaller. Its tectonic pattern is wide and gentle fold, and some early-middle Triassic sediments may exist in syncline region.

6. Discussion & conclusions

6.1 Stratigraphy of NCC in late Triassic

The residual stratigraphic succession of NCC in late Triassic has the characteristics of subareas: (1) Western Ordos in the interior-craton develps completely, to the southwest of which is the deposition and subsidence center of the basin. Upper Triassic strata extend eastwards to central Shanxi province and western Henan province in the hinterland of the basin, which are distributed in vast, continuous stretches. (2) In the north part of NCC, there is solitary distribution of the upper Triassic along west Liaoning-west Beijing-north Hebei. (3) In the southern part of the NCC, upper Triassic strata develop along north Qinling in west Henan-south Shaanxi area. Strata at the southern and northern edges of NCC are involved in late orogeny deformation, and now present like fragments between thrust faults.

6.2 Structural deformation of NCC in late Triassic

A tectonic-magmatic active belt formed during late Paleozoic is located in the northern Yanshan and western Liaoning province from east to west. The magmatic rock belt may be formed along the active continental side on the south of Soren suture when Siberia plate collided with NCC in Permian. During Mesozoic, NCC was developed in an intra-continental tectonic environment, which was a preexisting tectonic condition of Yanshan-Liaoning intra-continental orogeny deformation. In late Triassic, the northern NCC and Siberia plate collided and linked, making the paleo-Asian ocean disappear. On the south side, NCC collided with Yangtze plate, forming QDOB. QDOB formed earlier (during late middle Triassic) than Okhotsk structural belt in the north. Therefore, in late Triassic, the S-N compression on the north and south sides of NCC formed a "colliding" over-thrust tectonic pattern on the whole, which corresponded with WE striking structural belts pre-existed in intra-craton. In interior NCC, there is regional tectonic stability in Ordos basin in the west, which shows strata are better preserved with minor deformation. And there is strong deformation in the east, without upper Triassic. The boundary line of the two different areas runs approximately along Datong-Shijiazhuang–Ji'nan-Zhoukou.

6.3 Tectonics of proto-basin types in NCC

Analysis on syn-depositional structure records is very important for studying the process of regional structure deformation. Syn-depositional structure records keep the information of uplift, denudation and sediment process which are jointly controlled by deformation and other factors during the process of structure deformation in the geologic history. NCC region suffers from great reform due to Yanshan and Himalayan movements after the Triassic. And other questions have always been the hotspots under discussion. From the perspective of lithology, the remaining lower-middle Triassic can be tracked and contrasted from west to east, showing the same sediment background in the west and east of the NCC. From the perspective of lithofacies features, marginal facies sediment in the lower-middle Triassic are absent in many places. So it can be inferred that the area of proto-basin of NCC during the lower-middle Triassic should be more extensive than the current remaining area. In the late Triassic, NCC changes significantly. Firstly, the sediment boundary shrank compared with that in the early-middle Triassic. According to remaining distribution of the upper Triassic, the boundary was only limited in the western-southern region, including Ordos, middle and south Shanxi, west Henan, as well as Yanshan-Liaoning and north Qinling on the south and north sides. Secondly, the western and northern brinks of Ordos basin lack the upper Triassic, and marginal facies sediment develops in west-south brink; the late Triassic sediment is commonly absent in the east Bohai bay basin; the upper Triassic distribution in the southern North China is confined in Jiyuan region. All above explain that the late Triassic boundary of NCC retreated back to the west. The lower, middle and upper

Triassic in Ordos, Qinshui and Jiyuan basin are in conformity and unconformity contacts. Hence, the proto-basin of the west NCC in the late Triassic is inherited craton basin which overlaps the early-middle Triassic basin.

For the Mesozoic basin in the west of Beijing and Yanshan-Liaoning, because of prophase and anaphase structure reform and multiphase basin superposition and denudation, most of original patterns of the basin have been destroyed and divided into many small basins. In order to find out the sedimentary paleogeography, evolution and the relationship with tectonic effect, the late Triassic proto-basin is restored through extensively sediment section observation and contrast, paleocurrent measurement, gravel component contrast and structure deformation analysis. The upper Triassic Xingshikou, Laohugou Formations and time-equivalent strata has fewer outcrops, but can be infered to distributed in belt along Shangyi-Pingquan faults on the south side of Luanping and Chengde, extend eastward to Chaoyang region, and NEE striking distributed along the core of basin in the west of Beijing. The typical gravel molasse deposition in the two basin groups is correspondence for the north-side orogen thrust-nappe structure, with characteristics of intracontinental foreland basin. Similarly, the structures in the foreland basin are evident in the west segment of south brink of NCC. The foredeep zone is distributed along Shangzhou-Nanzhao in north Qinling, and touch Pingliang Foredeep zone of Ordos basin in the west. The foredeep zone consists of alluvial fan, delta and deep lacustrine deposit system. The fore-bulge zone is constituted by the basement uplift of Huanxian-Huoqiu. The back-bulge zone is distributed in NCC intra-basin, and subsidence center lies in Jiyuan-Tongchuan where deep lacustrine -delta system develops. "Foreland basin system" is reflected by north Qinling wedge-top zone, Pingliang-Nanzhao foredeep zone, Huanxian-Huoqiu Fore-bulge zone and Tongchuan-Jiyuan back-bulge zone from south to north.

7. References

[1] Burbank D W, Raynolds R G H. 1988. Stratigraphic keys to the Timing of Thrusting in Terrestrial Foreland Basins: Applications to the Northwestern Himalaya. In: Kleinsphn K L, Paola C (eds.), New Perspectives in Basin Analysis. Springer-Verlag New York Inc., 331-351

[2] Darby B J, Davis G A, Zheng Y, et al. 2001. Structural evolution of the southwestern Dagingshan, Yinshan belt, Inner Mongolia, China. GSA Mem, 194: 199-214

[3] Davis G A, Zheng Yadong, Wang Cong, et al. 2001. Mesozoic tectonic evolution of the Yanshan fold and thrust belt, with emphasis on Hebei and Liaoning Provinces, northern China. In: Hendrix M S, Davis G A, eds. Paleozoic and Mesozoic tectonic evolution of central Asia: From continental assembly to intracontinental deformation: Boulder, Colorado, Geological Society of America Memoir 194, 2001. 171-197

[4] Gao S, Rudnick R L, Carlson R W, Mc Donough W F, Liu Y S. 2002. Re-Os evidence for replacement of ancient mantle lithosphere beneath the North China craton. Earth and Planetary Science Letters, 198: 307-322

[5] Griffin W L, Zhang A D, O. Reilly S Y, Ryan G. 1998. Phanerozoic evolution of the lithosphere beneath the Sino-Korean craton. In: Flower M, Chung S L, Lo C H, Lee T Y, eds. Mantle dynamics and plate interactions in East Asia. Washington DC: American Geophysical Union, Geodynamic Series, 27: 107-126

[6] Jordan T E, Flemings P B, Beer J A. 1988. Dating Thruse-fault activity by use of foreland-basin strata. In: Kleinsphn K L, Paola C (eds.), New Perspectives in Basin Analysis. Springer-Verlag New York Inc., 307-330

[7] Menzies M A, Fan W M, Zhang M. 1993. Palaeozoic and Cenozoic lithoprobes and the loss of >120km of Archaean lithosphere Sino-Korean Craton, China magmatic processes and plate tectonics. In: Pritchard H M, Alabaster T, Harris N B W, Neary C R, Eds. Magmatic processes and plate tectonics. Geological Society London, 76: 71-81

[8] Menzies M A, Xu Y G.. 1998. Geodynamics of the North China craton. In: Flower M, Chung S L, Lo C H, Lee T Y, Eds. Mantle dynamics and plate interactions in East Asia. Washington DC: American Geophysical Union, Geodynamic Series, 27: 155-165

[9] Xu Yigang. 2001. Thermo-tectonic destruction of the Archaean lithospheric keel beneath eastern China: evidence, timing and mechanism. Physics and Chemistry of the Earth (A), 26: 747-757

[10] Cao Gaoshe, Li Xuetian, Liu Deliang, Zhou Songxing, Gao Yijun. 2003. Indosinian nappe structure and its hydrocarbon significance in Hefei basin and north Huaiyang tectonic zone. Oil & Gas Geology, 24(2):116-122

[11] Cao Hongxia. 2008. Research on the rule of depocenter migration and evolution of Late Triassic in Ordos basin. A doctoral dissertation submitted to Northwest University, 1-134

[12] Chen Chuanshi, Su Xianbo. 1995. Fault-block movement and evolution of the Jiyuan basin during the Mesozoic. Sendimentary Facies and Palaeogeography, 15(1):18-23

[13] Cheng Haihong, Sun Shu, Li Jiliang, et al. 1992. Tectonic constraint of foreland basin. In: Li J L. Research on the structure and evolution of the land-sea lithosphere in Southeast of China. Science and Technology of China Press, Beijing, 17-31.

[14] Chen Ruiyin, Luo Xiaorong, Chen Zhankun, Wang Zhaoming, Zhou Bo. 2006. Estimation of denudation thickness of Mesozoic stata in the Ordos basin and its geological significance. Acta Geological Sinica, 80(5): 685-693

[15] Chen Shiyue. 2000. Sedimentary tectonic evolution from Late Palaeozoic to Triassic in the South of North China block. Journal of China University of Mining & Technology, 29(5): 536-540

[16] Cui Shengqin, Li Jinrong, Sun Jiashu, Wang Jianping, Wu Zhenhan, Zhu Dagang. 2000. Sequence of tectonic movements and regional tectonic framework in the north edge of North China block. Geology Press, Beijing, 1-326.

[17] Cui Shengqin, Li Jinrong. 1983. On the Indosinian movement of China's Peri-Pacific tectonic belt. Acta Geological Sinica, 57(1): 51-62

[18] Deng Jinfu, Mo Xuanxue, Zhao Hailin, Luo Zhaohua, Du Yangsong. 1994. Lithosphere Root/De-Rooting and activation of the East China continent. Geoscience-Journal of China University of Geoscience, 8(3):349-356

[19] Deng Xiuqin, Lin Fangxiao, Liu Xianyang, Pang Jinlian, L Jianwen, Li Shixiang, Liu Xin. 2008. Discussion on relationship between sedimentary evolution of the Triassic Yanchang Formation and the Early Indosinian movement in Ordos basin. Journal of Palaeogeography, 10(2):159-166

[20] Du Xudong, Xue Linfu, Wu Guanghui. 1999. Distribution of Mesozoic basin and discussion geodymics of continent interior in the Eastern China. Journal of Changchun University of Science and Technology, 29(2):138-143
[21] Du Yuansheng, Yin Hongfu, Wang Zhiping. 1997. The late Caledonian-Early Hercynian basin's framework and tectonic evolution of Qinling orogenic belt. Earth Science, 22(4):401-405
[22] Fu Jinhua, Guo Zhengquan, Deng Xiuqin. 2005. Sedimentary facies of the Yanchang Formation of Upper Triassic and petroleum geological implication in southwestern Ordos basin. Journal of Palaeogeography, 7(1):34-44
[23] Guo Hua, Wu Zhengwen, Liu Hongxu, Wang Runhong. 2002. The overthrust-nappe tectonic framework of Yanshan intraplate orogenic belt. Geoscience, 16(4): 339-346
[24] Guo Hua, Wu Zhengwen, Chai Yucheng, Feng Ming. 2002. Mesozoic overthrust-nappe tectonic system in the Dabieshan orogenic belt. Geoscience, 16(2): 121-129
[25] Guo Xinian, Tang Zhonglin, Li Wancheng. 1991. Late Paleozoic coal accumulating regularities in Henan province. Wuhan: China University of Geosciences Press, 156-167
[26] He Zhengjun, Wang Zongqi, Ren Jishun. 1999. A preliminary research on sedimentary features and genetic mechanism of frontal basins before Jurassic large-scale Nappe in the Northern Region of North China. Scientia Geologica Sinica, 34(2):186-195
[27] Bureau of Geology and Mineral Resources of Henan Province. 1989. Regional geology in Henan province. Beijing: Geology Press
[28] Hu Bin, Huang Changchun. 2000. The types of fluvial deposits and its ichnofabric feature from the Upper Triassic epoch of Jiyuan basin. Western Henan. Henan Geology, 18(3):176-180
[29] Hu Jianmin, Liu Xiaowen, Xu Gang, Liu Jian, Zhang ShuanHong. 2005. The slidding-slump-mudflow sedimentation during the Late Late-Triassic to Middle Jurassic in Western Liaoning Province. China. Acta Geologica Sinica, 79(4):543-564
[30] Ji Youliang, Hu Guangming, Huang Jianjun, Wu Zhiping. 2006. Eroded strata thickness of Mesozoic and evolution of Mesozoic and Cenozoic basins in the Bohai Bay basin area. Acta Geologica Sinica, 80(3):351-358
[31] Jia Shixun, Zhang Xiankang. 2005. Crustal structure and comparison of different tectonic blocks in North China. Chinese Journal of Geophysics, 48(3):611-620
[32] The Studying Team of Mesozoic, Jiaozuo Mining Institute. 1982. The Mesozoic coal-bearing formations division and comparison in Henan. (Internal Report)
[33] Xie Dongning, He Mingxi, Zhou Lifa, Xie Qifeng. 2006. Characteristics of overthrust structures on northern edge of East Qinling-Dabie orogenic belt and hydrocarbon potentials. Oil & Gas Geology, 27(1):48-55
[34] Jin Chong. 2007. Structural pattern, evolution and dynamic mechanism of the Mesozoic strata in the Huanghua depression and Jiyang depression. A masteral dissertation Submitted to Ocean University of China, 1-70.
[35] Kang Ming. 1988. Mesozoic coal-bearing strata of Yima, Henan. Journal of Stratigraphy, (2):81-92
[36] Li Hongyan, Xu Yigang, Huang Xiaolong, He Bin, Luo Zhenyu, Yan Bin. 2009. Activation of north edge of the North China Craton in Late Paleozoic: Evidence from U-Pb dating and Hf isotopes of detrital zircons from the Upper Carboniferous

Taiyuan Formation in the Ningwu-Jingle Basin. Chinese Science Bulletin, 54(5):632-640

[37] Li Shuguang, Nie Yonghong, Liu Deliang, Zheng Shuanggen. 1998. Interaction between subducted continental crust and the mantle-I. Major and trace element geochemistry of the syncollisional mafic-ultramafic intrusions in the Dabie Mountains. Science in China,Ser.D,

[38] Li Shuguang, Nie Yonghong, S.R.Hart, Zhang Zongqing. 1998. Interaction between subducted continental crust and the mantle -II. Sr and Nd isotopic geochemistry of the syncollisional mafic-ultramafic intrusions in Dabie Mountains. Science in China,Ser.D,

[39] Li Shuangli, Ouyang Ziyuan. 1998. Tectonic framework and evolution of Xing'anling Mongolian Orogenic Belt (XMOB) and its adjacent region. Marine Geology & Quaternary Geology, 18(3):45-54

[40] Li Wenhou, Pang Jungang, Cao Hongxia, Xiao Li, Wang Ruogu. 2009. Depositional system and paleogeographic evolution of the late Triassic Yanchang Stage in Ordos Basin. Journal of Northwest University (Natural Science Edition), 39(3):501-506

[41] Li Zhong, Dong Renguo, Zheng Jianpin. 2007. Mesozoic volcanic-sedimentary configurations in north and south edges of the eastern North China Craton: Implications for tectonic transition mechanism. Journal of Palaeogeography, 19(3):227-242

[42] Li Zhong, Li Jiwei, Sun Su, Zhang Wenhua. 2003. Mesozoic basin-fill records in south foot of the Dabie Mountains: Implication for Dabie Orogenic attributes. Science in China, Ser.D,

[43] Liu Chiyang. 2007. Geologic characteristics and petroleum accumulation conditions of superimposed basin. Acta Petrolei Sinica, 28(1): 1-7

[44] Liu Chiyang, Zhao Hongge, Gui Xiaojun, Yue Leping, Zhao Junfeng, Wang Jianqiang. 2006. Space-Time coordinate of the evolution and reformation and mineralization response in Ordos Basin. Acta Geological Sinica. 80(5):617-638

[45] Liu Chiyang, Zhao Hongge, Wang Feng, Chen Hong. 2005. Attributes of the Mesozoic structure on the west margin of the Ordos Basin. Acta Geologica Sinica, 79(6):738-747

[46] Liu Chiyang. 1987. The tectonic evolution of Bohai Bay Basin and its characteristics. In: Collection on thesis of academic report conference for the 45th anniversary of the College of Geology, Northwest University. Xi'an: Shaanxi Science and Techology Press, 447-458

[47] Liu Le, Yang Minghui, Li Chunxia, Zheng Xiaofeng. 2009. Fractured reservoirs in metamorphic buried-hills in the Liaoxi low uplift and their hydrocarbon accumulation conditions. Oil & Gas Geology, 30(2):188-194

[48] Liu Shaofeng, Li Sitian, Zhuang Xinguo, Jiao Yangquan, Lu Zongsheng. 1996. Simulation of the subsidence and deposition of the foreland basin on the southwestern margin of Ordos. Acta Geologica Sinica, 70(1):12-22

[49] Liu Shaofeng, Li Zhong, Zhang Jinfang. 2004. Evolution and tectonic regime of the Mesozoic basins inYanshan area. Science in China, Ser.D, 34(zI):19-31

[50] Liu Shaofeng, Yang Shigong. 1997. The north-south difference and its formation mechanism in the west margin of Ordos Basin. Scientia Geologica Sinica, 32(3):397-408

[51] Liu Shaofeng, Zhang Guowei, Chen Shunyou, Yao Anping. 1999. Evolution of flexural basins and process of collision orogeny in East Qinling-Dabie Shan and its adjacent regions. Scientia Geologica Sinica, 34(3):336-346

[52] Liu Shaolong. 1986. The existence of a large-scale Triassic Sedimentary Basin in North China. Acta Geological Sinica, 60(2):128-138

[53] Liu Zhenghong, Xu Zhongyuan, Yang Zhensheng. 2003. Indosinian tectonic movement in the Daqingshan Region in Inner Mongolia. Geological Review, 49(5):457-463

[54] Liu Chengzhi, Ma Fengrong. 2006. The guide book for geological practice in the Beidaihe region. Beijing: Petroleum Industry Press, 80-84

[55] Lu Fengxiang, Zheng Jianping, Li Wupin, Chen Meihua, Yu Chunmei. 2000. The main evolution pattern of Phanerozoic mantle in the Eastern China: the "Mushroom Cloud" model. Earth Science Frontiers, 7(1):97-107

[56] Lu Fengxiang, Zheng Jianping, Shao Ji'an, Zhang Ruisheng, Yu Chunmei. 2006. Asthenospheric upwelling and lithospheric thinning in late Cretaceous-Cenozoic in eastern North China. Earth Science Frontiers, 13(2):86-92

[57] Lu□Dingyou, Yang Minghui, Zhou Xinhuai, Wei Gang, Liu Le, Li Chunxia, Zheng Xiaofeng. 2009. Structural characteristics and hydrocarbon accumulation in the buried-hills of the Liaoxi low-uplift, the Liaodong Bay Depression. Oil & Gas Geology, 30(4):490-496

[58] Liu Yuyan, Yang Weiran, Morinaga Hayaom, Adachi Yasuhisa, Yonezawa Takafumi, Yaskawa Katsumi. 1993. Paleogeographic history of North China continental block, Qinling geological block and Yangtze continent block since the Paleozoic Era. Geological Science and Technology Information, 12(4): 17-21

[59] Liu Yuyan, Yang Weiran, Morinaga Hayaom, Adachi Yasuhisa, Yonezawa Takafumi, Yaskawa Katsumi. 1993. Some paleomagnetic results on North China, Qinling and Yangtze blocks. Earth Science-Journal of China University of Geosciences, 18:635-641

[60] Ma Yinsheng. 2001. The evolution of Mesozoic-Cenozoic basin-mountain structure in the east Yanshan area and Xialiaohe basin. Journal of Geomechanics, 7(1):79-91

[61] Qi Yong'an, Su Xianbo. 1993. Trace fossils of Late Triassic meandering river sedimentary and their meanings, Jiyuan, Henan. Henan Geology 11(2), 119-122

[62] Qiu Ruizhao, Deng Jinfu, Zhou Su, Li Jinfa, Xiao Qinghui, Wu Zongxu, Liu Cui. 2004. The lithosphere type in the North China: Evidences on geology and geophysics. Science in China, Ser.D:34(8):698-711

[63] Ren Jishun. 1990. On the geotectonics of Southern China. Acta Geological Sinica, 64:275-288

[64] Ren Jishun. 1994. The continental tectonics of China. Acta Geoscientia Sinica, (3-4): 5-13

[65] Ren Zhanli, Xiao Hui, Liu Li, Zhang Sheng, Qin Yong, Wei Chongdao. 2005. Fission track evidences of tectonic-thermal evolution history in Qinshui basin. Cinese Science Bulletin, 50(z1): 87-92

[66] Shanxi bureau of geological and mineral. 1989. Regional geology in Shanxi province. Beijing: Geology Press, 673-674

[67] Shen Xiuzhi, Xue Aiming, Liu Deliang, et al. 1995. Relationship between tectonics and gas in basins, south of the North China. Hefei: University of Science and Technology of China Press

[68] Shi Quanzeng, Shang Yuzhong, Pang Jiqun, Tao Ziqiang. 1990. The nappe structure and distribution of coal field in the north slope of East Qinling Mountain, Henan province. Henan Geology, 8(4): 22-34

[69] Shi Quanzeng, Wei Xiangdong, Li Mingli, Pang Jiqun. 2004, The nappe structure and the strike-slip shear structure in the north edge of East Qinling Mountain, Henan province. Beijing: Geology Press, 15-101

[70] Tian Lifu, Hu Huabin, Hu Shengjun,Yang Zusen, Zhu Caifa. 1996. Discussion on the Indosinian Movement in the Western Yanshan Range. Journal of Shijiazhuang University of Economics, 19(6): 668-672

[71] Tian Lifu, Sun Liming, Hu Huabin, Zhou Pinwei, Hu Shengjun. 1996. Establishment of the Xingshikou Formation and its stratigraphic significance in the Western Yanshan Ranges. Journal of Shijiazhuang University of Economics, 19(6): 724-728

[72] Wan Tianfeng, Zhao Weiming. 2002. On the Mechanism of intraplate deformation in Chinese Continent. Earth Science Frontiers, 9(4): 451-463

[73] Wang Congfeng. 1987. Discovery of the Late Triassic strata in the North Henan Plain. Journal of Stratigraphy, (1): 48-52

[74] Wang Genhou, Zhang Changhou, Wang Guosheng, Wu Zhengwen. 2001. Tectonic framework of Western Liaoning Province and its evolution during Mesozoic. Geoscience, 15(1): 1-7

[75] Wang Guiliang, Yan Shouxun, Jiang Bo. 1992. The compound extension structure system of Meso-Cenozoic in West Shandong Province. Journal of China University of Mining & Technology, 21(3): 1-8

[76] Wang Guiliang, Liu Guijian, Zouhai, Li Chaofeng, Wang Jiyao. 1999. Coupling and transition of the Mesozoic basin mountain on the northern edge of North China Platform and its dynamics analysis. Coal Geology & Exploration, 27(6): 14-17

[77] Wang Lifeng, Zhou Pinwei, Hu Shengjun. 1998. The Late Triassic alluvial fan of Huailai county, Hebei and its structural environment. Journal Geology & Mineral Resources North China, 13(3): 230-235

[78] Wang Quan , Liu Xueya, Li Jinyi. 1991. Plate tectonics between cathaysia plate and angara paleo-landmass. Beijing: Press of PKU: 74-91

[79] Wang Qingchen, Sun Shu, Li jiliang, Zhou Da, Kenneth J. Hsu, Zhang Guowei. 1989. The tectonic evolution of the Qinling Mountain belt. Chinese Journal of Geology, (2): 129-142

[80] Wang Shide. 1987. Tectonics of Jiming Mountain and Yudai Mountain in XiaHuayuan Area. Journal of Hebei Geology College, (2): 1-15

[81] Wang Xin, Li Nan, Wang Yongdong, Zheng Shaolin. 2009. Discovery of huge complete cycads fossils in Upper Triassic strata, Western Liaoning and its significance. Chinese Science Bulletin, 54(13): 1937-1939

[82] Wang Yi, Lu Kezheng, Ren Anshen. 1994. Tectogenesis and volcanism and the basin evolution of the Mesozoic era in the Northeast of Jiyang depression. Journal of The University of Petroleum, China, 18(2): 1-8

[83] Wei Wenbo, Ye Gaofeng, Jin Sheng, Deng Ming, Jing jian'en, Peng Zhiqiang, Lin Xin, Song Shilei, Tang Baoshan, Qu Shuanzhu, Chen Kai, Yang Honhwei, Li Guoqiang. 2008. Geoelectric structure of lithosphere beneath eastern North China: features of a thinned lithosphere from magnetotelluric soundings. Earth Science Frontiers, 15(4): 204-216

[84] Wu Fuyuan, Sun Deyou. 1999. The Mesozoic magmatism and lithospheric thinning in Eastern China. Journal of Changchun University of Science and Technology, 29(4): 313-318

[85] Wu Hanning, Zhu Rixiang, Liu Chun, Chang Chengfa. 1990. Paleomagnetic observations in Norht China block: from Late Paleozoic to Triassic. Chinese Journal of Geophysics, 33(6): 694-701

[86] Wu Hanning, Zhu Rixiang, Liu Chun, Chang Chengfa. 1990. Paleomagnetic study on the Danfeng group ophiolite in Qinling area and its tectonic significance. Seismology and Geology, 12(1): 79-85

[87] Wu Hanning, Chang Chengfa, Liu Chun, Zhong Dalai. 1990. Evolution of Qinling fold belt and the North and South China blocks: The evidence of Geology and Paleomagnetism. Chinese Journal of Geology, (3): 201-214

[88] Wu Hanning, Chen Yan, Zhou Dinwu. 1992. A preliminary paleomagnetic result of the Danfeng ophiolite formation of the Shangdan zone in the Qinling belt (China). Chinese Journal of Geophysics, 35(3): 361-368

[89] Xia Bangdong, Li Peijun. 1996. Sedimentologic evidence on the Docking of Yangtse Plate and North China Plate in Middle-Late Triassic Epoch in East China. Acta Sedimentologica Sinca, 14(1): 12-21

[90] Xiao Yuanyuan, Ren Zhanli, Qin Jiangfeng. 2007. Geochemistry and Zircon LA-ICP MS U-Pb Dating of the Zijinshan Alkaline Complex in the Linxian County, Shanxi Province: Geological implication. Geological Review, 53(3): 433-440

[91] Xiao Zongzheng, Yang Honglian, Shan Qinsheng. 1995. On some Mesozoic geological problem in Beijing area. Journal of Stratigrphy, 19(4): 308-314

[92] Xing Zuoyun, Xing Jishan, Zhao Bin, Liu Jianhua, Yuan Xiaoping. 2006. Identification of Deep Structure of two episodes in North China and its significance-Yanshanian Movement and Deep Process. Geological Review, 52(4): 433-441

[93] Xu Bei, Chen Bin. 1997. The structure and evolution of the Middle Paleozoic orogenic belt between North China plate, north of Inner Mongolia and Siberia Plate. Science in China, Ser.D, 27(3): 227-232

[94] Xu Chunhua, Qiu Liangui, Lei Min, Li Xuetian. 2002. The evolution of the Dabie orogenic belt based on the sedimentary styles and tectonic patterns of the Hefei basin, Anhui. Sedimentary Geology and Tethyan Geology, 22(2): 91-98

[95] Xu Gang, Zhao Yue, Gao Rui, Li Qiusheng, Hu Jianmin, Liu Xiaowen, Wu Hai, Yang Fuquan, Zhang Shuanhong, Guan Ye, Zhang Jisheng, Bai Jin, Kuang Chaoyang, Wang Haiyan. 2006. Mesozoic basin deformation of Yanshan folded Fault Belt-Records of the intraplate deformation process: A case study of Xiabancheng, Chengde-Shangbancheng and Beitai Basins. Acta Geoscientica Sinica, 27(1): 1-12

[96] Xu Gang, Zhao Yue, Hu Jianming, Zeng Qingli, Liu Xiaowen, Wu Hai, Song Biao. 2003. Late Triassic thrusting in the Niuyingzi region, Western Liaoning, China. Acta Geologica Sinica, 77(1): 25-34

[97] Xu Hanlin, Feng Shiqi, Zhu Hongfa, Shen Yang, Zhao Ping. 2001. Formation and evolution of Late Triassic–Middle Jurassic foreland basins' tectonic framework in South China. Marine Origin Petroleum Geology, 6(1): 19-26

[98] Xu Hanlin, Zhao Zongju, Lu Fuliang, Yang Yilin, Tang Zuwei, Sun Guozhong, Xu Yunjun. 2004. Tectonic evolution of the Nanhuabei area and analysis about petroleum potential, 28(4): 450-463

[99] Xu Liming, Zhou Lifa, Zhang Yikai, Dang Ben. 2006. Characteristics and tectonic setting of Tectono-Stress field of Ordos Basin. Geotectonica et Metallogenia, 30(4): 455-462

[100] Xu Yigang, Li Hongyan, Pang Chongjin, He Bin. 2009. On the timing and duration of the destruction of the North China craton. Chinese Science Bulletin, 54(14): 1974-1989

[101] Xu Yigang. 2006. Formation of the Taihangshan Gravity lineament by the Diachronous lithospheric thinning of the North China craton. Earth Science-Journal of China University of Geosciences, 31(1): 14-22

[102] Xu Zhengyu, Lin Ge, Liu Chiyang, Wang Yuejun, Guo Feng. 2004. A discussion on amalgamation course between the South China and North China blocks: evidences from deformational characters in the Jianghan Superimposed Basin. Chinese Journal of Geology, 39(2): 284-295

[103] Xu Wenliang, Wang Dongyan, Wang Simin. 2000. p-T-t-c model of Mesozoic and Cenozoic volcanisms and lithospheric evolution in Eastern China. Journal of Changchun University of Science and Techonology, 30(4): 329-335

[104] Xu Shihong, Zhong Jianhua, Liu Zhongliang. 2007. Activity characteristics and controlling factors of the faults in the Hefei basin. Geotectonica et Metallogenia, 31(1): 31-36

[105] Yan Yi, Lin Ge, Li Zi'a, He Shijie, Xu Zhengyu. 2003. Detrital composition of Mesozoic sandstone and its implication for provenance and tectonic evolution of Beipiao (Jin-Yang) Basin, Western Liaoning Province. Acta Sedimentologica Sinica, 21(3): 441-447

[106] Yan Kongde. 1989. Tectonic conditions for formation of hypabyssal coal-generated gas deposits in Huainan-Huaibei Coalfield of Anhui. Journal of Anhui University of Science and Technology (Natural Science), 3: 1-8

[107] Yang Baojun, Zhang Mingsheng, Wang Pujun. 2003. Geological and geophysical analysis of the oil and gas region in China. Beijing: Science Press, 34-105

[108] Yang Geng, Chai Yucheng, Wu Zhengwen. 2001. Thin-skinned thrust nappe structures in Western Liaoning in the eastern sector of the Yanshan orogenic belt. Acta Geologica Sinica, 73(3): 322-332

[109] Yang Jinhui, Wu Fuyuan. 2009. Triassic magmatism and its relation to decratonization in the eastern North China Craton. Science in China, Ser.D, 39(7): 910-921

[110] Yang Minghui, Liu Chiyang, Zheng Menglin, Lan Chaoli, Tang Xuan. 2007. The two different types of marginal sequence architecture and their responses to the Middle-Late Triassic tectonic movement in Ordos basin. Science in China, Ser.D, 37(A01): 173-184

[111] Yang Minghui, Wang Simin, Chen Shanyong, Ming Haihui, Tang Xuan. 2005. Genetic classification and distribution of Ordovician carbonate buried-hills in Huanghua depression Bohai Bay basin. Oil & Gas Geology, 26(3): 310-316

[112] Yang Minghui, Wang Simin, Zheng Xiaofeng, Wang Jinxiu, Bai Hua, Zeng Peng. 2009. Aeromagnetic anomalies characteristics and tectonic subareas of Southern North China craton and adjacent regions. Geological Review, 55(6): 862-872

[113] Yang Minghui, Liu Chiyang, Sun Dongsheng, Cui Yongqian. 2002. Extensional tectonic system and its deep-seated setting of Jizhong basin, China. Geotectonica et Metallogenia, 26(2): 113-120

[114] Yang Minghui, Zhou Xinhuai, Wei Gang, Liu Le, Li Chunxia, Zheng Xiaofeng, Liu Xiaojian, Gao Libao. 2008. Segment, Linkage, and Extensional Fault-Related Fold in Western Liaodong Bay Subbasin, Northeastern Bohai Sea, China. Journal of China University of Geosciences, 19(6): 602-610

[115] Yang Nong, Chen Zhengle, Lei Weizhi, Zhang Huixu. 1996. Research on characteristics of Indosinian in Yanshan area of Norhtern Hebei. Beijing: Geology Press, 90

[116] Yang Shirong. 1994. Late Triassic-Jurassic sporopollen assemblages from Yima Area, Henan. Acta Palaeontologica Sinica, 33(6): 765-779

[117] Yin Hongfu, Du Yuansheng, Xu Jifeng, Sheng Jihu, Feng Qinglai. 1996. Radiolarian Fauna firstly discovered in Mian-Lüe paleo-suture belt of South Qinling Mountains. Earth Science-Journal of China University of Geosciences, 21: 184

[118] Yu Fusheng, Qi Jiafu, Wang Chunying. 2002. Tectonic deformation of Indosinian Period in eastern part of North China. Journal of China University of Mining & Technology, 31(4): 402-406

[119] Yu Hezhong, Han Shouhua, Xie Jinlong, Guo Qingxin, Wu Jinyun. 2006. Types of original sedimentary basins and tectonic evolution on southeastern margin of North China plate. Oil & Gas Geology, 27(2): 244-252

[120] Zhai Mingguo, Zhu Rixiang, Liu Jianming, Meng Qingren, Li Zhong, Zhang Hongfu, Hou Quanlin, Hu Shengbiao, Liu WEi. 2003. The vital transition period of Mesozoic tectonic regime in the eastern part of North China. Science in China, Ser.D, 33(10): 913-920

[121] Zhang Guoren. 2006. Mesozonic interplate orogenesis in Western Liaoning Province. A doctoral dissertation submitted to China University of Geosciences, 1-119

[122] Zhang Guowei, Dong Yunpeng, Lai Shaocong. 2003. The Mian-lue structural belt and suture zone in the south edge of Qinling-Dabie orogenic belt. Science in China, Ser.D, 33(12): 1121-1135

[123] Zhang Guowei, Dong Yunpeng, Yao Anping. 2001. Review on the development of studies on the tectonic and orogen process of orogenic belt, and discussing on some new key problems. Northwestern Geology, 34(1): 1-9

[124] Zhang Guowei, Meng Qingren, Lai Shaocong. 1995. Tectonics and structure of Qinling orogenic belt. Science in China (Series B), 25(9): 994-1003

[125] Zhang Hong. 1996. Mesozoic and Cenozoic Palaeotectono-stress field of Ordos basin. Journal Geology & Mineral Resources North China, 11(1): 87-92

[126] Zhang Hongbo. 2006. Stratigraphy and sedimentary facies study of Mesozoic in Dongpu and surrounding regions. A doctoral dissertation submitted to China University of Geosciences, 1-119

[127] Zhang Jinfang. 2002. Filling and evolving of Mesozonic basin in Chengde District and its relationship with tectonics. A doctoral dissertation submitted to China University of Geosciences, 1-63

[128] Zhang Yong. 2006. Mesozoic thrust tectonics around Cenozoic Yanqing-Huailai basin and their geological implications, Northwest of Beijing Municipality. A masteral dissertation submitted to China University of Geosciences, 1-74

[129] Editorial Committee of Petroleum Geology of the Changqing Oil Field. 1992. Petroleum Geology of China (Vol.12): the Changqing Oil Field. Beijing: Petroleum Industry Press

[130] Zhao Hongge, Liu Chiyang, Wang Jianqiang, Wang Feng, Yin Yan. 2007. Tectonic attribute of the western Ordos basin during the Late Triassic. Geology in China, 34(3): 384-391

[131] Zhao Wenzhi, Wang Xinmin, Guo Yanru, Liu Huaqing, Bai Yunlai. 2006. Restoration and tectonic reworking of the Late Triassic basin in western Ordos Basin. Petroleum Expoloration and Development, 33(1): 6-13

[132] Zhao Yue. 1990. The Mesozoic orogenies and tectonic evolution of the Yangshan Area. Geological Review, 36(l): l-13

[133] Zhao Chongyuan, Liu Chiyang. 1990. The regional tectonics of Mesozoic-Cenozoic in North China craton and the formation and evolution of petroliferous basin. In: Zhao C Y, Liu C Y, et al. The formation and evolution of the sedimentary basins in North China craton and its hydrocarbon occurrence. Northwest University Press, 10-21

[134] Zhao Chongyuan, Liu Chiyang. 1992. The remnant intracraton basin and its oil-bearing properties,Ordos Basin and Sichuan basin. In: Geological Society of China. Collection on thesis of "75" Academic Exchanges Conference about the important achievement of geological techology. Beijing: Science & Techology Press, 610-613

[135] Zhao Zongju, Yang Shufeng, Chen Hanlin, Zhu Guoqiang, Zhou Jingao. 2000. Tectonic attribute of basement in Hefei basin, Anhui Province, China. Scientia Geologica Sinica, 35(3): 288-296

[136] Zhao Zongju, Li Dacheng, Zhu Yan, Zhou Jingao, Feng Jialiang. 2001. The structure evolution and the petroleum system in Hefei basin. Petroleum Exploration and Development, 28(4): 8-13

[137] Zheng Qiugen, Zhang Yumin, Zhao Deyong, Liu Zhiqin. 1998. Tectonic evolution and Upper Triassic distribution in the Western Henan. Henan Petroleum, (2): 5-10

[138] Zheng Yadong, Davis G A, Wang Cong, Darby B J, Zhang Wenhou. 2000. The main questions of tectonic events and plate tectonics in Yanshan Belt. Acta Geologica Sinica, 74(4): 289-302

[139] Zheng Yadong, Wang Shizheng, Wang Yufang. 1990. The newly discovered oversize nappe structures and extentional metamorphic core complexes in the margin area between China and Mongolia. Science in China, Ser.B, (12): 1299-1305

[140] Zheng Yadong, Wang Tao. 2005. The kinematic and dynamic analysis of the Mesozoic nappe structures and extensional sloughing regime in the margin area between China and Mongolia. Science in China, Ser.D, 35(4): 291-303

[141] Editorial Committee of Petroleum Geology of the Central Plains Oil Field. 1993. Petroleum Geology of China (Vol.7): the Central Plains Oil Field. Beijing: Petroleum Industry Press, 250-256

[142] Zhou Xinke, Xu Huazheng, Hu Zongquan, Wang Chuangang. 2005. Characteristics of prototype basin and hydrocarbon potential evaluation of the Late Triassic in the west Henan Province. Petroleum Geology & Expeximent, 27(3): 211-217

[143] Zhou Jing'ao, Zhao Zongju, Deng Hongying. 1999. Tectonic evolution of the Hefei Basin and analysis of its petroleum potential. Acta Geologica Sinica, 73(1): 15-23

[144] Zhu Yanming, Qin Yong, Fan Bingheng, Li Tianzhong, Jiang Bo. 2001. Restoration and significance of the original thickness of Triassic system in Baohaiwan Bay basin. Journal of China University of Mining & Technology, 30(2): 195-200

[145] Zong Guohong, Shi Yangsheng, Wang Binghai, Wang Jie. 1998. Mesozoic structures and their relations to hydrocarbon traps in the Jiyang basin. Geological Review, 44(3): 289-294

7

Tectonic Implications of Stratigraphy Architecture in Distal Part of Foreland Basin, Southwestern Taiwan

Jong-Chang Wu[1], Kenn-Ming Yang[2*], Yi-Ru Chen[3] and Wen-Rong Chi[4]

[1]*Exploration and Development Research Institute, CPC Corporation, Taiwan, 1 Ta Yuan, Wen Sheng, Miaoli, 36010*

[2,3,4]*Department of Earth Sciences, National Cheng Kung University, 1 University Road, Tainan, 70101*

Taiwan

1. Introduction

A foreland basin system with its asymmetric basin geometry extends from the frontal areas of a mountain-building belt to the margin of craton and includes, as defined by DeCelles and Giles [1], wedge-top, foreland basin, forebulge and backbulge. Foreland basins have been well modeled based on the concept of flexed lithosphere under tectonic loading of thrust sheets and sediments deposited into the basin [2, 3, 4]. Rapid subsidence in a ancient foreland basin and the uplifted forebulge at its distal part has been viewed as the result of tectonic loading due to active thrust sheet emplacement [5, 6, 7, 8]. Corresponding to each episodic thrust belt advancement, a coarsening-upward clastic wedge would form and prograde into the newly subsided foreland basin [9, 10, 11].

Characteristics of sedimentary sequences in a foreland basin are, thus, related to dynamic and kinematic modes of orogen-foreland basin evolution. Inversely, such characteristics can be used to investigate tectonic motion of the adjacent orogen [12, 13]. Different models for the development of a foreland basin predict distinct stratigraphy architecture across the basin. An elastic plate model predicts that the asymmetric foreland basin with the uplifted forebulge would migrate toward the craton as the rising orogen advances [14, 15, 16] and that the continuous migration of the orogen-foreland basin pair would leave a regional unconformity in the basin, which is younger toward the craton [17, 18]. The strata sequences overlying and onlapping at the unconformity have been viewed as the initial deposits of the foreland basin development [19, 20, 21, 22, 23]. The elastic model can be applied to infer the scale of erosion at the forebulge unconformity and the rate of strata onlapping across the unconformity under the conditions of variable crustal rigidity and variable migration rate of the orogenic wedge [22].

If a visco-elastic model is applied, the stress imposed on a loaded lithosphere would be relaxed and the foreland basin would become narrower and deeper with the forebulge migrating

* Corresponding Author

toward the orogen [2]. This would cause lithofacies belts to migrate back-and-forth between the mountain-building belt and craton and form a more complicated stratigraphic architecture at the basin and superimposed unconformities at the distal part of the basin [24, 25].

In many studies of foreland basin, the addressed deflexed plates are characterized by continental crust with heterogeneous mechanical property; the heterogeneity is manifested by a torn plate as the base of a foreland basin [26] or by lateral variation in crustal rigidity [27, 28, 29, 30]. More importantly, the mechanical heterogeneity is due to the fact that the predecessor of a peripheral foreland basin is a rifted continental margin with large-scaled normal faults distributed within the stretched continental crust and, for such reason, the location of foreland basin would migrate from the most stretched crust with the least rigidity to a complete nonstretched crust [31, 32, 33]. Therefore, an initially narrow-deep foreland basin would evolve into a wide-shallow one. In addition, pre-existing and syntectonic normal faults would influence the development of foreland basin sequences and shape an atypical time-spatial distribution of lithofacies in the basin [22, 34, 35, 36, 37].

Before the 1990s in the last century, the proposed models for foreland basin sequences basically assume that a foreland basin is filled up with deposits right after it forms. Lately, time lag between erosion in the mountain side and deposition in the basin has been put into consideration when investigating and simulating a model of foreland basin sequences [38, 39, 40, 41, 42, 43]. In response to each episode of active thrust faulting, newly formed foreland basin by rapid subsidence would not be filled up with sediments in the early time; therefore, according to the theoretical model [4], the sediment-starved basin would be deep and narrow and accumulate deep water sediments in an "underfilled" state [39, 40, 41]. In the following stage, the orogenic wedge began to uplift and the consequential erosion provided the sediments to fill up the basin when the large-scale thrust faulting wane. From that moment, the basin gradually becomes widened and shallow and step into the "overfilled" state characterized by shallow water deposits [44]. Generally speaking, in response to each episode of large-scale thrust faulting in the orogenic belt, the subsidence rate in a basin, which is related to the rates of advancement of the front of orogenic belt and change of orogenic wedge morphology, the sediment supply rate and the global sea level change rate would all together affect the tectonostratigraphic architecture, time-spatial distribution of lithofacies and the superimposed unconformities in the basin [39, 40, 41, 42, 45, 46, 47, 48, 49]. Some models suggest that timing of rapid subsidence at different locations in a foreland basin has different tectonic implication [40, 41, 43].

In a young and on-going mountain building belt, such as that in Taiwan, the geological records in the foreland basin and fold-and-thrust belt are still well preserved and can be utilized to study the tectonostratigraphy problems and to test validity of different models for the problems. Taiwan is located on the convergent plate boundary between the Eurasian and Philippine Sea plates; its tectonic belts are aligned from east to west: the island arc in the Coastal Range, the suture zone along the Longitudinal Valley Fault Zone, the inner mountain-building belt of high-grade metamorphism in the Basement Complex, the outer mountain-building belt of low-grade metamorphism in the Backbone and Hsuehshan ranges, the imbricate fold-and-thrust belt in the Western Foothills, and the foreland basin (Figure 1, [50]). According to the model and definition of DeCelles and Giles [1], the coastal plain and its subsurface settings of southwestern Taiwan has been located in the distal part and the adjacent forebulge of the foreland basin system since the Late Neogene [23, 51, 52, 53, 50]. The present day forebulge is located on the central line of the Taiwan Strait [23, 51, 52]. Owing to the paragenetic relationship between the foreland basin and the adjacent mountain belt, the stratal sequences, lithofacies and subsidence history of the basin, especially those in the distal part and the forebulge of the basin system, would not only record the evolutionary history of the basin itself but also provide the most crucial indicators to infer the kinematics of the mountain-building process.

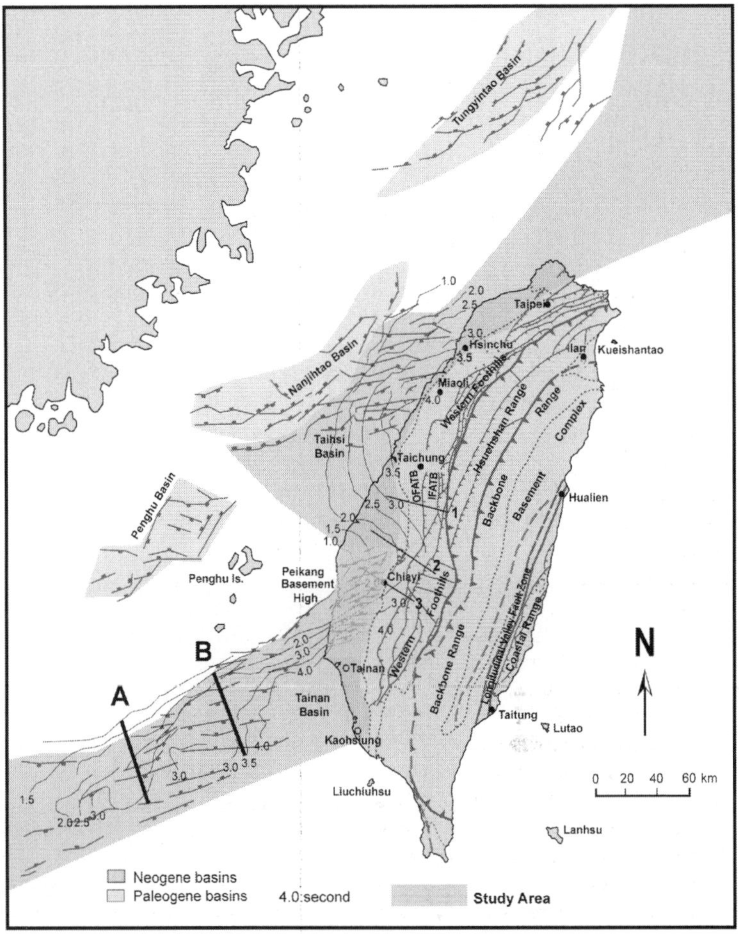

Fig. 1. Tectonic map of Taiwan and its adjacent areas. In the text of this article, the onshore foreland areas cover the alluvium and terrace gravel in the coastal plain, the outer fold-and-thrust belt (OFATB) and the inner fold-and-thrust belt (IFATB) in the Western Foothills. Note that, on this map, the Paleogene basin does not appear in the eastern part of the Taiwan Strait. This is because the Neogene settings are predominant in that area and almost overprint the Paleogene settings. WF, Western Foothills; HR, Hsuehshan Rang; BR, Backbone Range; BC, Basement Complex; LVFZ, Longitudinal Valley Fault Zone; CR, Coastal Range (From Yang et al. [50]). There are two rifted basins in the Neogene settings and the study area is in the northern margin of the southern basin.

The main purpose of this study is to integrate the well bore and seismic data to work on the analysis of subsidence history, stratal sequences and lithofacies in the distal part of the foreland basin in southwestern Taiwan (Figure 2). All the previous studies of the foreland basin in western Taiwan assumed that the mountain-building belt and foreland basin system have been continuously migrating westward to the craton since the initiation of the Taiwan orogeny [53, 54, 23, 50, 51]; they commonly regarded the deposits in the basin as a mega-sequence unit, based on which the time-spatial variation in lithofacies has been investigated and the geometry of the basin has been reconstructed. On the other hand, Yang et al. [55] analyzed the subsidence curves by Chi et al. [56] and pointed out that the tectonic development of the foreland basin and its counterpart in the mountain-building belt is episodic rather than continuous as proposed by Suppe [57]. Yang et al. [58] further proposed that the distal part of the foreland basin is characterized by superimposed unconformities in the subsurface and attributed the unconformities, with the bounded sequence units of third-order scale, and the following rapid subsidence to the episodic active thrust faulting in the mountain-building belt. The present article is the expansion of that proposed by Yang et al. [55, 58] and attempts to propose much more thorough descriptions and analysis of tectonostratigraphy in the distal part of the foreland basin in southwestern Taiwan. The first part of our works is to analyze in detail the subsidence curves in the distal part of the basin and to give a tectonic mode of epeirogenic movement in that area. We then propose two time-stratigraphy profiles based on the works of nanno-fossil age-dating from well bore data and seismic interpretation; we propose a more detailed and different tectonostratigraphic model from the previous studies. In the final part of this paper, we investigate the tectonic implications of the tectonostratigraphic model for the recent Taiwan orogeny.

2. Regional geology and previous studies

The Taiwan mountain-building belt is the result of the arc-continent collision between the Eurasia and Philippine Sea plates starting since the Pliocene (Figure 3; [59, 60, 61, 62]). During the orogeny, the westward vergent fault-and-thrust belt has also been propagating from the northeast toward the southwest owing to the oblique collision [63]. Age dating of nanno-fossil zones in the Coastal Range indicates that the age of the initial collision was 4 million years (My) [64]. Studies of sedimentary rock in the thrust belt show that the oldest sediments derived from the encroaching orogen are the Upper Pliocene [65]. However, Teng [66] suggested that the initial subsidence in the foreland basin due to tectonic loading of the growing orogenic wedge was in the Early Pliocene or at 5 Ma before present (bp). Teng [67] proposed that, according to the kinematic analysis of relative movement between the Eurasia and Philippine plates, the northern tip of the Philippine Sea plate started to collide with the boundary between continental and oceanic crusts in the Eurasian plate around 12 to 10 My bp and obducted over the passive margin of the Eurasian plate at 7 My bp. He also proposed that the mountain-building belt in northern Taiwan started to expose subaerially and propagate southward at 5 My and that a foreland basin in western Taiwan has developed since then. Huang et al. [68, 69] suggested that the timing of arc-continent collision should be at 6.5 My bp in northern Taiwan and 5 My bp in southern Taiwan.

Prior to the orogeny, the western Taiwan was located on the Eurasian passive continental margin (Figure 3) and had been went through two discernable major phases of Tertiary extensional tectonics (Sun, 1982; Yuan et al., 1989; Lin et al., 2003; Yang et al., 2006). The

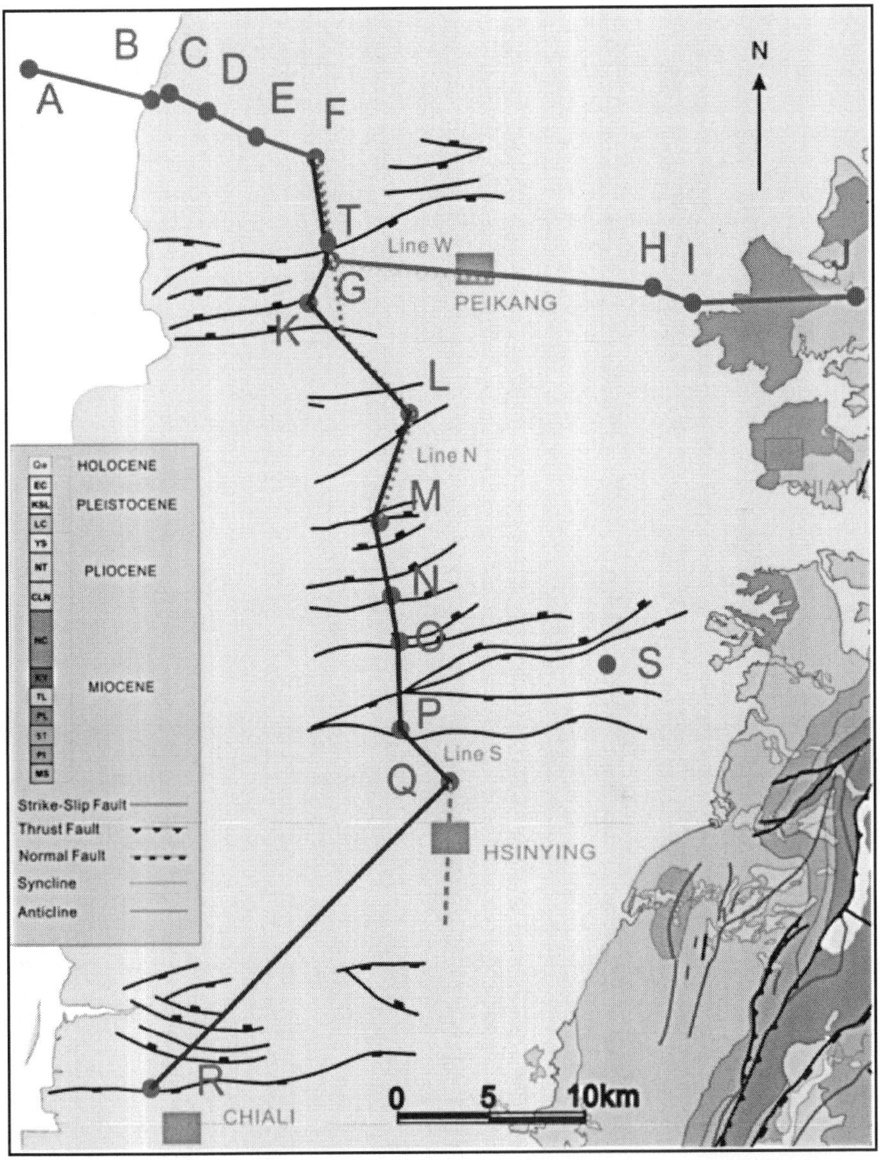

Fig. 2. Location map of well sites and seismic lines used for subsidence curve calculation and stratigraphy cross-section construction in this study. Locations of the east-west and north-south stratigraphy cross-sections are also shown in the map. The main subsurface structures in the study area are east-west striking normal faults. Part of foothills belt, which is characterized by fold and thrust structures, occurs to the east of the study area.

older phase primarily developed in the Paleogene period and ended in the Middle Oligocene [70, 71, 51, 52], forming a regional unconformity across the entire area of passive margin of southeastern China Mainland [70, 72]. However, the age of initiation of the post-rift sequences varies everywhere in the newly formed post-rift basin. In southwestern Taiwan, the age of the strata right overlying the regional unconformity is in NN1 zone [56]. The latest phase of the extensional tectonics initiated in the Middle to Late Miocene [73, 74, 70, 75 , 76, 77, 78, 71, 79, 51, 80] and formed two rifted basins separated by a basement high that stood nearly perpendicular to the front of thrust belt [80, 52]. The study area in this paper is located right on the northern margin of the southern rifted basins (Figure 1). The basin is composed of two opposite half-grabens (Figure 4), with its width increasing toward the northeast, indicating greater extension of the basin to the northeast (Figure 1; [80, 52]). The extensional tectonics is characterized by normal faults striking mainly east-west in the basin (Figures 1 and 2). The basin rifting has been propagating toward the southwest and the eastern part of the basin with greater extension now is buried beneath the fold-and-thrust belt [80]. The simple-shear model for the extension [81, 82] is applicable to interpreting the formation of the basin, because of not only the asymmetric feature across the basin but also its uplifting margins and subsiding center [70, 80]. This means that in the

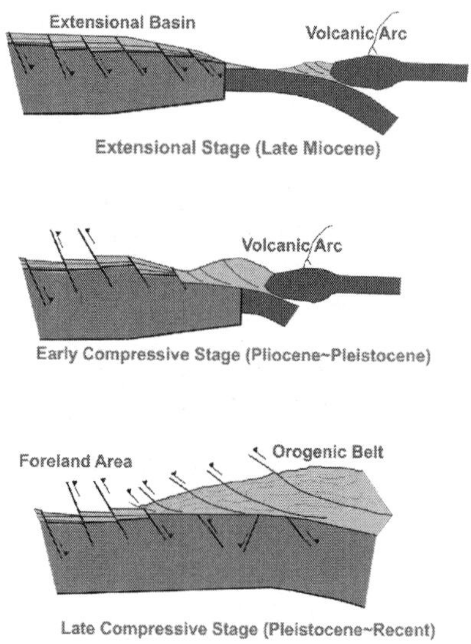

Fig. 3. Tectonic evolution of western Taiwan since the Late Miocene. Before the arc-continent collision, the passive margin of the Eurasia plate encountered the latest phase of extensional tectonics commencing from the Middle or Late Miocene. In the early stage of arc-continent collision, the preexisting normal faults that are close to the deformation front were reactivated while the extensional tectonics continued in the area to the west. In the latest stage, the inner part of the foothills belt have been dominated by low-angle thrusting while the reactivated normal faults and extensional tectonic stepped back to the foreland areas. More details about the structural setting in the foreland areas are given by Yang et al. [50].

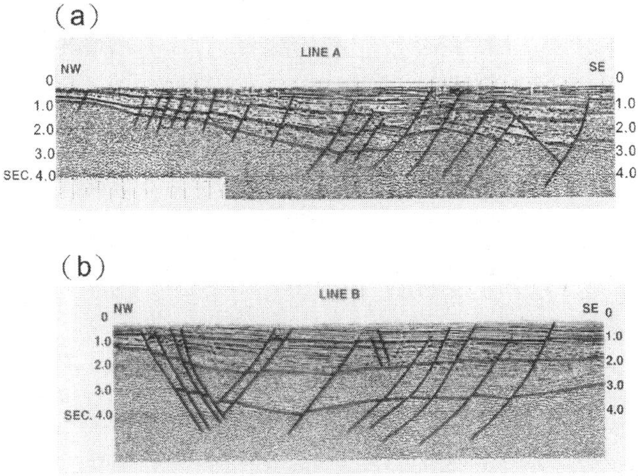

Fig. 4. Seismic sections showing half-grabens in the offshore area of southwestern Taiwan. The locations of the seismic lines are shown in Figure 1. The seismic lines run across the rifted basin, which formed during the latest phase of extensional tectonics. The rifted basin is composed of two opposed half-grabens with their widths widening to the east.

study area the foreland basin initially developed on the previously uplifted margin of the extensional basin [74, 70, 52]. Subsurface geology also indicates that normal faulting was still going on in the earlier stage of the foreland basin until the encroaching thrust sheets altered the local stress regime and caused the normal faulting to cease and step back cratonward [70, 52]. Although some studies attributed the latest stage of subsidence in the basin to the tectonic loading from the fold-and-thrust belt [83], so far no evidence has been found to indicate that the latest rifting and the recent collision were genetically related.

Referring to the geometric relationship between the passive continental margin and the thrust front, Suppe [57] assumed a constant migration rate for the westward advancement of the thrust belt and measured the rate of orogen growth and propagation along the passive continental margin. Based on the concept of critical taper of orogenic wedges, he also proposed that the mountain belt reached its today's elevation at about 3 My bp and that the cross-section width and area of the mountain range hadn't changed any longer since then [57].

The foreland areas in western Taiwan are equivalent to the areas covering the onshore fold-and-thrust belt (Western Foothills in Figure 1), the coastal plain outcropping with alluvial and terrace deposits, and the offshore Tungyintao, Nanjihtao, and Penghu basins developed in the Paleogene time (Figure 1). The fold-and-thrust belt is included as a part of the foreland area because its tectonic development has been strongly influenced by pre-existing normal faults. The coastal plain in southwestern Taiwan is located in front of the Taiwan mountain-building belt and the subsurface geological data from the coastal plain indicates that the thickness of upper Neogene, and depth of its correspondent sedimentary basin as well, increases dramatically toward the mountain-building belt to the east [70, 50, 52]. As for the offshore areas, structural cross sections [84, 85, 86, 49] through the Taiwan Strait show that the backbulge of DeCelles and Giles' [1] foreland basin system is almost equivalent to the linear zone connecting the Tungyintao, Nanjihtao, and Penghu basins; therefore, the entire Taiwan Strait can be considered as the western part of the foreland.

Covey [53] was the first study on the foreland basin in western Taiwan; he suggested that the strata of the upper Pliocene to Recent in the foothills belt and coastal plain are the foreland basin sequences, which are characterized by the sedimentary facies that are shallower upward and orogenward and record the history of westward migration of the orogenic belt. Based on Suppe's [57] kinematic model of orogenic evolution, Covey [54] proposed that an underfilled foreland basin (corresponding to deeper water facies) in the early stage evolved into an overfilled one (corresponding to shallow water and continental facies) in the later stage and then entered a steady state in the following stage, i.e., the subsidence was balanced by the sediment accumulation in the basin. During cratonward migration of the orogenic belt the sequences in the proximal part of the basin were cannibalized into the mountain-building belt and a new space was created for sediment accumulation at the same time in the distal part of the basin. Thus, cross-secton width and area of the basinal profile remained constant through the time. This may be observed in the sedimentary sequences which monotonously consist of prograding shallow marine, deltaic and fluvial deposits. Covey [53] suggested that the initiation of the foreland basin was at 3 My bp. Based on the analysis of subsidence history in the onshore and offshore of western Taiwan, Chou et al. [88] comes to the same conclusion.

However, some studies suggested that the initiation of foreland basin development is earlier. Sedimentary facies in the foothills belt give some signatures of tectonic loading from the mountain-building belt to the east at 5 My bp [66]. In their model for calculating the rigidity of crust in western Taiwan, Shiao and Teng [85] first proposed that the boundary between the Miocene and Pliocene is the base of the foreland basin. Yu and Chou [23], using subsurface well bore and seismic data in the offshore of northwestern Taiwan, identified a regional unconformity at the boundary between the Miocene and Pliocene, on which the Pliocene strata onlapping westward to the craton. In other words, the time gap between the overlying and underlying strata of the unconformity increases to the craton. They suggested that the characters of the unconformity indicate the erosion and uplifting in a forebulge [22]; therefore, they regarded the regional unconformity as the base of the foreland basin. The unconformity in southwestern Taiwan that was previously interpreted as the initiation of the latest stage of extensional tectonics was also regarded as the southward extending part of that in northwestern Taiwan [23]. The interpretation about the age of foreland basin initiation was followed by Lin and Watts [50] and Lin et al. [51], by whom the age were more definitely assigned as 6.5 Ma.

Among the most recent studies, Simoes and Avouac [89], using the isopach maps of each sequence of the Neogene in the offshore and onshore of western Taiwan by Shaw [90], calculated the rate of westward migration of the orogenic belt since the foreland basin initiated at 6.5 My bp. Tensi et al. [91] also used Shaw's [90] isopach maps to infer the location of the forebulge during the past 12 My. Both studies pointed out that the inferred migration rate of the mountain-building belt and its associated foreland basin is not coincident with the rate, which is well constrained, of relative motion between the Eurasia and Philippine Sea plates today. The rate of southward migration of the mountain-building belt by Simoes and Avouac [89] is far less than that of relative motion between the plates and the inferred locations of the forebulge before 5 My bp by Tensi et al. [91] which implies an unreasonable width of the foreland basin.

3. Analysis of subsidence history

Subsidence curves (Figure 5) from the distal part of the foreland basin of southwestern Taiwan were calculated using stratigraphic thicknesses from well bore data. Subsidence histories were determined by "backstripping" methods [92, 93]. Decompaction constants used to restore stratal thicknesses are from the empirical porosity-vs.-depth curve compiled by Sclater and Christie [93]. The thickness ratio of sandstone to shale was measured to decide the decompaction constant for each sequence unit. To acquire the original strata thickness for constructing the burial history and the subsidence curves, we adopt the biostratigraphic units of nonno-fossil zone, which are regarded as more corresponding to the chronostratigraphic units, as an interval to be decompacted. The correlation between absolute ages and nanno-fossil zones was based on the works of [64] and Okada and Bukry [94].

In addition to the decompaction corrections, which gave the curve of "total subsidence", effect of sediment loading, depositional water depths and eustatic sea level fluctuation were also corrected in order to obtain the subsidence which is caused by tectonic loading. Traditionally, lithofacies and their corresponding depositional environments have been used to infer the local relative sea level. Nevertheless, the controlling factors for lithofacies change are more complicate. In order to avoid the controversies in interpreting the depositional water depths, statistic method based on foraminiferal fossil assemblages was adopted to evaluate the paleo-water depths for subsidence curve calculation. The range of eustatic sea level fluctuations during the Neogene was less than 250 meters [95]. This component of accommodation was not corrected from the subsidence curves shown below but does not affect the position of major inflection points of the curves.

All the subsidence curves were calculated based on the strata overlying the regional unconformity that represents the end of the Paleogene extensional tectonics and the beginning of the post-rift stage in the Neogene time. As mentioned above, the latest phase of extensional tectonics initiated in Middle to Late Miocene with uplifting in the basin margin and a remarkably localized unconformity occurs in the subsurface of coastal plain of southwestern Taiwan.

The correlative conformity from a continuous succession of well bore data can be viewed as the onset of uplifting and can be obtained by comparing the complete succession with the nearby eroded ones. The onset of uplifting is around 10 My bp for all the calculated subsidence curves. The magnitude of erosion by uplifting were estimated and used to calculate a complete subsidence history prior to the onset of the uplifting. To estimate the thickness of the eroded strata and their sedimentation rate, we used some wells such as J and R, which preserve the continuous succession in the localities closer to the basin center. The sedimentation rate then was multiplied with the time gap between the age of strata right underneath the unconformity and that of correlative conformity to obtain the thickness of eroded strata for each drilled well. It must be emphasized that, since the drilled wells with thickness of complete succession are located closer to the basin center, where is characterized by greater magnitude of subsidence, the thickness of eroded strata for the other drilled wells should be the maximum estimation.

The subsidence curves indicate a gradually decreasing rate of subsidence during the post-rift phase prior to the onset of uplifting. The uplifting then was followed by another phase of rapid subsidence. The onset age of the intial rapid subsidence varies in the study area, but in general it is stepwisely younging to the northwest (Figure 5), indicating a progressive onlapping of deposits at the unconformity. However, the variation in age of initial rapid

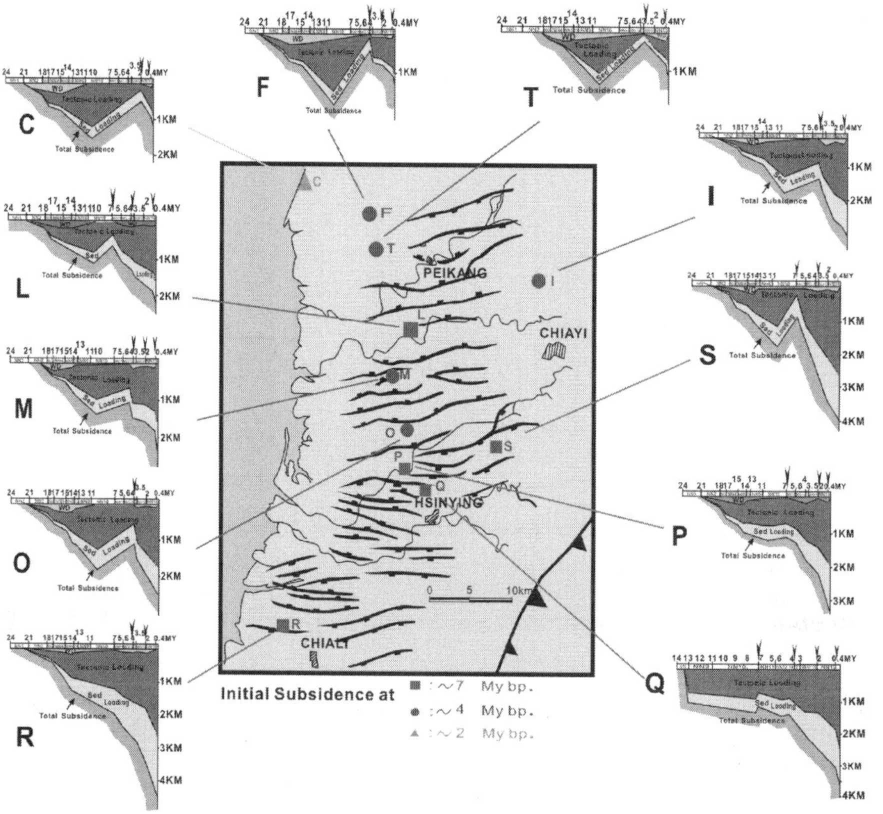

Fig. 5. Subsidence curves from drilled wells in the study area and location map of well sites with different ages of initial rapid subsidence. Arrow heads on each subsidence curve mark the ages of rapid subsidence initiation, which can be discerned as around 7, 5 to 4, 2 and 0.4 My. The well sites can be divided into three groups based on the age of initial rapid subsidence. Subsidence curves were modified from Chi et al. [56]. Detailed discussions of the subsidence curves are given in the text.

subsidence is not continuous across the entire study area; rather, the ages of onset of initial rapid subsidence can be discerned as around 7, 5 to 4, 2 and 0.4 My, although each age is within a small range of time. The well sites can be divided into three groups based on the age of initial rapid subsidence (Figure 5). In general, the well sites of younger ages of initial subsidence are located closer to the northwest. The first group of well sites with initial rapid subsidence onset at 7 My bp are located in the southeastern part of the study area. The subsidence curves show that rapid subsidence was followed by a period of slow subsidence, which in turn was followed by the next rapid subsidence with stepwise increasing rate (Figure 5). It is noticeable that even the subsidence curve from the drilled well of continuous succession shows the similar geometry of subsidence curve. The second group of well sites with initial rapid subsidence onset during 5 to 4 My bp are located to the northwest of the first group. Although the age of rapid subsidence is younger, the subsidence curves show the same geometry. The third group, which is represented by only one well site with initial rapid subsidence starting at 2 My bp, located at the northwestern part of the study area. The subsidence curve shows a convex-upward geometry, indicating an increasing rate of subsidence.

Comparison among the four groups of subsidence curves indicates that those subsidence curves that mark the earlier initial rapid subsidence also record the younger phases of rapid subsidence (Figure 5). Therefore, the entire area in the distal part of the foreland basin in southwestern Taiwan had underwent four discernible episodic phases of rapid subsidence starting at 7, 5 to 4, 2, and 0.4 My bp. The time-spatial distribution of the onset ages of rapid subsidence indicates that the area affected by the tectonic loading, which is manifested by rapid subsidence, have been progressively expanding to the northwest. Among the first group of well sites, well G is an exception; it is located among the well sites of younger rapid subsidence. Its tectonic implications will be deciphered in the next section after we demonstrate the subsurface structural and stratigraphic features from seismic interpretation.

4. Time-stratigraphy architecture

The local exploration geologists who work on the subsurface geology in the coastal plain of southwestern Taiwan used to adopt the boundaries of lithostratigraphic unit when working on the stratigraphic correlation. The lithofacies identification for each litho-stratigraphy unit is based on that of type section occurring in the foothills belt. Using the boundaries of litho-stratigraphy units for stratigraphic correlation is applicable to the strata deposited in the Miocene, during which the subsidence was slow in comparison with that in the later foreland basin period (Figure 5) and eustatic sea level fluctuation overwhelmed tectonics in shaping the stratigraphic architecture. Under such condition, the boundaries of litho- and chronostratigraphic units are nearly parallel or even identical [75, 77]. However, the situation is different for the foreland basin sequences, of which lithofacies change dramatically from that occurs in the foothills belt to that in the subsurface in the coastal plain [53, 54]. The lateral lithofacies changes even appear in local area; biostratigraphic study results from the southern part of this study area by Wu et al. [96] indicate that the sedimentary cycles are not coincident to the eustatic sea level fluctuation and, instead, strongly affected by tectonics. The study of sedimentary environments of the section in the foothills belt also gives the same conclusion [97]. In order to accurately illustrate time-spatial distribution of strata in the distal part of the foreland basin, we used the top of nanno-fossil zones from the drilled wells as the time line to construct a time-stratigraphic cross-section across the study area. Previous studies [74, 53, 77, 78, 23, 52] have shown that the base of the latest rift settings, and the foreland basin as well, in southwestern Taiwan deepens from the basement high in the northwestern part of the study area to the east and south. We constructed both north-south and east-west lines of time-stratigraphy cross-section; the former could be used to investigate the effect of the normal faulting on the stratigraphic architecture while the later would illustrate a typical asymmetric stratigraphic profile across a foreland basin.

The line of east–west stratigraphy cross-section extends from the near shore area in the west to the area very close to the thrust front of the foothills belt in the east (Figure 2). The well sites are mostly located in the western and eastern parts of the line. The stratigraphic architecture in the middle part is mainly constrained by seismic interpretation, which will be addressed below. The stratigraphy cross-section (Figure 6a) shows an eastward thickening succession, which is punctuated by two regional unconformities, which become conformities to the east. The localities of transition between the unconformities and their correlated conformities shift to the west. The time gap between the strata underlying and overlying the unconformities also increases to the west. Well J is at the eastern end of the line and its succession is continuous from the sequences of extensional basin to that of foreland basin. To its west, strata of NN11-13 onlap at the unconformity that is at the base of the foreland basin and pinch out at very short distance. In the middle part of the line the strata of NN11-13 appear again but terminate at a southeast-dipping normal fault that can be unequivocally interpreted on the seismic section line W (Figure 6b). The normal fault is the major boundary fault occurring along the northern margin of the latest extensional basin. Distribution of strata NN-14 continuously extends to the west until where they are truncated by another younger regional unconformity. The unconformity at the base of the strata NN 11-13 and 14 also merges with the younger unconformity to the west and forms another unconformity with large time gap between the underlying and overlying strata. Strata of NN 15-18 are truncated by the younger unconformity in the area between wells G and H; the stratigraphic architecture of transition between unconformity and the correlated conformity is constrained by the interpretation on seismic section line W (Figure 6b). Strata of N19 are ubiquitously overlying the younger unconformity and its correlated conformity across the entire cross-section. Comparing to entire foreland basin sequences, especially those between two unconformities and their correlated conformities, the eastward thickening of each unit of nanno-fossil zone, except strata of NN11, is not so remarkable. The most widespread units, NN14 and NN19, uniformly spread to the west where their thickness obviously decreases or they are truncated by the younger unconformity.

In order to illustrate the variation in stratal units across the boundary normal fault zone, part of the northern segment of north-south stratigraphy cross-section was designed to overlap with east-west cross-section (Figure 2). The stratigraphic architecture on north-south stratigraphy cross-section (Figure 7a) is similar to that on east-west cross-section; the foreland basin sequences thicken toward the basin center to the south and are truncated by two major regional unconformities, of which the spatial distribution of the younger one steps back to the north. Still, north-south cross-section shows some more complicated features of stratigraphic architecture, which are highly related to the normal faulting during the development of foreland basin. Thickening of the foreland basin sequences is not gradual but appears stepwise across several major normal faults of prominent displacement. In the northern part of the cross-section, stratal units of NN11-13 overlying the older unconformity are deposited in the downthrown side of the normal faults and separated from those deposited in the basin center where the units become conformable on the underlying strata. In the southern part of the cross-section, the strata of NN11, the oldest stratal unit overlying the older unconformity, are truncated by a local unconformity, which can be correlated to the stratal unit of NN12. Similarly, the stratal units of NN11-13, including the local unconformity between the units of NN11 and NN13, terminate at another major south-dipping normal fault. The existence of normal fault in shaping distribution of the stratal units of NN11-13 can be clearly illustrated by the interpretation on the seismic line N (Figure 7b). The boundary between the stratal units of NN11-13 and NN14, as shown on seismic section line N (Figure 7b), shows as an undulating feature, indicating erosion cutting down to the underlying strata. The scale of erosion is too significant

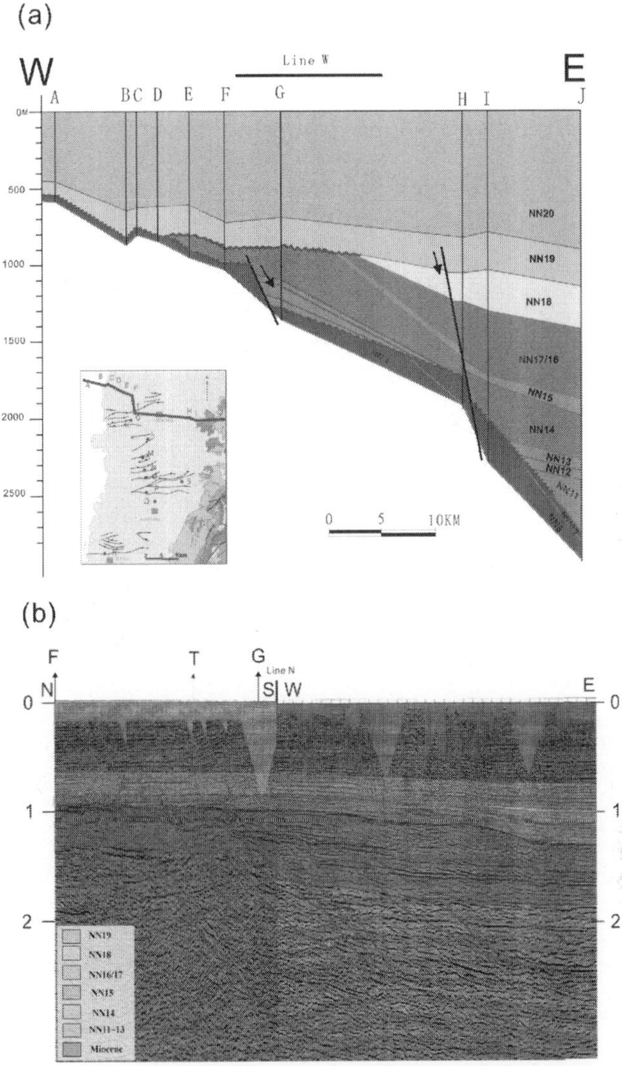

Fig. 6. Time-stratigraphic architecture of the foreland basin sequences shown by (a) east-west stratigraphy cross-section and (b) seismic line W for local stratigraphy constraint. The inset map shows location of the stratigraphy cross-section. The range of the seismic line is also marked on the stratigraphy cross-section. Although spatial distribution of strata is affected by normal faulting, the stratigraphic architecture illustrates westward younging stratigraphic settings, including the spatial distribution of the unconformities with their correlated conformities and the overlying strata. The normal fault that controls the deposition of localized strata of NN11-13 is constrained by the interpretation of the seismicc line W.

to be detected by nanno-fossil dating and, therefore, the corresponding unconformity does not appear on the stratigraphy cross-section. Nonetheless, its implication of sedimentation in the scheme of foreland basin will be addressed below. Localized distribution of the strata of NN15-17 underlying the younger unconformity is also shown on north-south stratigraphy cross- section (Figure 7a). Interpretation on the seismic section line N (Figure 7b) indicates that such distinct stratigraphic feature can be attributed to fault block tilting by normal faulting.

Wheeler diagrams (Figure 8) were constructed based on the stratigraphy cross-sections to illustrate time-spatial distribution of each nanno-fossil zone. On the east-west diagram (Figure 8a), the time gap of each regional unconformity increases to the west where the strata of NN19 directly overly the strata of NN6-7, the sequences prior to the foreland basin development. The diagram also shows that the basin margin, as defined by the boundary of eroded regime, has been migrating back-and-forth during the foreland basin development. The margin of the basin migrated toward the front of mountain-building belt to its farthest position during the NN11 zone and then started to retreat to the craton. The basin expanded cratonward rapidly to its maximum extent during the deposition of the strata of NN14. By the end of NN14, the basin margin migrated toward the front of mountain-building belt again and arrived at its easternmost position at NN8. Nonetheless, the easternmost position of the basin margin at this time is more cratonward than that at NN11, indicating cratonward shifting of the entire stratigraphic settings.

On the north-south diagram (Figure 8b), back-and-forth migration of basin margin still can be shown; however, the time-spatial distribution of strata of NN11-13 and NN15-18, which are coeval with the development of regional unconformities, are highly related to normal faulting of large displacement and occur in the downthrown side of the faults. Nonetheless, the ubiquitous unconformity-based stratal units, NN14 and NN19, basically are not affected by the normal faulting.

5. Discussions

Subsidence curves (Figure 5) and stratigraphy cross-sections (Figures 6, 7 and 8) in the study area give some important constraints for proposing any tectonostratigraphic models of foreland basin development in western Taiwan. In the sections below, we give some discussions regarding the controversies about several important related issues, including the initial time of foreland basin development, sedimentology of boundaries of third-order sequences and eustatic effects on the stratigraphy architecture. Once they are clarified, the tectonic implications of foreland basin sequences would be unraveled.

5.1 Onset time of foreland basin development

Rapid subsidence events starting at 5 to 4 and 2 My indicated by the subsidence curves can be well correlated to the unconformities and the overlying ubiquitous strata of NN14 and NN19. However, rapid subsidence event initiated at 7 My bp is not so well defined on the stratigraphy cross-sections (Figures 6, 7 and 8) because of onlapping of stratal units of NN11-13 at the unconformity occurs in the easternmost part of the study area. Since most part of the study area is located on the northern uplifting margin of the extensional basin, rapid subsidence following the uplifting at any places of the study area would be taken as the onset of tectonic loading during the foreland basin development. The onset age of the rapid subsidence at 7 My is very close to and might be correlated to that of the previously proposed initiation of the foreland basin in western Taiwan at the end of Miocene [85, 23, 49,

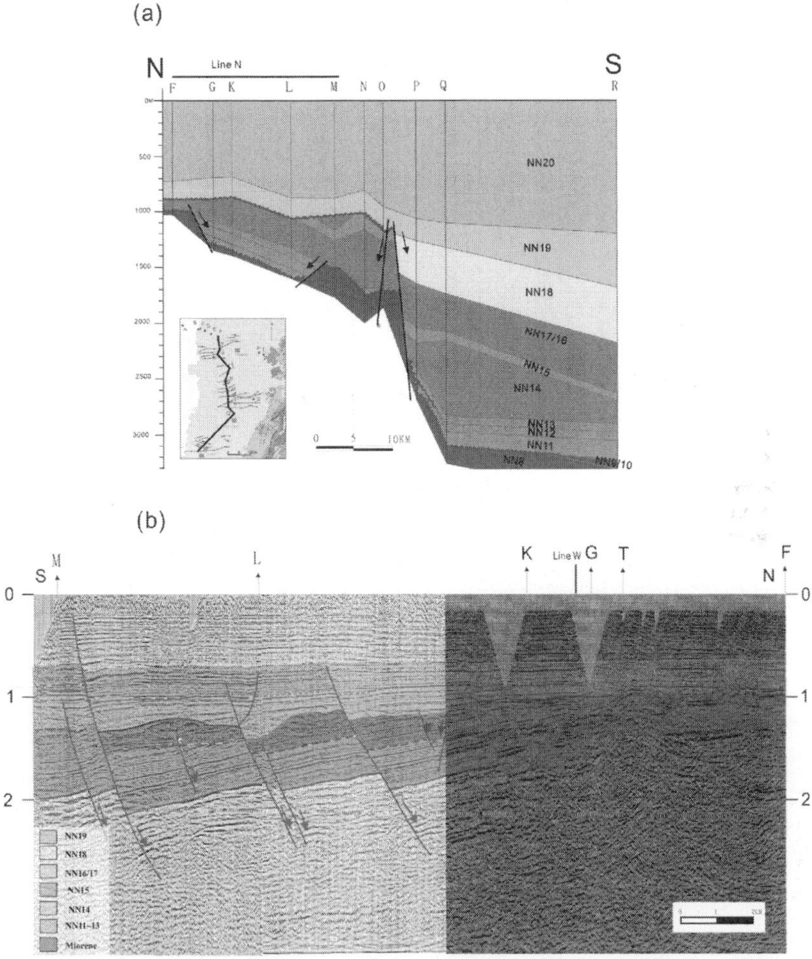

Fig. 7. Time-stratigraphic architecture of the foreland basin sequences shown by (a) north-south stratigraphy cross-section and (b) seismic line N for local stratigraphy constraint. The inset map shows location of the stratigraphy cross-section. The range of the seismic line is also marked on the stratigraphy cross-section. Spatial distribution of strata is highly affected by normal faulting; to the north of the normal fault located between wells O and P, the older unconformity is primarily at the base of strata of NN14 but is overlain by strata of NN11-13 in the downthrown side of normal fault of large displacement. To the south, foreland basin sequences are rather continuous, only disrupted by a local unconformity. Nonetheless, the stratigraphic architecture illustrates that stratigraphic settings, including the spatial distribution of the unconformities with their correlated conformities and the overlying strata, generally are younging to the north. Relationship of normal faulting with the irregular distribution of strata of NN11-13 can be illustrated on the interpreted seismic line N.

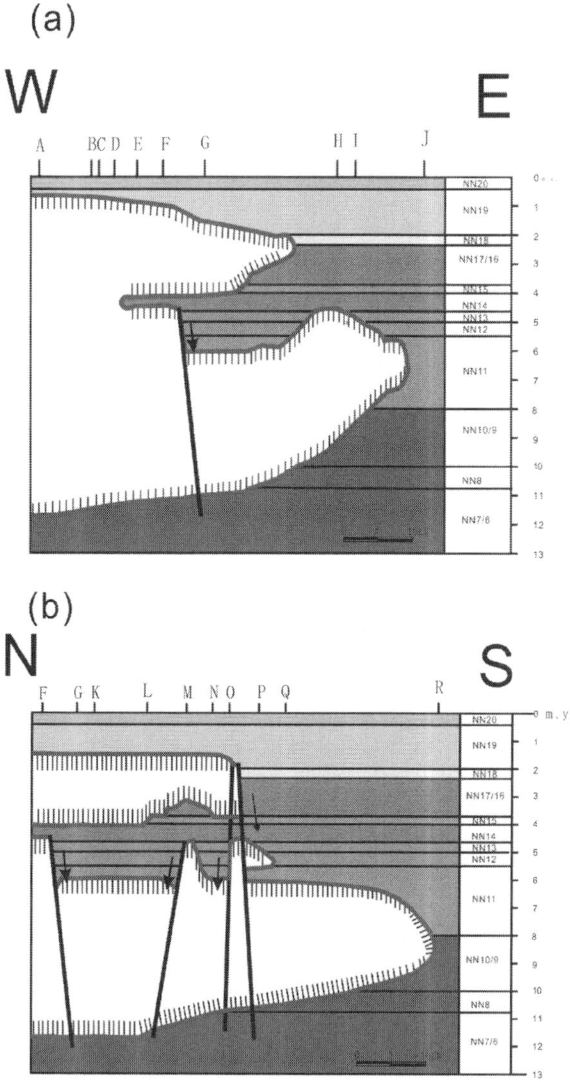

Fig. 8. Wheeler diagrams of (a) east-west and (b) north-south stratigraphy cross-sections, illustrating time-spatial distribution of the foreland basin sequences. On the east-west diagram, the time gap of each regional unconformity increases to the west where the strata of NN19 directly overly the strata of NN6-7. The basin margin, as defined by the boundary of eroded regime, has been migrating back-and-forth during the foreland basin development. On the north-south diagram, time-spatial distribution of strata of NN11-13 and NN15-18 are highly related to normal faulting and occur in the downthrown side of the faults.

50, 51, 89]. However, this contradicts some study results from northwestern Taiwan [66, 88], where the ages of initiation of foreland basin development have been dated younger than that in southwestern Taiwan. Considering the tectonic evolution of oblique collision for forming the mountain-building belt, the age of initiation of foreland basin development in the study area could only be younger than that in the northern areas. In addition, as pointed out by some previous studies [89, 91], the position of foreland basin margin in southwestern Taiwan at 6.5 My bp is not consistent with that deduced from the modern relative motion between the Eurasia and Philippine Sea plates. The occurrence of the stratal units of NN11-13 might give some answers to the controversy; the strata of the units, except that in the easternmost part of the study area where actually is in the basin center, were mainly accumulated in downthrown side of the major normal faults (Figures 6, 7 and 8), indicating that deposition of the units are highly related to the normal faulting. Therefore we suggest that the rapid subsidence onset at 7 My bp is the result of the normal faulting and rule out the possibility of foreland basin development onset at that time. We also suggest that the foreland basin had not commenced until the rapid subsidence onset at 5 to 4 My bp or the beginning of NN14.

5.2 Sedimentology of sequence boundaries

Subsidence curves (Figure 5) reveal four rapid subsidence events following uplifting and the causal unconformities. The resultant stratigraphic architecture can be divided into several sequence units of third-order scale bounded by the unconformities and their correlated conformities (Figures 6a and 7a). We demonstrate some prominent sedimentological features below to argue that the sequence boundaries are caused by or at least highly related to tectonic processes.

Well Q is located in the southern part of the study area. In terms of basin architecture, it is in the downthrown side of a major normal fault that is the demarcation between the uplifted margin and the basin center during the foreland basin development (Figures 2 and 7a). The subsurface succession from that well is continuous above the unconformity of 7 My bp between the stratal units of NN8 and NN11 (Figure 9). Sedimentary cycles of coarsening-upward can be identified as the sequence units of third-order scale and the sequence boundaries are at the unconformity-correlated conformities of 5 to 4, 2 and 0.2 My (Figure 9), distinctively. The characteristics of periodic coarsening upward sequences indicate a strong tectonic signature that can be comparable to other examples in some tectonically active belts [9, 10, 11]. The tectonic origin for the unconformities is also supported by the calculated magnitude of erosion of the unconformities in the Plio-Pleistocene [98], which scale is larger than that could be caused by eustatic sea level fall during the same period.

The well bore data was correlated to a seismic line tied to the well and the interpreted seismic profile shows that the sequence boundaries are characterized by prominent canyon morphology to the south of the well site (Figure 9). The largest scale of down-cutting morphology is the boundary at the base of the stratal units of NN19. The morphology of each submarine canyon has been studied and reconstructed in detail by Fuh et al.[99, 100] using a dense grid of seismic sections with some well bore data in the area between wells Q and R. The reconstructed morphology of submarine canyons shows that the regional trends of the axis of submarine canyons are parallel with that of the mountain-building belt, implying that the formation of the unconformities is related to the orogeny. Their studies also show that the submarine canyons gradually developed southward, consistent with the general trend of southward propagation of the mountain-building belt, but bounced back to the north relative to the preceding one when the most prominent canyon was developing at the base of NN19 [99]. Such back-and-forth pattern of submarine canyon migration is correlated to that shown on the stratigraphy cross-sections (Figures 6a and 7a) to the north.

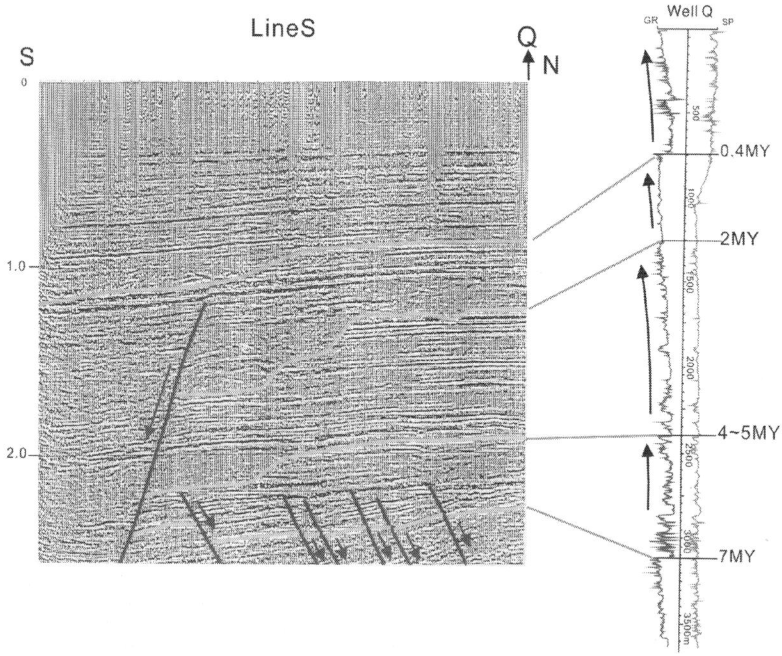

Fig. 9. Sedimentary cycles revealed by electric log data from well Q and their correlated sequences in the interpreted seismic line S. The sedimentary cycles are characterized by coarsening-upward sequences and are based by small-scaled unconformities, which turn into large-scaled down-cutting morphology of submarine canyons to the south.

5.3 Eustatic effects on stratigraphy architecture

A previous study of sequence stratigraphy of an outcropped section in the foothills belt [101] recognized several unconformities, including the one coeval with that of 2 My in the coastal plain, and suggested that the unconformities are the results of eustatic sea level fluctuation. In order to clarify the effect of eustasy on sedimentation of the foreland basin sequences in our study area, we compare the onset ages of rapid subsidence to the sea level fluctuation curves. The comparison demonstrates that the strata overlying the unconformities were deposited during the periods of sea level high stand or falling stage (Figure 10); the stratigraphy development is out-of-phase to the eustatic cycles. Therefore, the stratigraphic features in the foreland basin cannot be explained by eustatic sea-level fluctuation. Since the distribution of the strata directly overlying the unconformities is ubiquitous and not affected by the normal faults we suggest that such out-of-phase events should be related to subsidence induced by the telescopic effect of tectonic loading.

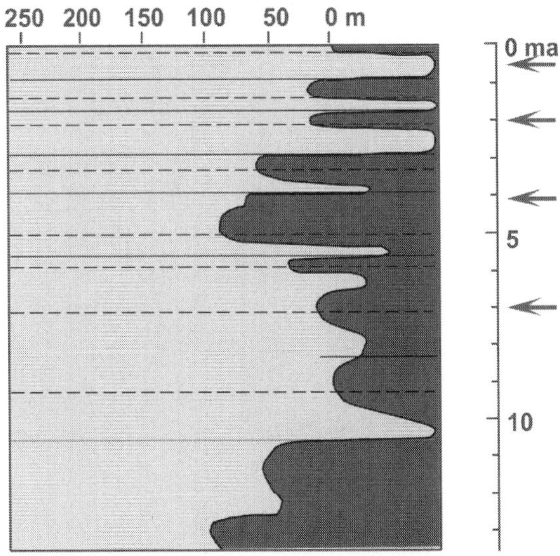

Fig. 10. Comparison of ages of rapid subsidence initiation to the eustatic sea level fluctuation [95] since the Late Miocene. All the rapid subsidence recorded in the foreland basin sequences happened in the sea level high stand or falling stages.

5.4 Tectonostratigraphic model for basin evolution

We have ruled out the possibility of eustatic effects on the stratigraphic architecture in the distal part of foreland basin. We suggest that the stratigraphic architecture, which is mainly shaped by the major unconformities, implies eastward migration of forebulge during episodic westward movement of the fold-and-thrust belt and later progradation of deposition toward the craton. Here, we propose a tectonostratigraphic model for the basin evolution in southwestern Taiwan since the beginning of latest extensional tectonics and through the foreland basin development.

The study area had been in the uplifted margin of the latest rifted basin up to 7 My bp (Figure 11a). The area encountered major normal faulting from 7 My bp to 4 My bp and syn-rift deposits were accumulated in downthrown side of the normal faults (Figure 11b), causing irregular spatial distribution of the stratal units of NN11-13 in the basin margin (Figures 7a and 8b). By the end of NN13 or at 5 to 4My bp, second phase of uplifting began (Figure 11c) and was followed by rapid subsidence and deposition of the ubiquitous strata of NN14 and the overlying strata onlapping toward the craton (Figure 11d). By the end of NN18 or at 2 My bp, the third phase of uplifting began, caused part of strata of NN14 and the overlying strata to be eroded (Figure 11e). From 2 My bp on, the uplifted area started to subside again and received another unit of ubiquitous strata of NN19 (Figure 11f.). The eroded area during the third phase of uplifting at 2 My bp, which boundary is defined by the transition between the unconformity and its correlated conformity, migrated cratonward relative to that of the preceding one.

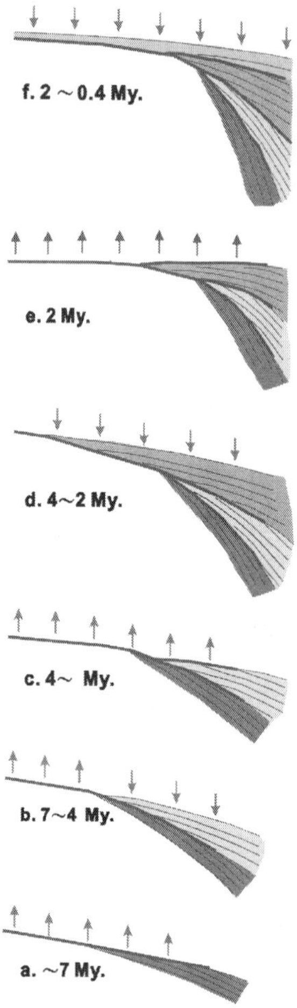

Fig. 11. Tectonostratigraphic model for the foreland basin evolution in southwestern Taiwan since the beginning of latest extensional tectonics and through the foreland basin development. Detailed descriptions for each stage are given in the text.

5.5 Tectonic implications

The characteristics of the subsidence history and stratigraphy architecture strongly imply that the foreland basin evolution favors the tectonostratigraphic model proposed by Flemings and Jordan [40] and Jordan and Flemings [41]. In their models, unconformity within the foreland basin sequences might represent active thrusting in the mountain-building belt, during which forebulge migrates toward the thrust front and causes uplifting in the distal part of the basin, while the rapid subsidence represents the later stage and quiescence of thrusting, during which sequences prograde toward the distal part of the basin and onto the unconformity. In southwestern Taiwan, sedimentation of the ubiquitous stratal units of NN14 and NN19 may represent the quiescent period of active thrusting in the mountain-building belt.

Migration rate of the forebulge in the distal part of the foreland basin can be approximately measured based on the eastern limit of each unconformity and the western limit of each westward onlapping sequences (Figure 12). The back-and-forth variation in positions of the above stratigraphic settings would provide some significant clues to infer the kinematics of recent orogen-foreland basin development. The result indicates that the migration rate was slower in the early stage but higher in the final stage than that derived from the previously proposed kinematic model of steady migration of the orogenic belt [57]. This implies that the pre-collision extensional tectonics might have caused weaker lithosphere beneath the foreland basin and that once the foreland basin migrated onto the less stretched lithosphere the basin would expand rapidly into the craton.

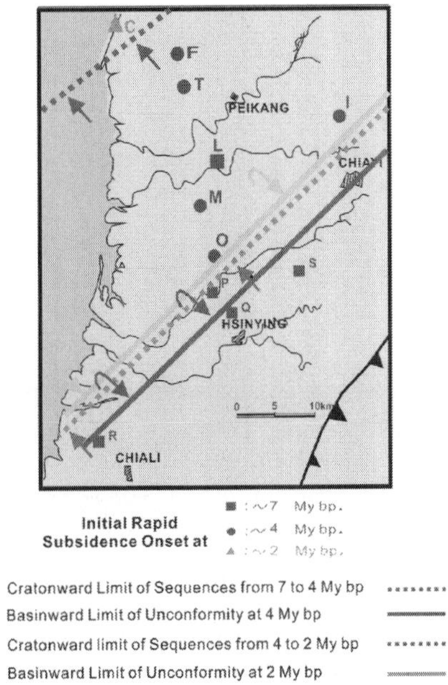

Fig. 12. Positions of the transition between the unconformities and their correlated conformities and the cratonward limitation of the distribution of unconformity-bounded sequences. The positions are determined based on the stratigraphy cross-sections (Figures 6a and 7a). Discussions of tectonic implication of variation in the positions are given in the text

6. Conclusions

1. Subsidence curves calculated from foreland basin sequences indicate that there are several rapid subsidence events at 7, 5 to 4, 2 and 0.2 My bp. The age of the initial rapid subsidence is younger toward the craton. The well sites that are characterized by earlier subsidence also record the other later events of rapid subsidence. Thus, the distal part of foreland basin encountered four discernible episodic events of rapid subsidence after the onset of rifting.

2. The stratigraphy cross-sections show that there are at least two unconformities in the foreland basin sequences, which divide the mega-sequence unit into several sequences of third-order scale. The characteristics of the unconformities are: 1, they merge into one unconformity toward the craton; 2, the time gap of each unconformity increases toward the craton, except where normal faulting created accommodation space for accumulation of older strata overlying the unconformity; 3, the spatial distribution of the younger unconformities with their correlated conformities shifts toward the craton.
3. Analysis of stratigraphy cross-sections indicates that the onset of foreland basin development was at 5 to 4 My bp, younger than that proposed in some previous studies.
4. The time-spatial distribution of the unconformities indicates back-and-forth migration of the basin margin in the distal part of foreland basin. The overlying sedimentary cycle of coarsening-upward sequences also implies the tectonic influence on the deposition of the foreland basin sequences in southwestern Taiwan.
5. The stratigraphic features in the foreland basin cannot be solely explained by eustatic sea-level fluctuation and the unconformities might result the episodic thrusting activity in the mountain-building belt to the east, which would cause eastward migration of forebulge and later progradation of deposition toward the craton.
6. The distal part of foreland basin in southwestern Taiwan had been in the uplifted margin of the latest rifted basin up to 7 My bp. The area encountered normal faulting from 7 My bp to 4 My bp. By the end of NN13 or at 5 to 4My bp, uplifting that corresponds to forebulge in the distal part of foreland basin began and was followed by rapid subsidence and deposition of the ubiquitous strata of NN14 and the overlying strata onlapping toward the craton. Close to 2 My bp, uplifting of the forebulge started again and caused part of strata of NN14 and the overlying strata to be eroded. From 2 My bp on, the uplifted area started to subside again and received another unit of ubiquitous strata of NN19.
7. The migration rate of the forebulge in the distal part of foreland basin was slower in the early stage than that derived from the previously proposed kinematic model for the steady migration of the orogenic belt. This implies that the pre-collision extensional tectonics might have caused weaker lithosphere beneath the foreland basin and that once the foreland basin migrated onto the less stretched lithosphere the basin would expand rapidly into the craton.

7. Acknowledgements

The authors like to thank Exploration and Development Research Institute, CPC Corporation, Taiwan, for providing invaluable well bore and seismic data and long-term supporting of the research works of this study. The authors thank Mr. Hsin-Hsiu Ting of Exploration and Development Research Institute, CPC Corporation, Taiwan, for reviewing and proof-reading the final manuscript. The authors also like to thank Ms. Yi-Jin Tang and Mr. Allen Zirei Yang from the Department of Earth Sciences, National Cheng Kung University, for preparing the figures and editing the text. Part of the research works of this study are the results of the project (NSC98-2116-M006-001) supported by National Science Council of Taiwan Government.

8. References

[1] DeCelles, P.G., and Giles, K.A., 1996. Foreland basin systems: Basin Research, 8: 105-123.

[2] Beaumount, C., 1981. Foreland basins: Geophysical Journal Royal Astronomical Society, 65: 291-329.

[3] Jordan, T.E., 1981. Thrust loads and foreland basin evolution, Cretaceous, Western United States: American Association of Petroleum Geologists Bulletin, 65: 2506-2520.

[4] Turcotte, D.L., and Schubert, G., 2002. Geodynamics (second edition). Cambridge University Press, Cambridge, 456 p.

[5] Shanmugam, G., and Walker, K.R., 1980. Sedimentation, subsidence, and evolution of foredeep basin in the Middle Ordovician, southern Appalachian: American Journal of Science, 180: 478-496.

[6] Heller, P.L., Bowdler, S.S., Chamber, H.P., Coogan, J.C., Hagan, E.S., Shuster, M.W. and Winslow, N.S., 1986. Time of initial thrusting in the Sevier orogenic belt, Idaho-Wyoming and Utah: Geology, 14: 388-391.

[7] Homewood, P., Allen, P.A. and Williams, G.D., 1986. Dynamics of the Molasse Basin of western Switzerland, in Allen, P.A., and Homewood, P., editors, Foreland Basins, Special Publication of the International Association of Sedimentologists, Blackwell Scientific Publication, Oxford, p. 199-218.

[8] Jordan, T.E., Flemings, P.B., and Beer, J.A., 1988. Dating thrust-fault activity by use of foreland basin strata, in Kleinspehn, K. L., and Paola, C., editors, New Perspectives in Basin Analysis, Springer-Verlag, New York, p. 307-330.

[9] Hayward, A.B., 1984. Sedimentation and basin formation related to ophiolite nappe emplacement, Miocene, SE. Turkey: Sedimentary Geology, 40: 105-191.

[10] Blair, T.C., and Bilodeau, W.L., 1988. Development of tectonic cyclothems in rift, pull-apart and foreland basins: Sedimentary response to episodic tectonism: Geology, 16: 517-520.

[11] Heller, P.L., Angevine, C.L. and Winslow, N. S., 1988. Two-phase stratigraphic model of foreland basin sequences: Geology, 16: 501-504.

[12] Allen, P.A., Homewood, P. and Williams, G.D., 1986. Foreland basin: An Introduction, in Allen, P. A., and Homewood, P., editors, Foreland Basin, Special Publication of the International Association of Sedimentologists, Blackwell Scientific Publication, Oxford, p. 3-12.

[13] Kleinspehn, P.B. and Poala, C., 1988, New perspectives in basin analysis, Springer-Verlag, New York, 453 p.

[14] Jacobi, R.D., 1981. Peripheral bulge-A causal mechanism for the Lower/Middle Ordovician unconformity along the western margin of the Northern Appalachian: Earth and Planet Science Letters, 56: 245-251.

[15] Speed, R.C., and Sleep, N.H., 1982. Antler orogeny and foreland basin: A model: Geological Society of American Bulletin, 93: 815-828.

[16] Mussman, W.J., and Read J.F., 1986. Sedimentology and development of a passive- to convergent-margin unconformity: Middle Ordovician Knox unconformity, Virginia Appalachians: Geological Society of American Bulletin, 97: 282-295.

[17] Cohen, C.R., 1982. Model for a passive to active continental margin transition: Implications for hydrocarbon exploration: American Association of Petroleum Geologists Bulletin, 66: 708-718.

[18] White, T., Furlong, K., and Arthur, M., 2002. Forebulge migration in the Cretaceous Western Interior basin of the central United States: Basin Research, 14: 43-54.

[19] Allen, P.A., and Allen, J.R., 1990. Basin Analysis: Principles and Applications, Blackwell Scientific Publication, 451 p.

[20] Allen, P.A., and Allen, J.R., 2005. Basin Analysis: Principles and Applications (second edition), Blackwell Scientific Publication, 560 p.

[21] Sinclair, H.D., Coakley, B.J., Allen, P.A. and Watts, A.B., 1991. Simulation of foreland basin stratigraphy using a diffusion model of mountain belt uplift and erosion: An example from the central Alps, Switzerland: Tectonics, 10: 599-620.

[22] Crampton, S.L., and Allen, P.A., 1995. Recognition of forebulge unconformities associated with early stage foreland basin development: Example from the North Alpine Foreland Basin: American Association of Petroleum Geologists Bulletin, 79: 1495-1514

[23] Yu, H.S., and Chou, Y.W., 2001. Characteristics and development of the flexural forebulge and basal unconformity of Western Taiwan Foreland Basin: Tectonophysics, 333: 277-291.

[24] Quinlan, G.M., and Beaumont, C., 1984. Appalachian thrusting, lithospheric flexure, and the Paleozoic stratigraphy of the Eastern Interior of North America: Canadian Journal of Earth Science, 21: 973-996.

[25] Tankard, A.J., 1986. On the depositional response to thrusting and lithospheric flexure: Examples from the Appalachian and Rocky Mountain basins, in Allen, P. A., and Homewood, P., editors, Foreland Basins, Special Publication of the International Association of Sedimentologists, Blackwell Scientific Publication, p. 369-392.

[26] Royden, L.H., Patacca, E., and Scandone, P., 1987. Segmentation and configuration of subducted lithosphere in Italy: An important control on thrust-belt and foredeep-basin evolution: Geology, 15: 714-717.

[27] Karner, G.D., and Watts, A.B., 1983. Gravity anomalies and flexure of the lithosphere at mountain ranges: Journal of Geophysical Research, 88: 10449-10477.

[28] Lyon-Caen, H., and Molnar, P., 1983. Constraints on the structure of the Himalayas from an analysis of gravity anomalies and a flexural model of the lithosphere: Journal of Geophysical Research, 88: 8171-8191.

[29] Wu, P., 1991. Flexure of lithosphere beneath the Alberta foreland basin: Evidence of an eastward stiffening continental lithosphere: Geophysical Research Letter, 18: 451-454.

[30] Waschbusch, P.J., and Royden, L.H., 1992. Episodicity in foredeep basins: Geology, 20: 915-918.

[31] Beaumont, C., Keen, C.E., and Boutilier, R., 1982. A Comparison of Foreland and Rift Margin Sedimentary Basins: Philosophical Transactions of the Royal Society of London Series a-Mathematical Physical and Engineering Sciences, 305: 295-317.

[32] Stockmal, G.S., Beaumont, C. and Boutilier, R., 1986. Geodynamic models of convergent tectonics: the transition from rifted margin to overthrust belt and consequences for foreland-basin development: American Association of Petroleum Geologists Bulletin, 70: 181-190.

[33] Watts, A.B., 1992. The effective elastic thickness of the lithosphere and the evolution of foreland basins: Basin Research, 4: 169-178.

[34] Lehmann, D., Brett, C.E., Cole, R. and Baird, G., 1995. Distal sedimentation in a peripheral foreland basin: Ordovician black shales and associated flysch of the western Taconic foreland, New York State and Ontario: Geological Society of American Bulletin, 107: 708-724.

[35] Lihou, J.C., and Allen, P.A., 1996. Importance of inherited rift margin structures in the early North Alpine Foreland Basin, Switzerland: Basin Research, 8: 425-442.

[36] Gupta, S., and Allen, P.A., 2000. Implications of foreland paleotopography for stratigraphic development in the Eocene distal Alpine foreland basin: Geological Society of American Bulletin, 112: 515-530.

[37] Bayona, G., and Thomas, W.A., 2003. Distinguishing fault reactivation from flexural deformation in the distal stratigraphy of the Peripheral Blountian Foreland Basin, southern Appalachians, USA: Basin Research, 15:503-526.

[38] Blair, T.C., and Bilodeau, W.L., 1988. Development of tectonic cyclothems in rift, pull-apart and foreland basins: Sedimentary response to episodic tectonism: Geology, 16: 517-520.

[39] Flemings, P.B., and Jordan, T.E., 1989. A synthetic stratigraphic model of foreland basin development: Journal of Geophysical Research, 94: 3851-3866.

[40] Flemings, P.B., and Jordan, T.E., 1990. Stratigraphic modeling of foreland basins: Interpreting thrust deformation and lithosphere rheology: Geology, 18: 430-434.

[41] Jordan, T.E., and Flemings, P.B., 1991. Large-scale stratigraphic architecture, eustatic variation, and unsteady tectonism: A theoretical evaluation: Journal Geophysical Research, 96: 6681-6699.

[42] Johnson, D.D., and Beaumont, C., 1995. Preliminary results from a planform kinematic model of orogeny evolution, surface processes and the development of clastic foreland basin stratigraphy, in Dorobek, S.L., and Ross, G.M, editors, Stratigraphic Evolution of Foreland Basin: *Society of Economic Paleontologists* and Mineralogists *Special Publication*, 52: 3-24.

[43] Jordan, T.E., 1995. Retroarc foreland and related basins, in Busby, C. J., and Ingersoll, R. V., editors, Tectonics of Sedimentary Basins, Blackwell Science, Cambridge, p. 331-362.

[44] Catuneanu, O., Beaumont, C. and Waschbusch, P., 1997. Interplay of static loads and subduction dynamics in foreland basins: Reciprocal stratigraphies and the "missing" peripheral bulge: Geology, 25: 1087-1090.

[45] Catuneanu, O., Hancox, P.J. and Rubidge, B.S., 1998. Reciprocal flexural behaviour and contrasting stratigraphies: A new basin development model for the Karoo retroarc foreland system, South Africa: Basin Research, 10: 417-439.

[46] Galewsky, J., 1998. The dynamics of foreland basin carbonate platforms: Tectonic and eustatic controls: Basin Research, 10: 409-416.

[47] Yong, L., Allen, P.A., Densmore, A.L. and Qiang, X., 2003. Evolution of the Longmen Shan foreland basin (western Sichuan, China) during the Late Triassic Indosinian Orogeny: Basin Research, 15: 117-138.

[48] Bayona, G., Cortes, M., Jaramillo, C., Ojeda, G., Aristizabal, J.J. and Reyes-Harker, A., 2008. An integrated analysis of an orogen-sedimentary basin pair: Latest Cretaceous-Cenozoic evolution of the linked Eastern Cordillera orogen and the Llanos foreland basin of Colombia: Geological Society of American Bulletin, 120: 1171-1197.

[49] Chou, Y.W., and Yu, H.S., 2002. Structural expressions of flexural extension in the arc-continent collisional foredeep of western Taiwan, in Byrne, T. B., and Liu, C. S., editors, Geology and geophysics of an arc-continent collision, Taiwan: Geological Society of America Special Papers, 358: 1-12.

[50] Lin, A.T., and Watts, A.B., 2002. Origin of the West Taiwan basin by orogenic loading and flexure of a rifted continental margin: Journal Geophysical Research, 107: 2-19.

[51] Lin, A.T., Watts, A.B. and Hesselbo, S.P., 2003. Cenozoic stratigraphy and subsidence history of the South China Sea margin in the Taiwan region: Basin Research, 15: 453-478.

[52] Yang, K.M., Huang, S.T., Wu, J.C., Ting, H.H. and Mei, W.W., 2006. Review and new insights on foreland tectonics in western Taiwan: International Geology Review, 48: 910-941.

[53] Covey, M., 1984. Lithofacies analysis and basin reconstruction, Plio-Pleistocene western Taiwan foredeep: Petroleum Geology of Taiwan, 20: 53-83.

[54] Covey, M., 1986. The evolution of foreland basins to steady state: evidence from the Western Taiwan foreland basin, in Allen, P., and Homewood, P., editors, Foreland Basins, Special Publications of the International Association Sedimentologists, 8: 77-90.

[55] Yang, K.M., Wu, J.C., and Chi, W.R., 1994. The tectonic implication of stratigraphy in the foreland basin: Program with abstracts, 1994 annual meetng, Geological Society of China, Taipei, p. 103-107.

[56] Chi, W.R., Yang, K.M., Wu, J.C., Chen, H.M., and Tang, H.M. 1987. The application of backstripping method in basin analysis and oil exploration of the Peikang area, Bulletin of Exploration and Development Research, 10: 63-88 (*in Chinese*).

[57] Suppe, J., 1981. Mechanics of mountain building and metamorphisms in Taiwan. *Memoir* of the *Geological Society of China*, 4: 67-89.

[58] Yang, K.M., Wu, J.C., Chi, W.R., and Ting, H.H., 2000. The tectonic implication of stratigraphy architecture in the foreland basin, western Taiwan (abstracts): American Association of Petroleum Geology 2000 Annual Convention, A162.

[59] Biq, C.C., 1972. Dual-trench structure in the Taiwan-Luzon region: Proceedings of the Geological Society of China, 15: 65-75.

[60] Chai, B.H.T., 1972. Structure and tectonic evolution of Taiwan: American Journal Science, 272: 389-422.

[61] Tsai, Y.B., 1978. Plate subduction and the Plio-Pleistocene orogeny in Taiwan: Petrology Geology of Taiwan, 15: 1-10.

[62] Teng, L.S. and Wang, Y., 1981. Island arc system of the Coastal Range, eastern Taiwan: Proceedings of the Geological Society of China, 24: 99-112.

[63] Suppe, J., 1984. Seismic interpretation of the compressively reactivated normal fault near Hsinchu, western Taiwan: Petroleum Geology of Taiwan, 20: 85-96.

[64] Chi, W.R., Namson, J., and Suppe, J., 1981. Record of plate interactions in the Coastal Range, Eastern Taiwan: *Memoir* of the *Geological Society of China,* 4: 155-194.

[65] Chou, J.T., 1973. Sedimentology and paleontology of the Upper Cenozoic system of Wewstern Taiwan: Proceedings of the Geological Society of China, 16: 111-143.

[66] Teng, L.S., 1987. Stratigraphic records of the Late Cenozoic Penlai Orogeny of Taiwan: Acta Geologica Taiwanica, 25: 205-224.

[67] Teng, L.S., 1990. Geotectonic evolution of late Cenozoic arc-continent collision in Taiwan: Tectonophysics, 183: 57-76.

[68] Huang, C.Y., Wu, W.Y., Chang C.P., Tsao, S., Yuan, P.B., Lin C.W. and Kuan-Yuan, X., 1997. Tectonic evolution of accretionary prism in the arc-continent collision terrane of Taiwan: Tectonophysics, 281: 31-51.

[69] Huang, C.Y., Yuan, P.B. and Tsao, S.J., 2006. Temporal and spatial records of active arc-continent collision in Taiwan: A synthesis: Geological Society of American Bulletin, 118: 274-288.

[70] Sun, S.C., 1982. The Tertiary basins of offshore Taiwan: Proceedings of the Second ASCOPT Conference and Exhibition, Manila, Philippine, p. 126-135.

[71] Yuan, J., Chen, J.T., Chow, C.T., Yang, K.M., Chen, H.M. and Lo, H.R., 1989. Evolution of Tertiary basins of western Taiwan and appraisal of hydrocarbon potentialities-Northern Basin and Peikang High region: Unpublished Report, CPC, MOEA, R.O.C. (*in Chinese*).

[72] Ru, K., and Pigott, J.D., 1986. Episodic Rifting and Subsidence in the South China Sea: American Association of Petroleum Geology, 70: 1136-1155.

[73] Hsiao, P.T., 1974. Subsurface geologic study of the Hsinying Coastal area, Taiwan: Petroleum Geology of Taiwan, 11: 27-39.

[74] Tang, C.H., 1977. Late Miocene erosional unconformity on the subsurface Peikang high beneath the Chiayi-Yunlin Coastal Plain, Taiwan: *Memoir* of the *Geological Society of China*, 2: 155-167.

[75] Leu, R.T., Chiang, C.L. and Huang, F.W.F., 1985. Study on the Pachangchi structure, Chaiyi, Taiwan: Petroleum Geology of Taiwan, 21: 13-32.

[76] Chow, J., Yuan, J., and Yang, K.M., 1986. Geological interpretation of the seismic in the Houpi area, Taiwan: Petroleum Geology of Taiwan, 22: 27-53.

[77] Chow, J., Yang, K.M., and Chen, H.M., 1987. Structural traps of the Paiho area, Southern Taiwan: Petroleum Geology of Taiwan, 23: 13-39.

[78] Chow, J., Yang, K.M., and Chen, H.M., 1988, Seismic Interpretation of the Subsurface Structures in the Yichu-Chiali Area, Southern Taiwan: Petroleum Geology of Taiwan, 24: 60-95.

[79] Yuan, J., Yang, K.M., Chi, W.R., Hu, C.C., and Chou, T.F., 1988. Episodic tectonism in the Tertiary continental margin of western Taiwan (abstracts): International symposium of Geodynamic evolution of eastern Eurasian margin, Paris, France.

[80] Yang, K.M., Ting, H.H., and Yuan, J., 1991. Structural styles and tectonic modes of Neogene extensional tectonics in southwestern Taiwan: Implications for hydrocarbon exploration: Petroleum Geology of Taiwan, 26: 1-31.

[81] Wernicke, B., 1985. Uniform-sense normal simple shear of the continental lithosphere: Canadian Journal Earth Science, 22: 8-125.

[82] Isller, D., McQueen, H., and Beaumont, C., 1989. Thermal and isostatic consequences of simple shear extension of the continental lithosphere: Earth and Planet Science Letters, 91: 341-358.

[83] Lee, T.Y., Tang, C.H., Ting, J.S., and Hsu, Y.Y, 1993. Sequence stratigraphy of the Tainan Basin, offshore southwestern Taiwan: Petroleum Geology of Taiwan, 28: 119-158.

[84] Sun, S.C., 1985. The Cenozoic tectonic evolution of offshore Taiwan: Energy, v. 10, p. 421-432.

[85] Shiao, T.W., and Teng, L.S., 1991. Flexural tectonics of Western Taiwan Foreland Basin: A preliminary study: Proceeding of the Third Taiwan Symposium on Geophysics, p. 437-446.

[86] Chou, Y.W., 1999. Tectonic framework, flexural uplift history and structural patterns of flexural extension in Western Taiwan Foreland Basin. Ph.D. dissertation, Taipei, National Taiwan University, 125 p (*in Chinese*).

[87] DeCelles, P.G., and Giles, K.A., 1996. Foreland basin systems: Basin Research, 8: 105-123.

[88] Chou, S.C., Teng, L.S., Chung, H.S., and Shiao, T.W., 1994. Geohistory of West Taiwan Foreland Basin: A preliminary study: Petroleum Geology of Taiwan, 29: 289-322 (*in Chinese*).

[89] Simoes, M., and Avouac, J.P., 2006. Investigating the kinematics of mountain building in Taiwan from the spatiotemporal evolution of the foreland basin and western foothills: Journal of Geophysical Research, 111: B10401.

[90] Shaw, C.L., 1996. Stratigraphic correlation and isopach maps of the wetsern Taiwan: TAO, 7: 333-359.

[91] Tensi, J., Mouthereau, F. and Lacomb, O., 2006, Lithospheric bulge in the West Taiwan Basin: Basin Research, 18: 277-299.

[92] Watts, A.B., and Ryan, W.B.F., 1976. Flexure of the lithosphere and continental margin basins, Tectonophysics, 36: 25-44.

[93] Sclater, J., and Christie, P.A.F., 1980. Continental stretching: An explanation of the post-mid Cretaceous subsidence of the central North Sea basin: Journal of Geophysical Research, 85: 3771-3739.

[94] Okada, H., and Bukry, D., 1980. Supplementary modficaton and introduction of code numbers to the low-latitude coccolth biostratigraphic zonation: Marine Paleontology, 5: 321-325.

[95] Haq, B.U., Hardonbol, J., Vail, P.R., 1988. Mesozoic and Cenozoic chronostratigraphy and cycles of sea-level change, in Wilgus, C.K., Ross, C.A., and Posamentier, H., editors, Sea-Level changes: an integrated approach: Society of Economic Paleontologists and Mineralogists, Special Publication, 42: 71-108.

[96] Wu, J.C., Wang, M.H., Tsai, C.C., Chi, W.R., Mei., W.W., 1994. Analysis and correlation of biostratigraphy, paleosedimentary environment in Hsinying, Tainan: Bulletin of Expoloration and Development Research, 26: 381-394 (*in Chinese*).

[97] Yeh, M.G., and Yang, C.Y., 1994, Depositional Environments of the Upper Miocene Area, Chaiyi, Taiwan: Petroleum Geolology of Taiwan, 29: 193-224.

[98] Fuh, S.C., 2000. Magnitude of Cenozoic erosion from mean sonic transit time, offshore Taiwan: Marine Petroleum Geology, 17: 1011-1028.

[99] Fuh, S.C., Liang, S.C., and Wu, M.S., 2003. Spatial and Temporal Evolution of the Plio-Pleistocene Submarine Canyons between Potzu and Tainan, Taiwan: Petroleum Geology of Taiwan, 36: 1-18.

[100] Fuh, S.C., Chern, C.C., Liang, S.C., Yang. Y.L., Wu, S.H., Chang T.Y. and Lin J.Y., 2009. The biogenic Gas Potential of the submarine canyon systems of Plio-Pleistocene foreland basin, southwestern Taiwan: Marine Petroleum Geology, 26: 1087-1099.

[101] Chen, W.S., Ridgway, K.D., Horng, C.S., Chen, Y.G., Shea, K.S. and Yeh, M.G., 2001. Stratigraphic architecture, magnetostratigraphy, and incised-valley systems of the Pliocene-Pleistocene collisional marine foreland basin of Taiwan: Geological Society of American Bulletin, 113: 1249-1271.

8

Seismic Hazard in Tien Shan: Basement Structure Control Over the Deformation Induced by Indo-Eurasia Collision

Buslov, Mikhail M.[1] and De Grave, Johan[2]
[1]*Institute of Geology and Mineralogy SB RAS, Novosibirsk,*
[2]*Department of Mineralogy & Petrology, Geological Institute, Ghent University, Gent,*
[1]*Russia,*
[2]*Belgium*

1. Introduction

The Tien Shan are an active intracontinental mountain belt that is part of the Central Asian Orogenic System (fig. 1). In general the mountain belt is composed of E-W trending ranges, separated by (sub)-parallel intramontane depressions. The western part of the Tien Shan is for a large extent situated in the Republic of Kyrgyzstan, while the eastern part is located in China. The Tien Shan can be regarded as a mobile belt between two rigid micro-continent or micro-plates: to the south this is the Tarim plate, to the north, the Aktyuz-Boordin (which in fact represents the southern part of Kazakhstan-Junggar plate) (fig. 2). The Tien Shan belt was formed during the Palaeozoic closure of the Turkestan Ocean [Allen et al., 1992; Gao et al., 1998; Chen et al., 1999; Brookfield, 2000]. Island-arc systems and micro-plates converged with the northern margin of Tarim and were accreted to Tarim in the Middle to Late Palaeozoic. Final closure of the Turkestan oceanic basin was accomplished in the Late Palaeozoic, in the Permian, when the Tarim and Tien Shan units collided with the Kazakhstan-Junggar plate [Laurent-Charvet et al., 2002; Van der Voo et al., 2006; Yang et al., 2007]. At this time, in the Permo-Triassic, this entire assemblage was also docked to the Siberian continent further north in present-day co-ordinates [Carroll et al., 1995; Levashova et al. 2007] and together these amalgamated terranes formed one of the nuclei of the current Asian continent. This Palaeozoic collision-accretion produced two ophiolite- and UHP-rock bearing suture zones [Allen et al., 1992; Zhang et al., 1993; Gao et al., 1998; Zhang et al 2002a; 2002b; Bazhenov et al., 2003; John et al., 2008]. During the various subduction and accretion episodes, the Tien Shan units were vastly intruded by granitoids [Solomovich & Trifonov, 2002; Mao et al., 2004; Konopelko et al., 2007; Solomovich, 2007; Pirajno et al., 2008; Van der Voo et al., 2006; Kröner et al., 2008]. An example of these is the extensive northern Kyrgyz granitoid batholith that is mainly of Ordovician - Silurian age, although smaller Permian plutons are found as well.
The basement of the Tian Shan consists of the Issyk-Kul (or Central Tien Shan or Ili) and Aktyuz-Boordin microcontinents. They are composed of Paleoproterozoic strongly metamorphosed rocks, covered by Neoproterozoic and Paleozoic volcanogenic-sedimentary and sedimentary rocks [Buslov et al., 2003; Mao et al., 2004; Konopelko et al., 2007; Solomovich, 2007; Wang et al., 2007a; 2007b]

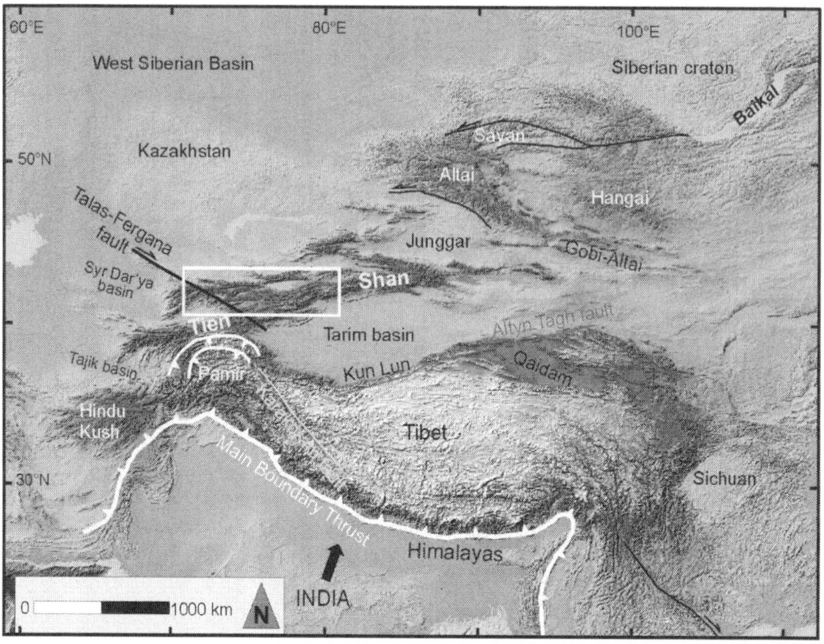

(a)

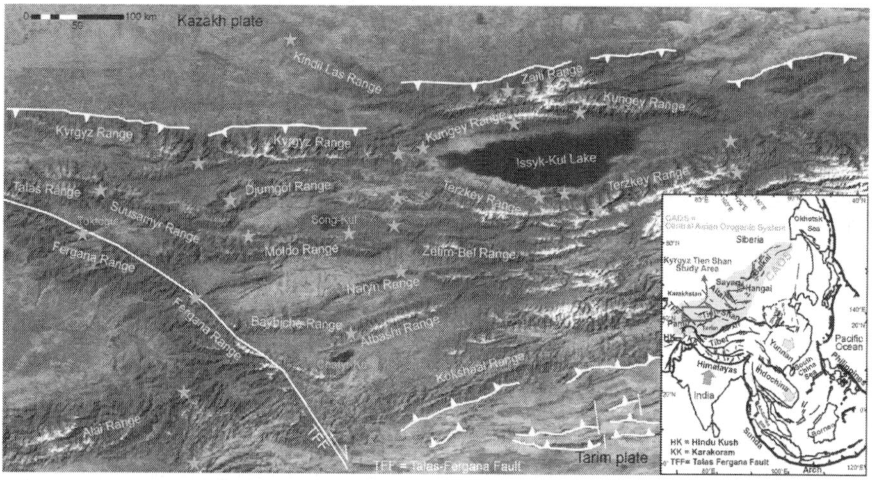

(b)

Fig. 1. (a) General location and setting of the (Kyrgyz) Tien Shan in the intracontinental Central Asian Orogenic System (CAOS). The white box corresponds to the position of Figure 1b; (b) Morphology and mountain ranges of the Kyrgyz Tien Shan. Stars indicate sample localities for geochronological investigations (see next chapter).

The ancestral Late Palaeozoic - Early Mesozoic Tien Shan belt was reactivated in the Late Mesozoic and a new mountain belt was built at this time along the inherited structural fabric. This is clearly illustrated by the foreland sediments of this Mesozoic Tien Shan range: Jurassic to Cretaceous sediments accumulated to several kilometre-thick foreland deposits in the Junggar (north) and Tarim (south) basins [Chen et al., 1991; Hendrix et al., 1992; Graham et al., 1993; Carroll et al., 1995; Cogné et al., 1995; Métivier & Gaudemer, 1997; Hendrix, 2000]. Also, the current fault-controlled intramontane basins - such as the Issyk-Kul basin - were formed as shown by their basal Jurassic and Cretaceous sequences that unconformably overlie the Palaeozoic basement [Cobbold et al., 1994]. After a period of Late Cretaceous and Early Cenozoic quiescence, the Tien Shan were again subjected to reactivation and finally evolved to the present-day intracontinental mountain belt that is still active in a mostly transpressive to purely compressive tectonic regime today [Allen & Vincent, 1997; Allen et al., 2001; Buslov et al., 2003].

These Late Cenozoic and active tectonics are associated with far-field effects of the India-Eurasia collision [e.g. De Grave et al., 2007a,b]. Erosion of the modern Tien Shan and the current transpressive to compressive tectonic regime are responsible for the exhumation of the basement-cored ranges as they appear today [Sobel et al; 2006; Oskin & Burbank, 2007], for the supply of the foreland sediments in the Tarim and Junggar basins and for the evolution of the Tien Shan intramontane basins [Heermance et al., 2007].

The most notable present-day crustal movements and their related strongest earthquakes (M>6) are concentrated at the boundaries of ancient micro-continents [Buslov et al., 2003, 2007]. The study of the Cenozoic structural position of Precambrian micro-continents, which are covered by think sedimentary units, would allow us to reconstruct the tectonic history of the Central Asian earth crust. Understanding the past tectonic events and processes would help us to study and potentially forecast modern catastrophic geological phenomena.

2. Problem formulation

2.1 Cenozoic tectonics and geodynamics of Central Asia

The main structure of the Central Asia orogenic system and of the Tien Shan in particular is formed by isometrically outlined Precambrian micro-continents (Tarim, Issyk-Kul, Aktyuz-Boordin, Junggar, Tuva-Mongolian) surrounded by accretion-collision zones. This complex crustal structure of the Tien Shan has a strong impact on the distribution of strain induced by the India-Eurasia collision. This strain is propagated northward as is indicated by the northward rejuvenation of intracontinental mountain ranges and intramontane basins (fig.2).

These micro-continents are submerged into (ancient) subduction zones and in the present-day structural pattern they appear as sub-vertically aligned rigid "roots" occurring as columns in the relatively softer matrix of the folded zones surrounding them. Since the Early Cenozoic initiation of the ongoing India-Eurasia collision, the Central Asian crust has been tectonically reactivated and its pre-Cenozoic structure appears to have been separated into 1) micro-plates or micro-continents overlain by Cenozoic sediments and 2) active fault zones in accretion-collision zones and their associated deformation systems or mountain ranges. The most important factors controlling the propagation of crustal deformation are firstly the presence of rigid micro-continents hosted by "soft" folded matrix and secondly of mantle plumes or at least diapirs of heated material and micro-plates rotating and migrating over so-called "plume heads" [Dobretsov et al., 1996; Buslov et al., 2003; Buslov, 2004].

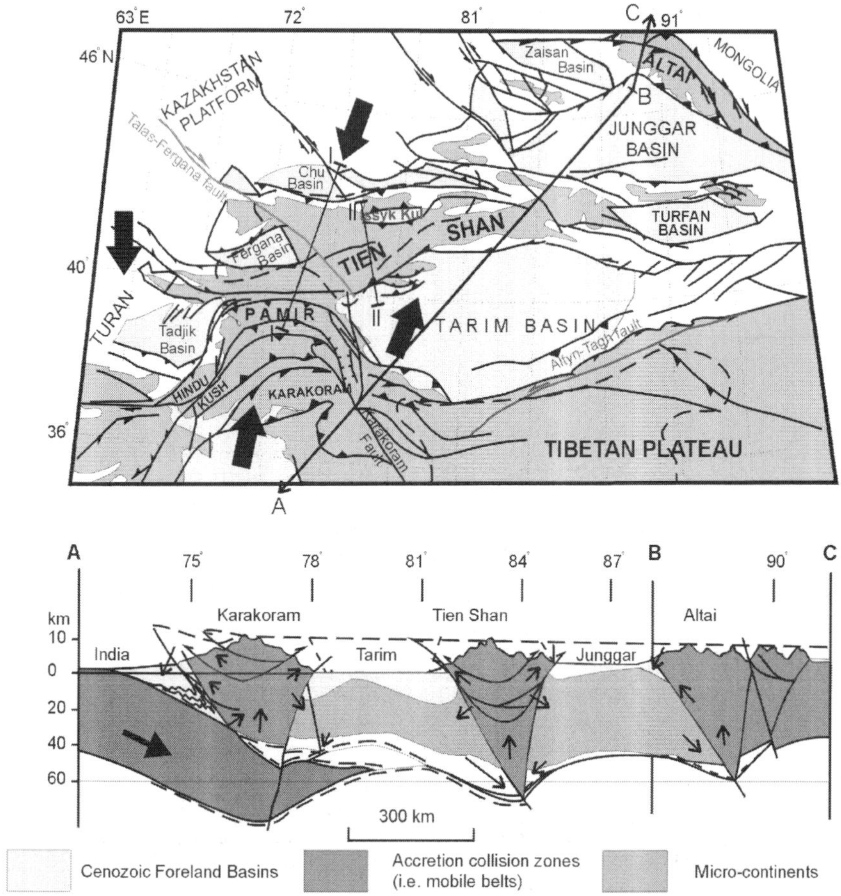

Fig. 2. Schematic Cenozoic tectonic situation and cross section (ABC) through the Central Asian Orogenic System, from India, over the Tien Shan, to the Altai region in South Siberia.

Two distinct deformation pulses have been recognized in the Cenozoic evolution of Central Asia for the period of 35-0 Ma based on new and published GPS, magneto-telluric, seismic, thermochronologic, structural and paleogeographic data on the regions of Altai, Transbaikalia, Tien Shan, Pamir, Himalaya and Tibet [Dobretsov et al., 1996; Buslov et al., 1999, 2003, 2007; Buslov, 2004; Abdrakhmatov et al., 1996, 2001; Aitchison et al., 2007; Avouac & Tapponnier, 1993; De Grave & Van den haute, 2002; Molnar & Tapponnier, 1975; De Grave et al., 2004, 2007a, b, 2008; Sobel et al., 2006; Tapponnier & Molnar, 1979].

The first deformation pulse (35-5 Ma) started relatively soon after the collision of India. This pulse is characterized by the indentation of the Pamir indenter resulting in the formation of high mountains in Tien Shan (in particular around the Issyk-Kul micro-continent) and around the Junggar plate. The reactivated basement faults and the mantle plume or diapir beneath the Tien Shan resulted in the tectonic layering of the lithosphere. During this pulse, the zone of compression and orogeny gradually extended northwards, away from the Indian collision zone and Pamir indenter to the Tien Shan, and then changing its direction somewhat to the northeast, extending over several thousands of kilometers from the Tien Shan, over Altai-Sayan to the Baikal rift zone. The deformation transmitted to such vast distances by the so-called "domino principle" (Dobretsov et al., 1996).

The second deformation pulse (5-0 Ma) was manifested in an EW-striking zone of ~600-1000 km wide extending from the Pamir and Tarim on one hand to the Siberian craton on the other. At this time, convergence and deformation induced mountain growth between these two rigid structures. At the present time this area is one of the most tectonically active parts of the earth's crust in Central Asia (Buslov et al., 2008).

In the latest Neogene - Early Quaternary mountain growth reached its peak over the whole territory of Central Asia resulting in the formation of continental molasse and catastrophic seismic events (earthquakes and landslides). Thus, the period of the NE gradual (semi-elastic or ductile) deformation transmitted from the Pamir and Tarim was followed by a period of brittle deformation and crustal hummocking or buckling over the whole territory of Central Asia between the Pamirs and the Baikal rift zone. This resulted in the formation of a strained zone between the active indenter of the Pamirs and the Siberian passive craton. The highest rates of displacement have been recorded within this zone: 20 mm/year in the Southern Tien Shan (northward migration), 10 mm/year in the Ukok plateau of the central Altai-Sayan (NE vector), 2-6 mm/year in the Northern Tien Shan (W and SE), 2 mm/year in the northern and southern parts of the Altai-Sayan area (northward and southward movements, respectively).

2.2 Tectonic layering of the Kyrgyz Tien Shan lithosphere and seismicity

The thrusting of the Tarim plate under the Southern Tien Shan and of the Pamirs onto the SW Kyrgyz Tien Shan results in the compression of the upper Tien Shan crust with a maximum velocity of 10-15 mm/yr (fig. 3), whereas India moves to the north with a velocity of 50 mm/yr relative to Central Asia and the Tien Shan in particular (Avouac et al., 1993; Avouac & Tapponnier, 1993). We suggest that a part of the convergence was accommodated by the formation of and flow in a viscous-elastic layer in the middle section of the Tien Shan crust. The deformation in the upper crust was mainly manifested along the Talass-Fergana fault and the strike-slip belts delineating and bounding the micro-continental fragments (fig. 3; cross-section I-I) as described for the Issyk-Kul area for example.

This tectonic layering of the Kyrgyz Tien Shan lithosphere implies the existence of plastic-viscous layers. Possibly, the layers can be related to the rotation of the SW Tien Shan block and the underthrusting of the Tarim plate and the indentation of its basement into the middle crust of the Tien Shan. This can explain the fast slip and deformation of the upper (20-30 km) crustal section.

Seismic S- and P-wave studies (fig. 3; cross-section I-I) show 10-20 km thick nearly horizontal wave-guide layers in the lower (35-50 km) and upper (10-20 km) crust of the Tien Shan (Bakirov et al., 1996; Sabitova and Adamova, 2001). These thick seismic wave-guide layers are however absent beneath the Fergana Basin and adjacent flat areas in the western Tien Shan. Similar wave-guide lenses are again found in the upper crust at a depth of 10-20 km, north of the Pamirs. The lower layer abruptly ascends to a shallower depth of 15-20 km near the Talass-Fergana fault. The southern part of the lower wave-guide is consequently uplifted 20 km relative to the northern part. Near the Tarim plate and NE of the Talass-Fergana fault, the wave-guide occurs at a depth of 20-40 km (fig. 3; cross-section I-I and II-

II). Near the southern margin of the Issyk-Kul microcontinent the northern extremity of the wave-guide is also located deeper, somewhat 15 km lower than its shallow southern occurrence (fig. 3; cross-section II-II).

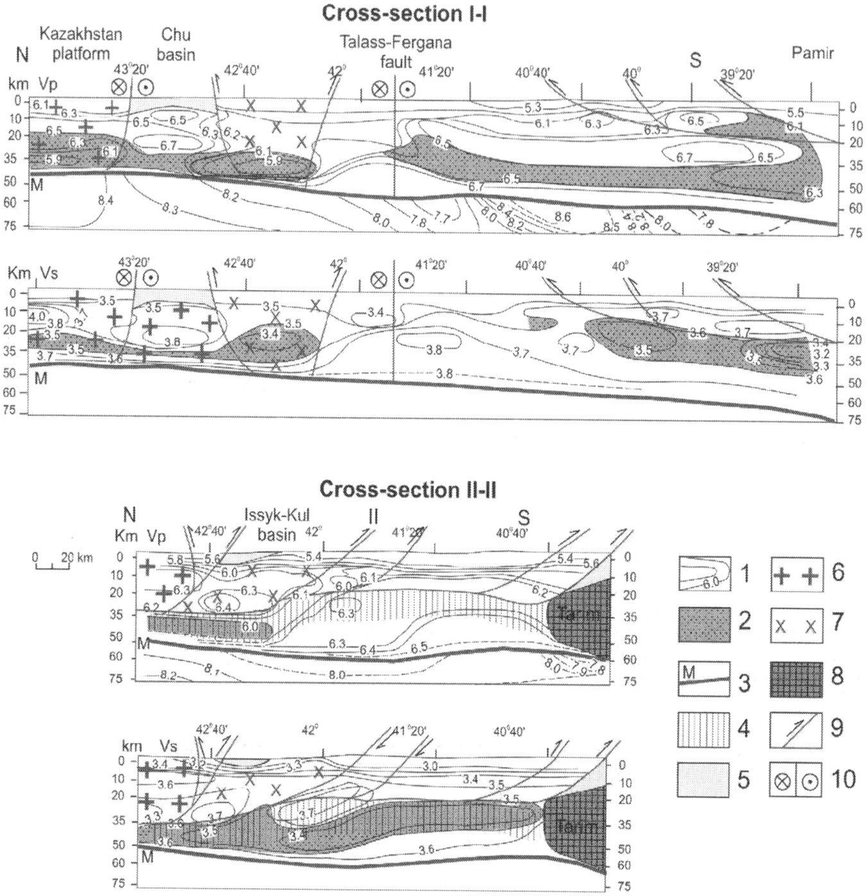

Fig. 3. Cross section I-I and II-II (locations indicated on figure 2) through the crust and upper mantle beneath the Kyrgyz Tien Shan (geophysical data after Bakirov et al., 1996): 1 - Seismic wave velocity iso-lines; 2- low seismic wave velocity zones; 3 - Moho discontinuity; 4 - high-conductive layers; 5 - Cenozoic basins; 6 - Kazakhstan platform (plate); 7 - Issyk-Kul micro-continent; 8 - Tarim micro-continent; 9 - thrust faults; strike-slip faults.

Magneto-telluric (MT) studies in the Tien Shan (Trapeznikov et al., 1997; Rybin et al., 2004, 2008) also indicate the presence of the aforementioned layers and lenses. Next to seismic wave-guide characteristics discussed above, these layers and lenses exhibit a high electric conductivity. The profiles in figure 3 show the position of the low seismic wave area (obtained by seismic data) and high conductivity layers (based on magneto-telluric data). There is a good correlation between the results of these different methods. A 15-25 km thick layer identified by MT, is located at a depth of 35-50 km to the north of the Issyk-Kul micro-continent and at a shallower depth of 20-35 km to its south. Oblique wave-guides mark the southern border of the Issyk-Kul micro-continent. Southwards, the wave-guide is located at the same depth as the basement of the Tarim micro-continent (20-50 km). The occurrence of these viscous layers would restrict the depth of faulting, and consequently the depth of earthquake hypocenters. Hypocenters of all Tien Shan earthquakes are indeed limited to a depth of less than 10-20 km (Sabitova and Adamova, 2001; Bragin et al., 2001) because obviously tectonic strain cannot exist in the viscous-plastic medium identified by the geophysical techniques.

There are three main seismically active regions in the western Tien Shan and Pamirs: the north Tien Shan (NTS), the south Tien Shan (STS) and the Pamir-Hindu Kush (PHK) (Fig. 4). The NTS generally is a low-active seismic zone, although large historical earthquakes as the 1885 Belovodskoye (M = 6.9), the 1887 Vernen (M = 7.3), the 1889 Chilik (M = 8.3), the 1911 Kebin (Kemin) (M = 8.2), the 1946 Chatkal (M = 7.5), and the 1992 Suusamyr (M = 7.5) have occurred. The 1902 Kashgar (M = 7.8), the 1907 Karatag (M = 7.3), the 1949 Khait (M = 7.4), and the 1974 Markansui (M = 7.3) earthquakes in the STS zone occurred against a background of high seismic activity. This is nicely illustrated by the recent (5 October 2008) M = 6.6 STS earthquake at the Pamir/Tien Shan thrust system near the Kyrgyz-Tajik border town of Nura that killed at least 72 people. The PHK is the most seismically active zone in Central Asia. It can be traced along the Pamir indenter front and has a characteristic S-shape. The PHK zone for example generated the 1909 Hindu Kush earthquake (M = 8.0) with a hypocenter at 230 km, several dozen 7.0 < M < 7.5 tremors with shallower focal points and thousands less strong events. The lithosphere beneath the PHK zone is unstable and is continuing to submerge into the upper mantle, providing very deep seismic sources at depths of up to 300 km. The seismic sources of the STS and NTS zones on the other hand are a lot more shallow and are located in the upper crust, mostly at a depth of 15 to 20 km, as they are constrained at depth by the viscous, plastic layering in the Tien Shan crust and lithosphere as described earlier.

Of special importance is the relation of the STS and NTS seismic zones to the mountain ranges bordering the Precambrian Issyk-Kul micro-continent. The aforementioned strong earthquakes (NTS) and high-velocity active fault movements are connected to the building of these mountain ranges. These events are clearly basement controlled as they are transpiring at the rigid body (micro-continent) borders. We can view the STS seismic events in the same context, with the Tarim microplate as rigid body thrusting underneath the Tien Shan.

3. Problem solution

3.1 Basement structure control over the deformation induced by Indo-Eurasia collision

As mentioned, micro-continents form an important feature in the tectonic collage of the Central Asian Orogenic System and were submerged into subduction zones during the formation and evolution of the Central Asian crust. In the present-day structural pattern of Central Asia, these micro-continents have sub-vertically aligned rigid roots occurring as columns in the relatively softer matrix of folded zones or mobile belts. The India-Eurasia collision, and the ongoing indentation of the Indian plate into the Eurasian continent,

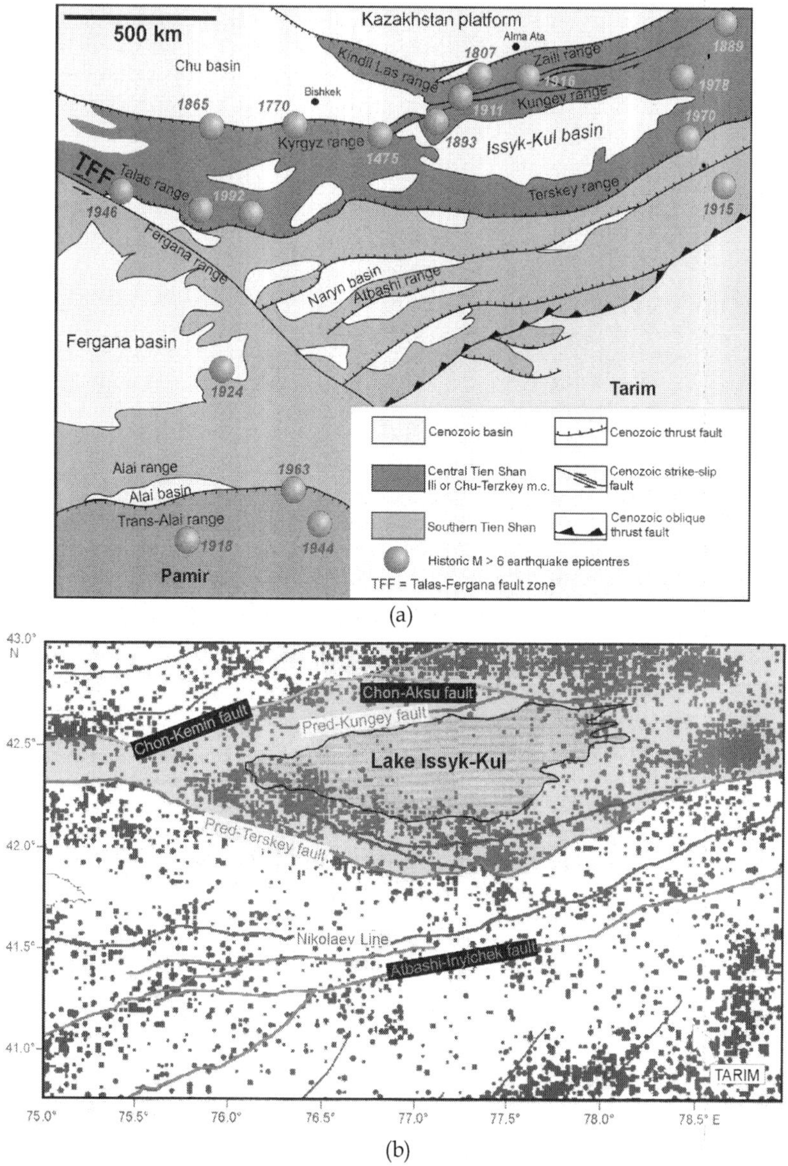

Fig. 4. (a) Present-day tectonic scheme of Kyrgyz Tien Shan with location of historic M > 6 earthquake epicentres. Most of these occurred around the edges of the Issyk-Kul (or Central Tien Shan or Ili) micro-continent. (b) M > 2 earthquake epicenters (black dots) since 1970 in the eastern Kyrgyz Tien Shan (modified after Tychkov et al., 2008). Again in this case, the epicenters cluster around the edges of the Issyk-Kul micro-continent (shaded pink).

reactivated the pre-Cenozoic structure of the Central Asian crust and separated it into (1) more or less undeformed micro-plates or micro-continents overlain by Cenozoic sediments and (2) active fault zones in accretion-collision zones and their related mountain systems (fig. 2). The most important factor controlling the propagation of crustal deformation into the Asian continental interior, away from the Indian convergence zone, is specifically the presence of these rigid micro-continents within the softer folded matrix [Dobretsov et al., 1996; Buslov et al., 2003; 2004; 2007].

In particular, during the Cenozoic, two factors affected the structure and geodynamics of the Tien Shan during its reactivation from the south: (1) thrusting of the Pamirs (west) and (2) under-thrusting of the Tarim plate (east) (fig. 2-4). These two processes are responsible for the formation of different structural-geodynamic Tien Shan provinces and are separated by the Talass-Fergana fault zone [Khudoley, 1993; Bakirov et al., 1996; Trapeznikov et al., 1997; Zubovich et al., 2001; Abad et al., 2003; Buslov et al., 2003, 2007]. These tectonic processes responsible for the complicated modern structure of the Tien Shan are still active. For example, the Issyk-Kul (or Central Tien Shan or Ili) micro-continent that hosts the Issyk-Kul basin, had been a homogenous structure during a long period prior to the latest Cenozoic, while there is evidence that young deformation reached its central part as late as the Quaternary [Trofimov, 1990]. At present, as in the case for many other Central Asian regions with rigid micro-plates in the local mobile belts, the strongest deformation processes in the Kyrgyz Tien Shan are recorded along the Issyk-Kul micro-continent margins. These margins are defined by a system of overriding thrusts that promote the further subsidence of the Issyk-Kul basin bottom and the formation of a pull-apart structure (fig. 5, 6) within the central Issyk-Kul basin [Buslov et al., 2003, 2007].

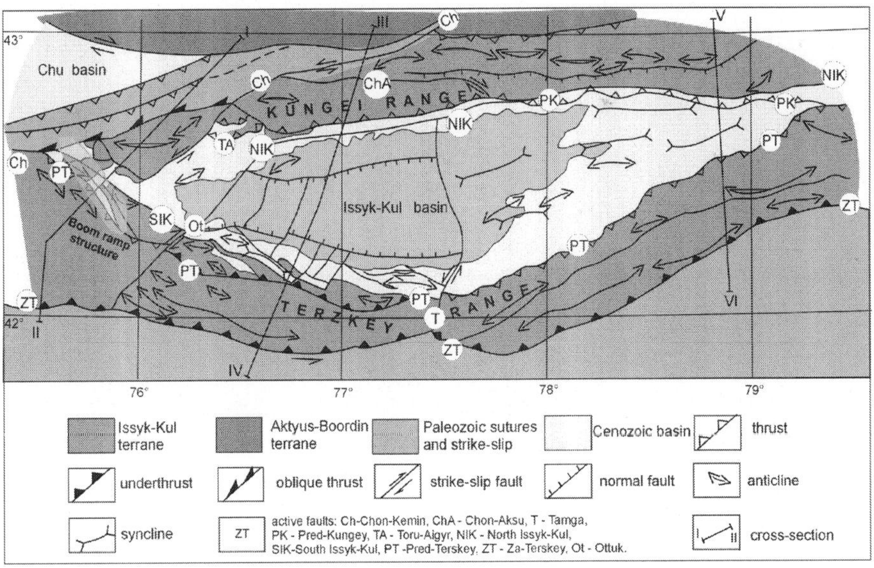

Fig. 5. Structural sketch map of the Issyk-Kul Basin and its basement, with positions of cross-sections from fig. 6.

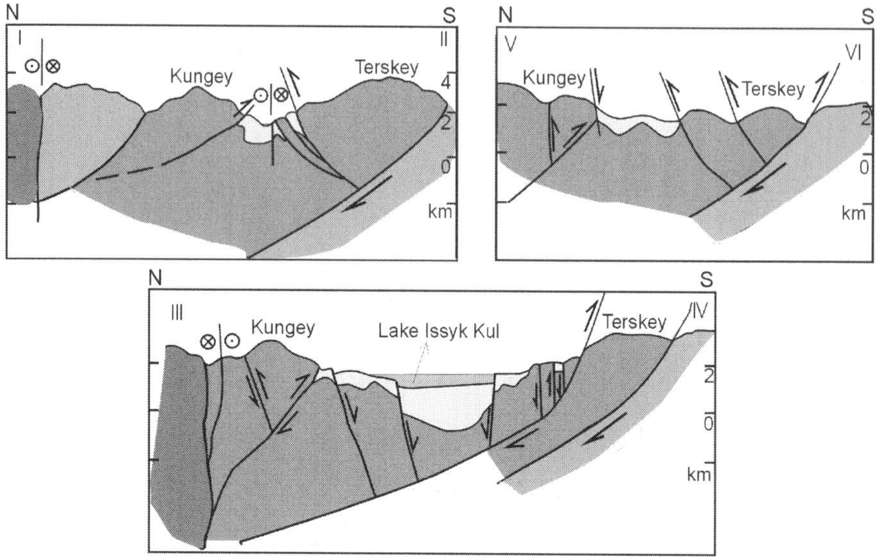

Fig. 6. Cross-sections across the western, central and eastern parts of the Issyk-Kul Basin (locations on figure 5).

The structural pattern of the pre-Mesozoic Tien Shan basement clearly influences the Mesozoic and Cenozoic reactivation of the Tien Shan orogen. This basement is highly heterogeneous and consists of the relatively rigid blocks of the Precambrian Tarim plate, the Issyk-Kul (or Central Tien Shan or Ili) micro-continent [Wang et al, 2007a; 2007b], and the Aktyuz-Boordin micro-continents (actually a part of the southern Kazakhstan platform: Zaili and Kindil Las ranges) [Djenchuraeva et al., 2008]. Their rigid bodies are enclosed in a more ductile, easier to deform matrix of Palaeozoic accretion-collision belts. It seems that these rigid blocks in the northern Tien Shan are displaced in the mobile matrix that surrounds them at variable velocities and directions with respect to each other. These displacements are considered an active far-field tectonic response to continued India-Eurasia convergence, and produces large earthquakes at the rigid body boundaries. All the epicentres of these large earthquakes are located within a narrow linear zone between the Aktyuz-Boordin and Issyk-Kul micro-continents, while other epicentres of M > 6 earthquakes [Wang et al., 2004; Tatevossian, 2007; Bourdeau & Havenith, 2008] mark the northern and southern margins of the Issyk-Kul microcontinent (Fig. 4). This indicates that crustal heterogeneity affects the formation of the active structures of the northern Kyrgyz Tien Shan [Buslov et al., 2003; 2004].

The Talass-Fergana fault zone, which is currently the site of dextral strike-slip movements, divides the Kyrgyz Tien Shan into a NE and SW part (fig. 4). As mentioned, two different tectonic mechanisms are responsible for this characteristic geometry: under-thrusting of the Tarim microcontinent under the southern Tien Shan [Neil & Houseman, 1997; Allen et al., 1999; Yang & Liu, 2002] and thrusting of the Pamir block onto the southwestern Kyrgyz Tien Shan [Lukk et al., 1995; Pavlis et al., 1997; Burtman, 2000; Coutand, 2002] respectively [Buslov et al., 2003]. The under-thrusting of Tarim and the thrusting of the Pamirs formed the Cenozoic structural pattern of the Tien Shan, but also resulted in the so-called tectonic layering of the upper lithosphere [Buslov et al., 2003; Lei & Zhao, 2007]. Different mechanisms of the Tarim and Pamir convergence were manifested in different time intervals and resulted in strike-slip movements along the Talass-Fergana fault. Of special interest is the fact that the displacement rate along the fault was 10 mm/yr during the Late Cenozoic and is as low as 2-3 mm/yr at present [Burtman et al., 1996; Meade and Hager, 2001]. This could be explained by the fact that the Tarim and Pamir convergent forces on the southern Tien Shan edges presently nearly cancel each other out (low slip rates), while earlier during the last 10 Ma the strain of the Pamir indentation dominated and resulted in the dextral strike-slip reactivation of the fault with higher displacement rates.

Given a total crustal shortening of 200 ± 50 km and the present-day shortening rate of 20 mm/yr, the northern Tien Shan could have been constructed within the last 10 Ma [Abdrakhmatov et al., 1996]. This latter study also suggest that the recent outspoken uplift of the Tien Shan could be the result of a marked increase in horizontal compressive forces following the abrupt rise of the Tibetan Plateau, as proposed by England and Houseman (1989). Apatite fission-track thermochronology, structural modelling and magneto-stratigraphy in the Kyrgyz Range (and adjoining ranges) and the Chu Basin in the northern Kyrgyz Tien Shan further showed that a first strong phase of rock uplift and deformation affected the northern Tien Shan 10-11 Ma ago, and a second accelerated phase occurred 5-3 Ma ago [Burbank et al. 1999; Bullen et al., 2001; 2003; De Grave et al., 2004; Sobel et al., 2006]. The latter acceleration is also expressed in the stratigraphy by a marked change in sedimentary environment in the Late Pliocene-Early Pleistocene: thick sequences of coarse conglomerates, sedimentary gaps and tectonic unconformities are observed [Cobbold et al., 1994; Abdrakhmatov et al., 2001].

As mentioned earlier, the Cenozoic tectonic activity of the Tien Shan is assumed to be the effect of the current indentation of India into the Eurasian continent [e.g. Molnar and Tapponnier, 1975; Cobbold and Davy, 1988]. Consequently the collision caused the propagation of deformation to the interior of the continent, resulting in crustal thickening and intracontinental mountain building. The Tien Shan consists of roughly EW-trending mountain ranges (with peaks exceeding 7000 m) alternating with sub-parallel sedimentary basins. Major Cenozoic faults mark the EW-trending boundaries between ranges and basins. However, the Talass-Fergana fault zone described above, strikes at an angle to the main structural direction and is oriented NW (fig. 1 trough 4).

The Cenozoic tectonic evolution of a vast part of Asia can be described in terms of the continuing convergence between India and Eurasia. After the initial collision [Dobretsov et al., 1996; Aitchison et al., 2007], India has continued its northward motion at reduced velocity and acts as a rigid indenter penetrating ~2000 km into Asia to cause postcollisional underplating (India) and uplift (Tibet) [Molnar & Tapponnier, 1975; Mercier et al., 1987; Cobbold and Davy, 1988; Le Pichon et al., 1992; Avouac & Tapponnier, 1993; Avouac et al., 1993; Dobretsov et al., 1996].

Orogeny in the Pamir and southern Tien Shan regions started later (post 35 Ma, Burtman, 2000) than the incipient Tibetan uplift during the collisional stage and was accompanied in the Late Oligocene by deposition of coarse-clastic, red continental molasse. The Miocene landscape of the Pamirs and the southern Tien Shan was dominated by < 3 km uplifts and intermittent depressions. In the Pliocene, red molasse gave way to grey molasse deposits as a result of climatic cooling, while the uplifts reached an altitude of 4-5 km. Further Quaternary uplift of the Pamirs and the Tien Shan produced the current typical glacial landscape [Chedia, 1986; Mikolaichuk, 2000; Abramowski et al., 2006; Zhao et al., 2006; Koppes et al., 2008].

Mechanisms of the formation of the Cenozoic structure of the Issyk-Kul micro-continental basement and its cover have been established via remote sensing, interpretation of satellite images, structural and geomorphologic studies and paleo-stress analyses. In the Palaeogene-Miocene the Issyk-Kul micro-continent remained relatively stable and about 4 km of sediments were deposited (Kokturpak, Kyrgyz, Issyk-Kul suites). Lacustrine sediments accumulated in the Palaeogene and lacustrine-proluvial ones in the Neogene-Quaternary mark the initiation of deformation. Clastic material was transported from the growing southern Tien Shan. Some fragments from the northern Tien Shan have been found in the Neogene-Quaternary deposits as well. These deposits point toward extensive Neogene erosion of the surrounding basement, and hence they imply significant denudation and exhumation of the basement. This exhumation is responsible for the cooling of the basement rocks that is observed through low-temperature thermochronologic techniques.

The tectonic activity and the deformation in the northern Tien Shan reached its peak in the Pliocene-Early Quaternary and resulted in the formation of the present-day topography and the strong deformation of the Issyk-Kul micro-continental basement and its Cenozoic cover. The direction of the regional compression changed from NW in the Late Miocene to NS in the Pliocene-Early Quaternary. The southern and northern margins of the micro-continent were hence gradually involved into thrusting and uplift and the surface area of the basin reduced. Ramp structures formed on its western and eastern ends as the mainly Ordovician - Silurian granitoid basement cored ranges were thrust onto the basin margins. The thrusting of these mountain structures over the basin was accompanied by the deposition of molasse-type sediments. Under-thrusting along the Kyrgyz-Terskey zone resulted in the splitting of the southern margin of the microcontinent into several blocks separated by oblique thrusts and strike-slip faults. Strike-slip movements along the Chon-Kemin fault resulted in thrusting and reverse faulting along the northern margin of the Issyk-Kul micro-continent (fig. 5, 6).

Although it has been intensively studied [Chedia, 1986; Trofimov, 1990; Mikolaichuk, 2000; Abdrakhmatov et al., 2002; De Batist et al., 2002; Buslov et al., 2003; Bowman et al., 2004], there is no unanimous view on the recent tectonics of the basin and its recent geodynamic evolution. The Issyk-Kul basin was formed along E-W striking normal faults and transverse faults in the Early and Middle Pleistocene. It had a much larger extension than at present, particularly to the East. It has been suggested that the lake basin originated as a pull-apart structure, which formed due to dextral displacements along strike-slip faults accompanied by regional NS compression [Klerkx et al., 1999]. About 4 km of sediments accumulated in the basin, the depocenters shifting progressively from east to west. At present, the basin has a trapezoid shape and is bounded by EW-NE-, and NW-striking faults (Fig. 5, 6). Trofimov (1990) argues that the structure of the Issyk-Kul basin evolved by the subsequent collapse of blocks in the west and east - from the periphery towards the center - while the faults and fault blocks in the south and north remained more stable. In an initial stage of recent basin evolution, during the Early and Middle Pleistocene, the displacement along the faults had reached 30-50 m. The normal faults were reactivated in the Late Pleistocene, the displacement reached 50-100 m. Catastrophic collapse of the central part of the lake, which subsided by 200 m, took place in the Middle Holocene. As a result a deep-water basin of almost 700 m depth formed during the last 10,000 years. This catastrophic collapse is supposed to be the direct cause to the regression of the Issyk-Kul in the Middle Holocene, when its water level rapidly lowered by 100 m [Trofimov, 1990]. De Batist et al. (2002) however argue that there is no evidence in the architecture of the off-shore sediment deposits to justify the hypothesis of a catastrophic collapse of the central basin floor.

The structural scheme is clearly different when considering different segments of the basin on a traverse from west, over central to east (fig. 5, 6). In the eastern part, which is well exemplified by the Boom syncline (section I-II in fig. 5), the basin has a well-defined ramp structure, delimited at both its northern and southern borders with the basement by thrust faults. Towards the east, the full ramp structure progressively evolves towards a half ramp structure. The northern area (in Russian literature, the western Pri-Issyk-Kul region), the Kungey range is thrust over the basin along the Toru-Aigyr fault [Mikolaichuk, 2000; Korzhenkov, 2000]. At the southern side the lacustrine sedimentary cover is in normal contact with the basement.

In the eastern part (section V-VI in fig. 5, 6), arched-up and linear folds widely affect the Cenozoic cover in the eastern Pri-Issyk-Kul region. The basin and basement are separated by thrusts and reverse faults on both the north and south sides. At the northern side, the Taldy-Sui oblique thrust delimits the basin. In the central part, the structural scheme is different. In the northern part, the Pred-Kungey oblique thrust joins with the Taldy-Sui thrust through the Aksui graben that constitutes the boundary between the east and central part. However, the most important structure of the northern Tien Shan is the NE-striking Chon-Kemin fault zone bounding the Issyk-Kul micro-continent on the north (fig. 5, 6). Recently, the presence of a significant sinistral strike-slip component has been evidenced in the kinematics of the Chon-Kemin fault zone [Delvaux et al., 2001, 2004], although previously it has been interpreted as a reverse fault [Chedia, 1986]. At the southern side, the Pred-Terskey fault zone bounding the Issyk-Kul micro-continent on the south (fig. 5, 6) is characterized by fault planes with complicated kinematics. The segments of the fault zone, which have different fault plane kinematics, are separated by the Tamga and Ottuk faults (fig. 5). The eastern segment - to the east of the Tamga fault - is a thrust: the Terskey range is thrust on the Cenozoic sediments. The middle segment - between the Tamga and Ottuk faults - is an under-thrust. The western segment is again a thrust. The eastern and middle segments are separated from the Pred-Terskey fault by the NS-striking Tamga fault, which sinistral offset is 700 m. The morphology of the Tamga fault is also complicated. At the junction point with the Pred-Terskey fault it is an oblique thrust. To the north, it transfers into a marginal fault (normal fault) of the subsiding trapezoid structure of the Issyk-Kul Basin and Lake. The Cenozoic sediments and Quaternary terraces, south of the lake, are affected by linear folds, as well as EW-striking flexures and reverse faults. Characteristic is the occurrence of numerous flexures and anticline folds with a gentle northern wing and steep-and-short southern wing often broken by faults, exhibiting a reverse sense of movement. The change of kinematic regime along the axis of the basin indicates various expressions of the impact of the NS compression on the Issyk-Kul microcontinent. We consequently propose that the thrusting of the Terskey range under the Issyk-Kul microcontinent resulted in the uplifting of its southern flank and subsidence of its central part. The northward dipping of erosion surfaces of the Terskey range and Cenozoic beds supports this suggestion. The transition zone between the uplifted part of the Issyk-Kul micro-continent (the western Terskey range) and the subsided part runs along the southern shoreline of the lake.

3.2 Termochronology

In order to constrain the aforementioned general Meso-Cenozoic geodynamic - tectonic history of the Kyrgyz Tien Shan in an absolute time frame, the crystalline basement rocks of several ranges were sampled for a thermochronological investigation as for example performed in the Siberian Altai Mountains [De Grave & Van den haute, 2002; De Grave et al., 2007b, 2008]. In most cases granitoid basement rocks were targeted for sampling, although a minor amount of metamorphic and metasedimentary samples were collected. Using the apatite fission-track (AFT) dating and modelling (~120-60°C) approach a continuous time-Temperature, tT-path for the different Kyrgyz Tien Shan mountain ranges can be constructed. All or some of these method were applied to the rocks of the following Kyrgyz Tien Shan mountain ranges (fig. 1b, more or less listed in a N-S sense): Kindil-Las, Zaili, Kyrgyz, Kungey, Terskey, Talas, Suusamyr, Fergana, Jumgöl, Moldo, Dzhetim Bel, Naryn, Atbashi, Alai, Trans-Alai.

In general terms, the data from most samples shows (fig. 7) that this Mesozoic event lasted until the Late Jurassic - Early Cretaceous (~150-100 Ma). Over the entire Kyrgyz Tien Shan territory the exact timing of this Mesozoic basement cooling shows some scatter, which indicates a possible punctuated or multi-phased driving force for this basement cooling. Given the ages, the specific low-T-sensitivity of the various thermochronologic methods applied, and the punctuated character, we interpret the Jurassic-Cretaceous Kyrgyz Tien Shan basement cooling as exhumation of this basement due to denudation of the overlying bedrock. The denudation and erosion can be associated with a phase of tectonic reactivation of the Tien Shan orogen and rejuvenated mountain building [Graham et al., 1993; Cobbold et al., 1994; Allen & Vincent, 1997; Métivier & Gaudemer, 1997; Hendrix, 2000; Li et al., 2004]. This Mesozoic tectonic activity and the resulting sediments are expressed in the field e.g. by Mesozoic embryonic intramontane basins (e.g. Issyk-Kul), still present in the current mountain belt, and their Mesozoic basal sediment load. Also, thick Mesozoic foreland deposits can be traced far into the Tarim, Junggar and Kazakh basins [Carroll et al., 1995; Hendrix, 2000]. In its turn, the Mesozoic reactivation and denudation can be further linked to the contemporaneous Cimmerian orogeny. This orogeny is connected to the convergence and collision of the Cimmerian blocks with the active southern margin of Eurasia. In particular, the punctuated collision of the Pamir-Tibetan blocks in the Jurassic and Early Cretaceous with Tarim/Tien Shan as a result of the closure of the Paleo-Tethys is the most probable cause of the Tien Shan reactivation. Distant effects of the collision events initiated Mesozoic movements along the inherited Tien Shan structures [Allen & Vincent, 1997].

When these movements ceased and tectonic quiescence was again installed in the Tien Shan region, the Mesozoic Tien Shan orogenic edifice was subjected to sustained erosion and peneplanation during which the underlying basement experienced only moderate exhumation and cooling. This episode is reflected in the regional thermal history reconstruction by modelled AFT data. These models (fig. 7) exhibit Late Cretaceous - Early Cenozoic near-horizontal tT-paths. The models indicate that most of the rocks sampled in the Kyrgyz Tien Shan remained at upper apatite annealing zone (APAZ) temperatures to lower AFT retention temperatures (~80-50°C), i.e. more or less the temperature interval corresponding to the crustal position the rocks reached at the cessation of Mesozoic tectonic activity and denudation. These tT-paths evoke quiescence and relaxation of the isotherms in the upper crust, and are hence suggestive of the aforementioned period of tectonic stability, peneplanation and red-bed formation.

In general the near-horizontal tT-paths are disturbed by a recent rapid cooling, corresponding to a last phase exhibited by the modelled tT-paths. This phase finally brings the samples to their present-day surface temperatures at their outcrop positions. In particular, this recent cooling brings the temperatures down from upper APAZ - lower AFT

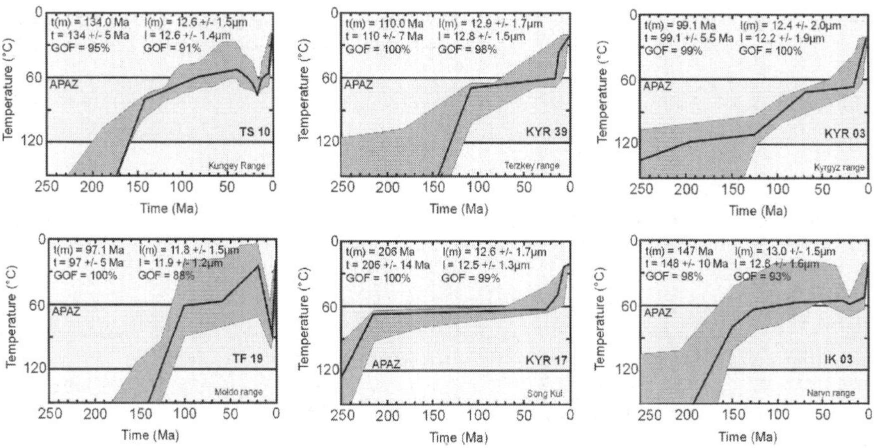

Fig. 7. Examples of thermal history models obtained on AFT age and length data from samples from the northern Kyrgyz Tien Shan batholith. For sample locations see figure 1b.

retention temperatures to ambient temperatures. The timing of this characteristic is restricted to the Late Cenozoic but shows some scatter between individual samples. Some samples exhibit this peak as early as 20 Ma ago, while others constrain the cooling to the last 5 Ma. However in most cases, it is safe to say that cooling initiated, seemingly in a somewhat poly-phased fashion, during the last 10 to 15 Ma. These cooling curves in the models might thus represent the timing of the denudation of the modern Tien Shan orogenic edifice and can be corroborated by several lines of independent geological evidence from such fields as sedimentology [Cobbold et al., 1994; Métivier & Gaudemer, 1997; Dill et al., 2007], magneto-stratigraphy [Sun et al, 2004; Charreau et al., 2005; 2006; Huang et al, 2006; Ji et al., 2008], geomorphology and structure [Tibaldi et al., 1997; Yin et al., 1998; Burbank et al., 1999; Abdrakhmatov et al., 2001; Thompson et al., 2002; Buslov et al., 2003; Fu et al., 2003; Hubert-Ferrari et al., 2007; Oskin & Burbank, 2007], geophysics [Trapeznikov et al., 1997; Bielinski et al., 2003; Rybin et al., 2004; 2008], geodesy [Abdrakhmatov et al., 1996; Reigber et al., 2001; Vinnik et al., 2004; Tychkov et al., 2008], and other geochronological studies [Sobel & Dumitru, 1997; Bullen et al., 2001; 2003; Sobel et al., 2006; De Grave et al., 2007a; Heermance et al., 2007]. We should caution the reader that this Late Cenozoic feature is mainly obtained by modelling the AFT data [Laslett et al., 1987; Ketcham et al., 2000]. Therefore a modelling artefact, recognized in AFT data modelling cannot be ruled out. However, at some places in the Kyrgyz and Terskey ranges reset, Late Cenozoic AFT ages are also found (fig. 8). For the central Terskey range for example, we find Late Cretaceous ages at higher elevations, but close to the Issyk-Kul basin, AFT ages between 17-38 Ma are obtained. In summary and given the strong independent geological evidence we interpret the Late Cenozoic, more specifically, Late Miocene to Plio-Pleistocene (15 - 5 Ma) and recent (5-0 Ma) cooling pulses as denudation of the current active Tien Shan mountain ranges as a far-field effect of ongoing India-Eurasia convergence. Moreover, this is further underscored by new U-Th-Sm/He dating results on the apatites from the Kyrgyz Tien Shan samples. Results from samples in the eastern Terskey Range yield AHe ages between 10-12 Ma and for the Zaili Range 10-33 Ma.

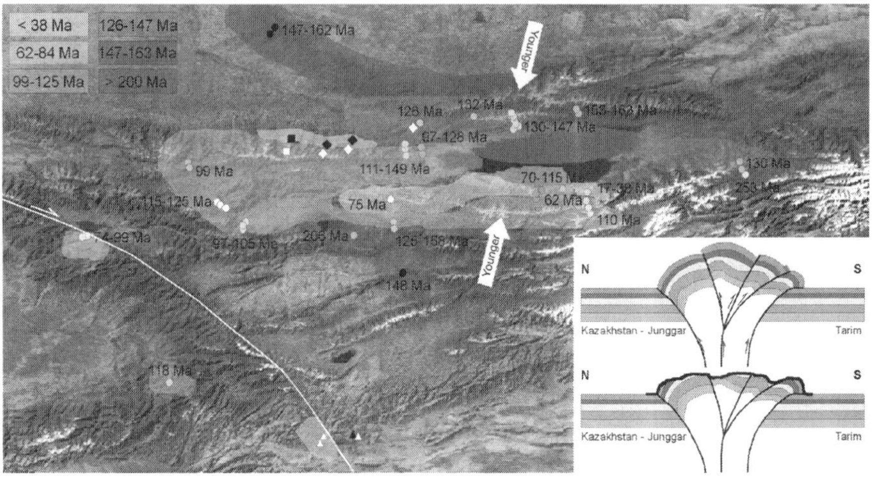

Fig. 8. AFT age zones in the Kyrgyz Tien Shan. Ages tend to get younger towards the more centrally located ranges.

Together with results from similar studies [Hendrix et al., 1994; Van der Beek et al., 1996; Sobel & Dumitru, 1997; Bullen et al., 2001; De Grave & Van den haute, 2002; De Grave et al., 2004; 2007a; 2007b; 2008; Sobel et al., 2006; Yuan et al., 2006; Jolivet et al., 2007; Vassallo et al., 2007], these observations seem to suggest that reactivation and deformation in the interior of the Eurasian continent is gradually rejuvenating northward through Central Asia since the Miocene as a distant effect of the ongoing indentation of India into Eurasia. This deformation resulted in transpressive mountain building and roughly 2 km of denudation of the northern Kyrgyz Tien Shan, in particular the mountain ranges around the Issyk-Kul micro-continent since the Late Miocene. The low-temperature thermochronology results also indicate a rejuvenation trend of ages from the north and south toward the central Kyrgyz Tien Shan mountain rages (fig. 8) that might point to an exhumed flower structure within the current transpressive to purely compressive tectonic framework.

3. Conclusions

In the Tien Shan, strong earthquakes and their triggered land-slides have been recorded within the fault zones which separate the ancient micro-continents presently migrating and moving in different directions and at different velocities. In the Cenozoic, two factors affected the structure and geodynamics of the Tien Shan: overthrusting of the Pamirs and underthrusting of the Tarim plate. These two processes were responsible for the formation of different structural-geodynamic provinces separated by the Talass-Fergana fault zone. The mantle plume activity beneath the Tien Shan resulted in the tectonic layering of the lithosphere and fast movement of the upper crust (10-15 km) over the plume head and its deformation. We believe that these tectonic and geodynamic processes for the complicated modern structure of the Northern Tien Shan and are still active. For example, the Issyk-Kul micro-continent had been a homogenous structure during a long time period, and the deformation reached its central part as late as Quaternary. At present, the strongest deformation processes are recorded along the micro-continent margins forming a system of thrusts and promoting the further subsidence of the Issyk-Kul lake bottom and formation of a pull-apart structure [Buslov et al., 2003, 2007].

Anticlockwise rotation of the Tien Shan block continued and dextral strike-slip displacement along the Talass-Fergana fault zone occurred at rates of about 10-15 mm/yr. Finally, interplay of the Tarim and Pamir convergence with the Tien Shan (~5 Ma ago to the present) gave rise to the maximum uplift and formation of the modern Tien Shan as they occur today, while the dextral strike-slip displacement along the Talass-Fergana fault zone fell to ~1 mm/yr [Zubovich et al., 2001].

The thrusting of the Tarim plate underneath the Tien Shan resulted in the shortening of the upper crust at a rate of ~10-15 mm/yr, whereas India moves northward at 50 mm/yr. A certain amount of the strain induced by the ongoing convergence may have been accommodated by the tectonic layering of the Tien Shan crust and lithosphere and the presence of viscous, plastic layers in the Tien Shan crust. The upper crustal strain has been buffered to a large extent by the Issyk-Kul micro-continent and the surrounding mobile belts. It therefore becomes clear that the recent tectonics of the northern Tien Shan is controlled by its pre-Cenozoic structure. Earthquakes of M > 6 clearly mark the northern and southern margins of the lens-shaped Issyk-Kul micro-continent, indicating that crustal heterogeneity affected the formation of the active northern Kyrgyz Tien Shan structures. Seismic and magneto-telluric studies show the tectonic layering of the Tien Shan lithosphere, with several nearly horizontal viscous layers. The lower layer is underthrust northward into the northern Tien Shan middle crust, as indicated by the seismic data shown in this paper. Tectonic layering of the lithosphere beneath the Tien Shan implies the existence of horizontal viscous layers, possibly related to the rotation of the SW Tien Shan block, and also to the underthrusting of the Tarim plate and indentation of its basement into the middle crust of Tien Shan. This caused high slip rates and failure in the upper crust to depths of 20-30 km.

Similarly to the Tien Shan, the strongest earthquakes and the highest-rate modern displacements are concentrated along the margins of micro-continents (micro-plates) in Altai as a result of the interaction of the Junggar and Tuva-Mongolian micro-continents.

4. Acknowledgments

The authors are supported by grants of the Fund for Scientific Research(FWO, Belgium) (JDG) and Russian Academy Scientist (ONZ-10.3)

5. References

Abad, I., Gutiérez-Alonso, G.Nieto, F.Gertner, I.Becker, A.Cabero , The structure and the phyllosilicates (chemistry, crystallinity and texture) of Talas Ala-tau (Tien Shan, Kyrgyz Republic): comparison with more recent subduction complexes, *Tectonophysics* , Vol.365, 2003, p.103-127.

Abdrakhmatov, K.E., Aldazhanov, S.A., Hager, B.H., Hamburger, M.W., Herring, T.A., Kalabaev, K.B., Makarov, V.I., Molnar, P., Panasyuk, S.V., Prilepin, M.T., Reilinger, R.E., Sadybakasov, I.S., Souter, B.J., Trapeznikov, Yu. A., Tsurkov, V.E., Zubovich, A.V., Relatively recent construction of the Tian Shan inferred from GPS measurements of present-day crustal deformation rates, *Nature*, Vol.384, 1996, p.450-453.

Abdrakhmatov K. Ye., Weldon R., Thompson S., Burbank D., Rubin C., Miller M., Molnar P., Onset, style and current rate of shortening in the central Tien Shan, Kyrgyz republic, *Russian Geology and Geophysics* ,Vol.42, 2001, p. 1585-1609.

Abdrakhmatov K.Y., Djanuzakov K.D., Delvaux D., Active tectonics and seizmic hazard of the Issyk-Kul basin in the Kyrgyz Tian-Shan. In: Klerkx J., Imanackunov B. (Eds), *Lake Issyk-Kul: its natural environment.* Kluwer Academic Publishers, Dordrecht, 2002, p. 147-160.

Abramowski U., Bergau A., Seebach D., Zech R., Glaser B., Sosin P., Kubik P.W., Zech W. , Pleistocene glaciations of Central Asia: results from ^{10}Be surface exposure ages of erratic boulders from the Pamir (Tajikistan), and the Alay-Turkestan range (Kyrgyzstan), *Quaternary Science Reviews*, Vol. 25, 2006, p.1080-1096.

Aitchison J.C., Ali J.R., Davis A.M., When and where did India and Asia collide?, *Journal of Geophysical Research*, vol. 212, BO5423, 2007, doi:10.1029/2006JB004706.

Allen M.B., Vincent S.J., Fault reactivation in the Junggar region, northwest China: the role of basement structures during Mesozoic-Cenozoic compression, *Journal of the Geological Society, London*, Vol. 154, 1997, p. 151-155.

Allen M.B., Windley B.F., Zhang Chi, Palaeozoic collisional tectonics and magmatism of the Chinese Tien Shan, Central Asia, *Tectonophysics*, Vol.220, 1992, p. 89-115.

Allen M.B., Vincent, S.J. Wheeler P.J. Late Cenozoic tectonics of the Kepingtage thrust zone: interactions of the Tien Shan and Tarim Basin, northwest China, *Tectonics* , Vol.18, 1999, p. 639-654.

Allen M.B., Alsop G.I., Zhemchuzhnikov V.G., Dome and basin refolding and transpressive inversion along the Karatau fault System, southern Kazakstan, *Journal of the Geological Society, London* 158, 2001, p.83-95.

Avouac J.P., Tapponnier P. Kinematic model of active deformation in Central Asia, *Geophysical Research Letters*, Vol. 20, 1993, p. 895-898.

Avouac J.P., Tapponnier P., Bai M., You H., Wang, G., Active thrusting and folding along the Northern Tien Shan, and Late Cenozoic rotation of the Tarim relative to Dzungaria and Kazakhstan, *Journal of Geophysical Research*, Vol. 98, 1993, p. 6755-6804.

Bakirov A.B., Lesik O.M., Lobanchenko A. P., Sabitova T.M., Evidences for the modern deep-seated magmatism in Tien Shan, *Russian Geology and Geophysics*, Vol. 37, 1996, p. 42-53.

Bazhenov M.L., Collins A.Q., Degtyarev K.E., Levashova N.M., Mikolaichuk A.V., Pavlov V.E., Van der Voo R., Paleozoic northward drift of the North Tien Shan (Central Asia) as revealed by Ordovician and Carboniferous paleomagnetism, *Tectonophysics*, Vol. 366, 2003, p. 113-141.

Bielinski R.A., Park S.K., Rybin A., Batalev V., Jun S., Sears, C., Lithospheric heterogeneity in the Kyrgyz Tien Shan imaged by magnetotelluric studies. Geophysical Research Letters, Vol.30, No.15, 2003, doi:10.1029/2003GL017455, p.4 .

Bourdeau C., Havenith H.B., Site effects modelling applied to the slope affected by the Suusamyr earthquake (Kyrgyzstan, 1992), *Engeneering Geology*, Vol.97, 2008, p. 126-145.

Bowman D., Korjenkov A., Porat N., Late-Pleistocene seismites from Lake Issyk-Kul, the Tien Shan range, Kyrghyzstan, *Sedimentary Geology* , Vol.163, 2004, p. 211-228.

Bragin V.D., Batalev V.Yu., Zubovich A.V., Lobanchenko A.N., Rybin A.K., Trapeznikov Yu. A., Qualitative bearing of modern movements on the deep geo-electric structure of the earth's crust in the Central Tian Shan and distribution of seicmicity, *Russian Geology and Geophysics*, Vol.42, 2001, p.1610-1629.

Brookfield M.E., Geological development and Phanerozoic crustal accretion in the western segment of the southern Tien Shan (Kyrgyzstan, Uzbekistan and Tajikistan), *Tectonophysics* , Vol. 328, 2000, p.1-14.

Bullen M.E., Burbank D.W., Garver J.I., Abdrakhmatov K,Ye., Late Cenozoic tectonic evolution of the northwestern Tien Shan: New age estimates for the initiation of mountain building, *Geological Society of America Bulletin*, Vol. 113, 2001, p.1544-1559.

Bullen M.E., Burbank D.W., Garver J.I., Building the Northern Tien Shan: integrated thermal, structural and topographic constraints, *Journal of Geology*, Vol.111, 2003, p. 149-165.

Burbank D.W., McLean J.K., Bullen M., Abdrakhmatov K.Y., Miller M.M., Partitioning of intramontane basins by thrust related folding, Tien Shan, *Kyrgyzstan. Basin Research*, Vol.11, 1999, p. 75- 92.

Burtman V.S., Cenozoic crustal shortening between the Pamir and Tien Shan and a reconstruction of the Pamir-Tien Shan transition zone for the Cretaceous and Palaeogene, *Tectonophysics*, Vol.319, 2000, p.69-92.

Burtman V.S., Skobelev S.F., Molnar P., Late Cenozoic slip on the Talas-Ferghana faults, the Tien Shan, Central Asia, *Geological Society of America Bulletin*, Vol.108, 1996, p. 1004-1021.

Buslov M.M. Cenozoic tectonics of Central Asia: basement control, *Himalayan Journal of Science*, Vol. 21, Issue 41, 2004, p. 104-105.

Buslov M.M., Zykin V.S., Novikov I.S., Delvaux D., Cenozoic history of the Chuya depression (Gorny Altai): Structure and Geodynamics, Russian Geology and Geophysics, Vol. 40, No.12, 1999, p. 1687-1701.

Buslov M.M., Klerkx, J., Abdrakhmatov K., Delvaux D., Batalev V.Yu., Kuchai O.A., Dehandschutter B., Muraliev A., Recent strike-slip deformation of the northern Tien Shan. In: Storti, F., Holdsworth, R.E., Salvini, F. (Eds). Intraplate strike-slip deformation belts. Geological Society of London, Special Publication, Vol. 210, 2003, p.53-64.

Buslov M.M., De Grave J., Bataleva E.V., Cenozoic tectonics and geodynamic evolution of the Tien Shan mountain belt, *Himalayan Journal of Sciences*, Vol. 2, 2004, p. 106-107.

Buslov M.M, De Grave J., Bataleva E.A.V., Batalev V.Yu., Cenozoic tectonic and geodynamics in the Tian Shan: synthesis of geology and geophysical data, *Journal of Asian Earth Sciences*, Vol. 29, 2007, p. 205-214.

Buslov M.M., Kokh D.A., De Grave J., Mesozoic-Cenozoic tectonics and geodynamics of the Altai, Tien Shan, and northern Kazkhstan, from apatite fission-track data, Russian geology and geophysical, Vol.49,No.9, 2008,p.648-654.

Carroll A.R., Graham S.A., Hendrix M.S., Ying D., Zhou,D., Late Paleozoic tectonic amalgamation of northwestern China: sedimentary record of the northern Tarim, northwestern Turpan and southern Junggar basins, *GSA Bulletin*, Vol. 107 , No.5, 1995, p. 571-594.

Charreau J., Chen Y., Gilder S., Dominguez S., Avouac J.-P., Sen S., Sun D., Li Y., Wang W.-M., Magnetostratigraphy and rock magnetism of the Neogene Kuitun He section (northwest China): implications for Late Cenozoic uplift of the Tianshan mountains, *Earth and Planetary Science Letters,* Vol. 230, 2005, p.177-192.

Charreau J., Gilder S., Chen Y., Dominguez S., Avouac J.-P., Sen S., Jolivet M., Li Y., Wang W., Magnetostratigraphy of the Yaha section, Tarim basin , China): 11 Ma acceleration in erosion and uplift of the Tian Shan mountains, *Geology,* Vol. 34, 2006, p.181-184.

Chedia O.K., *Morphology and neotectonics of the Tien Shan.* Ilim. Press, Frunze, 1986 (in Russian).

Chen C., Lu H., Jia, D., Cai D., Wu, S., Closing history of the southern Tianshan oceanic basin, western China: an oblique collisional orogeny, *Tectonophysics,* Vol. 302, 1999, p. 23-40.

Chen Y., Cogné J.-P., Courtillot V., Avouac J.-P., Tapponnier, P., Wang G., Bai M., You H., Li M., Wei C., Buffetaut E., Paleomagnetic study of Mesozoic continental sediments along the northern Tien Shan (China) and heterogeneous strain in Central Asia, *Journal of Geophysical Research* , Vol.96, 1991, p. 4065-4082.

Cobbold P.R., Davy P., Indentation tectonics in nature and experiment. 2. Central Asia. *Bulletin of the Geological Institutions of Uppsala,* Vol.14, 1988, p. 143-162.

Cobbold P.R., Sadybakasov E., Thomas J.C., Cenozoic transpression and basin development, Kyrghyz Tienshan, Central Asia. In: Roure, F., Ellouz, N., Shein, V.S., Skvortsov, I (Eds) *Geodynamic evolution of sedimentary basins, International Symposium.* Moscow, 1994, p. 181-202.

Cogné J.-P., Chen Y., Courtillot V., Rocher F., Wang G., Bai M., You H., A paleomagnetic study of Mesozoic sediments from the Junggar and Turfan basins, northwestern China., *Earth and Planetary Science Letters* , Vol.133, 1995, p.353-366.

Coutand I., Strecker M.R., Arrowsmith J.R., Hilley G. Thiede R.C., Korjenkov A., Omuraliev M., Late Cenozoic tectonic development of the intramontane Alai Valley, (Pamir-Tien Shan region, central Asia): an example of intracontinental deformation due to the Indo-Eurasian collision, *Tectonics,* Vol. 21, No. 6, 2002, 1053, doi:10.1029/2002TC001385, p.19.

De Batist M., Imbo Y., Vermeersch P., Klerkx J., Giralt S., Delvaux D., Lignier V., Beck C., Kalugin I., Abdrakhmatov K.Y., Bathymetry and sedimentary environments of Lake Issyk-Kul, Kyrgyz Republic (Central Asia): a large, high-altitude, tectonic lake. In: Klerkx, J., Imanackunov, B. (eds), *Lake Issyk-Kul: its natural environment. Kluwer Academic Publishers,* Dordrecht, 2002, p. 101-123.

De Grave J., Van den haute P., Denudation and cooling of the Lake Teletskoye region in the Altai Mountains (South Siberia) as revealed by apatite fission-track thermochronology, *Tectonophysics,* Vol. 349, 2002, p. 145-159.

De Grave J., Buslov M.M., Van den haute P., Intracontinental deformation in Central Asia: distant effects of India-Eurasia convergence revealed by apatite fission-track thermochronology, *Himalayan Journal of Sciences,* Vol. 2, 2004, p.121-122.

De Grave, J., Buslov, M.M., Van den haute, P., Distant effects of India–Eurasia convergence and Mesozoic intracontinental deformation in Central Asia: Constraints from apatite fission-track thermochronology, *Journal of Asian Earth Sciences* , Vol. 29, 2007a., p. 188-204.

De Grave J., Buslov M.M., Van den haute P., Dehandschutter B., Delvaux D., Meso-Cenozoic evolution of mountain range – intramontane basin systems in the southern Siberian Altai Mountains by apatite fission-track thermochronology. In: Lacombe, O., Lavé, J., Roure, F., Vergés, J. (Eds). *Thrust belts and foreland basins: from fold kinematics to hydrocarbon systems.* Frontiers in Earth Sciences, Springer Book Series, Berlin-Heidelberg, chapter 24, 2007b, p.457-470.

De Grave J., Van den haute P., Buslov M.M., Dehandschutter B., Apatite fission-track thermochronology applied to the Chulyshman Plateau, Siberian Altai Region, *Radiation Measurements* , Vol. 43, 2008, p. 38-42.

Delvaux D., Abdrakhmatov K., Lemzin I.N., Strom A.L., Landslides and surface breaks of the 1911 Ms = 8.2 Kemin earthquake, Kyrgyzstan, *Russian Geology and Geophysics* , Vol.42, 2001, p.1667-1677.

Delvaux D., Abdrakhmatov K., Lemzin I.N., Strom A.L., Kuchai O., Vittori T., Active tectonic deformation in the northern Tien-Shan of Kyrgyzstan. European Geosciences Union 1st General Assembly, 26-30 April, *Geophysical Research Abstracts* , Vol. 6, 2004, 06350.

Dill H.G., Kadirov O., Tsoy Y., Usmanov A. , Palaeogeography of Neogene red bed sequences along the Aksa-Ata river in the Parkent-Nurekata intramontane basin (Tien Shan Mountains, Uzbekistan): with special reference to the magnetic susceptibility of siliciclastic rocks, *Journal of Asian Earth Sciences,* Vol. 29, 2007, p. 960-977.

Djenchuraeva R.D., Borisov, F.I., Pak N.T., Malyukova N.N., Metallogeny and geodynamics of the Aktiuz-Boordu mining district, northern Tien Shan, Kyrgyzstan, *Journal of Asian Earth Sciences* , Vol.32, 2008, p. 280-299.

Dobretsov N.L., Buslov M.M., Delvaux D., Berzin N.A., Ermikov V.D., Meso- and Cenozoic tectonics of the Central Asian mountain belt: effects of lithospheric plate interaction and mantle plumes, *International Geology Review,* Vol. 38, 1996, p.430-466.

England P., Houseman G., Extension during continental convergence, with application to the Tibetan Plateau, *Journal of Geophysical Research* , Vol.94, 1989, p.17561-17579.

Fu B., Lin A., Kano K., Maruyama T., Guo J., Quaternary folding of the eastern Tien Shan, northwest China, *Tectonophysics* , Vol.369, 2003, p.79-101.

Gao Jun, Li Maosong, Xiao Xuchang, Tang Yaoqing, He Guoqi , Paleozoic tectonic evolution of the Tienshan orogen, northwestern China, *Tectonophysics* , Vol.287, 1998, p. 213-231.

Graham S.A., Hendrix M.S., Wang L.B., Carroll A.R., Collisional successor basins of western China: impact of tectonic inheritance on sand composition, *GSA Bulletin* , Vol.105, 1993, p. 323-344.

Heermance R.V., Chen J., Burbank D.W., Wang C., Chronology and tectonic controls of Late Tertiary deposition in the SW Tian Shan foreland, NW China, *Basin Reasearch,* Vol.19, 2007, p. 599-632.

Hendrix M.S., Evolution of Mesozoic sandstone compositions, southern Junggar, northern Tarim, and western Turpan basins, northwest China: a detrital record of the ancestral Tian Shan, *Journal of Sedimentary Research,* Vol.70, No. 3, 2000, p. 520-532.

Hendrix M.S., Graham S.A., Carroll A.R., Sobel E.R., McKnight C.L., Schulein B.J., Wang Z., Sedimentary record and climatic implications of recurrent deformation in the Tian

Shan: evidence from Mesozoic strata of the north Tarim, south Junggar, and Turpan basins, NW China, *GSA Bulletin* , Vol.104, 1992, p.53-79.

Hendrix M.S., Dumitru T.A., Graham S.A., Late Oligocene - Early Miocene unroofing in the Chinese Tien Shan: an early effect of the India-Asia collision, *Geology* , Vol.22, 1994, p. 487-490.

Huang B., Piper J.D.A., Peng S., Liu T., Li Z., Wang Q., Zhu, R., Magnetostratigraphic study of the Kuche depression; Tarim Basin, and Cenozoic uplift of the Tian Shan range, western China, *Earth and Planetary Science Letters*, Vol.251, 2006, p.346-364.

Hubert-Ferrari A., Suppe J., Gonzalez-Mieres R., Wang X., Mechanisms of active folding of the landscape (southern Tian Shan, China), *Journal of Geophysical Research* , Vol.112, B03S09, 2007, p. 39 .

Ji, J., Luo P., White P., Jiang H., Gao L., Ding Z., Episodic uplift of the Tianshan Mountains since the late Oligocene constrained by magnetostratigraphy of the Jingou River section, in the southern margin of the Junggar Basin, China. Journal of Geophysical Research 113, 2008, B05102.

John T., Klemd, R., Gao J., Garbe-Schönberg C.D., Trace-element mobilization in slabs due to non steady-state fluid-rock interaction: constraints from an eclogite-facies transport vein in blueschist (Tianshan, China), *Lithos,* Vol. 103, 2008, p.1-24.

Jolivet M., Ritz J.-F., Vassallo R., Larroque C., Braucher R., Todbileg M., Chauvet A., Sue C., Arnaud N., De Vicente R., Arzhanikova A., Arzhanikov S. , Mongolian summits: an uplifted, flat, old but still preserved erosion surface, *Geology* , Vol. 35, 2007, p. 871-874.

Ketcham R.A., Donelick R.A., Donelick M.B., AFTSolve: A program for multi-kinetic modeling of apatite fission-track data, *Geological Materials Research* , Vol. 2, 2000, p.1-32.

Khudoley A.K., Structural and strain analysis of the middle part of the Talassian Alatau ridge (Middle Asia, Kirgizstan), *Journal of Structural Geology* , Vol.15, 1993, p.693-706.

Klerkx J., Abdrakhmatov K., Buslov M., De Batist M., Vermeesch P., Imbo Y., Hus R. and Delvaux,D. , Active deformation in the Issyk-Kul basin (Kyrghyz Tian Shan): from pull-apart to transpressional ramp basin, *Second Intern. Congress of Limno-geology,* Brest, Abstract Book, 1999,T36.

Konopelko D., Biske G., Seltmann R., Eklund O., Belyatsky B., Hercynian post-collisional A-type granites of the Kokshaal Range, Southern Tien Shan, Kyrgyzstan, *Lithos,* Vol. 97, 2007, p.140-160.

Koppes M., Gillespie A.R., Burke R.M., Thompson S.C., Stone J., Late Quaternary glaciation in the Kyrgyz Tien Shan, *Quaternary Science Reviews* , Vol.27, 2008, p.46-866.

Korzhenkov A.M., Cenozoic tectonics and seismicity of the northwestern part of the Issyk-Kul basin (Tian Shan), *Russian Geology and Geophysics* , Vol. 41, 2000, p.971-982.

Kröner A., Hegner E., Lehmann B., Heinhorst J., Wingat, M.T.D., Liu D.Y., Ermelov, P., Palaeozoic arc magmatism in the Central Asian Orogenic Belt of Kazakhstan: SHRIMP zircon ages and whole-rock Nd isotopic systematics, *Journal of Asian Earth Sciences,* Vol. 32, 2008, p.118-130.

Laslett G.M., Green P.F., Duddy I.R., Gleadow A.J.W., Thermal annealing of fission tracks in apatite. 2. A quantitative analysis, *Chem. Geol.* , Vol. 65, 1987, p.1-13.

Laurent-Charvet S., Charvet J., Shu, L. Ma R., Lu H., Palaeozoic late collisional strike-slip deformations in Tianshan and Altay, Eastern Xinjiang, NW China, *Terra Nova* , Vol. 14, 2002, p. 249-256.

Lei J., Zhao D., Teleseismic P-wave tomography and the upper mantle structure of the Central Tien Shan orogenic belt, *Physics of the Earth and Planetary Interiors* , Vol.162, 2007, p.165-185.

Le Pichon, X., Fournier M., Jolivet, L., Kinematics, topography, shortening and extrusion in the India-Eurasia collision, *Tectonics* , Vol.11, 1992, p. 1085-1098.

Levashova N.M., Mikolaichuk A.V., McCausland P.J.A., Bazhenov M.L., Van der Voo, R., Devonian paleomagnetism of the North Tien Shan: implications for the middle-Late Paleozoic paleogeography of Eurasia, *Earth and Planetary Science Letters*, Vol. 257, 2007, p.104-120.

Li Z., Song, W. Peng, S., Wang,D., Zhang Z. , Mesozoic-Cenozoic tectonic relationships between the Kuqa subbasin and Tian Shan, northwest China: constraints from depositional records, *Sedimentary Geology*, Vol.172, 2004, p. 223-249.

Lukk A.A., Yunga S.L., Shevchenko V.I., Hamburger M.W., Earthquake focal mechanisms, deformation state, and seismotectonics of the Pamir-Tien Shan region, Central Asia, *Journal of Geophysical Research*, Vol.100, 1995, p.20321-20343.

Mao J., Konopelko D., Seltmann R., Lehmann B., Chen W., Wang Y., Eklund O., Usubaliev T., Postcollisional age of the Kumtor gold deposit and timing of Hercynian events in the Tien Shan, Kyrgyzstan, *Economic Geology* , Vol.99, 2004, p. 1771-1780.

Meade B.J., Hager B.H., The current distribution of deformation in the Western Tien Shan from block models constrained by geodetic data, *Russian Geology and Geophysics* , Vol.42, 2001, p.1622-1633.

Mercier J.L., Armijo R., Tapponnier P., Carey-Gailhardis E., Han T.L., Change from late Tertiary compression to Quaternary extension in southern Tibet during India-Asia collision, *Tectonics* , Vol.6, 1987, p.275-304.

Métivier F., Gaudemer Y., Mass transfer between eastern Tien Shan and adjacent basins (central Asia): constraints on regional tectonics and topography, *Geophysical Journal International* , Vol.128, 1997, p. 1-17.

Mikolaichuk A.V., The structural position of thrusts in the recent orogen of the central Tian Shan, *Russian Geology and Geophysics,* Vol.41, 2000, p. 961-970.

Molnar P., Tapponnier P., Cenozoic tectonics of Asia: Effects of a continental collision, *Science,* Vol.189, 1975, p.419-426.

Neil E.A., Houseman G.A., Geodynamics of the Tarim Basin and the Tian Shan in central Asia, *Tectonics,* Vol. 16, 1997, p.571-584.

Oskin M.E., Burbank D., Transient landscape evolution of basement-cored uplifts: example of the Kyrgyz Range, Tian Shan, *Journal of Geophysical Research* , Vol.112, F03S03, 2007, p. 20 .

Pavlis T.L., Hamburger M.W., Pavlis G.L., Erosional processes as a control on the structural evolution of an actively deforming fold and thrust belt: an example from the Pamir-Tien Shan region, central Asia, *Tectonics* , Vol.16, 1997, p.810-822.

Pirajno F., Mao J., Zhang,Zh., Zhang Zu., Chai F. , The association of mafic-ultramafic intrusions and A-type magmatism in the Tian Shan and Altay orogens, NW China: implications for geodynamic evolution and potential for the discovery of new ore deposits, *Journal of Asian Earth Sciences* , Vol.32, 2008, p. 165-183.

Reigber C., Michel G.W., Galas R., Angermann D., Klotz J., Chen J.Y., Papschev A., Arslanov R., Tzurkov V.E., Ishanov M.C. , New space geodetic constraints on the distribution of deformation in Central Asia, *Earth and Planetary Science Letters* , Vol.191, 2001, p.157-165.

Rybin A., Spichak V., Batalev V., Schelochkov G., Bataleva E., Safronov I., Magnetotelluric investigations of an active thrust fault in the Northern Tien Shan, Kyrgyzstan, Central Asia, *Proceedings of the 17th Workshop on Electromagnetic Induction in the Earth,* Hyderabad, India, 2004, p. 6 .

Rybin A., Spichak V., Batalev V., Bataleva E., Matyukov V.E., Array magnetotelluric soundings in the active seismic area of northern Tien Shan, *Russian Geology and Geophysics* , Vol.49, 2008, p. 337-349.

Sabitova, T.M., Adamova, A.A., Seismic tomography study of the Tien Shan crust: (results, problems, and prospects), *Russian Geology and Geophysics,* Vol.42, 2001, p.1543-1553.

Sobel E.R., Dumitru T.A. , Thrusting and exhumation around the margins of the western Tarim Basin during the India-Asia collision, *Journal of Geophysical Research* ,Vol.102, 1997, p. 5043-5064.

Sobel E., Chen J., Heermance R.V., Late Oligocene-Early Miocene initiation of shortening in the SW Chinese Tian Shan: implications for Neogene shortening rate variations, *Earth and Planetary Science Letters* , Vol.247, 2006, p.70-81.

Solomovich L.I., Postcollisional magmatism in the South Tien Shan Variscan orogenic belt, Kyrgyzstan: evidence for high-temperature and high-pressure collision, *Journal of Asian Earth Sciences* , Vol.30, 2007, p.142-153.

Solomovich L.I., Trifonov B.A., Postcollisional granites in the South Tien Shan Variscan collisional belt, Kyrgyzstan, *Journal of Asian Earth Sciences* , Vol.21, 2002, p. 7-21.

Sun J., Zhu, R., Bowler J., Timing of the Tianshan Mountains uplift constrained by magnetostratigraphic analysis of molasse deposits, *Earth and Planetary Science Letters* , Vol.219, 2004, p.239-253.

Tapponnier P. and P. Molnar, Active faulting and Cenozoic tectonics on the Tien Shan, Mongolia and Baikal regions, *J. Geophys. Res.*, Vol.84, 1979, p. 3425-3459.

Tatevossian R.E., The Verny, 1887, earthquake in Central Asia: application of the INQUA scale, based on coseismic environmental effects, *Quaternary International* , Vol.173-174, 2007, p. 23-29.

Thompson S.C., Weldon R.J., Rubin C.M., Abdrakhmatov K., Molna, P., Berger G.W., Late Quaternary slip rates across the central Tien Shan, Kyrgyzstan, Central Asia, *Journal of Geophysical Research* , Vol.107, B9, 2203, doi:10.1029/2001JB000596, 2002, p. 32 .

Tibaldi A., Graziotto E., Forcella F., Gapich V.H., Morphotectonic indicators of Holocene faulting in Central Tien Shan, Kazakstan, and geodynamic interpretations, *Journal of Geodynamics* , Vol.23, 1997, p.23-45.

Trapeznikov Yu.A., Andreeva E.V., Batalev V.Yu., Berdichevsky M.N., Van'jan L.L., Volykhin A.M., Golubtsova N.S., Rybin A.K., Magneto-telluric soundings in the Kyrgyz Tien Shan Mts, *Izvestija Academii Nauk SSSR, ser. Fizika Zemli* , Vol.1, 1997., p.3-20 (in Russian with English abstract).

Trofimov A.K., Quaternary deposits of the Issyk-Kul basin and their relation to tectonics, *Izvetija Academii Nauk Kyrgyzskoy SSR,* Vol.1, 1990, p.87-95 (in Russian with English abstract).

Tychkov S.A., Kuchai O.A., Bushenkova N.A., Bragin V.D., Kalmetieva Z.A. , Current crustal deformation in the northern Tien Shan: GPS and seismological data., *Russian Geology and Geophysics* , Vol.49, 2008, p.280-290.

Van der Beek P., Delvaux D., Andriessen P.A.M., Levi K.G., Early Cretaceous denudation related to convergent tectonics in the Baikal region, SE Siberia, *J. Geol. Soc. London* , Vol.153, 1996, p. 515-523.

Van der Voo, R., Levashova N.M., Skrinnik L.I., Kara T.V., Bazhenov M.L., Late orogenic large-scale rotations in the Tien Shan and adjacent mobile belts in Kyrgyzstan and Kazakhstan, *Tectonophysics* , Vol.426, 2006, p. 335-360.

Vassallo R., Jolivet M., Ritz J.-F., Brauchner R., Larroque C., Sue C., Todbileg M., Javkhlanbold D., Uplift age and rates of the Gurvan Bogd system (Gobi-Altay) by apatite fission track analysis, *Earth and Planetary Science Letters* , Vol.259, 2007, p. 333-346.

Vinnik L.P., Reigber C., Aleshin I.M., Kosarev G.L., Kaban M.K., Oreshin S.I., Roecker S.W., Receiver function tomography of the Central Tien Shan, *Earth and Planetary Science Letters* , Vol.225, 2004, p. 131-146.

Wang C.-Y., Yang Z.-E., Luo H., Mooney W.D. , Crustal structure of the northern margin of the eastern Tien Shan, China, and its tectonic implications for the 1906 M~7.7 Manas earthquake, *Earth and Planetary Science Letters* , Vol.223, 2004, p. 187-202.

Wang B., Chen Y., Zhan S., Shu L., Faure M., Cluzel D., Charvet J., Laurent-Charvet S., Primary Carboniferous and Permian paleomagnetic results from the Yili block (NW China) and their implications on the geodynamic evolution of Chinese Tianshan belt, *Earth and Planetary Science Letters* , Vol.263, 2007a, p.288-308.

Wang B., Shu L.S., Cluzel D., Faure M., Charvet J. , Geochemical constraints on Carboniferous volcanic rocks of the Yili block (Xinjiang, NW China): implications for the tectonic evolution of western Tianshan, *Journal of Asian Earth Sciences* , Vol.29, 2007b, p. 148-159.

Yang Y., Liu M., Cenozoic deformation of the Tarim plate and the implications for mountain building in the Tibetan Plateau and the Tian Shan, *Tectonics* , Vol.21, 1059, doi:10.1029/ 2001TC001300, 2002, p.17.

Yang T.N., Wang Y., Li, J.Y., Sun G.H., Vertical and horizontal strain partitioning of the central Tianshan (NW China): evidence from structures and $^{40}Ar/^{39}Ar$ geochronology, *Journal of Structural Geology* , Vol.29, 2007, p.1605-1621.

Yin A, Nie S., Craig P., Harrison T.M., Ryerson F.J., Qian X., Yang G. , Late Cenozoic tectonic evolution of the southern Chinese Tian Shan, *Tectonics* , Vol.17, No.1, 1998, p.1-27.

Yuan W.-M., Carter A., Dong J.-Q., Bao Z., An Y., Guo, Z., Mesozoic-Tertiary exhumation history of the Altai Mountains, northern Xinjiang, China: new constraints from apatite fission track data, *Tectonophysics* , Vol.412, 2006, p.183-193.

Zhang C., Zhai M., Allen M.B., Saunders A.D., Wang G., Huang X., Implications of Palaeozoic ophiolites from western Junggar, NW China, for the tectonics of Central Asia, *Journal of the Geological Society, London* , Vol.150, 1993, p.551-561.

Zhang, L., Ellis, D.J., Jiang, W. Ultra-high pressure metamorphism is western Tianshan, China: Part I. Evidence from inclusions of coesite pseudomorphs in garnet and from quartz exsolution lamellae in omphacite in eclogites. American Mineralogist 87, 2002a, 853-860.

Zhang L., Ellis D.J., Williams S., Jiang W., Ultra-high pressure metamorphism is western Tianshan, China: Part II. Evidence from magnesite in eclogite, *American Mineralogist*, Vol.87, 2002b, p.861-866.

Zhao J., Zhou S., He Y., Ye Y., Liu S. , ESR dating of glacial tills and glaciation in the Urumqi river headwaters, Tianshan Mountains, China, *Quaternary International* , Vol.144, 2006, p.61-67.

Zubovich A.V., Trapeznikov Yu.A., Bragin V.D., Mosienko O.I., Shchelochkov G.G., Rybin A.K. and Batalev V.Yu., Deformation field, Earth's crust deep structure, and spatial seismicity distribution in the Tian Shan, *Russian Geology and Geophysics* , Vol. 42, 2001, p.1634-1640.

Part 5

Advanced Concepts in Plate Tectonics

9

Lithosphere as a Nonlinear System: Geodynamic Consequences

E. G. Mirlin
Vernadsky State Geological Museum, Russian Academy of Sciences
Russia

1. Introduction

1.1 Nonlinear geological medium: theoretical aspect

The deterministic viewpoint for natural evolution was dominant in natural sciences for long time. This relates in full measure to geodynamics, and it seems that plate tectonics, being a basis for modern geodynamic conceptions for over 30 years, confirms such viewpoint. The plate tectonics, as a bottom of fact, presents a purely mechanical model based on a simple geometrical observation: the similarity of contours of continents gathered around the Atlantic Ocean. Formerly they composed a single whole, which was broken up and slid apart, to a first approximation, keeping their primary contours. Under this point of view it is meant that geologists deal with the geological environment, being a continuous monolith composed of various solid rocks. This monolith is mechanically divided into different in size fragments (a separate question is a cause of this fragmentation), and the task of tectonic geologists is just to collect the fragments, using their outlines, to reconstruct the history of continents' movements and development of the oceans and fold belts, and to forecast their further evolution. The plate-tectonic prognosis is taken to be sufficiently determined by the regularities of plate kinematics.

The progress of plate tectonics in explanations of many phenomena of the lithosphere structure is obvious, but the question is: what paradigm is going to take its place? This question is even more so appropriate as in the latest quarter of the 20th century the philosophic and methodological conceptions in the field of natural sciences changed from determinism to instability. Works by I. Prigozhin played a great role in the process. The basis of his «philosophy of instability» is dualism [1], stability and instability existing in the universe at the same time. The core of such perception is a thesis about development of highly organized structures in the conditions of nonequilibrium and nonlinear processes in different natural mediums [2, 3, 4]. Nearly three decades have passed since the appearance of instability philosophy, but it is obvious that the introduction of new ideas to geology goes rather slowly. At the same time, during the last two decades of the 20th century, i.e. practically simultaneously with the invention of this new philosophic conception, geophysicists (first of all, seismologists) carried out both experimental and theoretical investigations demonstrated that the medium, with which specialists in the field of the solid Earth deal with, belongs to the mediums characterized by instability and nonlinearity [5, 6, 7]. In essence, a crucially new view on rock characteristic and the whole lithosphere was established, in particular:

- hierarchical heterogeneity all over the scale range, from small mineral grains to planetary-scale irregularities;
- physical nonlinearity appearing in the interdependency of physical processes;
- energy activity, i.e. the ability to permanently produce energy as seismic, acoustic, and electromagnetic emission, as well as heat;
- changeability of physical properties in time, as result of activity and physical nonlinearity; and
- ability of geophysical processes for interaction.

Due to the above-indicated characteristics, the medium acquires properties of fluid; and, with that, self-similar processes, producing structures adapted for the consummation of incoming energy, take place in the medium. The Earth as a whole and its spheres represent the open systems, which convey and transform incoming energy.

An important point of the nonlinear-geophysical-medium conception relates with the fact that the engine of many geodynamic processes (splitting, rising of earthquake sources, rocks' melting, and others) is energy of rocks, which can be released under one or other type of actions [8]. In other words, these processes present a response of the medium to stress, and the more the medium deviates from equilibrium, the stronger is the response. In accordance with the nonequilibrium thermodynamics, the energy balance of an open system composed of rocks will be governed by two opposite entropy flows: d_i and d_e [9]. The first one depends on irreversible processes inside the system (as applied to rocks, the interatomic bond breakage in a crystal lattice) and the second one, on the energy exchange between the system and the environment. Under the low destruction speed, the system can be characterized by relatively stable conditions, under which $d_i = - d_e$, i.e. the entropy doesn't practically change with time. The most interesting and important properties characterize the nonequilibrium systems, when $d_i >> d_e$, i.e. the entropy inside the system sufficiently exceeds its energy losses through the exchange with the environment. In that case the self-organization process in the system leads to the formation of so-called dissipative structures, which existence is supported only due to the dispersion (dissipation) of energy incoming from the outside.

The development of methodology to solve real problems of geodynamics on the basis of instability and nonlinearity, undoubtedly, faces a lot of difficulties, as we are dealing with nonlinear physical systems, which adequate description requires an extremely complex apparatus. In this regard, a relatively rapid progress of plate tectonics is demonstrative. In many ways, it became possible due to the fact that both the main idea (detachment of lithosphere into plates shifting relative to each other) and well-composed methodology (analysis of plate-movement characteristics, quantitative assessments of the oceanic-lithosphere age, changes of submarine topography and geophysical fields with time, and etc.) appeared nearly at the same time. The problems, arising at the development of new methodological principles, are objective; therefore one wouldn't expect that the conception of nonlinearity would be so rapidly progressing as the paradigm of plate tectonics. Nevertheless, it is obvious that without its application the further progress in our understanding of nature of most geodynamic phenomenona becomes impossible.

In the last few years our efforts have been applied to the revelation of the data that could witness nonlinearity of lithosphere medium, as well as to interpretation of some geodynamic phenomenons in the context of a nonlinear medium. The goal of this article is to state (although in a condensed form) obtained results and to determine some of the problems appearing on the way.

2. Results and discussion

2.1 Fractal characteristics of geological mediums

The most important evidence of geological medium being nonlinear and unstable and of the self-organization processes in the medium is an extremely wide range of geomorphological, tectonic, and petrologic objects characterized by dimensional invariance of structural geometry, in fact, fractal organization of the objects. It includes geometrical pattern of seismic delamination of the Earth [9], submarine topography [10, 11], banks and channels of rivers [12], seismicity [13], fault systems and grains in rocks [14], lithospheric plates and blocks of various ranks [15], etc. These objects are quantitatively characterized by fractal dimension; furthermore, in many cases the scale self-similarity of structures is distinctly obvious under qualitative, phenomenological treatment. We would like to illustrate this by the most striking examples.

One of the most interesting in this regard objects is the Sea of Marmara basin, which recently has been the subject of comprehensive systematic investigations [16, 17]. A continental structural type characterizes the Earth's crust of the basin, and peculiar properties of its structure and evolution allow us to take it as an analogue of the early stage of continental breakup. The total opening amplitude nearby the Sea of Marmara reaches 100 km evidenced by from satellite geodetic measurements. The basin shape is an irregular rhomb, which allows us to characterize it as pull-apart, trans-displacement stretching structures (Fig.1a). A considerable thickness of the Quaternary sediments (on average 1-2 km up to 6 km in the northernmost and deepest depression) confirms that the structural depression of the Sea of Marmara was formed as a result of diagonal stretching.

Under more detailed investigations, it occurs that both branches, the North Anatolian and South Anatolian strike-slips, limiting a main pull-apart depression, are not uninterrupted and united. They also split along both diagonal and oblique faults regarding a general strike, into segments different in rank; each of the segments comprises rhomboid or oval in plane pull-apart depression smaller in size than a main basin. The nature of these structures is determined not only by general geometry, but by the analysis of morphotectonics of their edges, as well as by characteristics of layers' displacement within sedimentary series. The depression's slopes complicated by numerous benches (scarps) divided by inclined faults evidence their formation as result of diagonal stretching. Also, the depressions' flat bottoms are at 20-60 m deeper than the surface of the surrounding bottom. A major geometrical parameter of such depressions, a width between opposite sides, varies roughly within one order, from 10 km to 100 km.

Inside the Central Basin about 37 km wide (fig.1b-1c), a small rhomboid depression was discovered using side-looking sonar and seismic profiling. It is limited by numerous echeloned normal faults. The distance between opposite sides of this depression is not more than 10 km (fig.1d). Folded sediments both to the east and west from the depression can indicate compression, which is an important component of geodynamic settings.

Thus, analysis of available data, describing structure of the basin of Sea of Marmara in spread-spectrum scales, clearly shows scale invariance of pull-apart depressions, and indicates fractal divisibility of the Earth's crust. Structural depressions of various scales occurred in connection with horizontal tectonic displacements along strike-slips are widespread in the Earth's crust. In most cases they have a scale-invariant rhomb shape in plane [18]. So, different in rank pull-apart depressions in the Sea of Marmara have engaged our attention by their structural-geometry similarity to rhomb- and wedge-like depressions

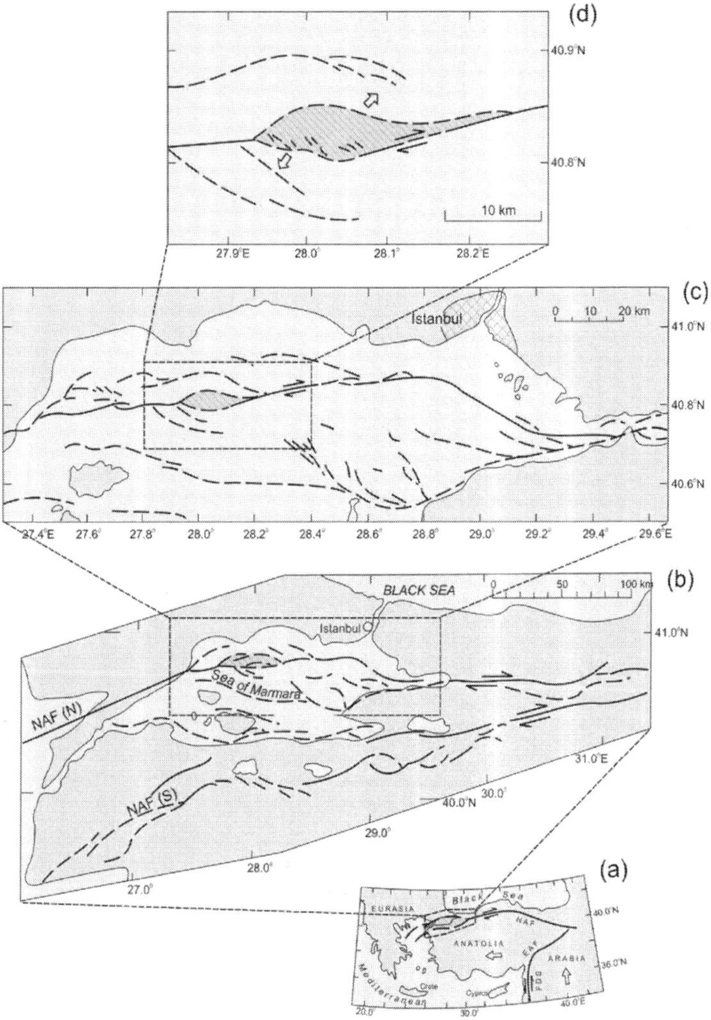

Fig. 1. Fractal structuring by the example of scale-invariant pull-apart structures of the Sea of Marmara and its framing after [16]. (a-d) sequential scale extension: (a) splitting of the North Anatolian Fault - NAF(N) and NAF (S) dividing the Anatolia Microplate from the Eurasian Plate. (FDS) Fault of the Dead Sea, Arrows show moving directions of plates and displacements along faults. (b) pull-apart basin of Sea of Marmara and different in rank rhomb-like or oval pull-apart structures along (N) northern and (S) southern branches of the North Anatolian Fault. Solid and dotted lines indicate main and general faults accordingly; and arrows, displacements along faults. (c) pull-apart structures of the northern branch; (d) one of rhomb-like local young depressions; arrows show displacements along the fault and the direction of the depression's opening.

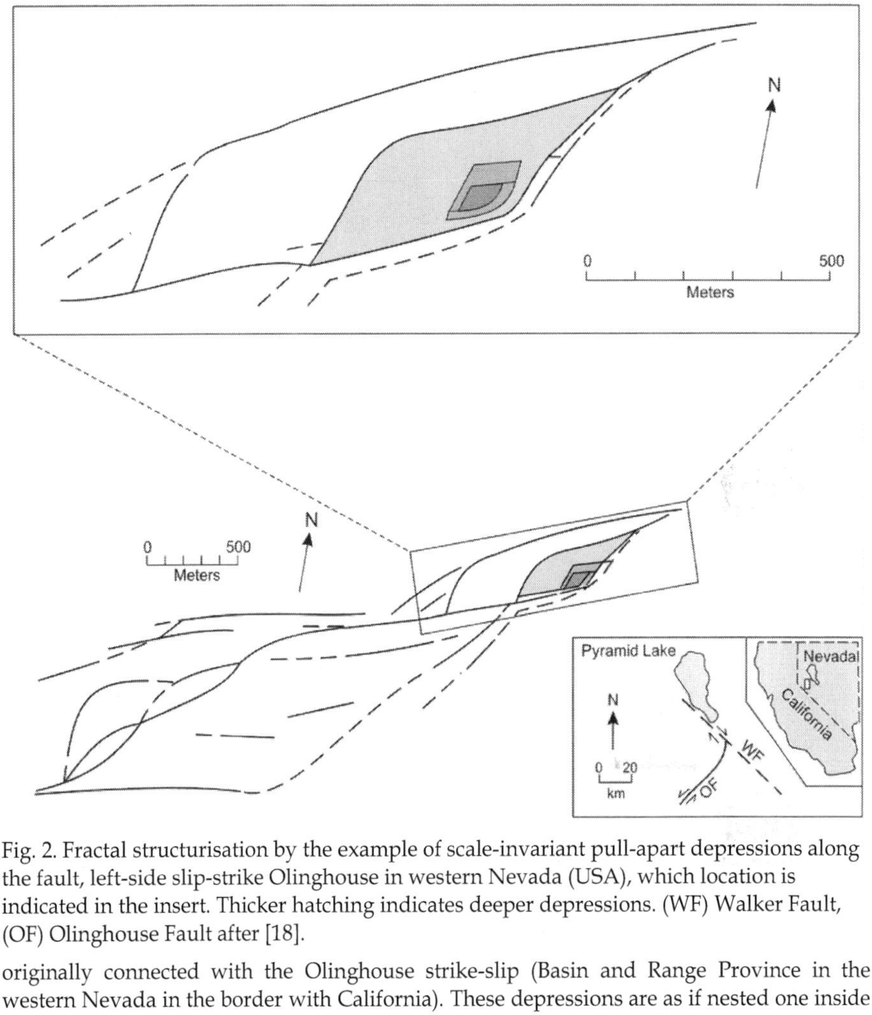

Fig. 2. Fractal structurisation by the example of scale-invariant pull-apart depressions along the fault, left-side slip-strike Olinghouse in western Nevada (USA), which location is indicated in the insert. Thicker hatching indicates deeper depressions. (WF) Walker Fault, (OF) Olinghouse Fault after [18].

originally connected with the Olinghouse strike-slip (Basin and Range Province in the western Nevada in the border with California). These depressions are as if nested one inside the other (fig.2), and their bottoms are dipped tens of meters beneath the adjacent surface [19]. Although the depressions have some differences in regional geodynamic settings, the Earth's crust structure, and depressions' sizes (they are rather smaller in Basin and Range Province than in the Sea of Marmara), they are similar in geometry and genesis, i.e. both examples demonstrate clear signs of structural self-similarity.

The next demonstrative example of scale isomorphism is related to spreading zones at the crests of the mid-oceanic ridges (MOR). A distinguishing characteristic of their structural geometry is segmentation different in rank, which itself indicates that the lithosphere is highly fractional, and for this reason the crests are good subjects for revelation of fractal structure formation. Echeloned in plan and various in rank geometry of spreading centers

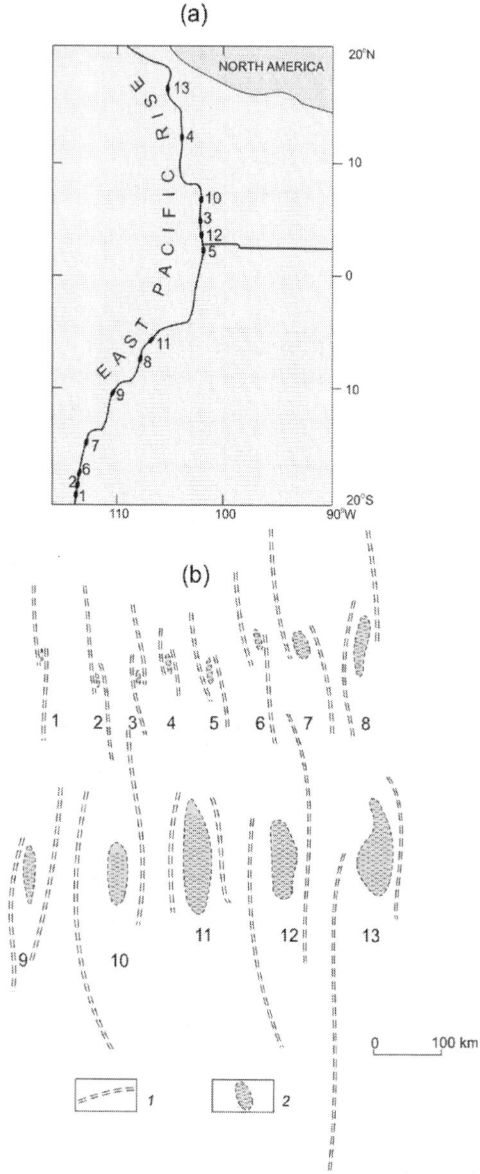

Fig. 3. Geometric self-similarity of overlapping spreading centres (OSC) at the ridge of the East Pacific Rise: (a) OSC locations; (b) their generalized morphotectonic schemes, based on comprehensive investigation of the submarine topography. (1) axes of spreading centers, (2) structural depressions between them, after [29].

indicates the presence of a shear component along the practically full length of the MΘR global system. The strike vector is oriented at different angles to the crests in the different sections of the ridge. In other words, in terms of geodynamics, the MOR crests present spreading zones with a shear component. The difference in the segments' geometry of slower and fast MOR is connected with the lithosphere plasticity, which depends of the lithospheric temperature regime, and, thus, geometry of spreading centers indicate the accretion regime of the Earth's crust [20].

Their structural-geometrical isomorphism can be clearly shown by the example of MOR crests with intermediate and high velocities (4-8 cm/year, 8-12 cm/year and more, accordingly). In such ridges, overlapping spreading centers (OSC) varied in size are widely known; most of them are mapped with high-resolution echo depth-sounders. The OSC geometry is quite simple and can be the subject of comparative studies. For this purpose, we analyzed bathymetry of 13 mapped in detail different in scale OSCs located at the East Pacific Rise (EPR) crest between 16°N and 20° S (fig.3a). Morphotectonic maps of these structures were built, and then simplified into easy-to-use general schemes for the comparison in a unified scale (fig.3b). In addition to echelon-like displaced spreading centers, the schemes show structural depressions, appearing as troughs elongated concordantly to the crest trend. Some variations in the OSC geometry are observed, but they are secondary in comparison with the indicated above general features.

Spreading velocities in the observed EPR segment vary from 9 cm/year to 16.2 cm/year, however a clear connection between morphometric parameters of OSC and velocities of the bottom growth has not been established. The most essential property of the observed structures is their geometrical self-similarity. The length of the shortest depression is about 15 km, and the longest one is 130 km, the length of the axis-rift segments limited by adjacent overlaps ranges from lesser than 15 km to more than 200 km, a relative depth of these depressions is first tens of meters for small overlaps and first hundreds of meters for large ones. In other words, quantitative adjectives change practically an order with constant geometry of these bottom forms, which makes a clear evidence for structural-geometrical isomorfism.

2.2 Vortex movements at the oceanic opening

Vortex structures and movements, accompanying the oceans' genesis, are the next important evidence for the nonlinearity of lithosphere medium. It is common knowledge that vortex movements various in rank are an essential feature of dynamics of the Earth's outer covers, atmosphere and hydrosphere, for which general features are instability and nonlinearity. For this reason we will pay more attention to the characteristics of vortex structures, as well as to possible nature of such movements.

As stated above, this medium has a block-hierarchical structure in every space-time scale, it is nonlinear, and energetically active, which provides the medium with properties of flowing fluid clouds; formation of vortex movements becomes high-probable and actual observations of the oceanic basins' evolution confirm that [21, 22]. Above all, this is the evolution of their structural geometry. It is characterized by two peculiarities: propagating of the spreading axis and its whirling. We can demonstrate this at the example of the North Atlantic, which is one of the well-studied areas of the World Ocean. The propagating of the spreading axes from the south (equatorial zone) to the north is clearly traced in its evolution since the Late Jurassic. In the Early Cretaceous (120 Ma) this propagating lasted that resulted in opening of the North Atlantic to the north of the Azores. However, this opening was complicated by the fact that secondary branches detached off a main opening trunk, were dying with time. Among such branches are the Rockall Trough, the Bay of Biscay, for which the opening was accompanied by rotation, the Iberian Peninsula, Basin of the Labrador Sea, as well as Porcupine and Baffin basins (fig. 4 a). By 90 Ma, the stretching had most likely stopped in the Bay of Biscay, Rockall Trough, and Porcupine Basin, but it continued in the Labrador Sea and presumably in the Baffin Basin (fig. 4b).

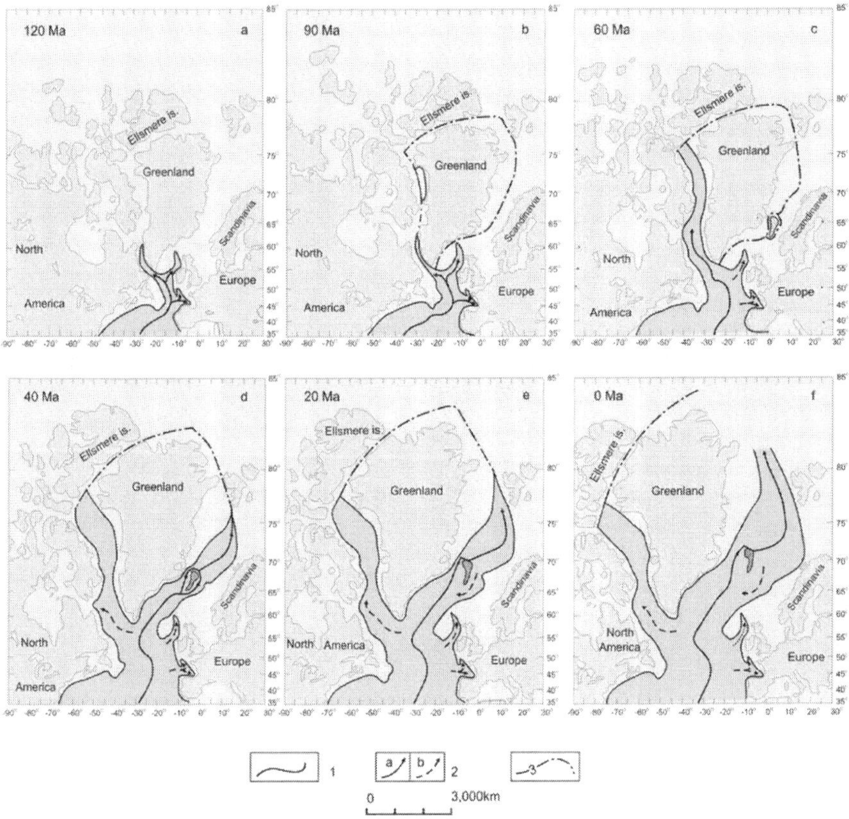

Fig. 4. Development of vortical spreading systems of the North Atlantic as determined from paleogeodynamic reconstructions [22, 39]. Time windows are shown in the figure. (1) Continental margin; (2) direction of spreading axis propagation: (a) active, (b) extinct; (3) line of initial opening of the ocean.

It is of interest that the separation of Greenland from Eurasia started about 60 Ma along not having existed by that time continental rifts, but noticeably to the west, along the axis of the mid-oceanic ridges of Reykjanes, Kolbeinsey, and Mona. At the same time spreading zones sprung to the southeast and northwest from the blocky Jan Mayen Ridge, which stayed apart and formed an independent, small in size continental lithosphere plate (fig. 4c).

By 40 Ma, the process of the oceanic formation had ended in the Labrador Sea and Baffin Basin and a general tendency to propagate the oceanic formation extended to the north as result of the separating Greenland from Eurasia. As before, the spreading axis propagating was not simple. In this case it was complicated by the presence of the Jan Mayen Microplate, as well as its rotation anticlockwise round a pole located in immediate proximity of Jan Mayen Island. This rotation was accompanied by the formation of two vortex-like spreading systems. With that, the northwest system propagated roughly in parallel with the general direction of the spreading axis of the North Atlantic, i.e. roughly to the north, and the development of the southeast branch (in the Norwegian Basin) is characterized by the propagating oriented about toward the North Atlantic system (fig. 4d).

At a period of 40 - 20 Ma, the Knipovich mid-oceanic ridge arose in the area of Spitsbergen Strike-Slip Zone as a result of the opening-axis propagating to the north. Due to the development of the above-mentioned MOR, a tendency of twisting the advancing opening general-zone of North Atlantic, which appeared at the previous stage, came to the accomplishment (fig.4e). Finally, the present structure of the North Atlantic (fig. 4f) clearly demonstrates results of the above-examined evolution: the development of the general spreading zone, as well as of the secondary branches, includes both propagating and simultaneous whirling. Because of this, the oceanic basin is not unified, but is constituted by the system of vortex-like depressions variable in size and age and having independent spreading systems. Accordingly, besides of large continental blocks of Eurasia, North America, and Greenland, there is a whole set of microcontinents and elevated blocks with subcontinental crust (Jan Mayen, Hatton, Rockall, etc.).

The above-considered basic tendencies in the North Atlantic evolution, propagating of the axis and their whirling, are known in the other spreading basins of the world ocean. At that, a size range of vortex systems is extremely wide. For example, at the EPR crest, which shape demonstrates a vortex 7.000 km long, there are microplates of Juan Fernandez and Easter Island, framed by pronounced vortex-like spreading zones with a length of 300 - 500 km (fig. 5). Their boundaries are presented as pseudo-faults formed with the propagating of spreading axis. Overlapping spreading centers at the EPR crest can be assign to vortex structures as well, as they show signs of vortex whirling in the propagating of spreading axis. In the Indian Ocean, the rift of Tadjoura together with the spreading zone of the Gulf of Aden and Arabian-Indian and Central Indian MORs presents a giant vortex about 8.000-km long, as if intruding into the continent of Africa. The tendency towards the vortex whirling of the West Indian MOR is clearly seen near the Rodriguez triple junction. Many of back-arc basins in the Pacific Ocean-Asian continent transition zone formed as a result of stretching also present vortex-like spreading systems various in size. As a whole, quantitative characteristics of such structural systems in the world ocean can change more than two orders, in other words, vortexes are characterized by different scale self-similarity. Continental massifs evolved from the Pangea breakup differ in size more than two orders as well; also, in detail studies microcontinents in their turn can be divided into individual blocks still smaller in size. Undoubtedly, the formation of self-similar elements during the lithosphere evolution presents an additional evidence for lithosphere to be considered as an open, nonlinear dynamic system, in which the processes of self-organization act [6] and where vortex movements of different scale may well be developed. This is also supported by the instability of the ocean formation - the displacement of the opening axis with space and time. The latest probably reflects the instability of convection-current dynamics, which is a major condition for vortex generation.

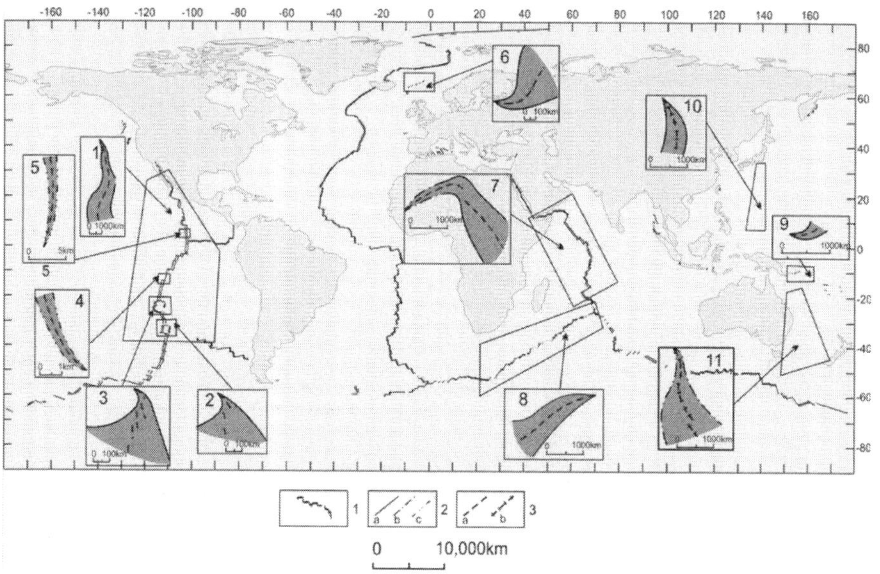

Fig. 5. Vortical systems in the World Ocean (numerals in figure): (1) East Pacific Rise (EPR); zones of microplate spreading: (2) Juan Fernandez; (3) Easter; branches of overlapping spreading axes at the EPR crest: (4) near 12°55' S; (5) near 5°30' N; (6) Norwegian Basin; (7) Tajura Trough, Gulf of Aden, and Central Indian Mid-Ocean Ridge; (8) West Indian Mid-Ocean Ridge; transition zone from Pacific Ocean to Asian continent: (9) Solomon Depression; (10) Shikoku and Parece Vela Basins; (11) Tasman Sea.
(1) Axes of the mid-ocean ridges; (2) boundaries of vortical systems: (a) along the isochron lines of the oceanic crust and pseudofaults, (b) along the ocean/continent boundary, (c) along the foot of the central plateau at the EPR crest; (3) spreading axes: (a) active, (b) extinct.

2.3 Vortex movements during the oceanic-basins' opening at the junction zone of Eurasia, Pacific, and Australia

As forming the Pacific margin of Eurasia, stretching efforts acted at regular intervals during relatively brief time (on average about 20 Ma) and were concentrated within lengthy band-like belts. Such periodicity appeared against the background of the successive migration of subduction zones (also intermittently) from Asiatic and Australian continents to the Pacific [23]. As result, since the Later Mesozoic to the present, marginal-continental volcano-plutonic belts different in age have been formed, as well as island arcs and marginal basins, altogether constituting the junction zone. Below we consider the spatially irregular variability of vortex-like spreading basins within this zone.

Within an active zone of the Eurasia junction, other oceanic basins are known besides the ones indicated at Fig. 5. Their opening is also characterized by the presence of a vortex component judging from their configuration in plane and from the structure of linear magnetic anomalies. Among these there are spreading zones of the West Philippines, Okinawa, the Mariana Trench, Caroline, North Fiji and South Fiji, Lau-Havre, Caroline-Manus, and, quite possibly, a row of others. They were developed under various geodynamic settings either in the immediate back region of island arcs in connection with their splitting or at a noticeable distance from subduction zones. Paleogeodynamic settings were reconstructed for some of marginal basins, and we discuss those of them, in which the vortex whirling phenomena appear most distinctly.

The vortex-like opening of the spreading basin generated at the Pacific-Australian plates' boundary to the west from New Zealand since Eocene is an exceptionally demonstrative example. In accordance with the names of adjacent oceanic basins, the authors of investigations devoted to this region [24] call it Tasman-Emerald Basin (TEB). Its evolutional scheme is based on combined tectonic analysis of both marine geophysical data and geological structure of South New Zealand Island.

The region under consideration covers the complex of underwater Macquarie Ridge and its extension on the island as Alpine Fault, underwater Campbell and Challenger plateaus, and deep Emerald and Tasman basins (inset in Fig.6). The latest ones are typical oceanic spreading basins, and, at the same time, a spreading zone of the Tasman Sea with the age of Later Cretaceous - Palaeocene is a branch of the Southwest Indian Mid Oceanic Ridge, like branches of spreading in the North Atlantic. Since Eocene (45 Ma), after spreading in the Tasman Sea ended, a new boundary between Australian lithospheric plate and Pacific one has been generated along the line separated earlier unified continent including Challenger and Campbell plateaus (Fig. 6). The fact that ophiolite belt of the Dan Ridge existed at that time in the south of Southern Island is fundamentally important. It is considered as an essential confirmation of relative motion between plates with the mentioned plateaus included to be not only spreading, but accompanied with whirling and compression in the area of the vortex closure. This is supported by the fact that the line of original opening was a broken saw tooth curve indicating the presence of a slide component at the early stage of plates' relative displacement, in accordance with a model of vortex kinematics.

For the next period (later Eocene-Oligocene; 30 Ma, fig. 6), formation of the Tauru overfault zone became a characteristic that can indicate the continuation of compression along with whirling. Moreover, a clearly echeloned structure of spreading zone was formed, which demonstrates the presence of a slide component during the basin opening. By the middle Miocene (15 Ma) both slide and shear components had become predominant, which was expressed by an abrupt change of transform faults' strikes, as well as by the appearance of a compression belt in the extreme north of the opening basin. This trend in the change of geodynamic settings lasted afterwards; as a result, at present the whole northern segment of the TEB axis zone is a compression belt (fig. 6d).

The other example of the vortex-systems' evolution relates with the Philippine Sea, where three extinct spreading centers are revealed: one, the most ancient, developed in Eocene, is located in the West Basin and two young centers (with Oligocene-Miocene age), in Shikoku and Parece Vela basins in the eastern part of the sea. The first one has the northwest–southeast strike and its evolution is divided into a row of stages with the gradual propagating of the spreading axis and the axis's whirling to the west-northwest. Figure 7 demonstrates both initial (a) and final (b) stages. At that, one can notice that at the final stage axes of linear magnetic anomalies are unconcordant in relation to the boundaries of the spreading basin and the axis of spreading after kinematics' change angularly cuts older one. This conforms to the location of oceanic-crust isochrones according to the above-considered kinematic scheme of vortex-like opening. As for Shikoku and Parece Vela basins, fan-shaped in plane even-aged systems of linear magnetic anomalies discovered in both are indicating the fact that the opening axis propagating was coming roughly towards each other [25].

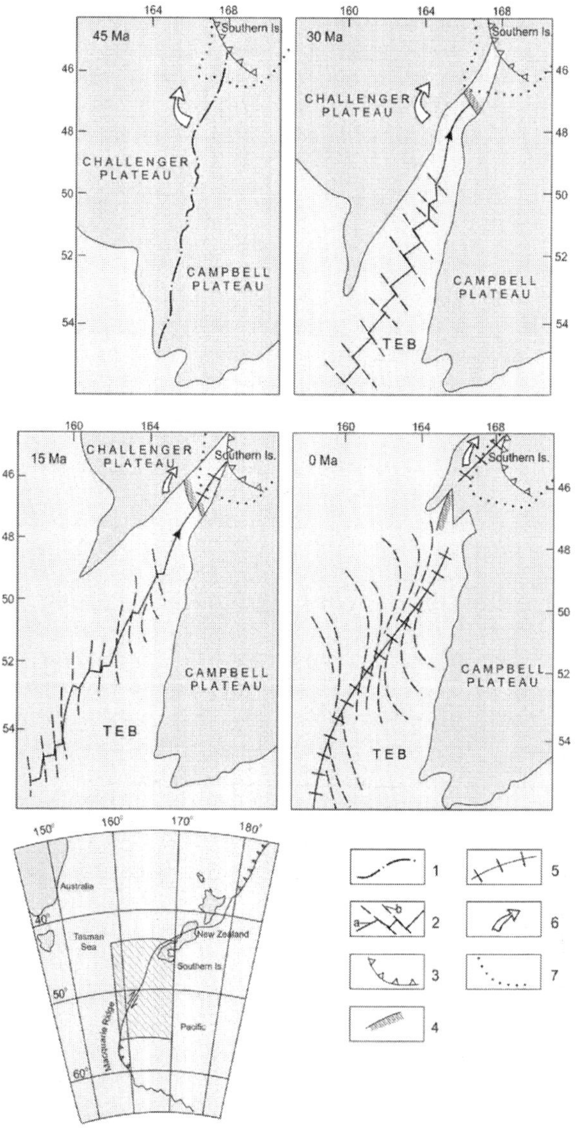

Fig. 6. Various time windows for the development of vortex-like spreading system of Tasman-Emerald Basin (TEB) after [24]. (1) a line of primary opening; (2) spreading axes (a) and transform faults (b), an arrow indicates the propagating direction; (3) ophiolite belt of Dan Ridge; (4) overfault zone; (5) compression belt, (6) direction of relative motion of Challenger Plateau relatively Campbell Plateau, (7) contours of Southeern Island. Hatching at the geographic scheme shows the position of the region under investigations.

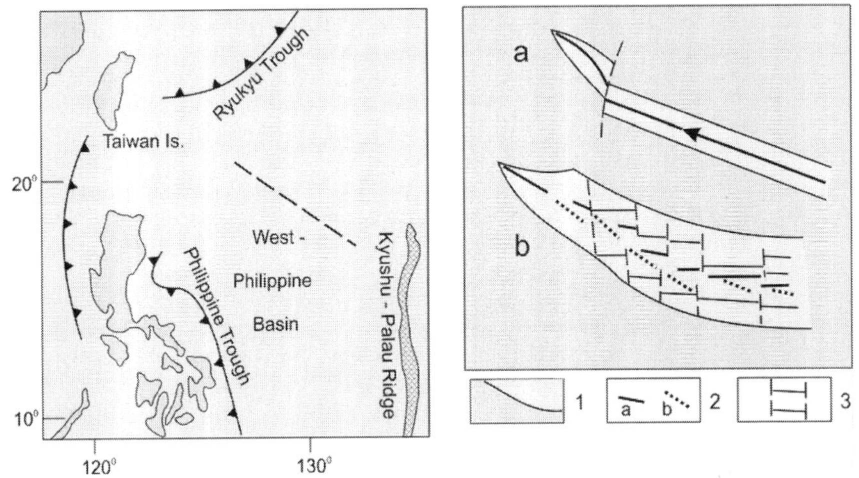

Fig. 7. The evolutionary stages of a vortex-like spreading system in the West Philippine Basin after [25]: (a) primary stage and (b) final stage. (1) boundary of the spreading system; (2) spreading axis: (a) before rebuilding, (b) after it; (3) axes of magnetic anomalies (solid lines) and transform faults (dotted lines). Hatching at the location scheme shows the position of the spreading zone.

The considered data witness to an important role played by the basins with vortex-like evolution in structure and development of the Pacific junction zone of Eurasia. At that, stretching belts, containing elements of these basins (subduction zones as well), show strong trend of migration from both Asian and Australian continents to Pacific, although isolated branches of non-contemporaneous stretching zones can be superimposed. Moreover, a doubtless connection is revealed between the spatial-temporal variability of stretching belts and age borders, corresponding to the change of kinematics of Eurasian, Pacific, and Kula lithospheric plates. The origin of new branches of vortex structures, which are sometimes characterized by opposite propagating of opening zones, are often confined to these age borders. This shows the similarity of evolution of the North Atlantic and stretching belts of the Pacific junction zone.

2.4 Possible nature of vortical movements

Essential difference in sizes of the oceanic vortex structures testifies that vortex motions are caused by dynamics of the Earth's covers essentially different in depth but with the same physical mechanism. Global vortexes (many thousands of km in size) can be resulted from the fact that at least some of the covers revolve around the central axis at a various speed and, consequently, move relatively each other. At present, most scientists are inclined to believe this idea, and it really looks probable, taking into consideration different physical specifications of the covers, especially, their integrated density and viscosity. If, for example, differential rotation of lower and upper mantle takes place, as well as upper mantle and lithosphere, in essence, this is equivalent to a flow in the mantle [26], which is likely to be unstable owing to reasons listed above. Instability of a border's dynamic surface originated from different rotational speeds of the covers generates a vortex component of motion and, as a consequence, a tendency to form vortex-like whirling of opening zones at the breakup of the megacontinent.

It is believed that the nature of a vortex component of motion when back-arc and inter-arc spreading basins were formed in the west and southwest Pacific margins is connected with subduction of relatively ancient, cold, and heavy lithosphere of Pacific under relatively hot and light lithosphere of Eurasia, as it is in the converging area of Pacific and Eurasian plates. A subduction zone is a dynamic interface where the relative motion of plates is dominanting along with simultaneous compression. Such phenomenon, essentially, may be considered as an analogue of a warm atmospheric front where cold relatively heavy air dives (subducts) under relatively light and hot air. A comparison of subduction with an atmospheric front is not a new idea; it exists in geological literature [27]. An atmospheric front is a dynamic interface between masses with different physical specifications (both a subduction zone and interface between covers). Atmospheric fronts present the exact cause for secondary atmospheric vortexes, arising at flexures of dynamic interfaces. Despite the distinctly different time scales, the physical basis of vortex formation is the same: flexures and dynamic instability of interfaces generate vortex motions both in atmosphere and in mantle in the area of plate converging.

Vortex motions caused the formation of small vortex-like structures, like overlapping spreading centers at the EPR crest, likely to be resulted from instability of flow of matter along the rise's axis. The existence of such flows is postulated based on the analysis of both structural development and interpretation of geophysical fields of oceanic rifts. It is significant that structural patterns like OSCs are discovered not only at fast spreading MOR in the Pacific but at the slow spreading Reykjanes and Kolbeinsey ridges in North Atlantic where the data on a flow of matter along the ridge's axis are very reliable.

2.5 Interpretation of some geodynamic phenomena with nonlinear medium model

The idea of both nonlinear geological medium and vortex motions in it allows us to suggest new interpretation of some geodynamic phenomena accompanying the ocean formation, which nature is hard to explain in the context of plate-tectonics' fixed paradigm.

Segmentation of passive continental margins and the oceanic bottom. The segmentation of passive continental margins and the oceanic bottom into sections of different orders has a global importance. It was repeatedly noticed that the configuration of continents (and, accordingly, the ocean) in plan are rounded. Broken stepwise contours connected with varied-order segmentation of passive continental margins and the oceanic bottom seem to be enclosed in smooth, rounded outlines of the oceanic basins. Undoubtedly, there is some in-depth meaning behind this geometrical enclosure. The progress in the study of physical specifications of the Earth's solid covers really resulted in the appearance of such concepts as the enclosure of deformation processes occurred in these covers [28]. The enclosure is generally meant the modification of continuum mechanic model, needed for description of mechanical difference in the crust and lithosphere for different spatial-temporal scales. As indicated above, for the oceanic lithosphere, the enclosure may be observed at a geometrical level.

Comprehensive geophysical researches of the ocean/continent boundaries in some regions of the world-ocean passive margins (for a example, at the South Atlantic) not only confirm similar peculiarity of structural geometry, but also reveal the self-similarity of stepwise configuration for the more detailed scale of researches [29]. The most of continental rifts, which usually are regarded as early evolutional stages of young oceans, are also characterized by both broken outlines in plane and separation of isolated structural segments. At that, the most of them demonstrate, in addition to stretching normal to the rift's axis, a slide component that characterize them as pull-apart structures. Together the data on configuration of continental rifts with geometry of the ocean/continent boundaries allow us to conclude that the breakup of Pangaea supercontinent took place along rounded lines, which at zooming occur to be stepwise ones, though they partly inherited suture zones of basement. It should be noted that in the limits of present plate-tectonic paradigm, a question about both enclosure of deformation processes and geometrical enclosure in the ocean formation is not considered at all; researchers are mostly deal with the analysis of structural links between continental rift zones and oceanic ones [30].

The specificity of vortex motions, occurring in unstable medium allows one to throw light on the nature of this phenomenon. Figure 8 presents a comparison of structural patterns of (a) a cyclonic synoptic vortex in the atmosphere, (b) the oceanic basin formed by vortical motion, and (c) the oceanic basin formed according to a plate tectonic model. If the continents' breakup and further evolution of the oceanic basin occur under effect of whirling, the opening line gets rounded in accordance with pattern of the vortex flow. The velocity of combined reciprocating and rotating motion of matter within it varies from its inner part to the outer zone (fig. 8a) This causes an appearance of the shear component in the lithosphere and reasons pull-apart features in continental rifts in the early continental stage of the ocean's opening. The line of initial opening of the ocean (at a later time, passive margins), though retains rounded shapes in general, but is divided into isolated segments, i.e. gets broken, saw-shaped. Geometry of such segmentation is characterized by

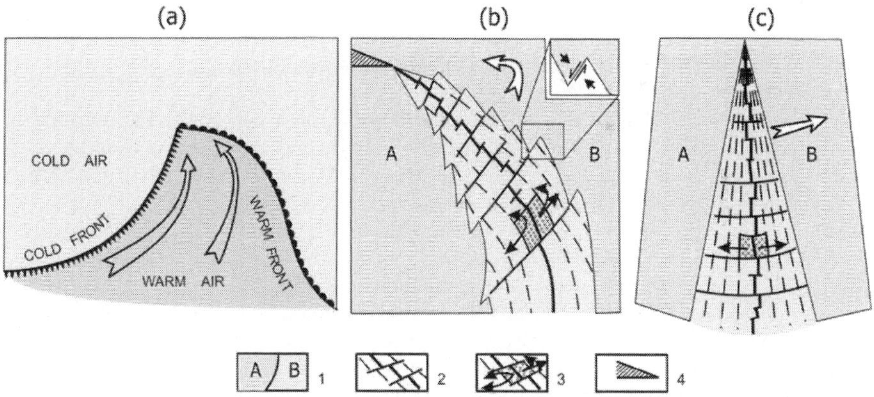

Fig. 8. Structural patterns of (a) a cyclonic synoptic vortex in the atmosphere, simplified after [40], (b) the oceanic basin formed by vortical movements, and **(c)** the oceanic basin formed according to a plate tectonic model. The arrows in Fig. 8a indicate the direction of warm air migration, while the arrows in Figs. 8b and 8c indicate the displacement of continental plate B relative to continental plate A. (1) Continental crust (A) and oceanic crust (B); (2) isochrons of the oceanic crust and transform faults; the heavy line is the spreading axis; (3) elementary bodies of plates and directions of their motion; (4) area of compression. The inset shows the orientation of stress at the boundary between the oceanic and the continental lithosphere during the opening of the oceanic basin.

self-similarity due to above-mentioned fractal properties of the medium. Moreover the change of velocities inside the vortex flow stimulates generation of dynamic separation surfaces within it and, accordingly, vortical structures lesser in size, which can also influence the configuration of the opening line.

At the stage of utter breakup of continental crust and of formation of spreading zones, vortical movements produce rotatory motions of varied volumes of matter around independent axes, which initiates twirling stresses in the new-formed thin oceanic lithosphere that is likely to be a major cause for origine of varied in rank structural echeloned segmentation.

Thus, both the geometrical roundness of continental contours and different in order segmentation of the oceanic bottom are connected with the specificity of vortical movements. But if the origin of the first phenomenon is apparently described by motion of viscous fluid (vortex flow), the nature of the second one in the best way should be interpreted on the basis of models of continuum mechanics (response of the breakable and elastic lithosphere to effect of vortex flows).

Tectonic delamination of the lithosphere. It was found that this phenomenon is widespread, and it is widely covered in the literature, including monographs [31]. As illustrated in many works, horizontal or close to horizontal displacements of deep or subsurface rock masses lay in the basis of origin of this phenomenon that results in the formation of roughly horizontal boundaries in the both crust and lithosphere, and tectonic stacking as well. From the analysis of high-tech seismic profiling results, numerous inclined reflectors were recognized in the oceanic crust. This fact, as well as results of both dredging and deep-water drilling, serves as the basis of new ideas of infrastructure and dynamics of the upper oceanic lithosphere. At the same time the tectonic nature of interpreted horizontal surfaces is doubtless. It was mentioned that tectonic delamination is peculiar not only to the modern oceanic lithosphere, but to ancient one as evidenced by structure of ophiolite complexes of different ages.

The authors of works devoted to this phenomenon note that a concept of the lithosphere's delamination is based on the data confirming a wide distribution of shear dislocations. It follows from the model of vortical movements that if a vortex flow affects the lithosphere's bottom, shear dislocations should occur there (subject to an upright position of the vortex axis, and, correspondingly, a horizontal orientation of the flow surface). Relatively independent rotation of varied in size masses in such flow should contribute to roughly horizontal displacements of some blocks of the crust and lithosphere and, hence, form sub-horizontal interfaces of tectonic nature. Moreover, rotation of varied in rank blocks of the crust appears independently one of another, around different axes, which inevitably result in the compression stress even in the crest zone of the mid-oceanic ridges. This was confirmed by comprehensive seismic observations in the axis of the Mid Atlantic Ridge where rupturing deformation reasoned by compression was revealed[32].

It is needless to say that, all presented above is only a qualitative scheme of origin of the oceanic lithosphere's delamination, but produces a basis for interpretation of the available data, as well as for carrying out quantitative assessment.

Fold deformations within passive margins. As follows from the model of the oceanic opening under the effect of vortical movements (fig.8b), geodynamic settings vary along the interface of continental plates. As the whirl develops, the compression component rises; the stronger is twirl, the stronger becomes compression that should be reflected in the structure of the Earth's crust. Inded, at the closure of strongly whirled vortical structures framing

Easter and Juan Fernández microplates at the EPR crest, underwater ridges resulted from compression are discovered [33]. The similar in nature of the crust deformations microplates of Gorda in the Pacific appear as both flexures and shortening of banded magnetic anomalies, and pattern of seismic activity [34]. Of special interest are fold deformations within passive continental margins, which don't correspond with widespread plate-tectonic evolutionary models, but can be quite well explained with the vortical motion.

Passive margins reflect the transfer from the ocean to the continent inside the lithosphere plate; they result from the continent's fragmentation during continental rifting, then, move aside of divergent boundaries, and with time change into regions with significant sag of the crust and deposition of a thick sedimentary stratum. They are not boundaries of plates, their name itself reflects their relatively tectonic inactivity: intensive sagging is dominant and other tectonic movements are excluded.

This roughly describes a plate-tectonic point of view on the origin and structure of these margins, whish is recognized by the most of researchers. However, intensive geophysical investigations carried out within such areas during the latest decades due to high oil-and-gas prospects revealed signs of fold deformations in some regions of passive margins. We consider them at the example of junction zones of passive type in the well-studied Norwegian-Greenland Basin [35; 36], which evolution from the position of vortical movements was considered above.

In contrary to the earlier ideas of the tectonic stability of the regions surrounding the basin, clear evidences of tectonic motions both vertical and horizontal during the Neogene time were revealed within them. Distinct signs of compression deformations of sedimentary stratum were discovered within the Atlantic margins of Norway, the Faroes, British and Shetland Islands, and along the southwestern margins of the Barents Sea; as well as at the eastern margins of Greenland (fig.9). In the middle of the Norway margins, they appear as local elevations framed by overfaults. Some of them are reflected in the bottom topography. Along the eastern segment of Jan Mayen Transform Fault, these elevations are characterized by clear echeloned pattern. The time of deformations was defined as Early Miocene to Later Eocene using analysis of seismic sections and drilling data. Some of positive folded structures were formed as result of both deformation stages.

Comprehensive three-dimensional seismic-stratigraphic researches of one of these elevations within passive margins of Norway brought out clearly that it was formed under tectonic compression normal (or roughly normal) to the rifting axis and spreading axis afterwards. The age of compression was defined: they were generated simultaneously or synchronously with the continental breakup at about 55 Ma. It is the researchers' opinion that folded structures could be result of both compression and shift. It is obvious that the initiation of this pair of forces fits with the idea of vortical nature of the spreading basin: at the initial stage of its opening, both components, compression and shift, appear in the area of the strongest whirling of vortical structure, which is being observed in reality.

In the eastern Greenland, folds and overfaults are found in the northern Jamson Land. They are of later Miocene age. Resent researches have shown that the folds have a wavelength of 5-10 km with total compression estimated as about 1%. This value agrees to the estimation of compression based on the evolutionary analysis of positive morphostructures at the western slope of Norway. Moreover, the continuation of the Greenlandian Transform Fault to the limits of the eastern Greenland is also accompanied by compression and folding.

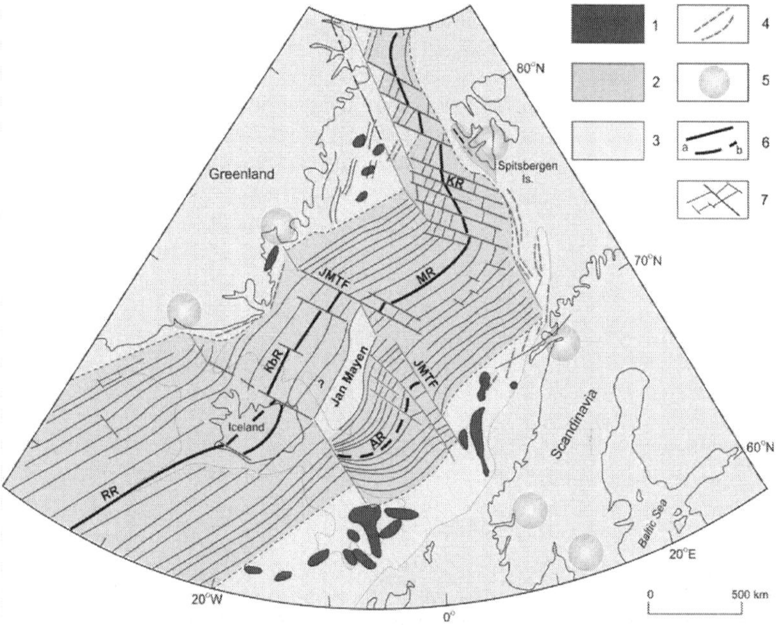

Fig. 9. Regions of passive margins of the Norwegian-Greenland Basin, where the folding of the sedimentary cover and the Neogene uplift are documented. Passive margins and the adjacent continental regions: (1) local uplifts formed by compression; (2) foldbelt west of Spitsbergen; (3) areas of the Late Cretaceous and Paleogene crustal extension; (4) shelf areas; (5) areas of intense Neogene uplifting. The ocean floor: (6) axes of (a) active and (b) extinct spreading; (7) axes of magnetic anomalies. (EJMFZ) East Jan Mayen Fracture Zone, (WJMFZ) West Jan Mayen Fracture Zone, (GFZ) Greenland Fracture Zone, (DSFZ) Denmark Strait Fracture Zone, (KnR) Knipovich Ridge, (MR) Mona Ridge, (KR) Kolbeinsey Ridge, (AR) Aegir Ridge, (RR) Reykjanes Ridge, modified after [16]

Within the southwest margins of the Barents Sea, tree major segments with different paleogeodynamic setting are known (from the south to the north): shift dislocations along the Senja Fault; oblique compression within margins of the Sørvestnaget Basin; and compression and shift along the zone of Hornsund faults. Such a change of geodynamic settings along the boundary of continental breakup can be connected with a vortical model of opening of this oceanic part.

The discovery and researching of domes in the sedimentary cover caused by tectonic compression in passive margins stimulated their further intensive study, and, in particular, works devoted to the comparison of these structures and ones long known within the ocean/continent junction zones of active type were carried out. One of such works presents the comparison of corresponding structures within Voring Plateau (Norway margins) and the northern Honshu Island (Japan) [37]. It was mentioned that tectonic evolution of so different global geostructures as passive margins of Norway and the active volcanic arc of Honshu (Japan) has a common feature, which consists in the fact that in both regions a long stretching period gave place to strong compression that caused generation of dome-like elevations.

It is obvious that although a problem of fold deformations of passive margins has aroused not a long time ago (because they have currently been found) it is of great importance from the position of both theoretic geotectonics and oil-and-gas prospecting. An applied aspect of this problem is evident and many specialists in oil-and-gas prospecting have realized it. In particular, disappointing results of prospecting at some districts of the Barents Sea are indeed connected with rising and erosion of deposits caused by compression.

Possible nature of mantle plumes. In a very nonequilibrium nonlinear medium of the Earth's solid covers, the various dynamically unstable interfaces can appear during the Earth's evolution. Correspondingly, this originates vortical movements of varied intensity. If we continue the analogy with processes in the Earth's outer covers, within its inner covers exceptionally strong (taking into account the medium properties) vortexes, tornados or twisters, can appear. In reality, among the Earth's lithosphere plates, there are small ones with angular velocity of rotation an order (or more) higher than the one of other plates. It can be suggested that they are rotated by intense vortical movements. Such fast-rotating (at a geological scale) plates are mostly restricted to weak permeable zones in the both crust and lithosphere where mantle masses with different physical characteristics interact similarly to the ones in the subduction zones. The formation of a fast-rotating vortex is most probable in such zones. A similar vortex in the atmosphere (tornado) is characterized by a strong vertical flow, which carries the matter from the bottom upwards. If such mechanism works in the inner covers, a narrow column of deep, hot, and softened matter with low seismic speeds should be found under fast-rotating plates. We consider some of the most striking examples from this standpoint.

Two typical hot spots, which origin by a common opinion is connected with upward mantle flow, are Iceland and Easter Island. Both of them are drawn towards tectonic junctions: Iceland is located at the intersection of the MOR's spreading zone and a transverse fault zone, and Easter Island is located not far off the triple junction of EPR and Chilean Rise. Both regions are characterized by the presence of very fast-rotating microplates. The Jan Mayen Microplate located north of Iceland during Palaeocene–late Miocene had a speed of rotation about 3 degree/Ma and the Easter Microplate is rotating still faster, about 15 degree/Ma. Cross-sections of the both crust and mantle up to 2,000 km were built for both regions based on the data of seismic tomography and varied in direction (fig.10). Beneath Iceland, a quite narrow roughly low-velocity vertical channel reaching the boundary of the lower mantle was revealed. Beneath Easter Island a channel with lower seismic velocities was also found in the middle and lower layers of the upper mantle (although it is not so clear as the one beneath Iceland). Both sections show distinct delamination of the upper mantle: in a vertical section, horizons with different seismic velocities are clearly distinguished. From the concept of vortical movements, it is known that such conditions are especially favorable for originating the fast rotating vortexes. We can notice that to explain magmatism of the Kamchatka region, as well as of some other regions with intraplate magmatism, a similar hypothesis was offered, suggesting a supposition of both, formation of strong vortical movements in the asthenosphere and «fluid-magmatic tornado» effecting the lithosphere and cause generation of hot spots [38].

The most of them are restricted to passive continental margins (for example, Deccan, Karoo, and Ferrar traps adjacent to the Indian Ocean, Etendeka, Parana, and Benue Trough adjacent to the Atlantic), in this connection it was repeatedly suggested that a causal relationship between plumes and continental breakup should exist. At the stage of continental breakup,

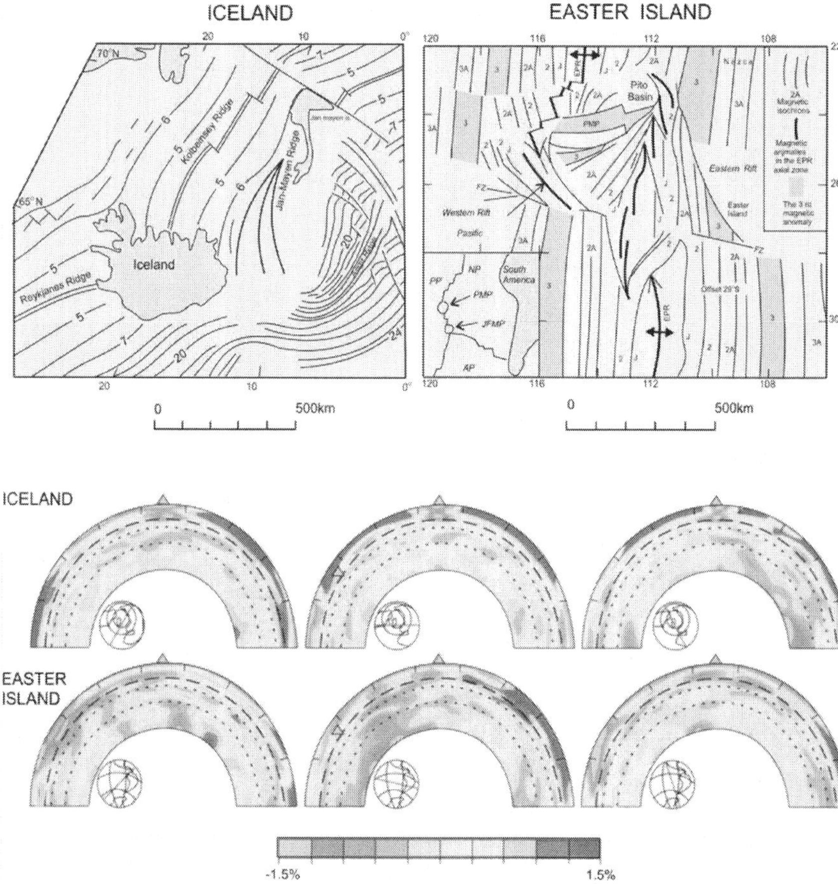

Fig. 10. Structural schemes of the Jan Mayen microplate, northeast of Iceland, and the Easter microplate at the crest of East Pacific Rise (upper panels) and tomographic sections along three profiles across these regions (lower panel), after the data reported in [41]. Axes of magnetic anomalies and their numbers are shown in the structural schemes. Location of the Easter and Juan Fernandez microplates is shown in the insets (see the upper right panel). (EPR) East Pacific Rise, (CR) Chile Rise, (NP) Nazca plate, (PP) Pacific plate, (AP) Antarctic plate, (PMP) Pacific microplate, (JFMP) Juan Fernandez microplate. Shades of gray show increased velocities of seismic wave propagation relative to the average (%): darker is faster and lighter is slower. The heavy dashed line in cross sections represents the boundary at a depth of 670 km, while the thin dashed lines represent the boundaries at depths of 1,000 km and 1,700 km.

the conditions favorable for generation of above considered fast-rotating vortexes could exist, and, correspondingly, for channels ascending deep matter in specific places of the breakup line. This line, as considered above, has always a saw-toothed shape, and at that, a stress field in the opposite parts of 'tooth' varies due to a specific character of vertical movements. To the explanation comes from Figure 8b, where the inset shows the orientation of stress vectors for one tooth at the moment of continental breakup. It is not difficult to see that a vortical component of the motion of the continental plate B relatively the plate A results in the generation of a dynamic subsurface where two forces are dominant, shift and compression. This is similar to dynamics of a subduction zone where conditions for the origin of strong vortexes exist. Turning to actual data we can see that at least some of the above mentioned trap provinces (for example, Etendeka and Benue Trough) are indeed restricted to those ledges of the African coastline where at the breakup stage similar geodynamic settings are known.

It is to be studied whether such hypothesis is correct to explain the nature of other hot sports and plumes. It is known that the mantle zones with lower velocities do not always correspond to them. However, a fast-rotating vortex is relatively short living thing (subject to the conditions of the medium where it forms). With this fact in mind we can suggest that the appearance of zones with lower velocities is dominantly governed by the evolutionary stage of one or other region of intraplate volcanism development.

4. Conclusion

Thus, the basis for the conception of nonlinear unstable geologic medium is an idea of opposite entropy flows influencing the energy balance of an open system. The use of the concept for geodynamic interpretations is not limited with the above shown examples. In particular, the origin of elastic vibrations in the Earth's crust, which is traditionally connected with earthquakes resulted from mechanical breaking of rocks is considered from the same standpoint. A thermodynamic scheme of rocks' destruction results in different interpretation of processes causing explosive-like energy output in the earthquake epicenter. In essence, we are dealing with a complete change of paradigm of seismology [9]. Other examples demonstrated self-organization of geological medium at various scales from micro level to mega level accompanied by formation of dissipative structures, i.e. the structures, in which the dissipation of endogenous energy is most efficient [38].

In the outer covers of the Earth's, atmosphere and hydrosphere, wide variety of vortical movements is observed, and they play a leading part in dynamics. The mentioned facts, as well as their interpretation, earnestly demonstrate existence of such motions in the inner solid covers, and they are of great importance in dynamics of these covers too. Suffice it to say that two thirds of the Earth's surface, to be exact - the oceanic lithosphere is formed under the direct impact of vortical movements. This conclusion is of fundamental importance for geotectonics, because it confirms above-mentioned theses about properties of the medium where processes defining the structural face of the Earth's crust take place, and these processes can not be considered as purely mechanical ones. Physical essence of these processes is a subject for special consideration, being beyond the scope of this article, however it is important to emphasize that vortical motions can be realized in a very nonequilibrium and nonlinear medium only. For one's turn, this implies that the epoch of domination of a plate tectonic paradigm comes to an end: its basis is mechanics of continuum, whereas the foundation of the future conception is nonlinear thermodynamics of a very nonequilibrium medium. Today's question is the revaluation of lithosphere as solid and fragile covers, because it is in inconsistent with ideas of medium fluidity and rank-invariant vortical movements in solid covers. Complex of lithospheric structures resulted from vortical movements should be considered in the frames of vortical tectonics, replacing new global tectonics. An additional illustration of this is the universality of vortical and spiral structures in micro-, meso-, and mega-world (spiral molecules, vortexes of both atmosphere and the ocean, vortical and spiral galaxies). The basis for a new geodynamic paradigm being formed at present is vortical tectonics, self-organization processes in nonlinear nonequilibrium mediums, and formation of fractal structures of lithosphere.

In summary it may be said that the switch to ideas of a nonlinear medium to explain the nature of most lithosphere structures is a general trend of geodynamics. Nowadays, the issue of the day and prerequisite to successful development is to work out and to start using the completely new (in comparison with fixed plate-tectonics' notion) method of resolving specific, practical problems.

5. References

[1] Prigozhin, I. R. Philosophy of Instability, *Philosophic matters* 1991, *6*, 46-57.

[2] Nikolis G. and Prigozhin, I. R. Self-Organization In the Nonequilibrium Systems. From Dissipative Structures Through Jitter To Order. Nauka: Mir, 1979; p 250 [in Russian].

[3] Haken H. Synergy. Mir: Moscow, 1980; p 185 [in Russian].

[4] Breakthrough to Synergy. Forward to the Third Millennium. Nauka: Moscow, 2002; p 478 [in Russian].

[5] Sadovsky M. A. Self-Similarity of Geodynamic Processes. *Herald of the Academy of Sciences of the USSR*, 1986, *8*, 3-11 [in Russian].

[6] Keilis-Borok V.I. Introduction: Non-linear systems in the problem of earthquake preduction. *Physics of the Earth and Planet. Interiors*, 1990, *61*, 1-7.

[7] Nikolaev A. V. Development of Methods of Nonlinear Geophysics. e-zine *Herald of the Academy of Sciences*, 2002, *1*, 20. Problems of Geophysics of the 21st Century. Vol. 1. Nauka: Moscow, 2003; p 311.

[8] Sharov, V. I. Tectonic Earthquake As a Nonequilibrium Thermodynamic Process of Rock Destruction, *Izv. Ross. Akad. Nauk. Fiz. Zemli*, 1992, *5*; pp121-127.

[9] Sharov, V. I. Study of Seismic Delamination of the Earth on the Threshold of the 21st Century. *Exploration and Conservation of Mineral Resources*, 1998, *2*; pp2-4 [in Russian].

[10] Barenblatt, G.I.; Zhivago, A.V.; Neprochnov, Yu.P.; Ostrovsky, A.A. Fractal Dimension: Quantitative Characteristic of Ocean Floor Topography. *Okeanologiya*, 1984, *24(6)*, 924-928.

[11] Malinverno, A. *Fractals and Ocean floor topography: A review and a model. In: Fractals in the Earth Science.* Plenum Press: New York, 1995; pp107-129.

[12] Nikora, V.I. *Fractal Dimension As a Quantitative Characteristic of Plain Geometry of Channels.* In: Dynamics and Thermodynamics of Rivers, Reservoir, and Marginal Seas: 3rd All-Union Conference. Abstracts. Nauka: Moscow,1989, 1, pp 46-47 [in Russian].

[13] Chelidze, T.A. Generalized Fractal Law of Seismicity. *Proc. of Acad. of Scie. of the USSR.* 1990, *314 (5)*, pp 1104-1105.

[14] Turcotte D.C. Fractals in petrology. *Lithos.* 2002. *65*.pp. 261-271.

[15] Sornet, D. and Pisarenko, V. Fractal Plate Tectonics. *Geoph. Res. Letters*, 2003, *30 (3):doi10.1029/2002GLO15043*

[16] Armijo, R., Meyer, B., Navarro, S. et al. Asymmetric Slip Partitioning in the Sea of Marmara Pull-Apart: A Clue to Propagation Processes of the North Anatolian Fault? *Terra Nova*, 2002, *14 (2)*, pp 80-86.

[17] Okay, A.I., Kashlar-Ozcan, A., Imren, C., Boztepe-Guney, A., Demirbag, E., Kuscu, I. Active Faults and Evolving Strike-Slip Basins in the Marmara Sea, Northwest Turkey: a Multichannel Seismic Reflection Study. *Tectonophysics*, 2000, *321*, pp189-218.

[18] Aydin, A. and Nur, A. Evolution of Pull-Apart Basins and Their Scale Independence. *Tectonics*, 1982, 1, pp 91-105.

[19] Sanders, C.O. and Slemmons, D.B. Recent Crustal Movements in the Central Sierra Nevada-Walker Lane Region of California - Nevada. Part III, The Olinghouse fault zone. *Tectonophysics*, 1979, *52*, pp 585-597.

[20] Dubinin, E. P. and Ushakov, S. A. *Oceanic Rifting*. GEOS: Moscow, 2001, 293 [in Russian].

[21] Mirlin, E. G. Fractal Formation of Structure at Different Stages of Oceanic Lithosphere Formation: Prerequisites, Examples, and Problems. *Okeanologiya*, 2006, *46 (1)*, pp 123-133.

[22] Mirlin, E. G. The Problem of Vortical Movements in the Solid Earth and Their Role in Geotectonics. *Geotektonika*, 2006, *4*, pp 282-296.

[23] Mirlin, E. G. and Zorina, Yu. G. The Openning of Oceanic and Marginal Spreading Basins: Contrastive Analysis. *Geotektonika*, 1992, *1*, pp21-33.

[24] Lebran, J-F., Lamarc, G., Collot, J-Y. Subduction Initiation At a Strike-Slip Boundary: The Cenozoic Pacific-Australian Plate Boundary, South of New Zealand. *Jour. of Geoph. Res.*, 2003, *108*, pp 15-1 - 15-8.

[25] Okino, K. and Fujioka, K. The Central Basin Spreading Center in the Phillipine Sea: Structure of an Exinct Spreading Center and Implications for Marginal Basin Formation. *Jour. of Geophys. Res.*, 2003, *108*, pp EPM 5-1 - 5-18.

[26] Khain, V. E. and Lomize, M. G. *Geotectonics with Principles of Geodynamics*. Universitet-Knizhnyi Dom: Moscow, 2005, 560 p [in Russian].

[27] *Vortices in Geological Processes*. Kamchatsky State Pedagogic Univ., Petropavlovsk-Kamchatsky, 2004; 297p [in Russian].

[28] Gol'din, S. V. *Physics of the Living Earth*. In: Problems of Geophysics of the 21st Century. Nauka: Moscow, 2003; 310 p [in Russian].

[29] Mirlin, E. G. , Kononov, M. V., Sushchevskaya, N. M. Vortical Spreading Systems in the Lithosphere and Upper Mantle. *Dokl. Earth Sci.*, 2005, *401A (3)*, pp 370-373.

[30] Kaz'min, V. G. and Borisova, I. A. Structural Relationships between Continental and Atlantic-Type Oceanic Rift Zones. *Geotektonika*, 1992, *27 (6)*, pp 80-98.

[31] Peive, A. A. Ruzhentsev, S. V. Sokolov, S. D., et al. *Tectonic Delamination of the Lithosphere*. Nauka: Moscow, 1980, 215 p [in Russian].

[32] Raznitsin, Yu. N. Tectonic Delamination of the Lithosphere in Young Oceans and Paleobasins. Nauka: Moscow, 2004, 270 p [in Russian].

[33] Larson R.L., Searl R.C., Kleinrock M.C., Shouten H., Bird R.T., Naar D.F., Reisby R.I., Hooft E.E., Lasthiotakis H. Roller-bearing tectonic evolution of the Juan Fernandez microplate. *Nature*, 1992, *356*, pp 571-576.

[34] Wilson, D.S. Deformation of the So-Called Gorda Plate. *J. Geophys. Res.*, 1989, *94(B3)*, pp 3065-3075.

[35] Boldreel, L.O. and Andersen, M.S. Tertiary Compressional Structures on the Faroe-Rockall Plateau in Relation to Northeast Atlantic Ridge-Push and Alpine Foreland Stresses. *Tectonophysics*, 1998, *300*, pp 13-28.

[36] Lundin, E. and Dore, A.G. Mid-Cenozoic Post-Breakup Deformation in the «Passive» Margins Bordering the Norvegian-Greenland Sea. *Marine and Petrol. Geol.*, 2002, *19*, pp 79-93.

[37] Mjelde, R., Iwasaki, T., Shimamura, H., Kanazawa, T.,.Kodaira, S., Raum, T., Shiobara, H. Spatial Relationship Between Recent Compressional Structures and Older High-Velocity Crustal Structures; Examples from the Voring Margin, NE Atlantic, and Northern Honshu, *Japan. J. of Geodynamics*, 2003, *36*, pp 537-562.

[38] Goryainov, P. M. and Ivanyuk, G. Yu. Self-Organization of Mineral Systems. Synergetic Principles of Geological Studies. GEOS: Moscow, 2001, 312 p [in Russian].

[39] Scotese, C.R. Paleomap Project. http://www.scotese.com 2002.

[40] Soldatenko, S. A. Synoptic Vortices in the Atmosphere and Ocean. *Soros. Obrazov. Zh.*, 1989, *2*, 78-84.

[41] Ritsema, J. and Allen, R.M. The Elusive Mantle Plume. *Earth Planet. Sci. Lett.*, 2003, *207*, pp1-12.

10

Layer-Block Tectonics, a New Concept of Plate Tectonics - An Example from Nansha Micro-Plate, Southern South China Sea

Hai-ling Liu, Hong-bo Zheng, Yan-Lin Wang,
Chao-Hua Wu, Mei-Song Zhao and Yun-Kong Du
Key Laboratory of marginal sea geology, South China Sea Institute of Oceanology, Chinese Academy of Sciences, Guangzhou 510301, China

1. Introduction

The Layer-block tectonics (LBT) is a new theory describing the layer-slip structure of lithosphere (Liu et al., 2002, 1999; Sun et al., 1991). According to this theory, a lithosphere plate, continental lithosphere plate in particular, is considered a composite of sub-plates connecting with each other horizontally and overlapping with each other vertically. The term "Layer" in the LBT emphasizes the rheological and stratifying characteristics of the lithosphere and the guiding and controlling role of mechanically "soft" layers with different deepness in the layer-slip movement of the lithosphere during the process of tectonic deformation. The term "block", on the other hand, emphasizes the discontinuity of various types of geological bodies segmentalized by dip-slip or strike-slip movement of lithosphere in horizontal direction. We use the concept of the LTB to cover the scientific thoughts of the other tectonic theories such as the gliding of layers (Mandle and Shippan, 1981), the flake tectonics (Oxburgh, 1972), the terrane tectonics (Irwin, 1972), the capped plates (Coleman, 1977), the extensional tectonics (Wernike, 1988, 1981), and so on. The obvious different of the LBT from the these tectonic theories, even from the plate tectonics theory, is that the LBT emphasizes the geotectonic effect of multi-levels of gliding surfaces within lithosphere-upper mantle including rheospheric top surface, Moho surface, mid-crust, top surface of sedimentary basement, and so on, rather than only emphasized singular Moho gliding surface in flake tectonics or the plate tectonics.

The lithosphere can be divided into different layers by the characteristics of material, energy, structure, rheological and chemical stratification at different depths (Su et al., 1996; Song et al., 1996; Wang et al., 1996; Wang, 1992; Rushentsev and Trifonov, 1985; Oxburgh, 1972). These layers interrelate with each other and stack-and-piece together to form an integral lithospheric aggregate. As the manifestation of this nature of stratification, the LBT is the result of bedding layer-slip, dip-slip (both in positive and negative direction) and strike-slip (in slant direction, sinistral or dextral) of geologic bodies under tectonic forces (vertically or horizontally).

All layer-block structures in various scales, whether large as the global lithosphere plate or small as a dislocation structure, have a common slip mechanism, and the "4-dimensional interrelated-action" faults enclosing them (Liu et al., 2002). The term "4-dimensional interrelated-action" means that dip-slip - layer-slip - strike-slip fault systems consisting of 3-dimensional edge faults which dynamically originated from the same layer-blocks should act jointly and synchronously (three dimensions of space plus one dimension of time), and the layer-blocks would hardly move when lacking any of these dimensions. This leads to the following principle of layer-block partition. Each level of layer-block is defined by slip surface and the "4-dimensional interrelated-action fault system" as boundaries. After each level of layer-block is defined, we will analyze the overlap and joint layer-blocks according to "Overpass-type" movement and the rules of multiple geodynamic systems, and finally draw an outline of the evolution of crust and upper mantle.

According to the depth of layer-slip surface, on which layer-block moves, and the depth of the boundary faults, there are 4 categories: ultra-crustal (incising depth is deeper than the base of lithosphere), crustal (incising depth reaches Moho discontinuity), basemental (incising depth reaches middle crust), and cover (incising depth is above the sedimentary basement) layer-blocks.

Then Nansha Micro-plate is located at the junction of modern Eurasian plate, the Pacific Plate, and the Indo-Australia Plate. Its geological structure is extremely complex. Based on comprehensive analysis of gravity, magnetic, seismic profiles with a total length of about 30,000 km (Yan and Liu, 2004; Liu et al., 2004, 2002, 1999; Schluter et al., 1996; Bai et al., 1996; Hinz et al., 1989; Hinz and Schuler, 1985) and relevant geological data, and in the light of the above-mentioned concepts and principle of partition of the LBT, we divided the Nansha Micro-plate into 6 layer-blocks, i.e., Nansha Ultra-crust Layer-block, Zengmu (Loconia, ZCL in Fig. 1), Nanwei (Riflemam Bank) - Andu (Ardasier Bank)(NCL in Fig. 1), and Liyue (Reed Bank) - North Palawan (LCL in Fig. 1) Crustal Layer-blocks, and Andu - Bisheng (Pearson Reef) (ABL in Fig. 1) and Liyue - Banyue (Half Moon Reef) (LBL in Fig. 1) Basemental Layer-blocks. We also discussed the forming mechanism of these layer-blocks.

Nansha plate is a Cenozoic micro-plate inherited with Mesozoic marine facies strata, magmatic rock and metamorphic rock basement (Liu et al., 2004; Kudrass et al., 1986). It is circled by several large plate-edge basins, such as, Wan'an (west Vanguard bank) basin, Zengmu basin, and Nansha (Borneo–Palawan) trough basin etc., and occupied by a number of intraplate basins. We give their initial division scheme on basal fault system and intra-plate basins in Nanwei - Palawan sea area in the east of Lizhun (Grainger bank)- Tinjar fault (LTf in Fig.1)in Nansha Micro-plate, and present an analysis on the forming mechanism of basal faults controlling the intra-plate basins, on the basis of geological and geophysical data, especially those of seismic profiles, of Nansha area, former research (Liu et al., 2002; Sun et al., 1991).

2. Characteristics of layer-block tectonic of the lithosphere of Nansha micro-plate

2.1 Nansha Ultra-crustal Layer-block

The Nansha Ultra-crustal Layer-block is defined by the ultra-crustal layer-slip surface, i.e. the bottom of Nansha lithosphere, and four boundary fault systems (Liu et al., 2004, 1999) as the controlling boundaries of its overall movement (Tab. 1 and Fig. 1). These boundary fault systems include the Kangtai - Xiongnan (Kangxiong for short, KXf in Fig. 1) ultra-crustal extending and slipping fault zone (in the north), Baxian (southwest Zengmu shoal)–Cuyo (Bacu for short, BCf in Fig. 1) ultra-crustal thrusting fault zone (in the south), Mindoro - Panay ultra-crustal strike-slip fault zone (Minpan for short, which is in the east and serves as a coordinating role, MPf in Fig.1) in the east, and the Wan'an - Natuna (Wanna for short, WNf in Fig.1) ultra-crustal strike-slip fault zone (in the west).

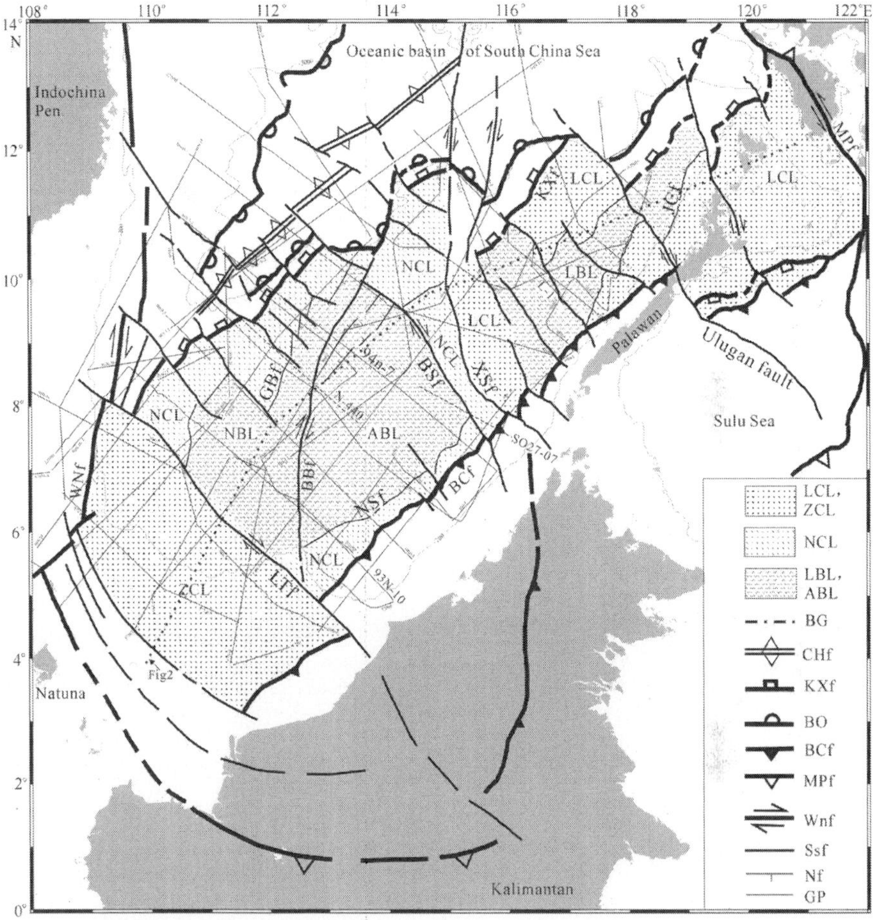

Fig. 1. Layer-block tectonics in Nansha micro-plate.
ABL= Andu - Bisheng basemental layer-block; BBf = Bisheng–Beikang fault zone; BCf = Baxian - Cuyo thrust-nappe fault zone; BO = Boundary of deep oceanic basin; BSf = Bisheng - Siling fault zone; CHf = Changlong - Huangyan (Scarborough Island) sea-floor spreading ridge fault zone; GBf = Guangya - Bisheng fault zone; GP=Geophysical profiles; JCf = Jianzhang - Calawit fault zone; KXf = Kangtai - Xiongnan extending and gliding fault zone; LBL = Liyue - Banyue Basemental Layer-block; LCL = Liyue - North Palawan Crustal Layer-block; LTf = Lizhun - Tinjar fault zone; MPf = Mindoro - Penay compressive strike-slip fault zone; NCL = Nanwei - Andu Crustal Layer-block; Nf = Normal fault; NSf = Nantong - Siling fault zone; Ssf = Secondary strike-slip fault; WNf = Wan'an - Natuna strike-slip fault zone; XSf = Xiyue - Siling fault zone; ZCL = Zengmu Crustal Layer-block.

<table>
<tr><th colspan="2">Name of layer-block</th><th colspan="2">Layer-slip - dip-slip - strike-slip 4-dimensional interrelated-action fault system</th></tr>
<tr><td colspan="2">Nansha ultra-crustal layer-block</td><td>Nansha ultra-crustal layer-slip - fault system</td><td>Nansha ultra-crustal layer-slip surface
Kangxiong ultra-crustal extending - gliding fault zone
Bacu ultra-crustal thrusting - nappe fault zone
Wanna ultra-crustal dextral pull-apart strike-slip fault zone
Minban ultra-crustal sinistral compression - strike-slip fault zone</td></tr>
<tr><td rowspan="3">Crustal Layer-block</td><td>Zengmu crustal block</td><td>Zengmu crustal layer-slip - fault system</td><td>Zengmu lower-crustal layer-slip surface
Wanna ultra-crustal dextral pull-apart strike-slip fault zone
Lizhun–Tinjar crustal strike-slip fault zone
Bacu ultra-crustal thrusting - nappe fault zone</td></tr>
<tr><td>Nanwei–Andu crustal layer-block</td><td>Nanwei–Andu crustal layer-slip - fault system</td><td>Nanwei–Andu lower-crustal layer-slip surface
Kangxiong ultra-crustal extending - gliding fault zone
Bacu ultra-crustal thrusting - nappe fault zone
Lizhu–Tinjar crustal strike-slip fault zone
Xiyue–Siling crustal dextral strike-slip fault zone</td></tr>
<tr><td>Liyue–North Palawan crustal layer-block</td><td>Liyue–North Palawan crustal layer-slip - fault system</td><td>Liyue–North Palawan lower-crustal layer-slip surface
Kangxiong ultra-crustal extending - gliding fault zone
Bacu ultra-crustal thrusting - nappe fault zone
Xiyue–Siling crustal dextral strike-slip fault zone
Minban ultra-crustal sinistral compression - strike-slip fault zone</td></tr>
<tr><td rowspan="2">Basemental Layer-block</td><td>Andu–Bisheng basemental layer-block</td><td>Andu–Bisheng basemental layer-slip - fault system</td><td>Andu–Bisheng middle crustal layer-slip surface
Nantong–Siling basemental extending - gliding fault zone
Guangya–Bisheng basemental extending - gliding fault zone
Lizhun–Tinjia crustal strike-slip fault zone
Bisheng–Siling basemental strike-slip fault zone</td></tr>
<tr><td>Liyue–Banyue basemental layer-block</td><td>Liyue–Banyue basemental layer-slip - fault system</td><td>Liyue–Banyue middle crustal layer-slip surface
Jianzhang–Calawit basemental extending - gliding fault zone
Xiyue–Siling crustal dextral strike-slip fault zone
Minban ultra-crustal sinistral compression - strike-slip fault zone</td></tr>
</table>

Table 1. 4-Dimension Interrelated-Action Fault Systems and Layer-blocks of Nansha Lithosphere

The geometry status of the Nansha ultra-crustal layer-slip surface can be imaged by the low speed layer (Vs<4.7km/s) of transverse wave shown in Fig. 2. The low speed layer of transverse wave can be considered as asthenosphere under Nansha lithosphere and lies in the depth between 58 to 148km. The depth of the top surface of the low speed layer reaches the maximum value 74km in the Nanwei bank - Taiping (Itu Aba) island area and rises gradually to the minimum value 58km in southwest and northwest area i. e. in Borneo and Indochina block. The difference between the maximum and minimum is up to 16km. The largest gradient is in the northwestern side of the Nansha Micro-plate (Fig. 3). In Nanwei bank - Taiping Island area, the thickness of this low-speed layer is 44km, and in Borneo area in southwest, the thickness increase to 68km. The thickness in northwest and Indochina area are up to 90km. This kind of coupling relation facilitated the southward-southeastward migration of ultra-crustal layer-block of Nansha lithosphere along with the migration of mantle asthenosphere (Liu et al., 1999).

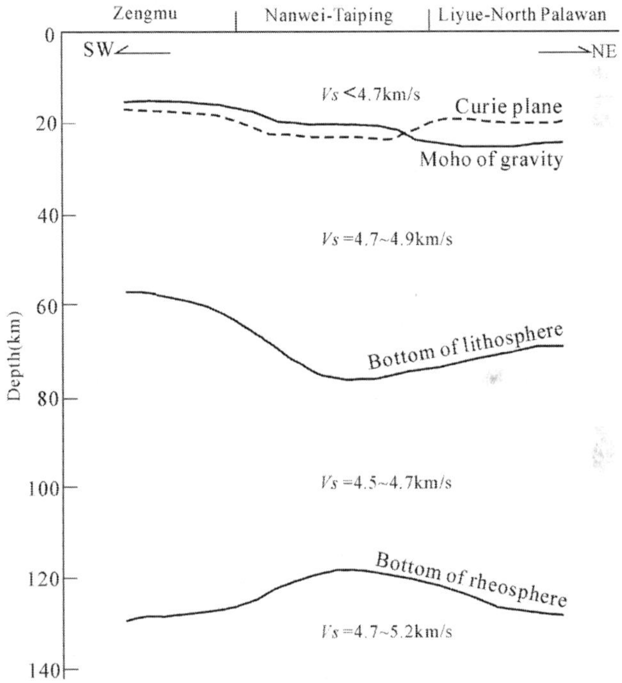

Fig. 2. Sketch map showing structural profile of crust—upper-mantle in southern South China Sea. After Wu et al. (1999), Xia (1997), Zhang et al. (1996), Yao et al. (1994), etc.. See Fig.1 for roughly location.

The above-mentioned 4 groups of dip-slip and strike-slip boundary fault zones, along with the ultra-crustal layer-glide surface, form an spatial and temporal kinematic system. This system constitutes a large-scale 4-dimensional interrelated-action fault system, i.e. the Nansha ultra-crustal layer-slip - dip-slip - strike-slip system, whose movement is unified in the overall southward drift.

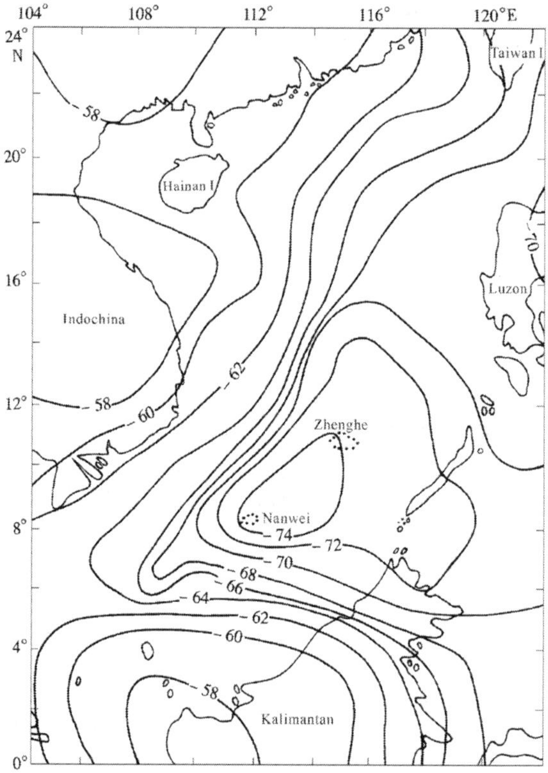

Fig. 3. Isobaths of lower boundary plane of Nansha lithosphere. The curves are drawn according to Vs=4.7km/s. The unit of the curves is km. Modified after Zeng et al. (1997).

2.2 Crustal layer-blocks in southern South China Sea

The Nansha Ultra-crustal Layer-block can be subdivided into 3 crustal layer-blocks, i.e., the Zengmu, Nanwei - Andu, and Liyue - North Palawan crustal layer-blocks, which are controlled by the lower crustal layer-slip surface and crustal fault systems.

The lower crustal layer-slip surface is the relative gliding plane between crust and upper mantle. Generally, it is a sharp seismic wave velocity interface corresponding Moho discontinuity. Sometimes it can be a transitional thin layer with gradually changing of wave velocity, or an obscuring interface, or a composite layer alternated with high and low velocity fine layers. Considering its property, it can be a chemical interface, or mineral phase changing interface, or even a mechanically non-capable layer, which is the reflection of tensional cracks of rocks extensively developed under super-high static pressure. The depth of Moho is an important parameter to decide the rheological characteristics of lithosphere and intra-plate strain caused by plate boundary forces. It indicates crustal maturity, crustal type, and isostasy degree, and controls the crustal layer-blocks formed by tensional breakup, separating, and downward sliding of crust.

In the southern area of the South China Sea, the depth fluctuating of Moho discontinuity is more complex than the ultra-crustal layer-slip surface. However, its general tendency of change is similar to that of ultra-crustal layer-slip surface as described above, and rises to SW-S. From Liyue bank in the NE, via Nanwei and Andu banks in the middle, to Zengmu basin in the southwest, the Moho changes roughly in three steps, from 24km (Liyue bank in the northeast) to 20km (in the middle) and to 16km (Zengmu basin)(Fig. 2). The magnetic survey (Fig. 4) also reflect the similar characteristics of portioned crustal blocks, which is supported by terrestrial heat flow field (Fig. 2). Each step platform can be considered as an independent layer-slip plane. In this way, we can think that there are three lower crustal layer-slip planes, the Liyue - North Palawan, Nanwei - Andu, and Zengmu lower crustal layer-slip planes, from northeast to southwest. The transitional slope zones are the fault zones, i.e. Xiyue (Yiyue or West York Island)- Siling (Commodore reef) (XSf in Fig.1) and Lizhun - Tinjar (LTf in Fig. 1) crustal strike-slip fault zones incising the crust. These fault zones separate the three crustal layer-blocks above-mentioned.

2.3 Basemental layer-blocks in the southern area of the South China Sea

Generally, basemental layer-blocks are partitioned by middle crustal layer-slip surface and dip-slip or strike-slip boundary fault zones that incise only to middle crustal layer-slip surface. The middle crustal layer-slip surface generally develops in middle-crustal layer. The middle crust is usually consisted of granite and dioritoid rocks, and is 8~20km in thickness and 10~15km in buried depth (Huang et al., 1994). Under normal geothermal conditions, this depth is adaptable for greenschist metamorphic process of quartz deformed from ductility to plasticity. In addition, since there are abundant of radioactive elements concentrated in this depth, local melting may occur. Therefore, it behaves plastically and rheologically. Above this layer-slip surface, there is the relatively brittle rigid layer of upper-crustal crystalline basement, which is consisted of granitoid intrusive rocks and metamorphic rocks; and below this layer-slip surface, there is the lower-crust consisted of relatively strong gabbroic rocks. The middle crust layer-slip plane can provide a space for the concentrated releasing of gravity energy and horizontal stress energy, downsliding, and inner-crust diving of upper crust. Because of the existence of this surface, the upper crust loses its tectonic deformation energy during the plastic flow process and becomes too weak to dive into lower crust due to insufficient energy. Most thick-skinned tectonics, such as Basin - Range Province and thrust-superimposed orogenic zone, are controlled by this layer-slip surface (Li et al., 1996).

Some signs of sliding of middle crust layer-slip surfaces have been revealed in the southern area of the South China Sea. By upward extrapolation of magnetic data with the steps of 5km, 10km, and 15km, respectively. We analyze the space characteristics of middle crustal layer-slip surface. As shown in Fig. 4, in the south of the area of Lizhun (Grainger) bank - Yinqing (London) reefs - Feixin (Flat) island, the zero-contours of upward extrapolation of magnetic data show a SE-migration tendency of increasing with the step of upward extrapolation. According to this result, it can be concluded that there is a detachment surface with depth comparable to that of middle crust (about 10km in depth). The Andu–Bisheng and Liyue–Banyue basemental layer-blocks slided and tilted prominently into southeast direction along this surface.

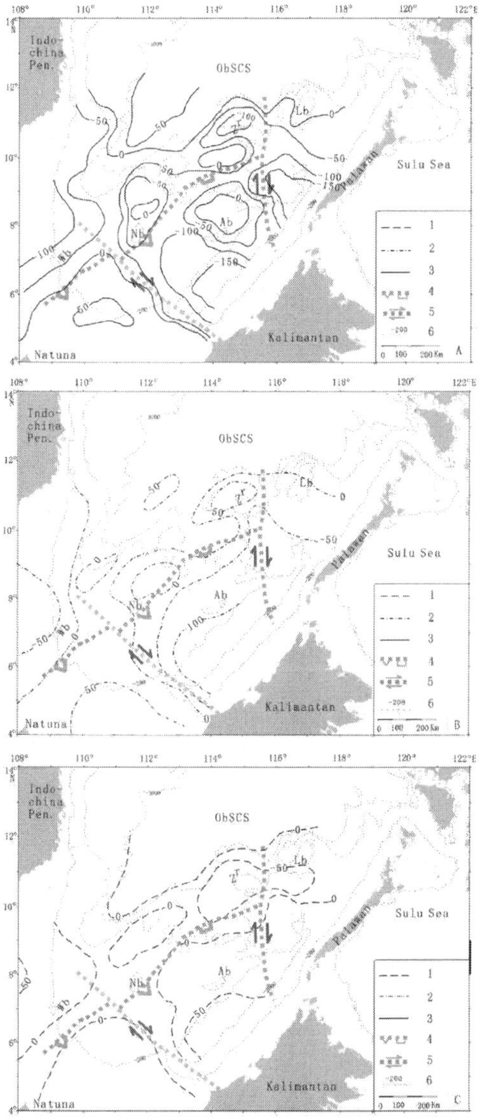

Fig. 4. Map showing magnetic anomaly curves of upward extrapolation for steps of 5, 10 and 15km in turn in southern South China Sea. A is for the step of 5 km, B is for 10km, and C is for 15km. 1=magnetic anomaly curves of upward extrapolation for the step of 15km; 2=magnetic anomaly curves of upward extrapolation for the step of 10km; 3=magnetic anomaly curves of upward extrapolation for the step of 5km; 4=Mid-crustal layer-slip plane; 5=Strike-slip fault zone' 6=Water isobaths; Ab=Andu bank; Lb=Liyue bank; Nb=Nanwei bank; OBSCS=Oceanic basin of South China Sea; Wb=WanAn bank; TP=Taiping island.

3. Characteristics of main intra-plate basins in Nansha area

Most basins, in particular those featuring of stretching, are caused by tilting or subsiding of the basement.This requires the existence of a "4-dimensional interrelated-action" (Liu et al., 2002)between boundary faults. Basemental layer-blocks control the formation of intra-plate basins. As shown in Fig. 5, we recognized 3 basin groups controlled by the basemental layer-blocks, i.e., Nanwei–Andu basin group (NBG in Fig. 5) in southwest, Liyue–Palawan basin group (LBG in Fig. 5) in east, and the Feixin–Nanhua (Cornwallis South Reef) basin group (FBG in Fig. 5) in between. The Nanwei–Andu basin group includes the well-know west Nanwei, east Nanwei, Beikang, and Andu basins etc. (Adb, Bkb, Wnb, and Enb in Fig. 5, in turn). The Liyue - Palawan basin group includes north Palawan, west Palawan and Liyue basins etc. (Npb, Wpb, and Lyb in Fig. 5, in turn) The Feixin–Nanhua basin group is a large-scale strike-slip basin in a whole. From Fig. 5, we can see that all these basin groups developed in different crustal layer-blocks, and are directly controlled by their basal layer-blocks respectively.

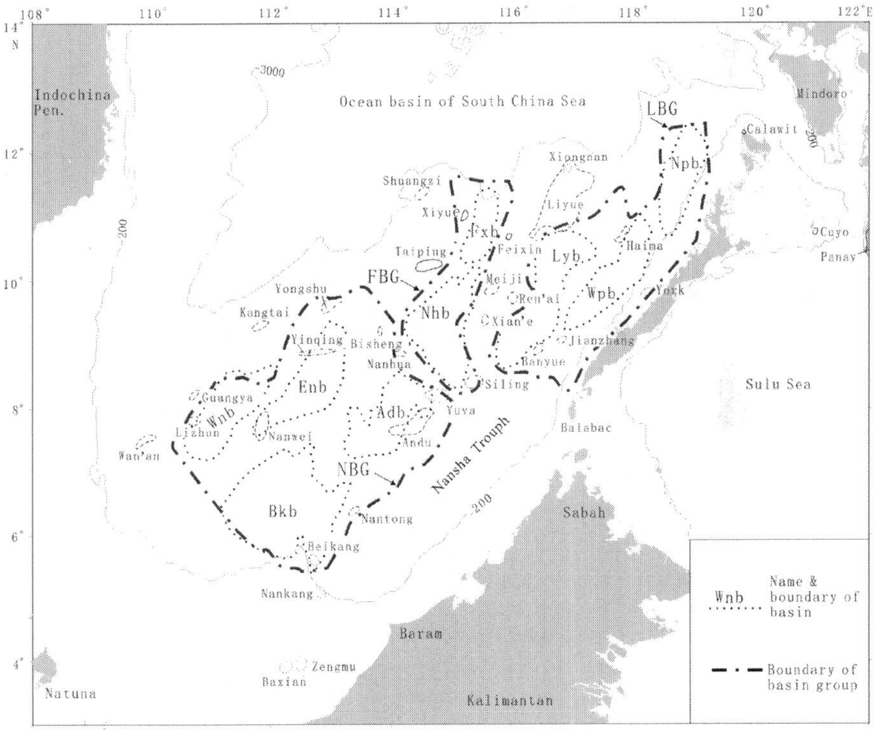

Fig. 5. Distribution of intra-plate basins in Nansha micro-plate.
Adb = Andu basin; Bkb = Beikang basin; Enb = East Nanwei basin; FBG = Feixin–Siling basin group; Fxb = Feixin basin; Lyb = Liyue basin; NBG = Nanwei–Andu basin group; Nhb = Nanhua basin; Wnb = West Nanwei basin; Wpb = West Palawan basin; Npb = North Palawan basin; LBG = Liyue–Palawan basin group.

3.1 Characteristics of Nanwei–Andu basin group and its basin-controlled faults

Nanwei–Andu basin group was developed in Nanwei - Andu crustal layer-block, and is controlled by the movement of Nanwei - Andu basemental layer-block. It is enclosed by Bisheng - Siling strike-slip fault zone (BSf in Fig. 1) in the northeast and Lizhun - Tinjar strike-slip fault zone (LTf in Fig. 1) in the southwest, and Nantong - Siling fault zone (NSf in Fig. 1) in the southeast and Guangya - Bisheng fault zone (GBf in Fig. 1) in the northwest. These fault zones share the Nanwei - Andu upper-crust layer-sliding surface, form a dip-slip - layer-slip - strike-slip system with "4-dimensional interrelated-action", and control the formation of Nanwei - Audu basin group.

3.1.1 Lizhun – Tinjar fault zone

It is a major NW-strike deformation zone in Nansha Islands. It is an apparent boundary line for both topography and geophysics. The magnetic field (Zhang et al., 1996), gravity field (Su et al., 1996a), and geothermal field (Ru and Pigott, 1986) across this line show great differences. It is a crustal fault deep through Moho. In Early Miocene, it was a dextral fault zone, with 100km horizontal offset (Young, 1976). In Late Miocene, it became sinistral. And from the Quaternary Period, it became dextral again. Its activities influenced the formation of the Andu–Bisheng crustal layer-block, and the Nanwei - Andu basin group.

3.1.2 Bisheng – Siling basal strike-slip fault zone

The northwestern section of this NW direction fault zone passes the east of Bisheng Island, and its southeastern section runs to the east of Siling Reef, where it can be traced by observing the activities of Xiyue - Siling fault in its later period. It is located largely in Nanhua (Pigeon) waterway, and is consisted of several nearly parallel faults. Seismic profile shows a negative flower structure (Liu et al., 2002).

3.1.3 Nantong – Siling basal extensional sliding fault zone

Basically, this zone is an extensional fault zone developed along the southeast of Nantong Reef - Siling Reef and the northern edge of Nansha Trough. It is consisted of several nearly parallel normal faults (Liu et al., 2002). Most of these faults are dip NW direction, except that the southwestern section runs in NEE-strike direction and the northeastern section runs in NE-strike direction. Its middle section is cut by several NW-strike translation faults. In the half graben formed through activities of this fault zone, the sedimentary covers started to develop in Paleocene Epoch. The half graben is filled with Paleocene to Early Oligocene clastic deposit. The activity of the faults was stronger during Late Oligocene to Early Miocene. From gravity and magnetism profile (Cui, 1996), it can be seen that this fault zone only disturbed into middle crust with the depth of 6~8km,. Its detachment surface is near ductile bed in the middle crust. The rock density above this depth is extremely inhomogenous but rather homogenous at 2.7 g/cm^3 in deep (see Fig. 6).

3.1.4 Guangya – Bisheng basemental extending dip-slip fault zone

This fault zone controls the northwestern edge of Nanwei - Andu basin group. It dip to SSE or SE direction, and is intersected into sections by several NW strike-slip faults. In magnetic field map, this fault zone is located right at the transitional zone between a dome and a depression of the top-interface of the magnetic basement. The Nanwei - Andu basin group developed in the depression in the southeast of Guangya - Bisheng basemental extensional dip-slip fault zone.

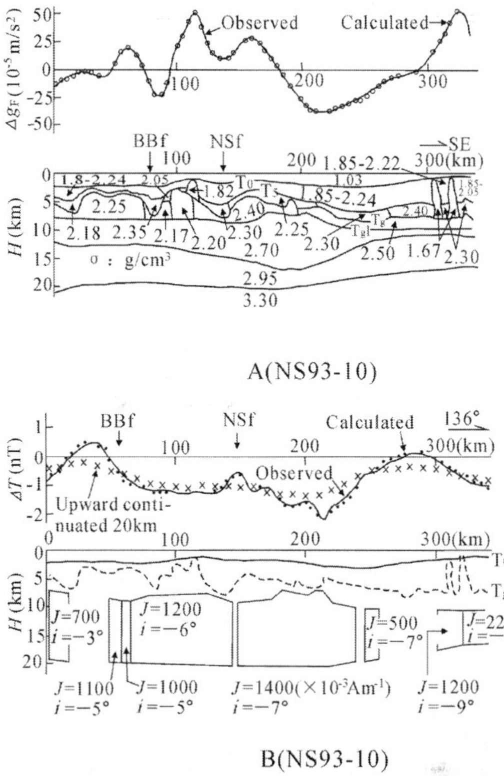

Fig. 6. Gravity and magnetic profiles crossing Nantong-Siling basemental extension fault zone. A = Gravity profile; B = Magnetic profile. BBf = Bisheng–Beikang fault zone; NSf = Nantong - Siling fault zone; Packed up from Cui (1996). See NS93-10 of Fig.1 for location of the profile.

3.1.5 Andu–Bisheng upper crust layer-slide surface

In the middle of Nansha Micro-plate and in the south side of Lizhun bank - Yinqing reefs - Feixin reef line, the zero contours of upward extrapolation of magnetic anomaly shows a tendency of southeastward drifting with increasing of the upward extrapolation steps (Liu et al., 2002). From this, it can be presumed that there exists a detachment surface extending along the layer in deep, which is called Andu–Bisheng upper crustal layer-slide surface, in the middle crust (about 10km in depth). Andu–Bisheng upper crustal block (i.e. Andu-Bisheng basemental layer-block) slips and tilts along this surface southeastward, and formed a series of NE basemental depressions in the northwestern edge. The aerial magnetometer measurement conducted by Aero-Geophysical Prospecting and Remote Sensing Center of Chinese Ministry of Geology and Mineral Resources, also revealed the basemental depressions and shows that the central basement of the depressions is 2~4km deeper than that of the two sides (Liu et al., 2002).

3.1.6 Nanwei – Andu Basin Group

From seismic profiles (Fig. 7), this group can be easily seen a complex, a face-to-face dip-slip faulted-block group, and is consisted of multiple half grabens and horsts controlled by primary face-to-face tilting boundary faults. Two primary boundary-controlled faults, Guangya - Bisheng and Nantong - Siling positive dip-slip - extending faults (Fig. 7, 8), extend into depth, decouple, flatten gradually, and merged into the Andu–Bisheng middle crustal layer-slide surface. The main activities of the faults happened between Early Tertiary and Early Miocene. They started to subside in late Early Paleocene, earlier than that in Zengmu basin (Zhong et al., 1996, 1991). Among the basins, the Beikang and West Nanwei basins reached their peaks of tectonic subsidence at middle Eocene period, with 0.2~0.3km/Ma and 0.6km/Ma subsidence rate respectively. The extension coefficient of the Beikang basin is 1.4~1.72. In Late Eocene, the thermal subsidence became stronger, and steep tectonic subsidence appeared in Pliocene.

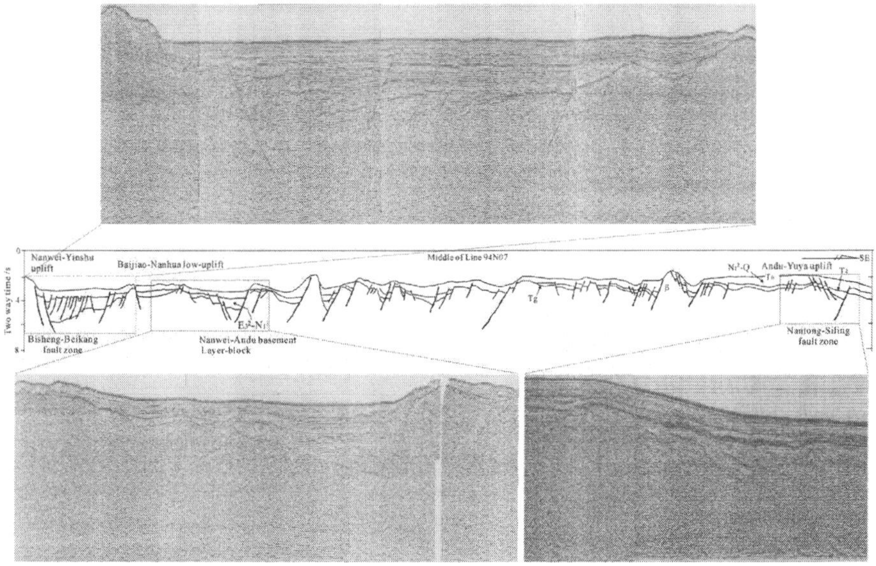

Fig. 7. Interpreted seismic profile 94N07 transecting Nanwei-Andu basemental layer-block. See Fig.1 for the location of the profile.

3.2 Basic characteristics of Liyue – Palawan basin group and its basin-controlling faults

Liyue - Palawan basin group developed in Liyue - North Palawan crustal layer-block, and is controlled by Liyue - Banyue basemental layer-block. The dip-slip - layer-slip - strike-slip "4-dimensional interrelated-action" fault system, which encloses the basemental layer-block and controls the formation of this basin group, is consisted of Minpan ultra-crustal sinistral compression and strike-slip zone in east and Xiyue–Siling crustal dextral strike-slip zone (XSf in Fig. 1) in west, Jianzhang (Royal Captain Shoal) - Calawit (Island) basemental extending positive fault zone (JCf in Fig. 1) in south, and Liyue - Banyue middle crustal layer-slide surface.

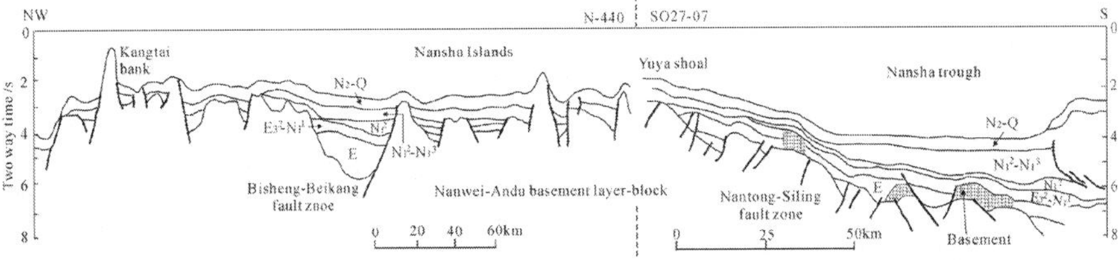

Fig. 8. Interpreted seismic profiles transecting Nanwei-Andu basemental layer-block
A = profile transecting eastern part (SO27-07) and western part (N-440) of Nanwei-Andu basemental layer-block. See Fig.1 for locations of the profiles.

3.2.1 Minpan ultra-crustal sinistral compression – strike-slip zone

This zone connects northwards with the passive subduction zone in the east of South China Sea oceanic crust (i.e. Manila trench), and extends to Taiwan Island. In the south, it extends to Negros and Cotabato trenches, subduction zones of Sulu and Sulawesi oceanic crusts. These two trenches started their arc activities in Late Miocene (7Ma) and Early Pliocene (Pubellier et al., 1991). From Taiwan Longitudinal Valley to Cotabato trench, the entire fault zone becomes the boundary between Eurasian Plate and West Philippine Oceanic Plate. It is obvious that this fault zone has been active for a long time. It is a ultra-crust strike-slip fault zone cutting deep into lithosphere, and significantly influenced the development of Nansha Micro-plate.

3.2.2 Xiyue – Siling crustal dextral strike-slip fault zone

This fault runs along a NS-strike trough to the west of Liyue Bank. Its southern section is merged into the southeastern section of NW-strike Nanhua waterway and turned into SE direction. The Xiyue - Siling crustal strike-slip fault zone starts from the southern edge of the oceanic basin of he South China Sea, passing through the west of Liyue Bank and east of Siling Reef, and enters into the lowest section of Nansha Trough (water depth is larger than 3300m). Southward, it extends into Sabah area, and separates the EW-strike structure of northeastern Sabah and NE-strike structure of southwestern Sabah (Yao, 1995; Tongkul, 1990). This fault zone is reflected on both magnetic and gravity fields. Together with Lizhun - Tinjar crust strike-slip fault, it cut into Nansha lower crust layer-slide surface and made it a three-level step-like structure. The southern section of this fault zone is rather steep, with deeper sections inclining to the east and converging into the layer-slide surface of Liyue - North Palawan lower curst.

The strata on each side of the fault are different. The lower structural layer in the east is Early Jurassic delta -shallow-marine facies sandstone-mudstone to Early Cretaceous littoral to shallow-marine facies coal-bearing clastic rock series (Kudrass et al., 1986; Taylor and Hayes, 1980). In the west, the lower structural layer of northwest part of Nansha Islands is even older, and may be the Triassic marine sedimentation. The strata of the middle structural layer in the east are thin in Ren'ai (Second Thomas) Reef - Liyue Reef area, to less than 1km mostly. They are a set of unmetamorphic Paleocene - Eocene delta facies and open shallow sea - half deep-marine facies clastic deposit. The age of strata filling in the bottom of half grabens lasted into Late Cretaceous. The middle structural layers in the west, however, are rather thick to 1~3km. The thickest layer appears near Nankang (South Luconia) shoal, to 4.5km in thickness, with half graben deposits filled in its lower part, and sheet-like draping layers in its upper part (Liu et al, 2007). The Late Oligocene to Early Miocene shallow-sea platform layered carbonate rock of the upper structural layer just distributes evenly in the areas on the two sides of Xiyue - Siling fault zone.

There are at least two apparent tectonic events. The earlier one cut into lower Miocene series, and the later one cut into Pliocene series to Quaternary system (Hinz and Schuler, 1985). From seismic profiles (Hinz and Schuler, 1985), it can be seen that strike-slip and extension events cause the formation of apparent half graben structures, and the throw of the faults are 1.7~3.0s (two way time). The events of the north part stopped at the end of middle Miocene; the south part, however, continued till Recent due to the Sabah thrust (Yao, 1995). Its activity is directly related to the formation of Nanwei - Andu and Liyue - North Palawan crustal layer-blocks, Liyue - Banyue basemental layer-block, Liyue - Palawan basin group, and the Feixin - Nanhua basin group.

3.2.3 Jianzhang – Calawit extending – dip-slipping fault zone

This fault zone starts from the northwest of Calawit Island of Calamian Islands (Fig.5). In the north, it extends to Minpan ultra-crustal sinistral trans-compression - strike-slip zone. To the east of Jianzhang shoal in southwest, it submerges under the progressive thrusting and mélange wedge of South Palawan. The fault zone ends at Balabac - Balukelo regional shear fault, which is out of the eastern beach of Sabah. This fault zone is cut into several sections by a series of NW or near NS-strike transcurrent faults. The major transcurrent faults include the Ulugan dextral strike-slip fault (Fig.1). Numerous profiles show that the extension - detachment actions of this fault zone occurred from the Late Cretaceous to early Early Tertiary, and these tectonic activities caused the formation of asymmetrical half graben sedimentary basin (which is deep in southeast and shallow in northwest). The extension fault developed in Pre-Oligocene stratum sequence. Only a few faults in southwestern sections cut into the overlapping carbonate sequence (Late Oligocene-to-Early Miocene Nido formation) or Quaternary system. The extension - detachment surface, which dips in NW direction, is steep at upper part and gentle at lower part. It converges into the plastic layer-slide surface in middle crust, and extends to Moho at some extremely thin sectors (Schluter et al., 1996). It seems that there were at least twice compressive thrusts occurred in the northeast of the fault zone. The earlier one was in about Paleocene-to-Middle Eocene (E_1 - E_2^2) and the later was after Early Miocene, caused an extensive faulting in northeast of Liyue - North Palawan area and in the Pre-Oligocene sedimentary delta wedge in the south of Liyue bank.

One of the results of the event of this fault zone is the formation of a complex half graben structure in NE-SW-strike direction, with its southeastern part subsided and northwestern part uplift (i.e. deep in southeast and shallow in northwest). This makes the platform-like top of Liyue - Banyue basal layer-block inclines to southeast as a whole, and deepens gradually as it runs into Palawan trough (Fig. 9).

3.2.4 Liyue – Palawan basin group

It includes North Palawan, West Palawan and Liyue basins etc. It developed in the middle-south of Liyue - Banyue basal layer-block, and its long-axis is generally in NE direction. It began to subside as Nanwei - Andu basin group (Zhong et al., 1991; Ru and Pigott, 1986) at the Late Paleocene (B. P. 55Ma). In Early Eocene, it subsided rapidly to 1.1km, and then subsiding process slowed down (Zhong et al., 1991).

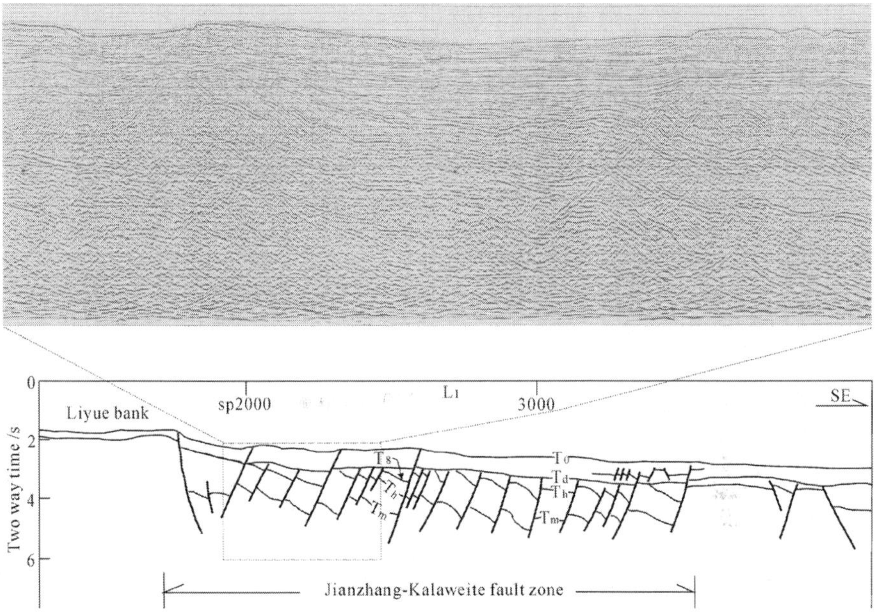

Fig. 9. Seismic profile L_1 transecting western Liyue-Banyue basemental layer-block (See Fig.1 for location) . Td, T_8, Th, Tm are the reflect horizons between lower Miocene and mid-Miocene, upper Eocene and upper Cretaceous, lower and lower Cretaceous, and lower Cretaceous and pre-Cretaceous, respectively.

3.3 Feixin – Nanhua basin group

This strike-slip and pull-apart basin group is mainly controlled by Bisheng - Siling basal strike-slip fault zone and Xiyue - Siling crust strike-slip fault zone. It is formed as the result of relative dextral strike-slip between Nanwei - Andu crustal layer-block and Liyue - North Palawan crustal layer-block, which are located in the east and west sides of the basins. It is formed mainly in Eocene and Miocene Epoch.

4. Forming mechanisms of the main cenozoic sedimentary basins within Nansha Micro-plate

The key condition for the movement and migration of a layer-block is the formation of a transformation mechanism that controls the three-dimensional boundary fault system of a layer-block. Layer-blocks in geodynamical system will show the tendency of overall movement when they are applied with sufficient tectonic forces. The system transition and conversion between three-dimensional boundary faults interrelating with the whole layer-block but with different properties of movement, is the prerequisite for realizing the movement of the whole layer-block. According to the multiple dynamics principles, the formation of layer-block structure is controlled by multiple-geodynamics, and the driving mechanism for layer-block of diverse levels is distinct.

There is a direct genetic relation between basemental layer-blocks and intra-plate basins in Nansha Micro-plate. The genesis of intra-plate basin is different from that of plate-edge basin which energy comes from mantle convection. The genesis of intra-plate basin, however, is not only influenced by movement of plate, but also by intra-plate force. The basins inside Nansha Micro-plate mainly received its energy from rheomorphism of middle crust. Depending on different ways of action of basin-forming force, the basins can be divided into three types as mentioned above: Feixin - Nanhua basal strike-slip - pull-apart basin, Andu - Bisheng basemental face-to-face dip-slip - detachment basin, and Liyue - Banyue basemental unidirectional dip-slip - detachment basin. Since the first type has the same strike-slip - pull-apart mechanism as that of ordinary strike-slip fault, we will not give further discussion. In the following paragraphs, we will focus on the basin-forming model of the other two types.

The genesis of Andu–Bisheng basal block is in the following procedure: The thermal uplift of mantle of South Chin Sea in late Mesozoic caused the lithosphere pure shear extension of South China Sea. As a result, the lithosphere mantle and lower crust of Nansha occurred extension and rheomorphism, and caused the destabilization of gravity of upper crust. The upper crust then used middle crust layer as the layer-slide surface, and started dip-slip - tilting movement along the pre-existed NE-strike Guangya - Bisheng fault and NE-strike Nantong - Siling fault. As the face-to-face dip-slip of the two faults continued, the strata near fault-side along northern and southern boundaries of hanging walls kept descending. The underlying plastic substances were squeezed to the bottom of upper crust of lower walls, and caused the uplifting and denudation of upper cursts of lower walls. Meanwhile, some plastic substances in middle crust were squeezed to the bottom of middle hanging walls, which enhanced the low-uplifting effects of internal part of Nanwei - Andu basemental layer-block, and caused the formation of complex graben structure (Fig. 10A). This procedure, if considering its mechanics of deformation, has a mechanism very similar to the mechanism of cantilever beam on elastic foundation (Li et al., 1995) (Fig. 10C).

The other type of basemental layer-block is the Liyue - Banyue basal layer-block. Its main structural feature is concurrent-direction tilting and uplifting fault block group (Fig. 10B). The extending - dip-slip movement of the NE-strike Jianzhang - Calawit extending - gliding fault zone caused the tilting–sliding of the layer-block along the underlying middle crustal layer-slide surface. As a result, there formed the North Palawan, West Palawan and other fault basins, whose main axes are all in NE direction, in the southeast of the Liyue–Banyue basal layer-block; and the Liyue bank - Haima (Seahorse) bank area in the northwest of the Liyue - Banyue basal layer-block began to rise.

The cause why the upper crust can easily slide on the middle crust surface is that there are nano-sized particle layers developped between the upper-crust and mid-crust. The nano-particles are characteried by higher density, higher strength, and lower rolling friction force (f_2 in Fig.11); and exist in almostly all faults and layer-slip surfaces in natural world. In the case with nano-particles, the friction is rolling friction. Under the same normal pressure force (P in Fig.11), the rolling friction force is far less than the sliding friction force (f1/f2 can be up to 18) (f_1 in Fig.11). So, the upper crust can easily move along the mid-crust layer.

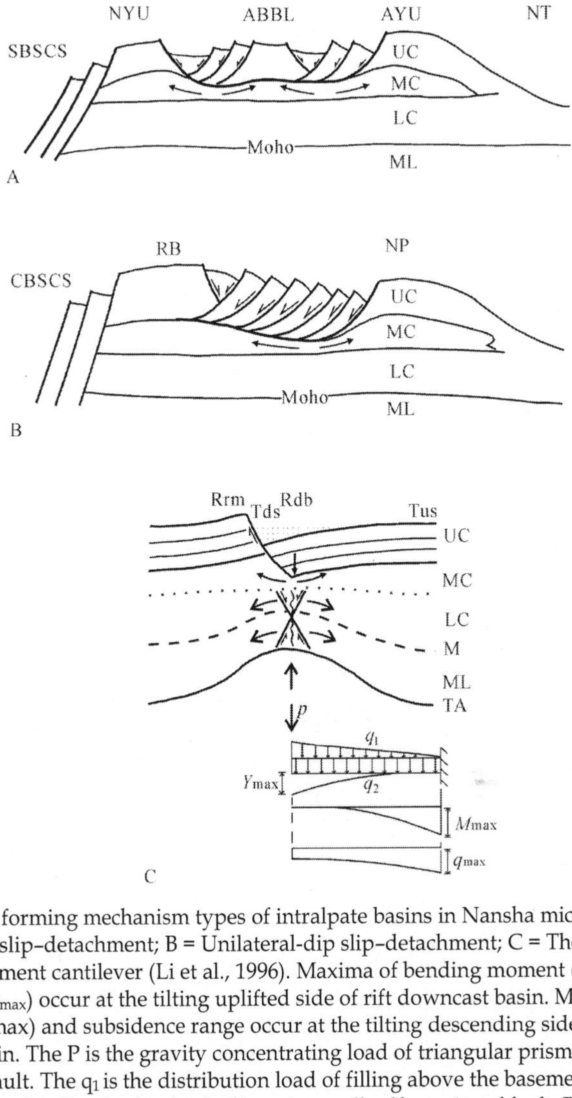

Fig. 10. Major forming mechanism types of intralpate basins in Nansha micro-plate. A = Opposite-dip slip–detachment; B = Unilateral-dip slip–detachment; C = The flexural model of elastic basement cantilever (Li et al., 1996). Maxima of bending moment (Mmax) and shear stress (q_{max}) occur at the tilting uplifted side of rift downcast basin. Maxima of deflection (Ymax) and subsidence range occur at the tilting descending side of the rift downcast basin. The P is the gravity concentrating load of triangular prism covering above the plane of fault. The q_1 is the distribution load of filling above the basement of basin. The q_2 is the uniformity distribution load of hanging wall self-gravity of fault. Both M and Q positively correlate with the length and gravity of girder. ABBL = Andu–Bisheng basemental layer-block; AYU = Andu–Yuya (Investigation shoal) uplift; CBSCS = Central basin of South China Sea; LC = lower crust; M=Moho; MC = mid-crust; NP = North Palawan; NT = Nansha trough; NYU = Nanwei–Yongshu (Fiery Cross or N. W. Investigtor reef) uplift; RB = Reed bank; Rdb = Rift downcast basin; Rrm = Rift rising mountain; SBSCS = Southwest basin of South China Sea; TA = top of asthenosphere; Tds = Tilting descending side; Tus = tilting uplifted side; UC = upper crust.

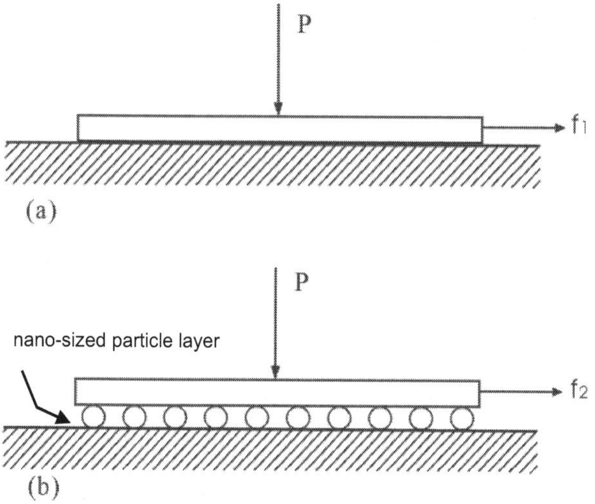

Fig. 11. Rolling friction comparing with sliding friction. (a) f_1: sliding friction force in the case without nano-particles; (b) f_2: rolling friction force in the case with nano-particles. Generally, the f_2 is far smaller than the f_1.

5. Conclusions and discussions

The geological and geophysical data from Nansha area show the characteristics of the LBT. From the data, we identify the Nansha ultra-crustal layer-block, Zengmu, Nanwei - Andu, Liyue - North Palawan crustal layer-blocks, and Andu - Bisheng and Liyue - Banyue basemental layer-blocks. They are products of multiple geodynamic systems.

The fault structure of Cenozoic sedimentary basement inside Nansha Micro-plate is a dip-slip - layer-slip - strike-slip "4-dimensional interrelated-action" fault system. The Nanwei - Andu and Liyue - Banyue basal dip-slip - layer-slip - strike-slip fault systems controlled the development of Nanwei - Andu, Liyue - North Palawan, and Feixin - Nanhua basin groups in Nansha Micro-plate. The formation of basins in Nansha Micro-plate is the result of multiple dynamical forces. It is influenced by the tectonic evolution of Nansha Micro-plate since Cenozoic era, but most importantly, it is directly controlled by the plastic rheomorphism effect of middle crust. The basin-forming mechanism of basins in Nansha Micro-plate shows a great diversity. The Nanwei - Andu basin group has a uniformed basemental face-to-face dip-slip - detachment mode, while a basemental single-direction dip-slip - detachment model is applicable for Liyue - North Palawan basin group.

The proposal of these models for the forming mechanism of the intra-plate basins is provided with guiding implication for the exploration of oil and gas or gas hydrate resources inside Nansha Micro-plate. They should give us some elicitations as follows: zones which are in the hanging walls of basin-controlled boundary faults and near the faults, should be remunerative for oil-gas exploration. Along these zones, with the gliding - tilting movement of these basin-controlled growth boundary faults, not only some conditions of "generation - movement - reservoir - preservation" of oil-gas resources could be formed in the neonatal Cenozoic sedimentary strata, but also possible oil-gas within the pre-Cenozoic marine-facies strata underlaid the Cenozoic Erathem could remove upwards along neonatal faults and form oil-gas accumulations. In the concrete, the secondary structures, such as rolling anticlines and tilted fault blocks, which developed on the hanging walls of Nantong - Siling and Guangya - Bisheng faults in Nanwei - Andu basin group, as well as paleo-buried hills structures developed around low uplifts between these two fault zones should be advantaged zones for oil-gas accumulations in Nanwei - Andu basin group, while the zone along southern side of the Jianzhang - Calawit Extending - Dip-slipping Fault Zone i.e. the northwestern shelf of Palawan Island should be regarded as better belt for oil-gas accumulations in Liyue - Palawan basin group. Some oil-gas fields discovered in above-mentioned advantageous belts, such as the Crestone exploration area of China in Beikang basin, Thanh Long or Blue Dragon oil-and-gas fields in west Nanwei basin and active gas/oil fields or new discovered fields in northwest Palawan basin, and so on, are very good exemplifications.

6. Acknowledgement

This study is jointly funded by the National Basic Research Program of China (973) (2009CB2194 and 2007CB411700), the State Fund for Natural Science of China (40976029).

7. References

Bai Zhiling, Zhang Guangxue, Zeng Xianghui, et al. (1996). The geological and geophysical comprehensive research symposium of the southeastern Nansha sea area. Wuhan, China: China University of Geosciences Press (in Chinese), 1~89

Coleman R G. (1977). Ophiolites, Ancient Oceanic Lithosphere. Mineral and Rocks, 12, Springerverleg, Berlin, Heid lberg, New York, 147~158

Cui Ruyao.(1996). Gravity and magnetic anomalies characteristics and its interpretation in the Southeastern Nansha sea area. In: The geological and geophysical comprehensive research symposium of the southeastern Nansha sea area. Wuhan, China: China University of Geosciences Press (in Chinese), 14~25

Hinz K and Schuler H U. (1985). Geology of the dangerous ground, South China Sea, and the continental Margin off southwast Palawan: Results of SONNE cruises SO-23 and SO-27. Energy, 10(3/4): 297~315

Hinz K, Fritsch J,Kempter EHK *et al*. (1989). Thrust tectonics along the north-western continental margin of Sabah/Borneo.Geologische Rundschau, 78(3):705-730

Huang Huaizeng, Wu Gongjian, Zhu Ying, et al. (1994). Study on dynamics of Lithosphere. Beijing: Geological Press (in Chinese), 1~131

Irwin W P. (1972). Terranes of the western Paleozoic and Triassic belt in the southern Klamath Mountains, California. U.S. Geological Survey Professional Paper 800C:C103~C111

Kudrass H R, Wiedicke M, Cepek P, et al. (1986). Mesozoic and Cenozoic rocks dredged from the South China Sea (Reed Bank area) and Sulu Sea and their Significance for plate tectonic reconstructions. Marine and Petroleum Geology, 13: 19~30

Li Yangjian, Zhang Xingliang, Chen Yancheng. (1996). Continental layer-controlled tectonics' introduction. Geological Publishing House, Beijing, China (in Chinese), 1~164

Liu Hailing, Sun Yan, Guo Lingzhi, et al. (1999). On the boundary faults' kinematic characteristics and dynamic process of Nansha ultra-crust layer-block. Acta Geologica Sinica (English edition), 73(4): 452~463

Liu Hailing, Guo Lingzhi, Sun Yan, et al.(2002). Study on fault system in Nansha block (South China Sea) and the block's lithospheric dynamics. Scientific Press, Beijing, China (in Chinese), 1~123

Liu Hailing, Yan Pin, Zhang Boyou, et al. (2004). Role of the Wan-na Fault System in the Western Nansha Islands (Southern South China Sea) Waters Area. Journal of Asian Earth Sciences, 23(2): 221~233

Liu Hailing, Xie Guofa, Yan Pin, et al. (2007). Tectonic implication of mesozoic marine deposits in the Nansha Islands of the South China Sea. Oceanologia et Limnologia Sinica (in Chinese), 38(3): 272~278

Mandle G and Shippan G K. (1981). Mechanical model of thrust sheet gliding and imbrication. In: K. R. McClay and N. J. Price (eds.), Thrust and Nappe Tectonics. Blackwell Scientific Publication, London, 38~46

Oxburgh E R. (1972). Flake tectonics and continental collision. Nature, 239: 202~204

Pubellier M, Quebral R, Rangin C. et al. (1991). The Mindanao collision zone, a soft collision event within a continuous Neogene strike-slip setting. Journal of SE Asian Earth Sciences. Sp. Iss, 6(3/4): 239~248

Ru Ke and Pigott J D. (1986). Episodic rifting and subsidence in the South China Sea. AAPG Bull, 70(9): 1136~1155

Rushentsev S V and Trifonov V G. (1985). Tectonic layering of the lithosphere. Episodes, 7: 44~48

Schluter H U, Hinz K, Block M. (1996). Tectono-stratigraphic terranes and detachment faulting of the South China Sea and Sulu Sea. Marine Geology, 130: 39~78

Song Xiaodong and Richards P G. (1996). Seismological evidence for differential rotation of the Earth's inner core. Nature, 382(6588): 221~224

Su Weijian, Dziewonski A M, Jeanloz R. (1996). Planet within a planet: rotation of the inner core of earth. Science, 276:1883~1887

Su Daquan, Huang Ciliu, Xia Kanyuan. 1996a. The crust in the Nansha trough. Scientia Geologica Sinica (in Chinese), 31(4):409~415

Sun Yan, Shi Zejin, Shu Liangshu, et al. (1991). Studies on the layer slip-dip slip fault structures and the petroleum Geology (Taking the Middle-lower Yangtze area as an example), A research monograph of the oil and nature gas (2). Publishing House of Nanjing University, Nanjing, China (in Chinese), 1~181

Taylor B and Hayes D E. (1980). The tectonic evolution of South China Sea Basin. In: Hayes D E (ed). The Tectonic and Geologic Evolution of Southeast Asian Seas and Islands Part 1. Geophysics Monoger, 23: 89~104

Tongkul F. (1990). Structural style and tectonics of western and northern Sabah. Bulletin of the Geological Society of Malaysia, 27:227~239

Wang Shengzu. (1992). A multi-layer Tectonic model for interior deformation of continental plate. In: MA Zong-jin. On multiple esrthquake laters in continent—florilegium of international congress of multiple esrthquake later sciences. Beijing: Earthquake Press (in Chinese), 191~203

Wang Xiaofeng, Li Zhongjian, Chen Boling, et al. (1996). The Evolution of Tan-Lu Fault Zone and its Geological Significance. <Origin and Evolution of the Tan-Lu fault system, eastern China> Workshop of the 30th IGC, Beijing, 12~20

Wernike B. (1981). Low-angle normal faults in the Basin and Range Province: nappe tectonics in an extending orogen. Nature, 291(5817):645~648

Wernike B. (1988). Basin and Range Extensional Tectonics. Dept of Earth and Planetary Sciences, Harvard University, 23~33

Wu Nengyou, Zeng Weijun,. Li Zhenwu, et al. (1999). Characteristics of Upper Mantle Activity in the South China Sea region and the Indochina Mantle Plume. ACTA GEOLOGICA SINICA, 73(4):464~476

Xia Kanyuan. (1997). Structure of the oceanic crust and its spreading history in the South China Sea. In: Gong Zai-sheng et al. (eds), Continental margin Basin analysis and Hydrocarbon Accumulation of the Northern South China Sea. Science Press, Beijing, China (in Chinese), 12~26

Yan Pin and Liu Hailing. (2004). Tectonic-stratigraphic division and blind fold structures in Nansha Waters, South China Sea. Journal of Asian Earth Sciences, 24:37~348

Yao Bochu. (1995). Characteristics of Zhongnan—Liyue fault and its significance for tectonics. In: South China Sea Geology Research. Wuhan, China, China University of Geosciences Press (in Chinese), 7: 1-14

Yao Bochu, Zeng Weijun, Hayes D E. et al. (1994). The geological memoir of South China Sea surveyed jointly by China & USA. Wuhan, China: China University of Geosciences Press (in Chinese), 1~204

Young H A. (1976). The geothermal gradient map of Southeast Asia—A progress report. SEAPEX program, offshore South East Asia conference, paper 6, 1~5

Zeng Weijun, Li Zhenwu, Wu Nengyou, et al. (1997). The upper mantle activation in South China Sea and the Indosinian mantle plume, In: Geological Research of South China Sea (9). China University of Geosciences Press, Wuhan, China (in Chinese), 1~19.

Zhang Yixiang, Deng Chuanming, Zhou Di. (1996). Characteristics of magnetic abnormity. In: Kan-yuan Xia et al.(eds), Geology - Geophysics and Oil-gas resources in Nansha Islands and its neighborhood waters. Science Press, Beijing, China (in Chinese)., 65~83

Zhong Jianqiang, Huang Ciliu, Zhang Wenhuan. (1996). On Tectonic Subsidence of North Zengmu Basin Since Oligocene. In Xia Kan-yuan, et al. (eds), Geology - Geophysics and Oil-gas resources in Nansha Islands and its neighborhood waters. Beijing, Science Press (in Chinese), 120~125

Zhong Jianqiang, Zhou Di, Zhang Wenhuan. (1991). Primary analysis on tectonic subsidence of Reed Bank basin and North Palawan basin. In: Nansha comprehensive science investigate team (eds), Research Papers on Geology-geophysics and Islands-Reefs in Nansha Islands and its neighborhood waters (1). Beijing, Ocean Press (in Chinese), 93~98

Part 6

Neotectonics: Advanced Techniques of Investigation

11

The Role of Geoelectrical DC Methods in Determining the Subsurface Tectonics Features. Case Studies from Syria

Jamal Asfahani[1]
Atomic Energy Commission
P.O. Box 6091, Damascus,
Syria

1. Introduction

The tectonic and faulted zones are characterized by a pronounced change in rock physical properties, where the application of geophysical methods allows directly detecting and delineating such zones. The geoelectrical DC methods are the most appropriate for characterizing the tectonic and faulted zones. Different DC configurations have been developed for the location of tectonic and faulted zones, such as combined resistivity profiling, (Mares *et al.*, 1984).

In the last few years, a large number of high-resolution seismic reflection surveys have been conducted (e.g., Williams et al.,1995; Palmer et al., 1997; Van Arsdale et al., 1998 to provide information on Quaternary fault geometry and timing. For very shallow investigation, ground-penetrating radar (GPR), which can bridge the gap between high-resolution seismic surveys and trenching, has been applied by Cai et al. (1996) in the San Francisco Bay region. Although, the GPR yields a high- resolution picture down to 4 to 6 m, but the high number of GPR reflections and diffractions resulting from complex sedimentary and tectonic features does not usually permit an unambiguous location of fault, (Demanet et al. (2001). However, when the fault is delineated by other geophysical methods, the interpretation of GPR data gives valuable information on the deformation pattern close to the fault and its position.

At the border of Nevada and California, Shields et al. (1998) have used several geophysical techniques (seismic reflection, magnetic, and electromagnetic) to locate the extension of the Parhump Valley fault zone. Demanet et al. (2001) have also applied various geophysical techniques (electrical profiling, electromagnetic, GPR, seismic reflection) along the Bree fault scrap (Western border of the Roer Graben) in order to locate and image an active fault zone in a depth range between a few decimeters to a few tens of meters. These acquired geophysical data are considered as a reconnaissance tool prior to trenching. Parrales et al., 2003 has executed a site investigation by using combined geophysical methods in a faulted area in Managua, Nicaragua. A mapping of active capable faults by high resolution geophysical methods has been carried out by Chwatel et al., 2005, where several examples from the Central Vienna basin were provided. Caputo et al., 2007, 2003 have characterized

[1] jasfahani@aec.org.sy

the late Quaternary activity along the Scorciabuoi fault (Southern Italy) as inferred from electrical resistivity tomographies. Piscitelli et al., 2009 showed tomography examples in studying active tectonics from the Tyrnavos basin, Greece. The characterization of Quaternary faults by electric resistivity tomography in the Andean Precordillera of Western Argentina has also been shown by Sabrina et al., 2009.

More recently, Massoud *et al.*, 2009 have applied the directional azimuthal resistivity sounding and joint inversion of VES-TEM data to delineate the shallow subsurface structure near Lake Qaroun, El Fayoum in Egypt.

Asfahani, 2007 (a, and b), Asfahani, 2010-a has applied the geoelectrical DC method, particularly VES soundings in the Khanaser valley in Northern Syria to estimate the water resources. He has delineated subsurface structures and outlined fresh, brackish and saline water accumulations through treating the VES data by applying Pichugin Habibulaiev technique (1985). Asfahani in the paper of (Asfahani and Radwan, 2007) enhanced later the proved efficiency of the Pichugin Habibulaiev technique and made it applicable even in pronounced topography and relief areas. Accordingly, shallow and young subsurface structures were delineated and used thereafter as a basis for deducing the tectonic origin of Khanaser Valley. Asfahani in the paper of Asfahani *et al.*, 2010 has successfully applied different geoelectrical methods in Al-Lujj area, Northwestern Syria in order to explain the subsurface tectonic origin of the Kastoon Dam in the Ghab trough depression.

This chapter mainly concentrates on the use of DC methods, particularly the vertical electrical sounding technique (VES), and the interpretation of the VES data by applying the enhanced Pichugin$ Habibuleav approach for determining the subsurface tectonics in different case problems taken from Syria.

2. Vertical Electrical Sounding (VES)

Vertical Electrical Resistivity Sounding (VES) is generally used to determine vertical variations in electrical resistivity. In this technique, an electrical current is imposed by a pair of electrodes at a varying spacing expanding symmetrically from a central point, while measuring the surface expression of the resulting potential field with an additional pair of electrodes at appropriate spacing, (Fig.1). For any array of current electrodes A and B and potential electrodes M and N, apparent resistivity ρa is expressed according to Dobrin (1976) by:

$$\rho a = K.\frac{\Delta V}{I} \tag{1}$$

Where

$$K = \frac{2\pi}{\frac{1}{AM} - \frac{1}{BM} - \frac{1}{AN} + \frac{1}{BN}} \tag{2}$$

In the equation [1], I is the current introduced into the earth and ΔV is the potential measured between the potential electrodes. The instrument used in gathering the field data measures directly the resistance $\frac{\Delta V}{I}$, and apparent resistivity ρa is subsequently obtained after computing the geometric coefficient (K) for a given position of current and potential electrodes (Dobrin 1976).

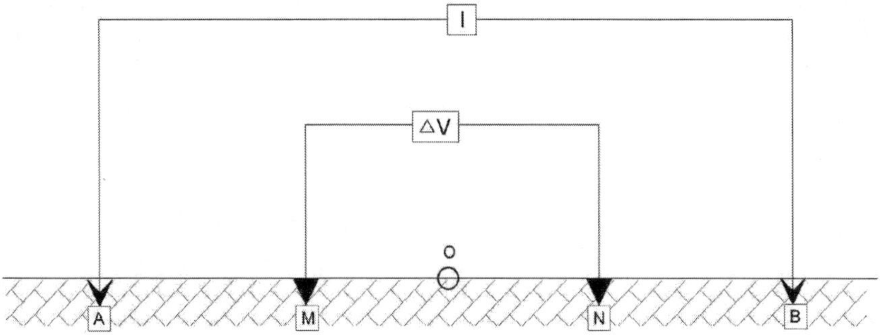

Fig. 1. Schlumberger Configuration in the field.

According to Asfahani; 2010-d a new modification and development on the described traditional Schlumberger configuration consists of using two kinds of AB/2 spacings: The first spacings are purposely designed in order to obtain reliable detailed data on the shallow depths, not exceeding 50m.

The new added AB/2 spacings are:

1, 1.3, 1.68, 2.18, 2.82, 3.66, 4.74, 6.15, 7.97, 10.33, 13.38, 17.35, 22.49, 29.15, 37.78, 48.87, 63.48, 82.27 and 106.6m.

The second traditional AB/2 spacings are:

3,5, 7, 10, 15, 20, 30, 40, 50, 70, 100, 150, 200, 300, 400, 500, 750, 1000, 1500 and 2000m.

In the field application, the two shallow and deep spacings are executed in the same time, such as two-geoelectrical field curves are obtained at a given VES location.

Each of the two obtained field resistivity curves is separately interpreted according to the new proposed interpretation approach consisting of the following three distinguished steps, Asfahani;2010-d:

1. The first interpretation step is achieved by applying the traditional curve matching technique and using the master curves (Orellana and Mooney 1966).
   Accordingly, corresponding subsurface layers thicknesses and resistivities are approximated. The approximate models are thereafter accurately interpreted using an inverse technique program (Zohdy 1989; Zohdy and Bisdorf 1989), until good and reasonable fitness between field and regenerated theoretical curves are obtained.
2. The second interpretation step is achieved by using the technique of PichuginHabibuleav (1985), Asfahani *et al.*, 2007, 2010.
   This interpretative technique is considered as the most sophisticated for distinguishing fractured zones and dipping contacts between different rock types and can be easily applied only on a series of VES soundings distributed on a given oriented profile.
   The superimposing of the results provided by applying the two mentioned approaches allows resistivity, thicknesses and structural model to be established along the studied profiles.
3. The third step is achieved by using the available geological information in order to finally establish a geological cross- section along the studied profiles.

Applicability of the described geoelectrical configurations is raised by the fact that they provide detailed tectonic information about shallow and deep pentration depths. The advantages of such configurations are demonstrated through several field application examples presented with their analysis and interpretations. Therefore, the specific configuration with its particular shallow design is recommended to be applied for detecting and imaging faulted and fractured zones in Quaternary and Recent deposits and for exploration and mining geology, where shallow information depths are required.

3. Pichugin $ Habibulaev technique 1985

This technique is considered as the most sophisticated one for distinguishing tectonically fractured zones and oblique contacts between different rock types, and permits the determination of faults direction and dip amounts.

When an electrical current passes through a plane contact between two outcropping formations of different resistivities ρ1 and ρ2, (Fig.2) then electrical field boundary conditions at this contact, are characterized by the following:

- If the center of vertical electrical sounding is exactly located over a vertical contact between two formations of different resistivities ρ1, ρ2, and the configuration is perpendicular to this contact, then the resulting measured resistivity ρK is given by the following equation:

$$\rho K = (\rho 1 + \rho 2) / 2 \quad (3)$$

- and if the configuration is parallel to such a contact, then the resulting measured resistivity ρ'K is given by the following equation:

$$\rho' K = 2\,\rho 1\,\rho 2 / (\rho 1 + \rho 2) \quad (4)$$

  In both cases, the resistivity does not depend on AB or MN lengths.
- If there are two vertical electrical soundings VES1 and VES2, performed on either side of a vertical contact, then all profiling curves for every given AB/2 will be intersected in one point located over this vertical contact. The data of vertical electrical soundings is therefore converted to be represented by the form of horizontal curves as multi-depths profiling curves for every given AB/2. The locations of vertical electrical soundings, realized on a given profile, are plotted on abscissa using a linear scale, while the corresponding apparent resistivities (ρK or ρ`K) for each given AB/2 are plotted on ordinate using a logarithmic scale. The intersection points of all horizontal curves termed as "Points of Non-Homogeneity", (PNH) and labeled by (+) symbol are plotted on a 2D (x, z) geological section. The depth z of each of them can be determined according to the following equation:

$$Z = [(AB / 2)_i + (AB / 2)_j] / 2 \quad (5)$$

  Where $(AB / 2)_i$ and $(AB / 2)_j$ are the half separations between the electrodes A and B, for which two horizontal curves are intersected. The fractured zones are identified and determined by the presence of (PNH) on vertical pseudo vertical lines.

The geological interpretation of the PNH is based on the following assumptions:

- When the (PNH) are distributed according to the oblique lines located at shallow depths, they point to the presence of an inhomogeneous lithological contact;
- If they are arranged along oblique lines dipping down at an angle exceeding 30° in depth, they represent a tectonic fractured zone;

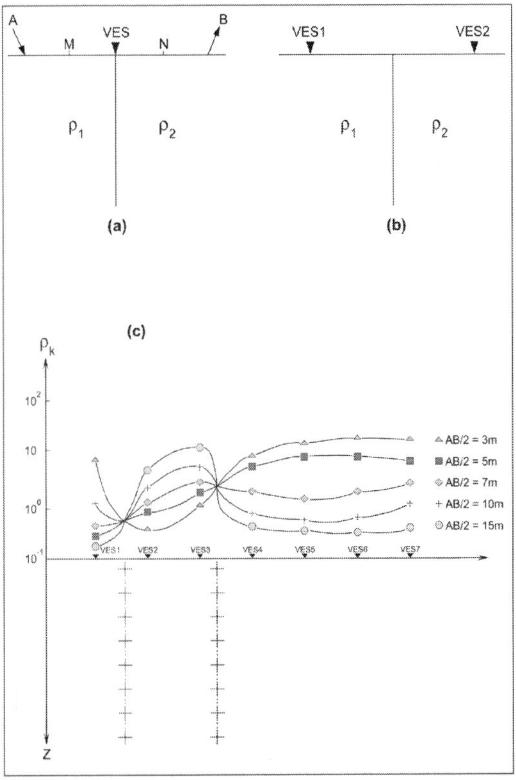

Fig. 2. The principle of Pichugin and habibuleav technique (1985).

- If they are randomly scattered near the surface, they indicate an homogeneous lithology;
- If they are arranged in a regular form, they reflect may be the presence of geological structures in the region of study (syncline, anticline, or simply horizontally layered strata).

More recently, Asfahani in the paper of Asfahani and Radwan (2007) enhanced the technique by taking into consideration the real topographic variations of the studied soundings (VES) distributed along a given profile, in order to accurately acquire reliable subsurface structures. This enhanced technique is widely used for identification of the subsurface tectonics as will be shown in the following five case problems presented in this chapter:

3.1 Characterizing the active tectonic of the Quaternary and Recent deposits in Al-Ghab depression region

Quaternary and Recent sediments represent indeed, a valuable record of the active tectonic events which have happened in the last 10.000 years. Being mainly composed of soft sediments covered usually by top soil, they are very susceptible to be weathered and eroded even in a short time, where their traces of surface expressions containing the records of the occurred tectonic deformation are partially or completely eradicated. Additional short term factors that either damage, bury or wipe away the remaining rest of this natural record, are represented by escalating diversified human activities, started from the dawn of civilization, ranging from land cultivation to giant dams construction.

The need to support the morphotectonic mapping of surface active tectonic features with subsurface data in order to acquire a reliable, intact, and well-preserved a 3-D image of Quaternary sediments deformation is one of the main challenge in active tectonic studies. The need to implement adequate methods able to provide a truthful insight in such sediments becomes more and more pressing.

Most of the geophysical methods used today in many applications have been invented and developed to meet the increasing demand of the society for new water, mineral and energy resources, which usually lie under considerable depth. Accordingly, geophysicists strive to devise much deeper-penetrating methods and contrive new data interpreting methods and models.

In active tectonic researches, the scope is reversed, since data on shallow depth subsurface structures are strongly needed to map and image the near surface fault trace with great accuracy.

Syria lies at the northern margin of the Arabian plate, and to the south of the East Anatolian Fault (EAF) that separates the Arabian plate from the Anatolian plate. The N-S trending Dead Sea Fault System (DSFS), which crosses through the western parts of Syria is a major sinistral transform plate boundary between the African (Levantine sub-plate) and the Arabian plates (Fig.3). It accommodates their differential movement, linking the Red Sea /

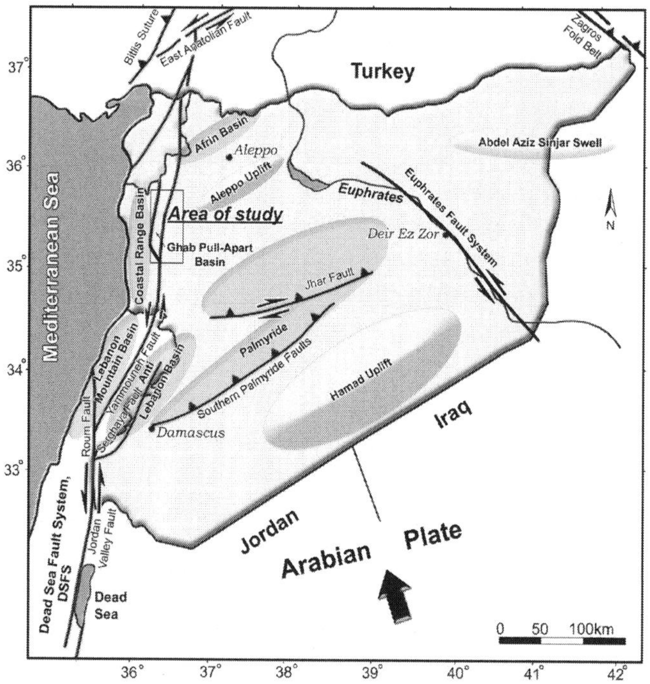

Fig. 3. The main tectonic features of Syria and the location of the study area.

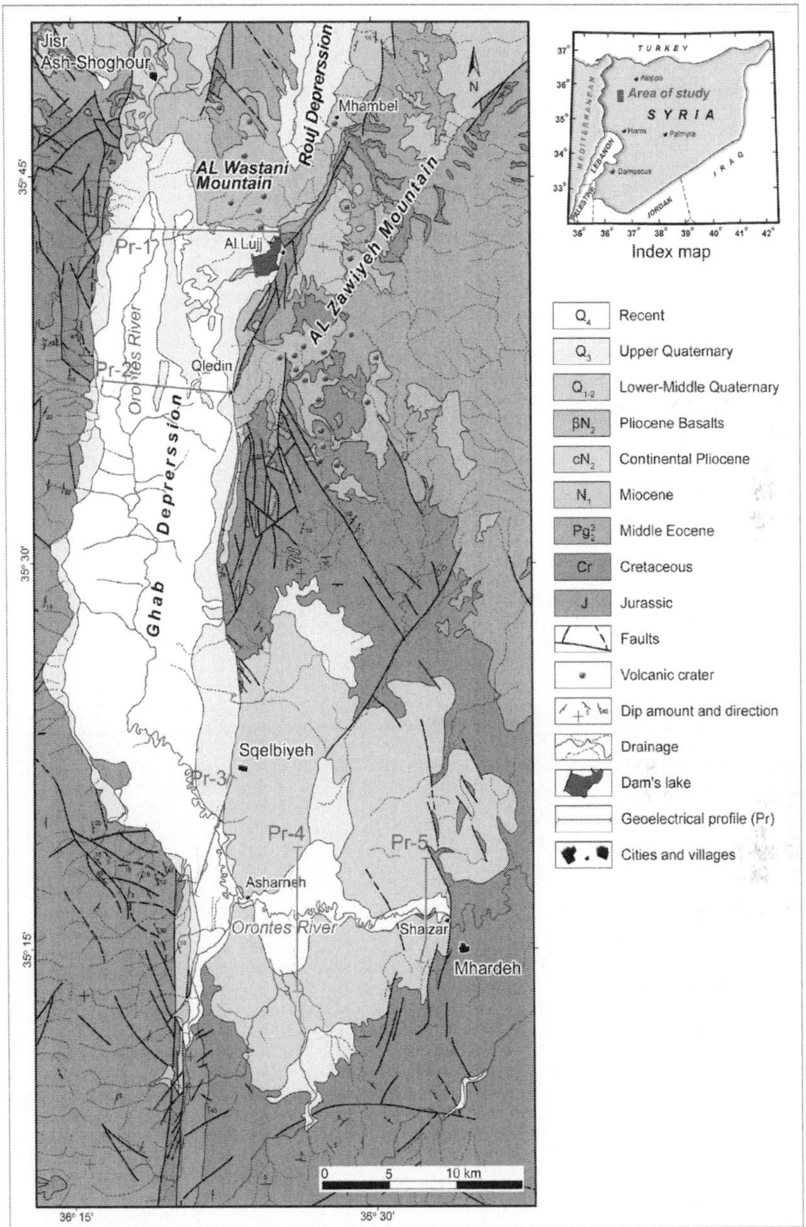

Fig. 4. Simplified geological map of the Al Ghab area, with the locations of the executed geoelectrical profiles.

Gulf of Aqaba seafloor spreading to Neo-Tethyan collision in Turkey, involving complex structural deformation controlled by the prevailing stress field along it.

Ghab rift valley is developed along the northern parts of the DSFS in Syria. Brew et al. (2001a, 2001b) consider this pull-apart basin as a deep structure opened in response to a left-step in the DSFS during Pliocene to Holocene. It is bounded from the north and northeast by the NNE elongated Al Rouj Valley and Al Wastani Mountain. It is separated from Al Zawiyeh Mountain to the east and from the Coastal Chain to the west by bounding linear faults and to the south by a NW normal fault, (Fig.4). The Ghab depression is filled by horizontal, 90-150 m thick Pliocene lacustrine sediments, covered by a thin sheet of Quaternary lacustrine sediments (Ponikarov 1966).

The Asharneh depression is of an elliptical shape, stretching in a meridian direction. It boards the Ghab depression from the East. The East of Asharneh depression is limited by a system of small echelon- like faults.

Pliocene continental deposits fill the tectonic basins of Ghab and Asharneh. It appears on the marginal edges of those tectonic depressions margin deposits in contact with the older rocks of different ages.

The Pliocene effusives composed of basalt and tuff cover have been developed in the northern part of the Ghab depression. The basalt cover is slightly inclined in southward direction, where it is buried under Upper Quaternary and Recent deposits.

Shallow and deep vertical electrical soundings (VES) have been applied in Al-Ghab depression region in order to characterize the Quaternary and Recent deposits.

More than 70 VES soundings distributed on five profiles are carried out, where the locations of those profiles are shown in Fig.4. The acquired VES data have been interpreted by the new proposed approach constructed by the three previously described steps. The second tectonic step which consists of finding the points of non homogeneity (PNH) is the most important in such a proposed approach, and allows the subsurface tectonic features along a given profile to be easily traced. This chapter shows just an example of the interpretation results obtained by applying the proposed interpretative approach on the shallow and deep Zaizon profile as follows:

3.2 Pr-1 shallow and deep Zaizon profile

This profile of a length of 13 Km, was executed in E-w direction and has 13 VES data points. The interpretation of this profile by the new proposed interpretative protocol allows very detailed information to be obtained for the shallow depth not exceeding 50m as shown in Figure.5-a. In accordance with the available geological information and the multi-layer models obtained by the inversion, a geoelectrical cross section was established along this profile. The cross section revealed in general four to five layered subsurface mediums as follows:

- The first uppermost, geoelectrical layer is of a thicknesse varying between 0.3 and 1.8m, with an average thickness of 0.7m. It exhibits resistivity values ranging between 4.4 and 33 Ohm.m. It is composed of salty marshs and eolain sands.
- The Recent geoelectrical layer is relatively thin (0.8- 5.7m), and exhibits resistivity values ranging between 1.7 and 144 Ohm.m reflecting the effect of surface conditions and/ or the very lithological variations. It is composed of peat, clayey loams, sands, and pebbles.
- The Quaternary geoelectrical layer is relatively thin (15.5- 60m), and exhibits resistivity values ranging between 0.7 and 60 Ohm.m. It is composed of alluvium, proluvium, clay and lacustrine limestone.

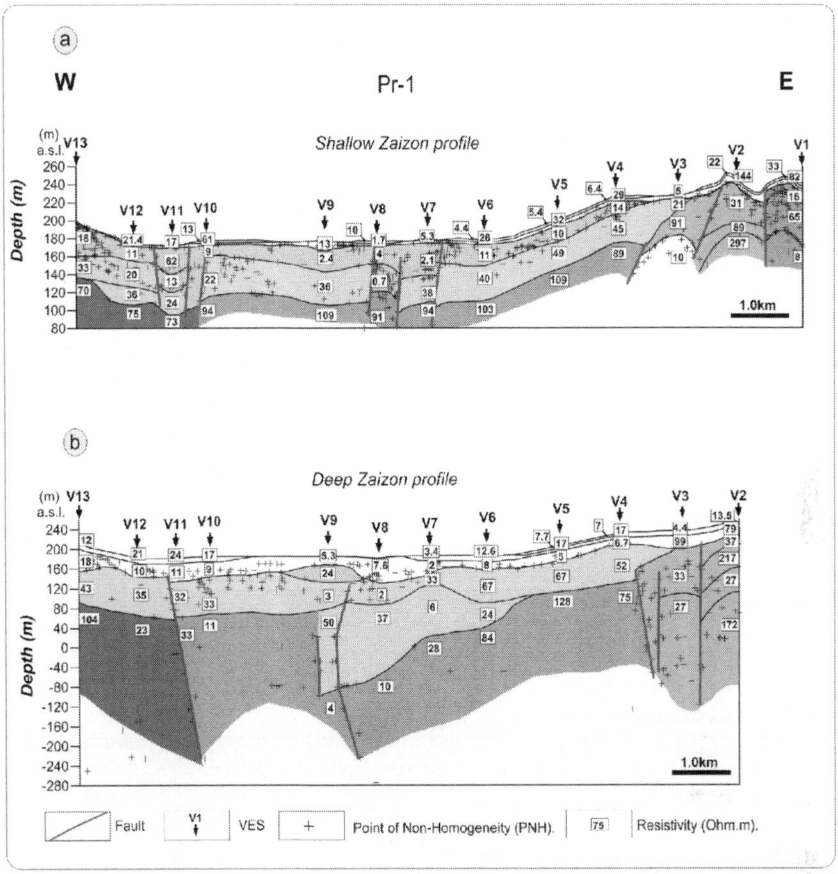

Geological legend are the same shown in geological map in Figure 3

Fig. 5. Shallow and deep geoelectrical cross section on Pr-1 Zaizon profile.

The Neogene (Pliocene) geoelectrical layer is of a thicknesse varying between 2 and 31m with an average thicknesse of 24m. Its resistivity values are ranging between 21 and 49Ohm.m with an average of 35Ohm.m.

- The Neogene rocks are composed of lacustrine limestones, sandstones, coarse sands and conglomerates, and the effusive rocks which are composed of extinct volcanic vents filled with volcanic breccia and erosion product.
- The Paleogene geoelectrical layer is of a resistivity values ranging between 89 and 109Ohm.m with an average of 98Ohm.m. This Paleogene layer is shown between VES-4 and VES-10 (Fig.5). It is composed of clayey limestone, and organic limestone.
- The Cretaceous geoelectrical layer is of a resistivity values ranging between 70 and 75Ohm.m with an average of 73Ohm.m. This Cretaceous layer is shown only under the VES-11,12 and VES-13 (Fig.5). It is composed of blocks of limestone, dolomite, clay and flint.

The distribution of the PNH provides an idea about the subsurface structures and the sedimentological deposition along the studied profile. The boundary between the Quaternary and Paleogene geoelectrical layers is marked by the distribution of such points.
The positions of the faults along this profile have been well located. To the west of VES.2, one can easily notice the presence of a fault, which penetrates the Pliocene lava.
Three faults have been identified between VES6 and VES8, the first one nearest to VES7 cuts the Quaternary and Recent deposit cover. The second one located between VES7 and VES8 cuts only the Quaternary deposits. The third one beneath VES8 cuts the Quaternary and Recent deposits. The area between VES6 and VES8 could be considered as a faulted zone.
Another faulted zone composed of three faults is located between VES10 and VES12 and characterized by gradual decline, where a large thickness of Recent deposits have been accumulated as shown in Fig.5.
The interpretation of deep profile shown in Fig.5-b revealed the presence of the same detected layers described by the shallow array. The investigation depth obtained by applying the deep array is more than 250m, where a general description of the different geological layers is established, but there are no details about the Recent and Quaternary layers as obtained by the shallow array. The deep subsurface tectonic is also determined where the locations of the faults are also shown in Fig.5-b.
The fractured zones determined along shallow and deep profiles are in an acceptable concordance. However, some shallow faults are absent in the deep profile as shown under the deep VES7, where such faults only affect the shallow layers, or swallow up with each other along the deep profile as shown under the deep VES10 and VES11.
Similar results have been obtained while interpreting the other four shallow profiles (Qledin (Pr-2), Sqelbiyeh (Pr-3), Asharneh (Pr-4), and Shazar (Pr-5)), where the resistivities and thicknesses of the different geological formations have been well determined as shown in Table [1].

Profile Name	Recent		Qunternary		Neogene		Paleogene		Cretaceous	
	AV Resis (Ohm.m)	AV Thickness (m)	AV Resis (Ohm.m)	AV Thickness (m)	AV Resis (Ohm.m)	AV Thickness (m)	AV Resis (Ohm.m)	AV Thickness (m)	AV Resis (Ohm.m)	AV Thickness (m)
Pr-1	44	1.6	12	29	35	24	-	-	73	-
Pr-2	34	1.2	25	10	56	21	50	-	36	-
Pr-3	25	6	20	17	27	31	53	-	-	-
pr-4	40	3.5	18	27	21	16	-		-	-
Pr-5	22	2	34	9	32	20	-	-	190	-

Table 1. The geometric and resistivity characteristics of different geological ages along the five executed profiles.

3.3 Delineation of the impact of Ghab extensional tectonics on the Qastoon Dam in Northwestern of Syria

Zaizoon and Qastoon earth fill dams, distanced 5 km from each other at the northern parts of Ghab pull apart, were constructed during 1990-1996 and set in operation in 1996. Zaizoon dam collapsed in 2002, cosing 20 lives, flooding 5 villages and submerging 8,000 ha of fertile agricultural land.
The latent impact of the Ghab extensional basin tectonic setting, and associated deformations resulted by active tectonics on the Qastoon dam in northern Ghab in Syria

have been evaluated, (Asfahani *et al.*, 2010). This was achieved by applying an appropriate methodology essentially based on morphotectonic mapping and integrated geophysical surveys, consisting of electrical resistivity profiling, vertical electrical sounding and self-potential techniques. The integrated interpretation of the acquired morphotectonic and geophysical data allowed the detection of subsurface deformed structures, either underlying the Qastoon dam lake floor, or close to it. It is believed that these active structures were developed through the ongoing active tectonic processes occurring in the Northern Arabian plate. The tectonic survey proved that the N66.5°E striking Wadi Al Mashta fault, extending beneath the Qastoon dam lake floor, is one of the youngest active structures, and that the intersection of the fault with the Qastoon dam prism is a water-leaking point. Dam supporting measures, continuous monitoring and precautious disaster management are therefore recommended to be urgently adopted and practiced.
Asfahani in the paper of (Asfahani *et al.*, 2010) has applied the electrical resistivity profiling (ERP) to image faults and fractures for penetration depths ranging between 10 and 20 m,. The ERP was performed along two profiles at the Qastoon-Al Lujj site, the first one, labeled Pr-1, is 435 m long and oriented E21°S, the second one, labeled Pr-2, is 330 m long and E-W oriented (Figs. 6, 7a, and 7b). Traditional Schlumberger configuration with two fixed AB constant spacings of 30 and 60 m (Figs. 6, 7a, and 7b) was used for measuring the resistivity. A distance of 3 m between potential electrodes M and N is chosen. In fact, the choice of such current and potential spacings in this configuration has proven its efficacy in providing resistivity measurements from the required active tectonic depths. The computed geometric

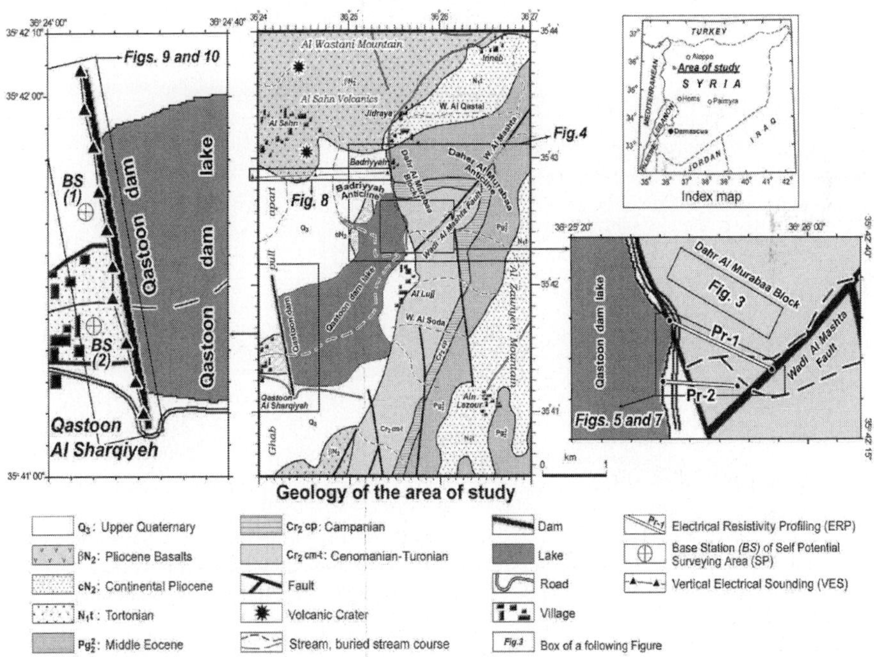

Fig. 6. Geology of the Al-Ghabe and location of Qastoon dam.

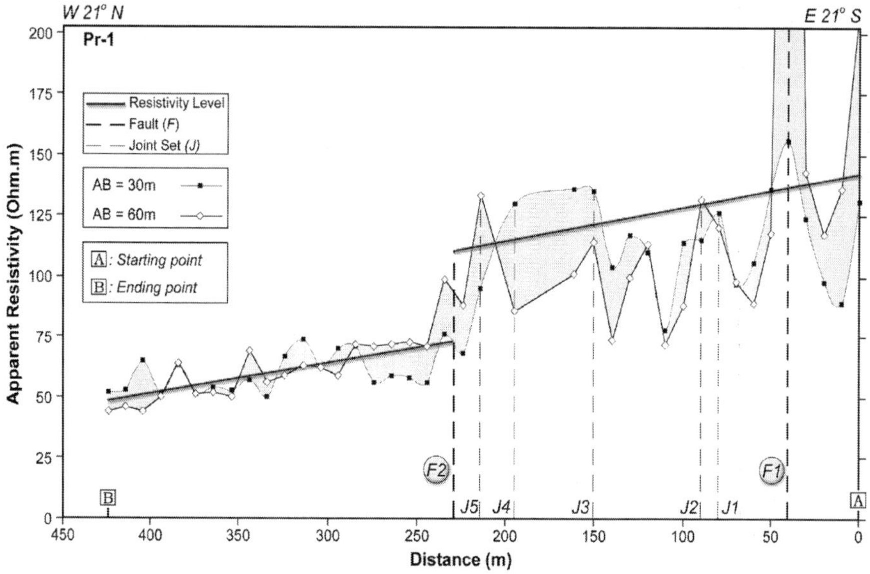

(a)

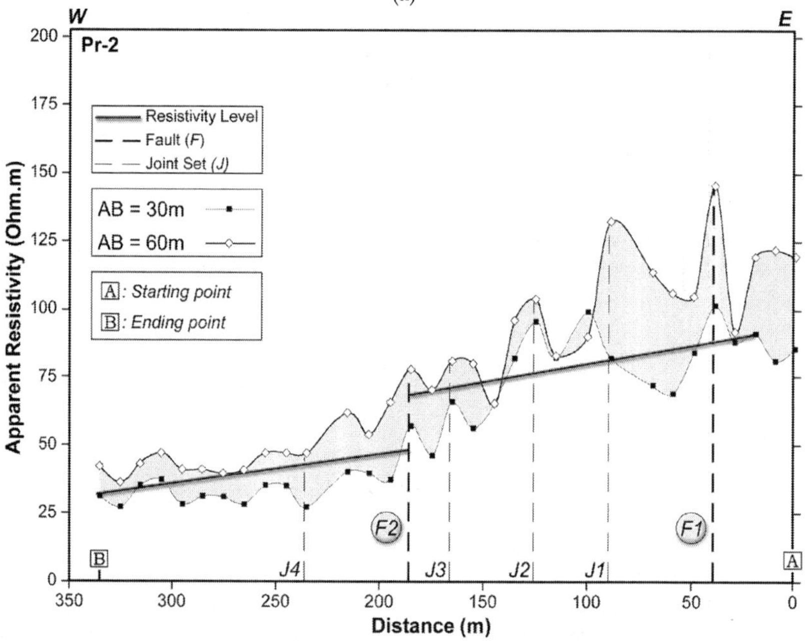

(b)

Fig. 7. (a) Apparent resistivity along Pr-1. (b) Apparent resistivity along Pr-2 coefficients for AB of 30 m and 60 m are 233 m and 940 m respectively. Forty points along the Pr-1, and thirty two points along Pr-2, were measured with an interval of 10 m between every two successive measurements.

ERP was also carried out using Lee configuration to detect the locations of PNH which could be interpreted and attributed to the presence of faults and fractures. Lee configuration includes the use of five electrodes AMONB placed on a straight line on the ground (where OM=ON), as shown in Fig. 8.

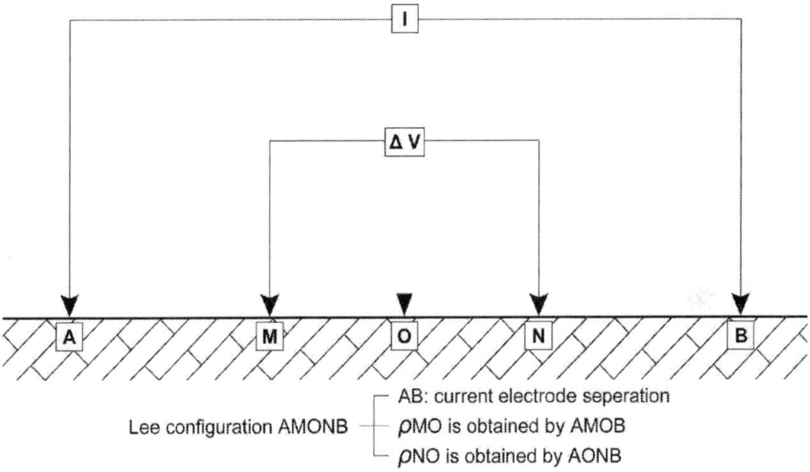

Fig. 8. Lee configuration in the field.

Three resistivities are measured by using such configuration as follows:

- Traditional ρa is results through A and B and measuring potential difference between M and N;
- ρ_{OM} resistivity results through electrical current between A and B and measuring potential difference between M and O;
- ρ_{ON} resistivity results through current between A and B and measuring potential difference between N and O.

The geometric coefficients of Lee configuration are computed for current electrode AB spacings of 30 and 60 m to be 466 m and 1880 m respectively, (the potential electrodes MN separation is constant and equals to 3m). The resistivity ratio variations ρ_{OM}/ρ_{ON}, for Pr-1 and Pr-2 profiles for both AB of 30 and 60 m, were followed, traced and interpreted (Figs. 9a, and 9b).

The tectonically-oriented interpretation of ERP (Pr-1) indicates the presence of two different apparent resistivity regions traceable at both 10 and 20 m penetrating depths for AB of 30 m and 60 m respectively (Fig. 7a).

- The first region extends 230 m westward from profile Pr-1, and is characterized by an average resistivity of ~130 Ohm.m. Within this region, a sharp resistivity peak (more than 200 Ohm.m). At 40 m from the profile starting point A, was interpreted as a sub-vertical fault (F1). Less pronounced resistivity peaks were interpreted as possible parallel joint sets (J).

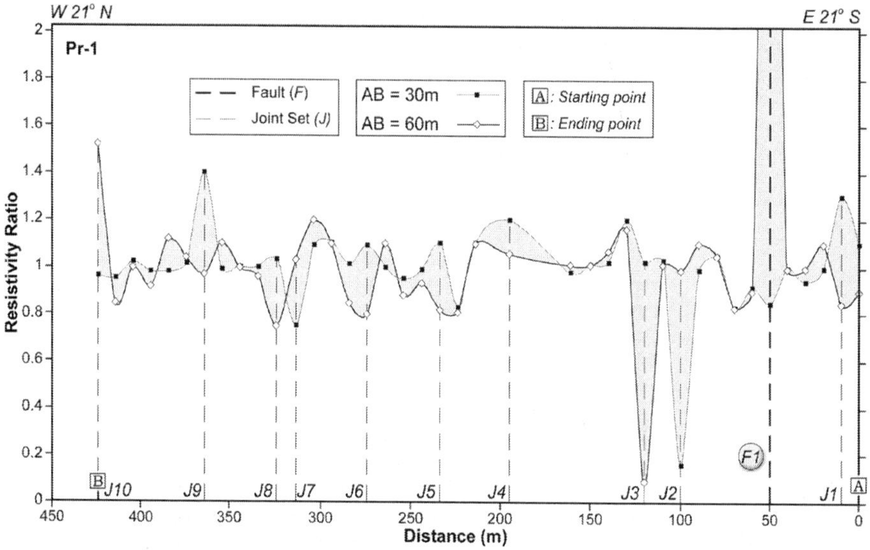

(a)

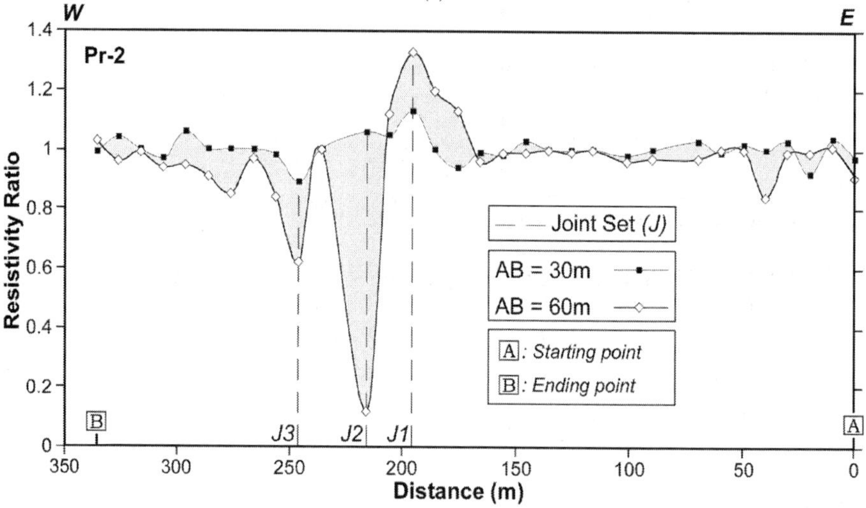

(b)

Fig. 9. (a) Apparent resistivity ratio along Pr-1. (b) Apparent resistivity ratio along Pr-2.

- The second resistivity region extends from the 230 m point on the profile to the profile endpoint B, with a 60 Ohm.m average resistivity. The deflection between these two resistivity levels was interpreted as another sub-vertical fault (F2). The weak resistivity peaks along this profile might be considered as minor deformational features.

Analysis of ERP (Pr-2) reveals a similar case, with two regions of average resistivities 75 and 40 Ohm.m values (Fig. 7b). The slope between them is traceable at both 10 and 20 m penetrating depths for AB of 30 m and 60 m respectively, marking a sub-vertical fault labeled (F2), distanced 185 m from the profile's starting point. A moderate peak is recognized in the 60 m-spaced AB, interpreted as (F1) fault distanced 40 m from Pr-2 profile starting point. Less pronounced peaks were interpreted as joint sets (J). F1 fault, traced on both profiles Pr-1 and Pr-2 was interpreted as the extension of the N66.5°E Wadi Al Mashta fault in depth, while F2 fault was interpreted as a parallel fault to F1.

The interpretation of resistivity ratios profiles (Lee Configuration) is based on the studying of the variations of the ratio ρ_{OM}/ρ_{ON}. As a rule, resistivity ratio value of ρ_{OM}/ρ_{ON} = ~1, indicates an homogenous medium between M and N electrodes. Ratio values, deviating from 1, reflect a sharp geoelectrical contrast between adjacent media of different electrical characteristics, caused mainly by fractures and faults. Ratio variations and characteristics of Pr-1 and Pr-2 profiles were shown in Figures 9a, and 9b. The value that differs from 1 is interpreted accordingly as potential shallow active faults.

The interpretation of the *Badriyyah* VES profile by the inverse modeling revealed the presence of a lithological sequence that contains Middle Eocene chalky limestone and limestone, followed by Pliocene lacustrine marl and chalky limestone, caped by Pliocene basaltic flows and ending up by Quaternary alluvial sediments, (Fig.10). The interpretation of this VES profile by the Pichugin and Habibullaev method shows a clear deformation represented by faulting and folding of the previously mentioned sediments, where these results are in accordance with surface morphotectonic mapping. Integration of the results suggests the presence of a vertical, ~50 m deep normal fault, defining the western side of the 170 m wide Badriyyah graben. This finding is confirmed by the morphotectonic surface mapping of two normal faults striking N15°W and N15°E . Folding of the Eocene and Pliocene sediments, confined between the Badriyyah graben western fault and Al Sahn volcanics, have been observed through the interpretation of VES data. The interpretation findings were in concordance with surface morphotectonic field mapping that yields to detect the Badiryyah anticline (Fig. 10). The geoelectrical interpretation also detected faults that bound and penetrate Al Sahn basaltic volcanics, erupted high probably in Pliocene and post-Pliocene i.e. Quaternary, due to active tectonic processes.

The geoelectrical interpretation of Badriyyah VES profile confirms the presence of meaningful active tectonic deformation. The Badiryyah anticline and graben, folding and normal faulting of Pliocene sediments and normal faulting of Quaternary sediments those fill the aforesaid graben are evident examples of such deformation (Fig. 10). A southward draining paleo channel incising through them, and the presence of younger sediments burying this channel point to ongoing tectonics reflected by late steady subsidence of Badriyyah graben's central part, making it a potential water leakage site from Qastoon dam lake.

The interpretation of the N12°W shallow *Qastoon dam* VES profile of 1750 m long, measured parallel to the Qastoon dam, allowed the obtaining of a geoelectrical subsurface image beneath the dam prism (Fig. 11). It defined three pronounced spots with clear alignments of the Point of Non-Homogeneity (PNH) beneath V1, V5 and V8 soundings. The (PNH) alignment beneath V1 is relatively less significant since it lies beneath the dam northern end. The other two PNH alignments under V5 and V8 are alarming since they underlie the central part of Qastoon dam prism. They have been interpreted by considering the distribution of the measured resistivities values, as a SW extension of the NE striking Wadi

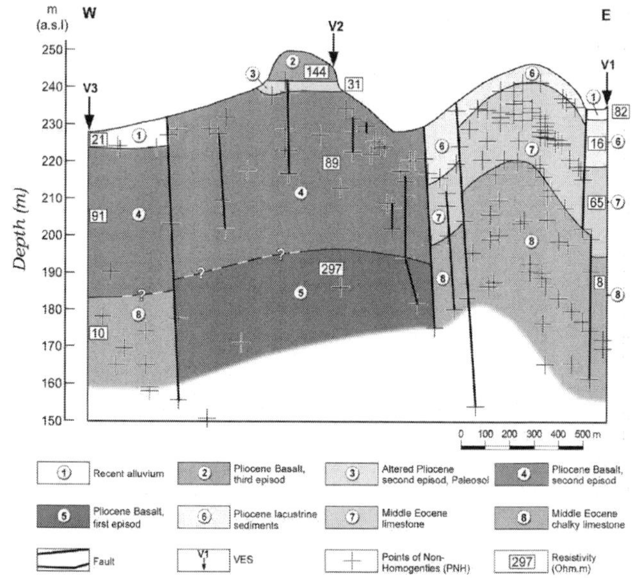

Fig. 10. Shallow subsurface geoelectrical cross section obtained along Badriyyah VES profile.

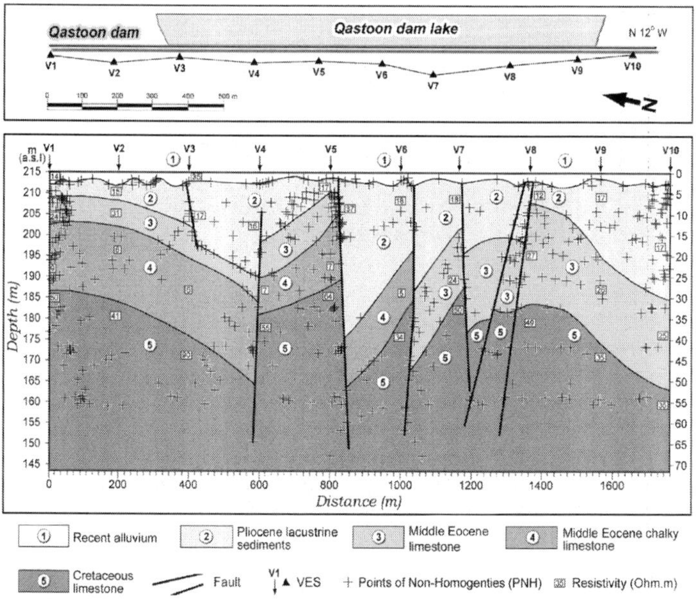

Fig. 11. Shallow subsurface geoelectrical cross section obtained along Qastoon Dam VES profile.

Al Mashta fault. Accordingly, V5 and V8 sites were selected to conduct Self potential (SP) measurements for detecting any possible water leakage at them from the dam lake. It is worth mentioning that after Zaizoon dam collapse, Qastoon Dam Monitoring Authority, installed 18 geodetic monitoring points on the dam prism and other 15 points at the dam's backside, to monitor land deformation in the area surrounding the dam. The authority conducts also continuous leveling using three-primary fixed benchmarks, seven secondary benchmarks and eight roving ones. Qastoon dam water controlling team, responsible for monitoring lake water level, spotted three water-leaking points. The first one is situated at the dam prism, the second one is located at the basaltic rocks that outcrop at the dam lake northern bank, while the third is located 200 m west of the dam. These observations are important, particularly the presence of the second leaking point, since it coincides with the N15°E striking normal fault, that defines the western border of the above mentioned Badriyyah graben. The first leaking point will be discussed while interpreting the SP data.

Two SP grids were established over area 1 and area 2, where the locations of V5 and V8 are respectively covered (Fig. 12). The map of SP for area 2 clearly reveals a distinguished N65°E SP elongated anomaly. The elongation axe is distanced 325 m from the southern end of the dam prism, and indicates a possible water leakage from Qastoon dam along it.

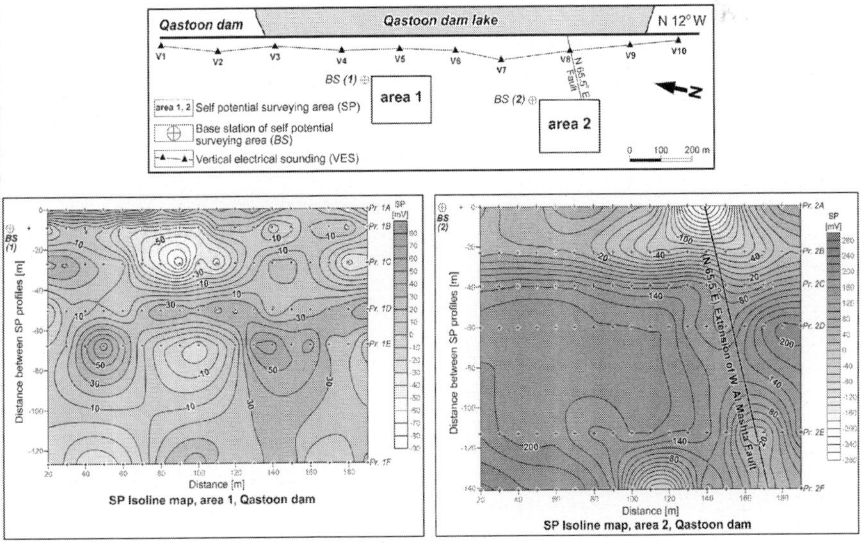

Fig. 12. SP maps in Areas 1 and 2 at the Qastoon Dam.

Remarkably, this elongation coincides with the direction of the NE striking Wadi Al Mashta fault. This strengthens the interpretation of a southwestern extension of Wadi Al Mashta fault beneath the dam lake's floor and the dam itself. Croot relationships, indicate that Wadi Al Mashta fault is either younger than the N15°W fault, which bounds the Badriyyah graben eastern end, or it is an older fault that than underwent a later reactivation. Accordingly, Wadi Al Mashta fault's formation and/or reactivation is indeed very recent, and its future reactivation is quite probable representing a real adjourned danger on Qastoon dam, (Asfahani *et al.*, 2010).

The map of SP established in area 1 could be interpreted as shallow jointing set conjugate to Wadi Al Mashta fault (Fig. 12). The presence of 10° and 20° dipping Pliocene marly clay, clayey marl and clay horizons, forming the limbs of Badriyyah anticline, are hazardously lubricant horizons able to trigger slumping within Qastoon dam lake in response to a probable future earthquake. This bears an additional latent danger pending on Qastoon dam itself and on the surrounding villages.

3.4 Delineation of the tectonic of the Khanasser Valley in Northwestern of Syria

In Khanasser valley, considered as a semi arid region in Syria, the shallow groundwater presents electrical conductivities ranging from 0.1 to 20 mS/cm. In order to study the hydro geological conditions of such region, a good knowledge is required of the geometry of the aquifer at depth.

Khanaser Valley was therefore geoelectrically thoroughly surveyed through a grid consisted of twelve VES profiles as shown in Figs, 13, and 14). The tectonically-oriented Pichugin & Habibullaev method was enhanced to be applicable in areas of rugged relief and topography. The enhanced profiles were tectonically interpreted and subsurface structures within Khanaser Valley were delineated. Accordingly, a tectonic evolutional scenario of the valley was established and its hydrogeological characteristics were derived.

The established iso-apparent resistivity maps for different AB/2 depth penetrations indicate the presence of two different geological structures; characterized by very conductive zones of a resistivity less than 4 Ohm.m related to the intrusion of salt water in Quaternary and

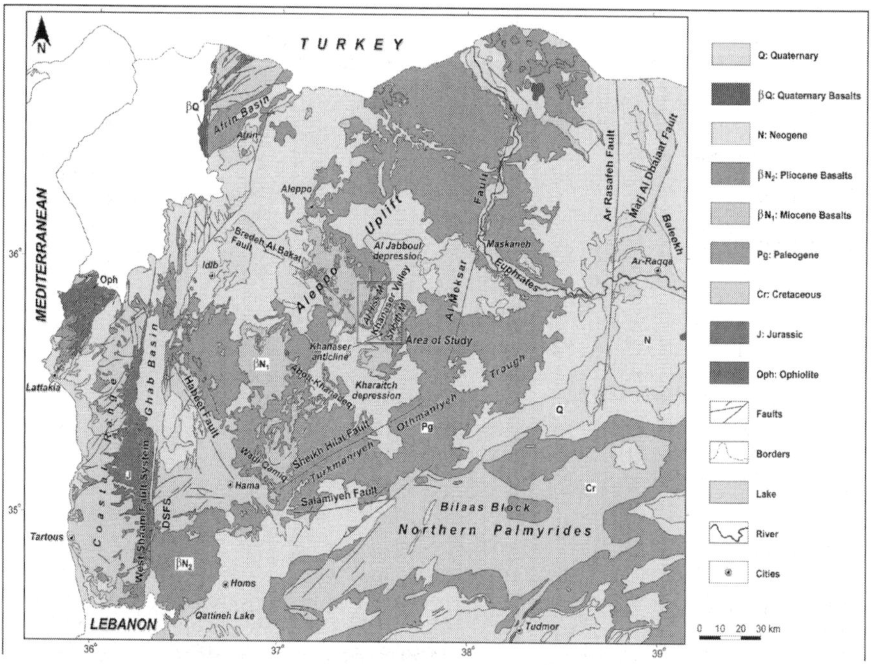

Fig. 13. Location of the Khanasser Valley.

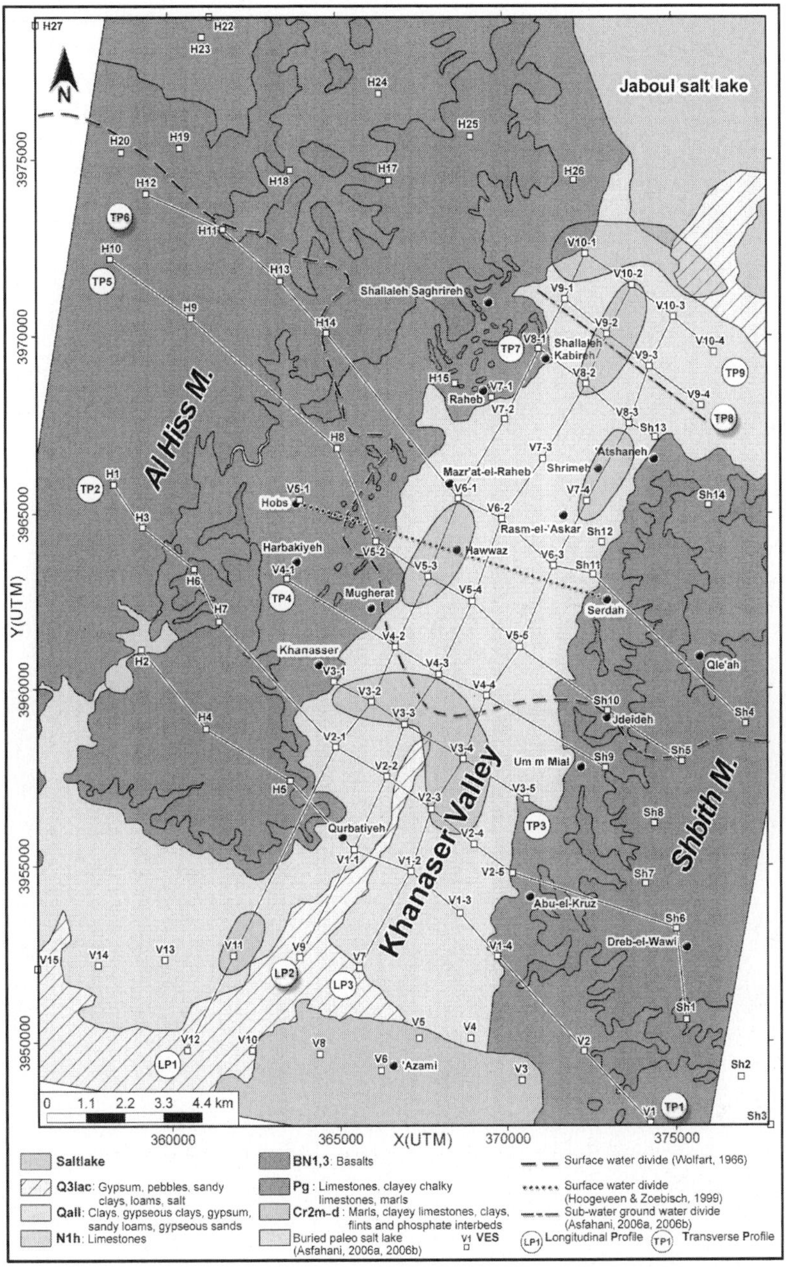

Fig. 14. Geology of the Khanasser Valley with the location of measured VES profiles.

Paleogene deposits. Resistive zones have been signaled in Jebel Al Hass in the west and Jebel Shbith in the east, characterized by a resistivity exceeding 300 Ohm.m, due to the presence of basalt formation of "BN1" age. The thickness of Quaternary, Paleogene and their electrical characteristics have been precisely determined. The top of Maistrechtian and its electrical characteristics have been also well established, Asfahani, 2007-a. Quaternary paleosabkhas were delineated through the studying of three longitudinal profiles a long the valley itself (LP1, LP2, and LP3), Asfahani, 2007-a.

Figures.15 and 16 show just two examples of the results interpretation of the transverse cross-section TP1 and TP6 by the application of both traditional and enhanced Pichugin $ Habibulaev method. The results of this interpretative method are presented by the PNH (+), which mainly show the locations of fractured zones. A clear concaved shape of the (+) with a width of 7Km was observed along the profile TP1 west of the Qurbatiyeh village. Clear normal faults, underlying the concaved shape bound sharply the Khanasser valley, and the field check confirms the presence of a volcanic crater to the northwest of Qurbatiyeh village, (Asfahani and Radwan, 2007).

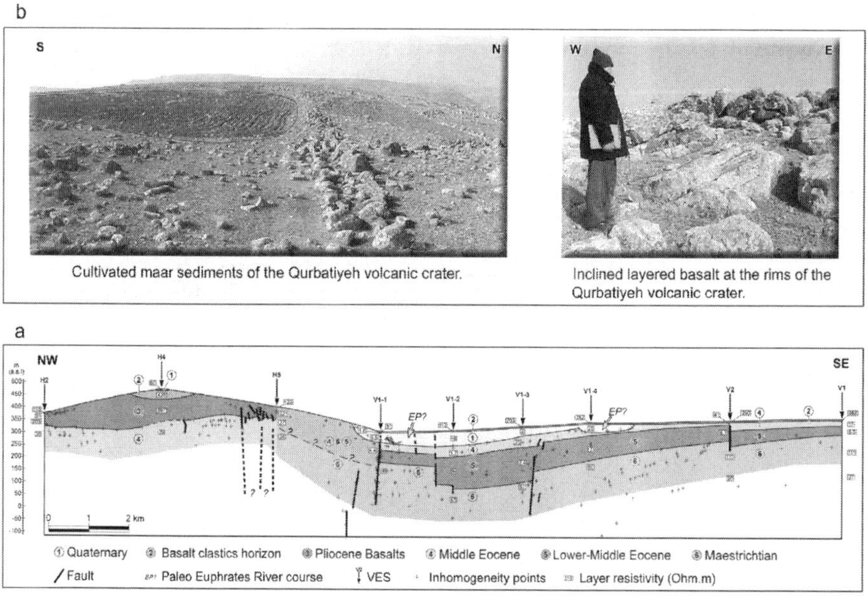

Fig. 15. Interpretation of transverse TP1 profile in the Khanasser Valley.

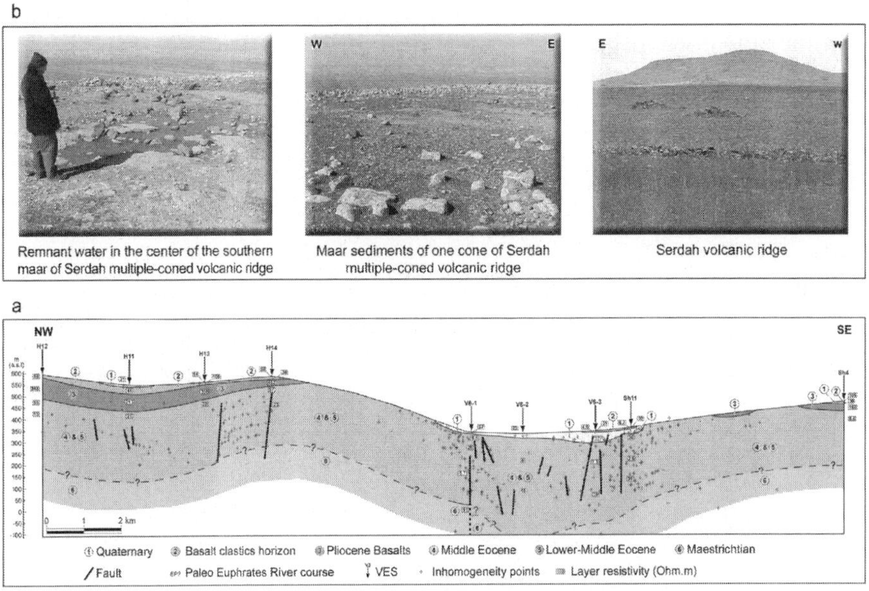

Fig. 16. Interpretation of transverse TP6 profile in the Khanasser Valley.

3.5 Application of the adapted shallow geoelectrical configuration in exploration and mining geology of phosphate deposits

The phosphatic deposits in Syria are actually mined in two main locations, Khneiffis and Al-Sharquieh phosphate mines as shown in Fig.17-a.

The Al-Sharquieh mine develops gradually to the south and southwest, where the phosphatic layers are approximately deposited in a horizontal position.

The geoelectrical and radioactive signatures of phosphate deposits in Al-Sharquieh mine in Syria have been identified through an extensive research work already published, (Asfahani and Mohamad, 2000). During this work, different geoelectrical methods have been successfully introduced and advised to be applicable while prospecting for phosphate resources.

The lithological and radioactive data of more than 40 drilled pits are used together in order to support the results obtained by the application of geoelectrical methods, where the location of those pits is shown in Fig.17-b.

On each of the drilled pits, a shallow vertical electrical sounding has been already carried out, where a total of 45 VES distributed on a regular grid was interpreted. The same VES measurements have been recently reinterpreted in order to clarify the subsurface tectonics, where seven field examples have been herewith presented with their interpretation, (Figs. 18, and 19), Asfahani, 2010-b. The structural picture of the study region has been well established, where a geoelectrical properties difference between Northern west and Southern east directions has been noticed. The Northernwest direction is characterized by uplifting structure where phosphatic beds are exposed at or near the surface. Profile-4 obviously shows this uplifting structure which is well indicated by the PNH distribution.

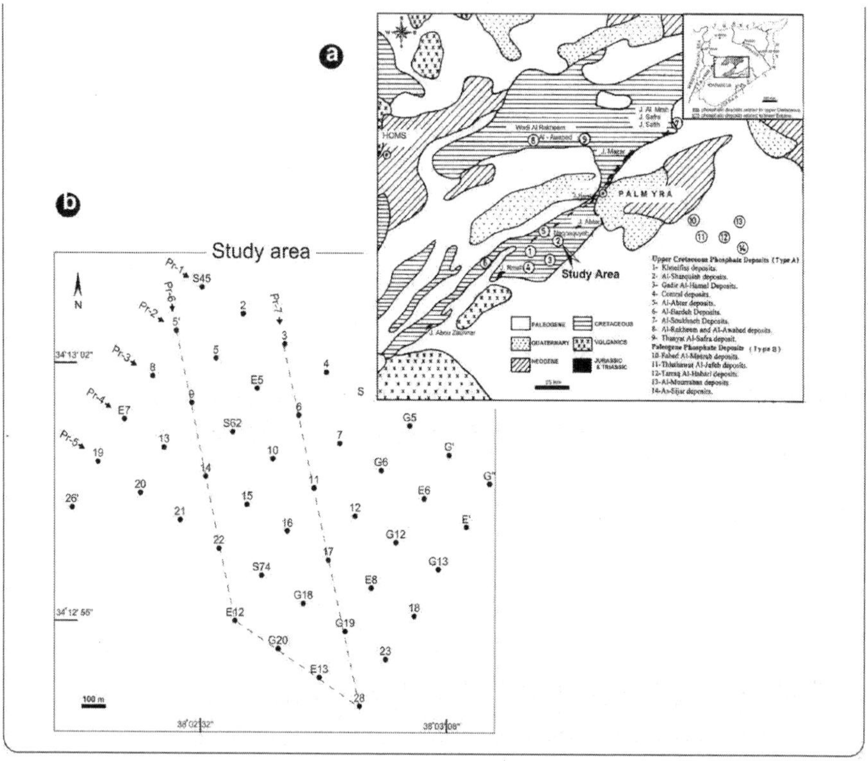

Fig. 17. a: Geology of the phosphate mines in Syria. b: Location of VES profiles in Al-Sharquieh mine.

The Southern east direction is characterized by low resistivity values and deepening of the phosphatic layers which deposited in a negative topographic structure such as subsiding sedimentary basin. This basined structure is beautifully indicated by the non homogeneity points distribution while applying the technique of Pichugin$Habibulaev (1985). The very condense clustering of non homogeneity points observed in many locations along the most of the studied profiles at depths ranging between 5 and 15 m are due to rapid lateral lithological variations normally observed in such sedimentological environments. The distribution of non homogeneity points largely contributes in delineating the positions of the faults in both Northwest and Southeast directions. The locations of the identified faults are characterized by high radioactivity due to the uranium concentrations in the phosphatic rocks. It appears quite clear that the maximum uranium concentration is confined to zones of tectonic weakening and to zone of fracturing.

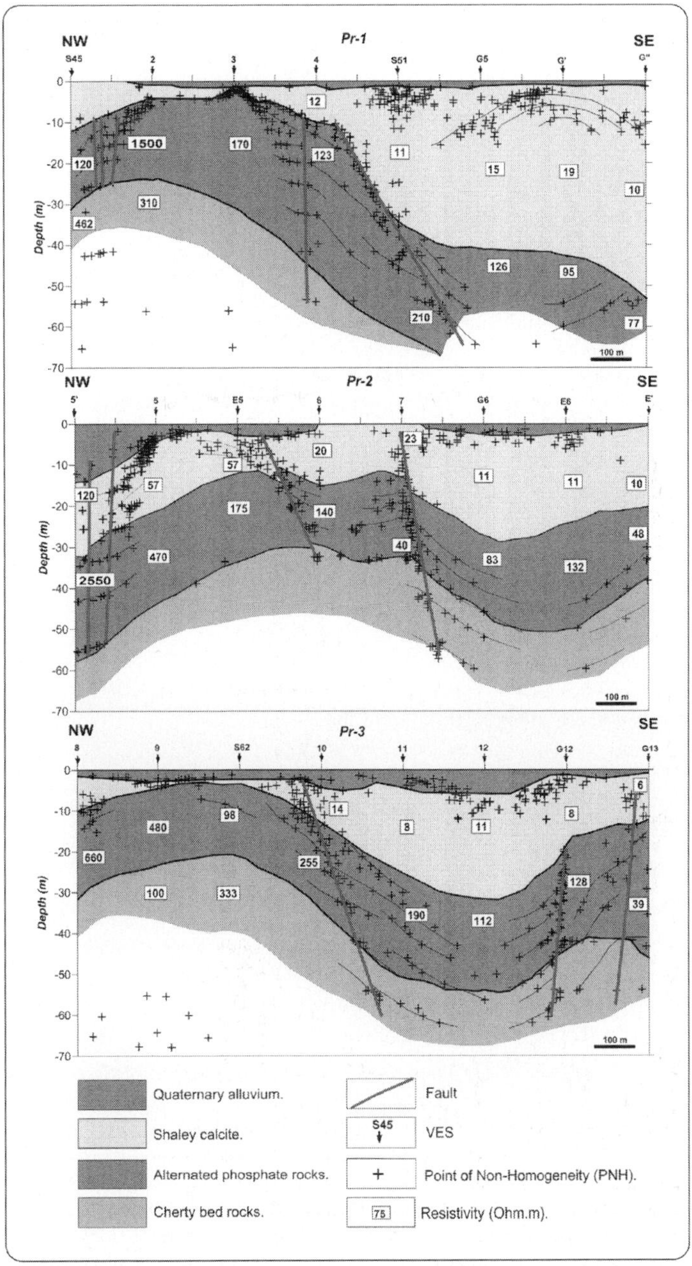

Fig. 18. Interpreted geological cross-sections of Pr-1, Pr-2,Pr-3 and Pr-4.

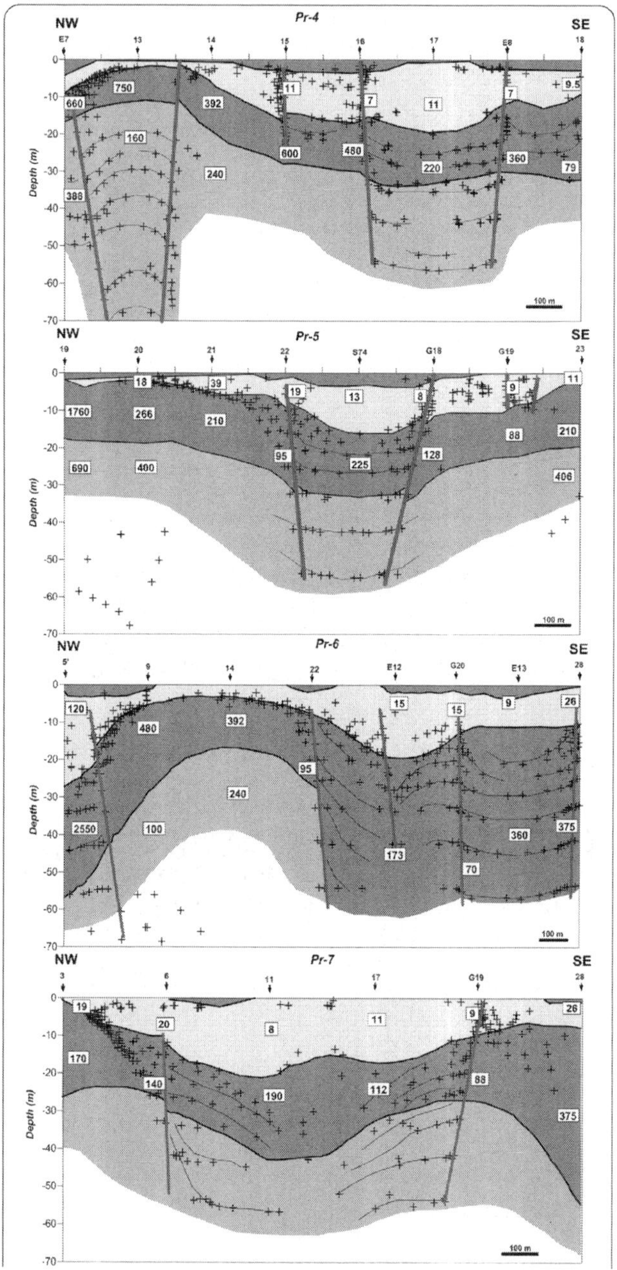

Fig. 19. Interpreted geological cross-sections of Pr-5, Pr-6, and Pr-7.

4. Application of deep Schlumberger configuration in Northeastern of Syria for the *determination of the favorable structures for sulpher prospecting*

Electrical and structural characterstics of formations favorable for sulpher occurrence in northeast of Syria are described using geoelectrical prospecting methods. Simple (VES) and combined (CVES) Schlumberger vertical electrical soundings and geoelectrical profiling using Wenner configuration were applied to Techreen structure, Asfahani and Mohamad, 2002. The geoelectrical research has been concentrated on the studying of six profiles (A,B,C,D,E, and F) shown in Fig.20 and located at the borders of anticlines, where positive and negative structures are joined and salt formations have a tendency to disappear. Secondary structures characterized by high apparent resistivity exceeding 3000Ohm.m were located at each of the studied profil using Wenner profiling configuration. These secondary structures are demonstrated to be favorable for sulpher prospecting by both drilled wells and vertical electrical soundings. More than 84 VES measurements were carried out in the study area, where thicknesses and resistivities of the Lower Al-Fares, Al-Garibeh and Al-Dibbaneh formations were determined. The interpretation of those VES distributed along the studied profiles by the Pichugin $ Habibulaev technique allows the subsurface tectonic to be precisely determined for a depth penetration corresponding to AB/2 of 1000m. Figure.21 shows a beautiful example of such results interpretation carried out along the profile A. It was found that sulfur occurrences in the research area are controlled by tectonic paths that are well defined by geoelectrical methods.

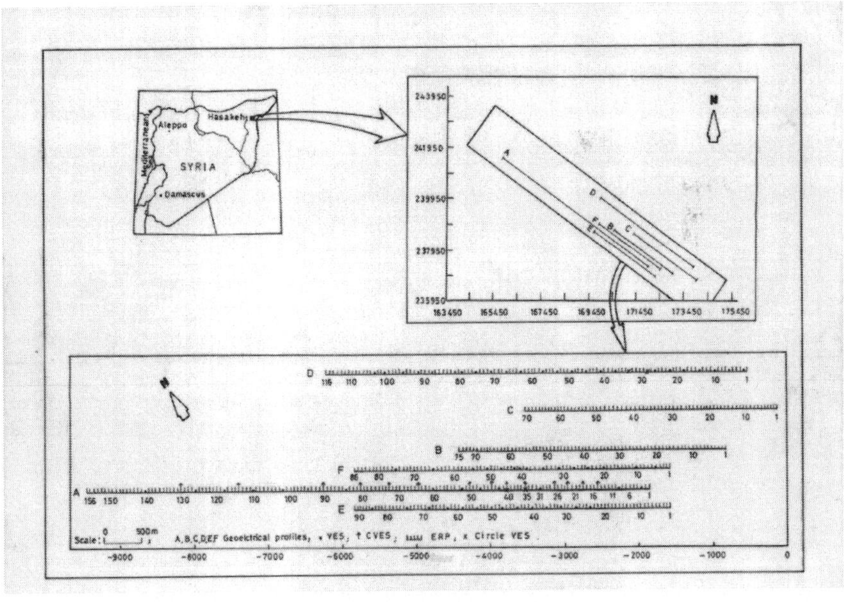

Fig. 20. Location of Techreen structure for sulpher prospecting.

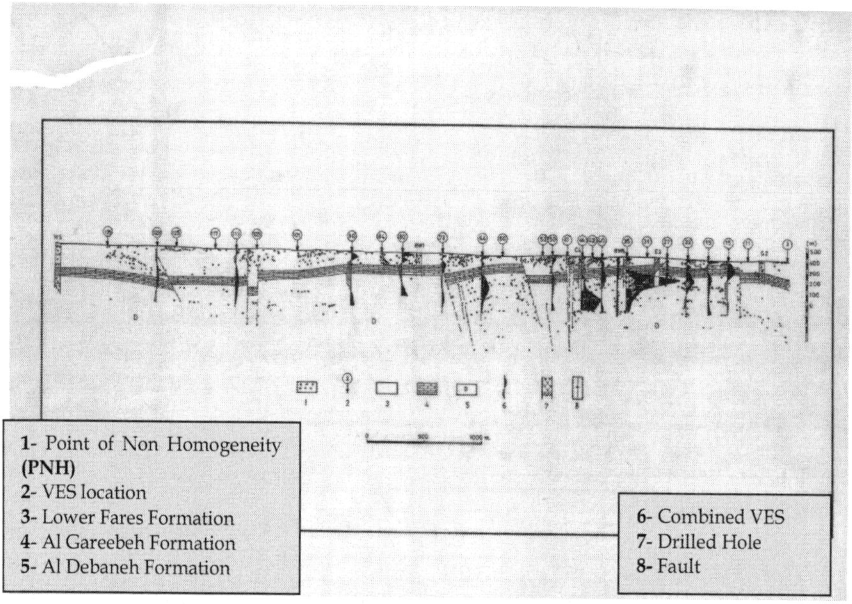

Fig. 21. Interpretation of VES distribution along profile-A.

5. Conclusion

Useful and important applications of the geoelectrical methods in delineating the subsurface tectonics features for solving different geological problems from Syria have been presented. The DC geoelectrical resistivity Schlumberger configuration is mainly developed and modified in order to simultaneously obtain reliable data for the shallow and deep penetration depths. The interpretation of the VES data by the enhanced Pichugin $ Habibulaev technique allows an integrated subsurface tectonic to be established along the studied VES profile, where typical selected examples have demonstrated the efficiency of such an interpretative technique and the role of DC geoelectrical methods in determining the subsurface tectonic features. This integrated interpretative technique with the application of modified Schlumberger configuration is therefore strongly recommended when tectonic subsurface information is required.

6. Acknowledgment

Author would like to thank Dr. I. Othman, General Director of Syrian Atomic Energy Commission for permission to publish the materials presented in this chapter.

7. References

Asfahani, J. (2007-a), Geoelectrical Investigation for Characterizing the Hydrogeological Conditions in Semi Arid Region in Khanasser Valley, Syria. Journal of Arid Environnments 68, PP 31-52.

Asfahani, J. (2007-b), Electrical Earth Resistivity surveying for delineating the characteristics of ground water in semiarid region in Khanaser Valley, Northern Syria, Hydrological Processes 21, 1085-1097.

Asfahani, J. (2010-a), *Electrical Resistivity Investigations for Guiding and Controlling Fresh Water Wells Drilling in Semi Arid Region in Khanasser Valley, Northern Syria, Accepte for publication in Acta Geophysica.*

Asfahani, J. (2010-b), A New interpretation approach of shallow and deep geoelectrical configuration for application in tectonic and exploration mining geology, Submitted for publication in Exploration and Mining Geology.

Asfahani, J., and Radwan, Y. (2007), Tectonic Evolution and Hydrogeological Characteristics of Khanaser Valley, Northern Syria, Derived from the Interpretation of Vertical Electrical Soundings, Pure Appl. Geophys. *164*, 2291-2311.

Asfahani, J., Radwan, Y., and Layyous, I., (2010), Integrated Geophysical and Morphotectonic Survey of Impact of Ghab Extensional Tectonics on the Qastoon Dam, Northwestern Syria, Pure Appl. Geophys. *167*, 323-338.

Asfahani, J., Mohamad, R. (2000), Investigation of electrical properties of radioactive phosphatic layers in Al-Sharquieh mine, Syria, Exploration and Mining Geology., Vol.9, 141-148.

Asfahani, J., Mohamad, R. (2002), Geoelectrical Investigation for sulfur prospecting in Techreen structure in northern Syria, Exploration and Mining Geology., Vo.11, 49-59.

Brew, G., Barazangi, M., Al-Maleh, K., and Sawaf, T. (2001a), Tectonic and Geologic Evolution of Syria, GeoArabia *6(3)*, 573-616.

Brew, G., Lupa, J., Barazangi, M., Sawaf, T., Al-Imam, A., and Zaza, T. (2001b), Structure and Tectonic Development of Al-Ghab Basin and the Dead Sea Fault System, Syria, Journal of the Geological Society *158*, 665-674.

Cai, J.,McMechan, A., and Fisher,M.A., (1996) , Application of ground penetrating radar for investigation of near surface fault properties in the San Francisco Bay rgion. Bull. Seis Soc.Am., 86, 1459-1470.

Caputo, R., Salviulo, L., Piscitelli, S., and Loperte, A., (2007) , Late Quaternary activity along the Scorciabuoi fault (Southern Italy) as inferred from electrical resistivity tomographies: Annals of Geophysics, Vol. 50, N. 2, P 213-224.

Caputo, R., Piscitelli, S., Oliveto, A., Rizzo, E., and Lapenna, V., (2003) , The use of electrical resistivity tomographies in active tectonic: Examples from the Tyrnavos basin, Greece, Available online 5 June 2003, ArchEnviMat.

Chwatel, W., Decker, K., and Roch, K., (2005) , Mapping of active capable faults by high-resolution geophysical methods: Examples from the central Vienna basin: Austrian Journal of Earth Sciences, Volume, 97.

Dobrin, M. (1976), Introduction to Geophysical Prospecting, Mc Graw-Hill, New York.

Demanet, D., Renardy, F., Vanneste, K., Jongmans, D., Camelbeeck, T., and Meghraoui, M., (2001), The use of geophysical prospecting for imaging active faults in the Roer Graben, Belgium: Geophysics, Vol. 66, No.1, P. 78-89.

Sabrina Y.F., Augusto E.R., Jose M. Cortes and Carla M., (2009), Characterization of Quaternary faults by electric resistivity tomography in the Andean Precordillera of Western Argentina, Journal of South American Earth Sciences, Vol 28, Issue 3, PP: 217-228.

Mares, S., ed., (1984), Introduction to Applied Geophysics, (Reidel Publ Co., Dordrecht, Boston, Lancaster) P. 581.

Massoud, U., El Qady, G., Metwaly, M, and Santos, F. (2009), Delineation of shallow subsurface structure by azimuthal resistivity sounding and joint inversion of VES-TEM data: case study near Lake Qaroun, El Fayoum, Egypt, Pure Appl. Geophys. *166*, 701-719.

Orellana, E., and Mooney, H.M. (1966), Master Tables and Curves for Vertical Electrical Sounding Over Layered Structures, Interciencia, Madrid.

Palmer, J.R., Shoemaker, M., Hoffman, D., Anderson, N. L., Vaughn, J. D., and Harrison, R. W., (1997), Seismic evidence of quaternary faulting in the Benton hills area, southeast Missouri: Seis. Res. Lett., 68, 650-661.

Pichugin, N.I., and Habibullaev, I.K.H. (1985), Methodological Recomendations in Studying Geo-Tectonic Conditions of Vertical Electrical Soundings Data with Application of EC Computer for Solving Hydrogeological and Geo-Engineering Problems, Tashkend, (in Russian).

Parrales, R., Dahlin,T., and Rubi, C., (2003) , Site investigation with combined methods in a faulted area in Managua, Nicaragua- A pre-study: Procs. 9th Meeting of Environmental and Engineering Geophysics, Prague, Czech Republic, P-076.

Piscitelli, S., Caputo, R., Lapenna, V., Oliveto, A., and Rizzo, E., (2009), Electrical imaging survey across activity faults: Examples from the Tyrnavos basin, Greec, GNGTS-Atti del 21° Convegno Nazional/01.01

Ponikarov, V.P. (ed.) (1966), The Geological Map of Syria, scale 1:200000, sheets I-37-XIX and I-36-XXIV, Ministry of Industry, Damascus.

Shields, G., Allander, K., Brigham, R., Crosbie, R., Trimble, L., Sleeman, M., Tucker, R., Zhan, H., and Louie, J.N., (1998), Shallow geophysical survey across the Pahrump Valley fault zone. California- Nevada border: Bull. Seis Soc.Am., 88, 270-275.

Van Ardsal, R., Purser, J., Stephenson, W., and Odum, J., (1998), Faulting along the southern margin of Reeflot Lake, Tennesse: Bull. Seis Soc.Am., 88, 131-139.

Williams, R., Luzietti, E, A., and Carver, D, L., (1995), High- resolution imaging of Quaternary faulting on the Crittenden County fault zone, New Madrid seismic zone, northeastern Arkansa: Seis. Res. Lett., 66, 42-57.

Zohdy, A.A.R. (1989), A New Method for the Automatic Interpretation of Schlumberger and Wenner Sounding Curves, Geophysics *54*, 245-253.

Zohdy, A.A.R., and Bisdorf, R.J. (1989), Schlumberger Sounding Data Processing and Interpretation Program, U. S. Geological Survey, Denver.

12

Salt Tectonics of the Lisan Diapir Revealed by Synthetic Aperture Radar Images

Damien Closson[1], Najib Abou Karaki[2], Nada Milisavljević[3], Frédéric Hallot[3], and Marc Acheroy[3]

[1]Signal and Image Centre. Avenue de la Renaissance 30, Brussels 1000,
[2]University of Jordan, Environmental and Applied Geology Department, Amman,
[3]Signal and Image Centre. Avenue de la Renaissance 30, Brussels 1000,1
[1,3]Belgium
[2]Jordan

1. Introduction

1.1 General setting

In the Oligocene, the Africa-Arabia plate broke up separating Arabia and Sinai as individual plate and sub-plate. Since then, the Arabian plate moves northward along the Jordan - Dead Sea Transform (JDST) fault more rapidly than the Sinai sub-plate. This left-lateral strike slip movement (Figure 1) had displaced Early Miocene dykes across the fault zone up to 100 km (Quennell, 1958; Freund et al., 1970; Garfunkel et al., 1981) and had resulted in the development of rhomb-shaped grabens such as the Dead Sea pull-apart basin along the main fault (Figure 1, inset).

The Dead Sea Basin is the largest pull-apart along the JDST fault. It is about 150 km x 15 km. Repeated structural subsidence resulted in the accumulation of sedimentary rocks as much as 10 km thick (Garfunkel and Ben Avraham, 1996). Inside the basin, two sets of faults, both oriented roughly N-S, can be recognized (Figure 1, inset). The first set is the extension of the northern and southern segments of the JDST fault, forming the pull-apart basin (Ben Avraham, 1997; Garfunkel, 1997). These faults are accommodating most of the horizontal motion of the JDST fault. The second set is constituted by the transverse faults, oriented NNW-SSE, that cross obliquely the basin at interval of 20-30 km. Between these faults, the basin infill is slightly back-tilted toward the south with no large deformations (Gardosh et al., 1997; Ben Avraham, 1997; Al Zoubi and ten Brink, 2001). Strong deformations are known only near the diapirs formed by the salt of the Sedom Formation (2 km thick). The Sedom formation formed from the late Miocene to the Pliocene (5.3 - 2.5 Ma). It is composed of 75% rocksalt which arrived via marine ingression from the Mediterranean and Red Seas. This flooding has ceased, however, with a rise of intrusive rock in the Araba and Jezreel valleys (Figure 1). The rocksalt is interbeded with anhydrite and gypsum, reddish dolomite, silt, sand and clay. Since the early Miocene, the center of sedimentary deposition existed where the Lisan area (Figure 1, inset) is currently located.

During the Pliocene excessive accumulation of sediment caused a diapiric upward movement of accumulated lower-density sediment to begin, thus forming several salt domes structures. Among these diapirs, the one located under the Lisan area is the largest (Figure 1, inset; Figure 2) and constitutes the element under investigation.

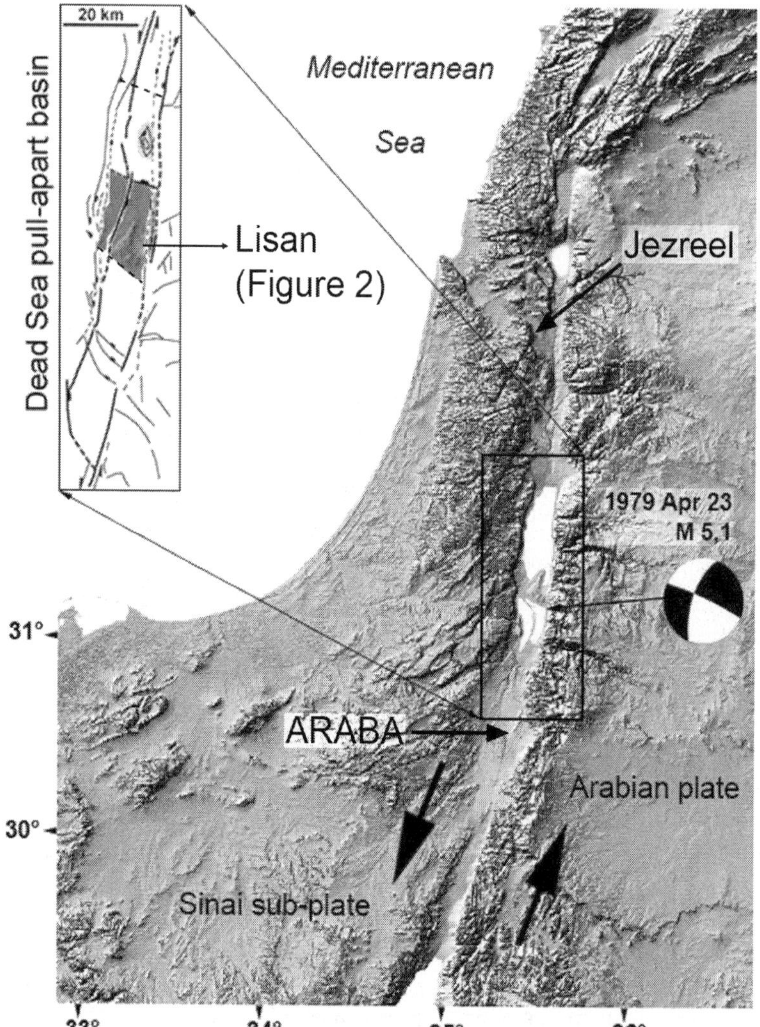

Fig. 1. The area of interest in its global setting. The shaded relief image (Space Radar Topography Mission, Feb. 2000) focuses on the Jordan – Dead Sea transform fault zone. Inset shows the main tectonic elements of the Dead Sea pull-apart basin (after Ben Avraham, 1997; Ben Avraham and Lazar, 2006). The focal mechanisms associated with the earthquake of April 23, 1979 (Arieh et al., 1982) is representative of all focal mechanisms calculated on a fault plane compatible with the general direction of the Jordan - Dead Sea Transform.

1.2 Goal of this work

This chapter presents and discusses ground surface displacements affecting the Lisan area. The displacement fields have been investigated owing to the processing of 27 radar images acquired by C- and L-band sensors onboard satellites (ALOS, ENVISAT, and ERS 1&2). The time period concerned ranges from 1992 to 2008. Uplift and subsidence have been mapped and analyzed inside a geographical information system. Two "tandem" pairs of ERS 1&2 satellite have been processed to create two digital surface models of the studied area. It compensates for the lack of updated information in the fields of topography and geomorphology. The work on these models targeted the mapping of lineaments, the detection drainage pattern anomalies, the definition of the drainage texture, and the extraction of topographic features which can be explained by structural or stratigraphic conditions.

2. Background

2.1 Tectonic setting

At regional scale, the Lisan appears to be a structurally complex rhomboid-shaped tectonic unit (Figure 2 inset). Parallel and curvilinear faulted zones define either elongated or almond-shaped grabens on its eastern and western sides (Lynch Strait, Ghor Al Mazra'a, Ghor Al Haditha - Figure 2). North and south, the Lisan is respectively bounded by a faulted zone separating the northern to the southern basin and by a discontinuity (Figure 2, inset, dashed line) representing the edge of the Lisan salt diapir (Ben Avraham and Lazar, 2006).

The top of the salt dome (Figure 2) is estimated to be at 125 m below the surface based on drill holes that have penetrated over 100 m of deformed lake deposits (Neev and Emery, 1967; Zak, 1967, Bender, 1974). With respect to the Lisan peninsula, the main out-cropping formation is the Lisan Marl (chalk, marly chalk, gypsum, salt and conglomerate). Based on photogeological interpretation and field observations, Sunna (1986) subdivided this formation into two Members (Lower and Upper) while Bartov et al. (2006) subdivided the marls into three Members (Lower: 60-40 Kyr, Middle: 40-32 Kyr, and Upper: 32-15 Kyr) based on U-Th dating method.

Al Zoubi and ten Brink (2001) used seismic data correlated to several deep wells (Zak, 1967; Bender, 1974) to determine the size of the elongated Lisan diapir: 13 km x 10 km x 7.2 km (Figure 2, top salt contours). Its geometry is tectonically controlled. At the ground surface, fault escarpments of 20 to 40 m (Figure 3, see location on Figure 2) indicate the limit of the former peninsula. On the western side it is difficult to determine the orientation of individual faults due to insufficient exposure caused by the setting up of saltpans (Figure 2) in the 1990s. Bartov et al. (2006) observed 1-3 m of displacements of the exposed faults along the SW-NE escarpment. They displaced the upper part of the Middle Member (~35kyr) and perhaps younger Lisan deposits. Bartov (1999), Al-Zoubi and ten-Brink (2001) related the western fault boundary (Figure 4, from 729-3459 to 737-3472) to the ongoing salt tectonics underneath the Lisan wave-cut platform (Figure 3).

On Figure 2, the topographic contour lines -395m and -422m corresponding respectively to the shoreline in 1963 and in 2009 have been drawn. They underline that during the last fifty years, the Dead Sea level had decreased. From 1977, the Lisan peninsula became a land bridge between the eastern and the western coasts because of the drying up of the Lynch Strait. The Dead Sea split in two parts. The salt remains soak the recently emerged lands.

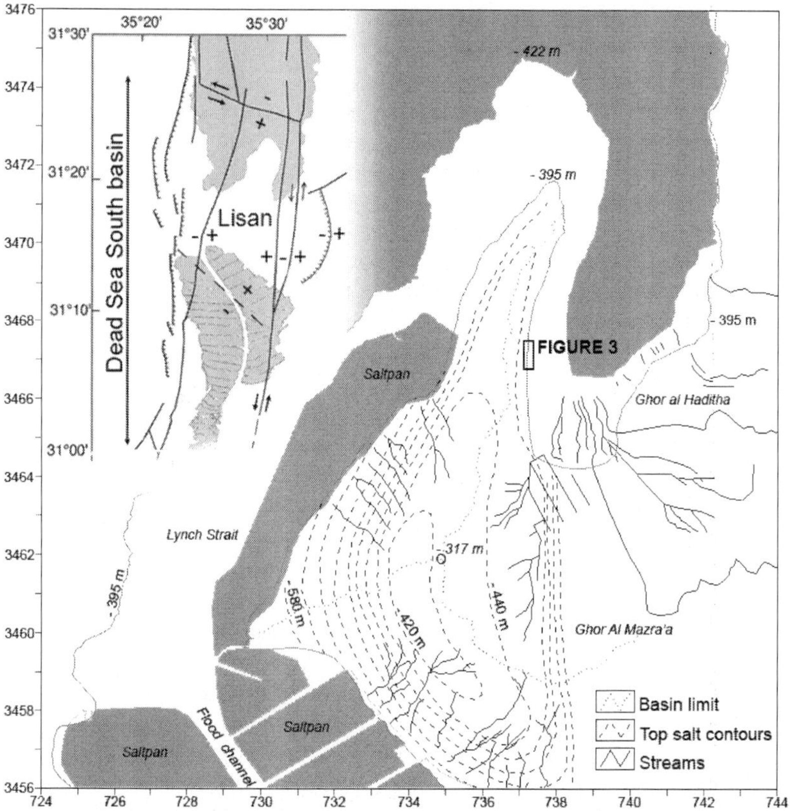

Fig. 2. Top salt of the Lisan diapir (contours from "The salt upwarp of El Lisan", map compiled by Bender, 1967, and Abu-Ajamieh, 1987). The radial network of ephemeral streams emphasizes the location of maximum thrust (see elevation point -317m, top of the Lisan). Upper right inset displays the major faulted zones and discontinuities as well as the upthrown and downthrown fault blocks in the Dead Sea southern basin (based on Ben Avraham 1997, Ben Avraham and Lazar, 2006). Coordinate are expressed in UTM km, 36, WGS84.

They are thus susceptible for dissolution and compaction. The Lisan tectonic block (Figure 2, inset) gathers the highest concentration of sinkholes and subsidence in the whole Dead Sea area (Closson et al., 2009a).

Along the eastern side, Sunna (1986) and Bartov et al. (2006) have interpreted the fault network (Figure 4, from 737-3456 to 737-3470), as one of the strike-slip faults that bound the Dead Sea pull-apart structure. Sub-parallel normal faults trending N-S have formed a 50 m stepped escarpment toward the flat Ghor Al Haditha - Mazra'a graben. This fault zone displaces post-Lisan (<15kyr) deposits. Its northern part is manifested by normal faults dipping to the east (Figure 3), while its southern segment shows smaller displacements with extensive folding.

Fig. 3. Vestiges of a fault plane preserved in an alignment of triangular facets. The fault scarp had been produced by active normal faulting along the eastern margin of the former Lisan peninsula. Repeated faulting has produced a rock cliff tens of meters high. Erosion modified the fault scarp but, because the fault plane extends hundreds of meters down into the bedrock, its effects on erosional landforms persisted for several thousand years.

Folds in the Lisan formation delineate the structures of some sub-domes (e.g. Figure 4, lineaments at 734-3460 and at 736-3457) and indicate that it is still rising. Sunna (1986) and Bartov et al. (2006) studied the overall peninsula at the exception of the emerged lands (between the elevations of -395 m and -422 m, see Figure 4). This surface had been investigated more recently, mainly by using remote sensing techniques (Closson, 2005a, 2005b; Closson et al., 2007). Figure 4 shows a major elongated dome with smaller sub-domes located at its southern edge. The structural high of the center of the main dome is located at coordinates 735-3462 (Figure 4, lineaments). Localized joints radiating from this center are found in many places. They are the consequences of the flexure of the dome structure. In the vicinity, the folds have a wavelength of 50-100 m and their amplitude is 10-20 m (734-3462). The folding postdates the base of Upper Member of the Lisan formation. (~32 kyr) and was active during the deposition of the Upper Member (32-15 kyr) due to thickness variation near the folds (Bartov et al., 2006). Southward, the major southern dome (Figure 4, lineaments at 736-3458) has a structural peak of -323 m, it is about 3 km long and 2 km wide with an axis orienting at N45E. The dome is flanked by minor folds dipping 3°- 10° with a wavelength of about 100 m. A few hundred meters southeastward, another dome has a peak of -340 m (Figure 4, faults at 738–3457). It is 3 km long and 1.5 km wide with an axis oriented at N30E. It is characterized by secondary folds crossing the structure and trending 90E-100E with flank dipping up to 40°.

A depression separates these two very different structural domains and supports the location of a major fault zone in between. In this area one fault expose strikes N23E and displaced the top of the Lisan formation (738-3458).

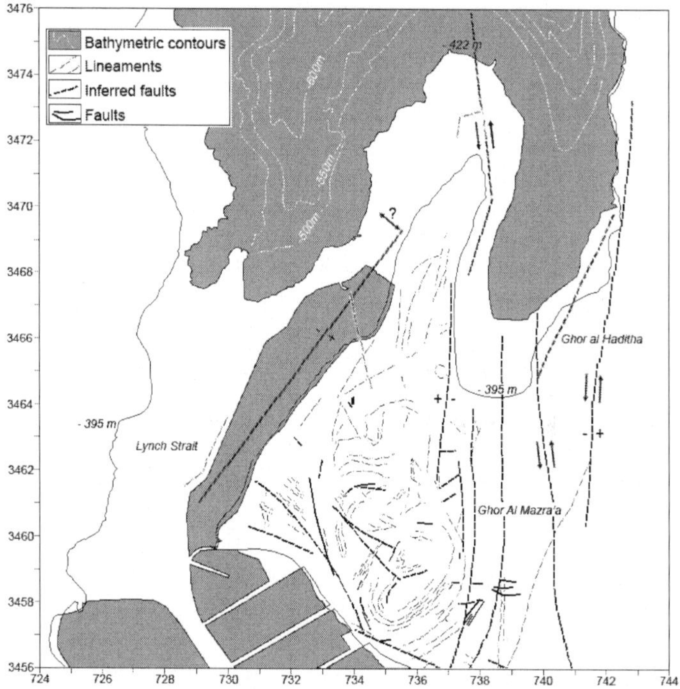

Fig. 4. Lineaments and (inferred) faults of the Lisan, Mazra'a and Ghor Al Haditha area, Jordan, based on the work done by Sunna, 1986; Closson et al., 2003a; Closson et al, 2003b; Closson et al., 2005; Closson, 2005a, b; Diabat, 2005, Bartov et al., 2006, Closson et al, 2007; Closson & Abou karaki, 2009a, b; Closson et al., 2010a, b; Bathymetric contours (Hall, 1997) allow identification of deep and wide submarine canyons that could result from the erosion contemporary to a near complete drying up of the Dead Sea (Neev and Emery, 1967). Topographic contour lines of -395m and -422m correspond respectively to the shoreline in 1963 and in 2009.

2.2 Current rates of deformation

When dealing with ground displacements in the Lisan area, six major constraints have to be taken into account:

- the horizontal strike slip movement of the JDST;
- the subsidence of the pull-apart basin;
- the rising of the salt diapir;
- the compaction of the sediments in the zones that have emerged during the last 50 years;
- the possible (but never detected) isostatic rebound in the same zone in relation with the rapid disappearance of about 28 m of Dead Sea brine (-395m in the mid-1960s to -423m in 2010) (E. Salameh, personal communication);
- the dynamic of the halokarst in relation with the Dead Sea level lowering.

The first four are briefly summarized below. The fifth is hypothetical and had not yet been taken into consideration. The latest point had been tackled in Closson et al. (2007).

2.2.1 The horizontal strike-slip movement of the JDST and the subsidence of the pull-apart basin

The modern slip rate is still under debate. Geological observations, however (mostly south of 33°N) suggest nearly pure left-lateral strike-slip faulting and estimated slip rates range between 1 and 20 mm/yr with preferred values around 5 to 7 mm/yr (Freund et al., 1968; Garfunkel et al., 1981; Gardosh et al., 1990; Ginat et al., 1998; Klinger et al., 2000; Niemi et al., 2001; Pe'eri et al., 2002; Ferry et al., 2007). The central part of the JDST is an active fault zone. The historical and instrumental seismicity data suggest that it has generated magnitude 7 (magnitude equivalent) or even larger earthquakes, approximately once every few hundred years. In the XX century, three well documented instrumental earthquakes $5 < M < 6$ occurred in or very close to the Dead Sea proper in 1956, 1979 (Figure 1, focal mechanism), and 2004 (Abou Karaki, 1987). The distribution of late Pleistocene subsidence in the central Dead Sea Basin is controlled by the local tectonic regime. The normal border faults show a maximum subsidence rate of 0.3 mm/yr of the rim or median block. This block extends basinward up to about 2.5 km where an oblique fault related to the strike-slip segment causes a 0.6 mm/yr subsidence (Bartov and Sagy, 2004).

2.2.2 Lisan diapir uplift and subsidence

Bartov et al. (2006) postulated a long-term uplift rate of 2 mm/yr, based on the depth of the Pliocene salt (>3500 m). The Holocene dome structure indicates ongoing deformation. The center and margins of the dome are undergoing different geomorphologic process. Channels deeply incised the margins but no significant incision is found in the central area.

The upper member of the Lisan Formation is truncated due a transgressive episode of the lake during the Holocene. It occurred ~6000 years B.P. and reached of -370 m. Bartov et al. (2004) deduced that the dome raised and passed over the elevation of -370 m. Taking the present datum of the truncated top of the structure (-315 m) as a marker, the dome was raised 55 m during the last ~6000 years. This indicates an average uplift rate of 9 mm/yr.

From interferometric processing applied to 16 ERS scenes with the JPL Sar processor, Baer et al. (2002) observed that "change interferograms show no sign of uplift in the southern or central domes where the diapir is closest to the surface. Possible evidence for uplift is seen in the northern part of the peninsula. Here, 10 mm and 20 mm of uplift are observed in the 22- and 50-month interferograms, respectively, suggesting an average uplift rate of 4–5 mm/yr". However, Shimoni et al. (2002) published a cumulated 93 months vertical deformation map from different profiles (11-Jun-92 to 21-Mar-99) where two uplift areas characterized by a maximum of 30 mm (4 mm/yr) had been delineated. Subsidence zones were also located north and south of the Lisan with value of 150 mm (22 mm/yr). Here, the DORIS software was used to process a dataset of 17 ERS scenes.

3. Material and methods

3.1 Satellite images and software

Radar interferometric techniques have been applied to 27 scenes (single look complex format) acquired by three different sensors onboard satellites: 14 ERS AMI-SAR (8 descending; 6 ascending), 2 ENVISAT ASAR, and 11 ALOS PALSAR (4 descending; 7

ascending). Two interferometric processors have been used to perform the study: Sarscape module (version 4.2) of Envi and Radar Mapping Suite (version 2010) of Erdas Imagine. Table 1 lists the images based on the date of acquisition.

ERS	ERS	Envisat	Alos	Alos
ASC	DSC	DSC	ASC	DSC
29-May-99	10-Dec-00	27-Apr-08	20-Dec-08	17-May-08
11-Oct-97	4-Jul-99	23-Mar-08	4-Nov-08	1-Apr-08
10-Oct-97	11-Oct-97		4-May-08	15-Feb-08
16-Dec-95	14-Jul-96		19-Mar-08	15-Nov-07
15-Dec-95	29-Jul-95		2-Feb-08	
30-Jun-93	30-Jul-95		18-Dec-07	
	5-Aug-93		2-Nov-07	
	11-Jun-92			

Table 1. Image dataset

3.2 Radar interfrometry techniques

Radar interfrometry techniques exploit the phase information of the signal emitted and received by a sensor onboard a satellite to generate digital surface models (Interferometric Synthetic Aperture Radar or InSAR technique) and/or ground motion images (Differential InSAR or DInSAR technique). These techniques have been successfully used for years to study the topsoil deformations in the Dead Sea area (e.g. Derauw 1999, Baer et al. 2002, Shimoni et al. 2002, Closson et al., 2003, closson et al., 2010). However, practically all published results were obtained from the processing of ERS-1 and ERS-2 satellite data. In this work we exploit the images acquired from the following sensors: ERS AMI-SAR, ENVISAT ASAR (both characterized by C-band, wavelength = 5.6 cm), and ALOS PALSAR which uses L-band (wavelength = 23.62 cm). Table 2 brings an overview of the sensors and their satellites.

The principle of the InSAR and DInSAR techniques is briefly summarized. Two images acquired from the same orbit but at different time (46 days apart for ALOS and 35 days for ERS and ENVISAT - Table 2), and with a similar incidence angle are necessary to generate an interferogram (image of the phase difference).

As an example, Figure 5 displays the principle of InSAR for digital surface model generation. In T_0, a radar antenna onboard a satellite acquires one image of a point P at the ground surface. The measure of the distance MP leads to the phase Φ_M. In T_1, the same or a twin satellite passes over the area of interest but not exactly at the same position. The measure of the distance SP leads to the phase Φ_S.

The InSAR techniques exploit the phase difference ($\Delta\Phi_{INT} = \Phi_S - \Phi_M$) which is related to the distance (SP – MP). Considering a single pixel footprint:

In T_0: $\Phi_M = 2 * MP * (2\pi / \lambda) + \Phi_0$

In T_1: $\Phi_S = 2 * SP * (2\pi / \lambda) + \Phi_0$

InSAR equation: $\Delta\Phi_{INT} = \Phi_S - \Phi_M = (SP - MP) * (4\pi / \lambda)$

λ = radar wavelength (5.6 cm for ERS and ENVISAT, 23.6 cm for ALOS)

	ERS (C band)	ENVISAT (C band)	ALOS (L band)
Altitude of the satellite	~790 km	~790 km	~700 km
Central frequency band	5.3 GHz	5.331 GHz	1.27 GHz
Wavelength	5.6 cm	5.6 cm	23.6 cm
Polarization	VV	VV HH HH/HV VV/VH HH/VV	VV HH HH/HV VV/VH
Incidence angle	~23.2°	15° - 45°	~38.7°
Maximum perpendicular baseline	1100 m	1100 m	15380 m
Resolution (range)	~24.5 m	30 - 150 m	~8.6 m
Resolution (azimuth)	~5 m	~5 m	~5 m
Swath	~100 km	58 - 110 km	~70 km
Repeat cycle	35 days	35 days	46 days
phase noise (*)	2.1 mm	2.1 mm	3.3 mm
(*) i.e., line of sight (LOS) precision of over the 100 m to 5000 m wavelength band			
fringe rate	4	4	1
Coherence	less	less	better
Tropospheric effects	Identical	Identical	Identical
Ionospheric effects	better	better	~17 times greater than C-band
Phase gradients/Greater deformation	less	less	better
Measurement accuracy	better	better	Less (~1.6 times)

Table 2. List of some parameters allowing a comparison between sensors used in this work.

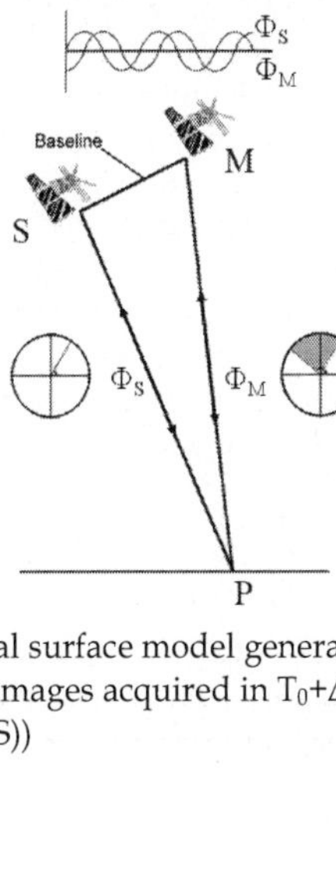

Fig. 5. Principle of InSAR for digital surface model generation. M = "Master" or reference image acquired in T_0. S = "Slave" images acquired in $T_0+\Delta T$ (ΔT = 1 day (ERS), n*35 days (ERS, ENVISAT), n*46 days (ALOS))

Among others, $\Delta\Phi_{INT}$ gathers information about the topography ("topographic fringes"). An analogy can be done between the "fringes" that appear over an interferogram and the surface between two contours of a topographic map. In the same way, the "altitude of ambiguity" of an interferogram is similar to the contour interval.
The sensitivity to the topography of an interferogram ("altitude of ambiguity") depends on the relative position of the two sensors (distance MS). Among others, it depends on the "perpendicular" baseline parameter. The value of the "altitude of ambiguity" increases drastically when the distance of the perpendicular baseline tends to zero. An interferogram generated from two acquisitions having a very short perpendicular baseline will present only a few "topographic fringes". In other words, a short perpendicular baseline means an interferogram equivalent to a topographic map with a few contours and a large contour interval. Therefore, if a concentration of fringes is observed over such an interferogram, then ground displacements are suspected.
Conversely, when the perpendicular baseline increases, the "altitude of ambiguity" decreases so that the interfrogram will reveal the most subtle topographic variations of the ground surface. A limitation exists for the length of the "perpendicular baseline". Indeed, the more the perpendicular baseline increases, the more the image of point P (Figure 5) will differ so that it become impossible to coregistrate the two acquisitions and compute the phase difference. Table 2 gives the maximum perpendicular baseline allowed (in theory).
The more the temporal baseline increases, the more the possibility of ground movements increases. Surface displacements happen for various reasons (e.g. water or oil extraction lead to subsidence while volcanoes inflation leads to uplift). In this case, $\Delta\Phi_{INT}$ gathers information about the topography and ground movements ("displacement fringes") that have occurred between the two acquisitions. Figure 6 displays the principle of DInSAR for deformation measurement: a subsidence occurred between the two acquisitions and the point P dropped to the position P'.

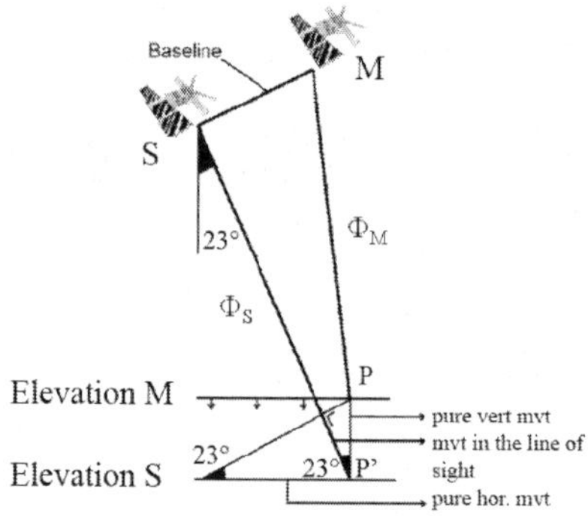

Fig. 6. Principle of DInSAR for deformation measurement. As an example, 23° is the incidence angle for ERS satellites.

In T_0: $\Phi_M = 2 * MP * (2\pi / \lambda) + \Phi_0$

In T_1: $\Phi_S = 2 * SP' * (2\pi / \lambda) + \Phi_0$

DInSAR equation: $\Delta\Phi_{INT} = \Phi_S - \Phi_M = (SP - MP) * (4\pi / \lambda) + (SP' - SP) * (4\pi / \lambda)$

$$\Delta\Phi_{INT} = \Delta\Phi_{Topo} + \Delta\Phi_{Mov}$$

If one focuses only on the ground movements ($\Delta\Phi_{Mov}$), the "topographic fringes" have to be retrieved using the technique of "differential interferometry". By simulating $\Delta\Phi_{Topo}$ from an available digital elevation model of the area of interest, it is possible to obtain $\Delta\Phi_{Mov} = \Delta\Phi_{INT} - \Delta\Phi_{Topo\_simulated}$

From the DInSAR equation one can deduce that $\Delta\Phi_{Topo} = (SP - MP) * (4\pi / \lambda)$ is a function of the distance between M and S while $\Delta\Phi_{Mov} = (SP' - SP) * (4\pi / \lambda)$ is independent of the baseline. $\Delta\Phi_{INT}$ is therefore more sensitive to the ground displacement than the topography. From Figure 6, one can deduce that $\Delta\Phi_{Mov\_pure\ vertical} = (\Delta\Phi_{Mov} / \cos 23°) * (\lambda / 4\pi)$. One centimenter of ground displacement along the line of sight will lead to $\Delta\Phi_{Mov} \sim 127°$. Considering a typical noise of an interferometric phase of the order of $\pi/6 = 30°$, movements of about 2.5 mm can be detected. With respect to the topography, if a difference of elevation of 20 m is considered, then the $\Delta\Phi_{Topo}$ will be about 43° with a baseline of 50 m and about 8.6° for a baseline of 10 m (in the case of ERS satellite).

Table 3 shows an example of the spatial and temporal baselines available for a set of 6 ERS acquisitions in ascending mode. Each image is chosen as the reference and spatial and temporal characteristics of the other scenes calculated respectively.

Spatial

	29-May-99	11-Oct-97	10-Oct-97	16-Dec-95	15-Dec-95	30-Jun-93
29-May-99		75	294	81	253	86
11-Oct-97			369	156	178	151
10-Oct-97				213	547	218
16-Dec-95					334	5
15-Dec-95						339
30-Jun-93						

Temporal

	29-May-99	11-Oct-97	10-Oct-97	16-Dec-95	15-Dec-95	30-Jun-93
29-May-99		595	596	1260	1261	2159
11-Oct-97			1	665	666	1564
10-Oct-97				664	665	1563
16-Dec-95					1	899
15-Dec-95						898
30-Jun-93						

Table 3. Spatial and temporal baselines of 6 ERS images acquired in ascending mode. Grey cells are discussed in the text.

The pairs 10/11-Oct-97 and 15/16-Dec-95 have been acquired at one day apart, when the ERS 1 and 2 satellites were orbiting together. Because of this short delay, the scatters on the ground are generally not disturbed (no displacement such as displayed on Figure 5) and thus allow an accurate measurement. The baselines are wide (369 m and 334 m) allowing a detailed description of the topography. One pair is characterized by a very short baseline:

30-Jun-93 and 16-Dec-95 = 5 m. In this case, the interferogram displays essentially ground displacements over a period of 899 days (if they exist).

4. Results

4.1 Digital surface model of the Lisan and features extraction

Figure 7 displays a digital surface model realized by applying InSAR techniques to a "tandem" pair of ERS images acquired in ascending mode 15/16-Dec-95. The perpendicular baseline is quite high with 334 m leading to an "altitude of ambiguity" of 31.5 m. No atmospheric artifact has been found on this pair (Derauw 1999).

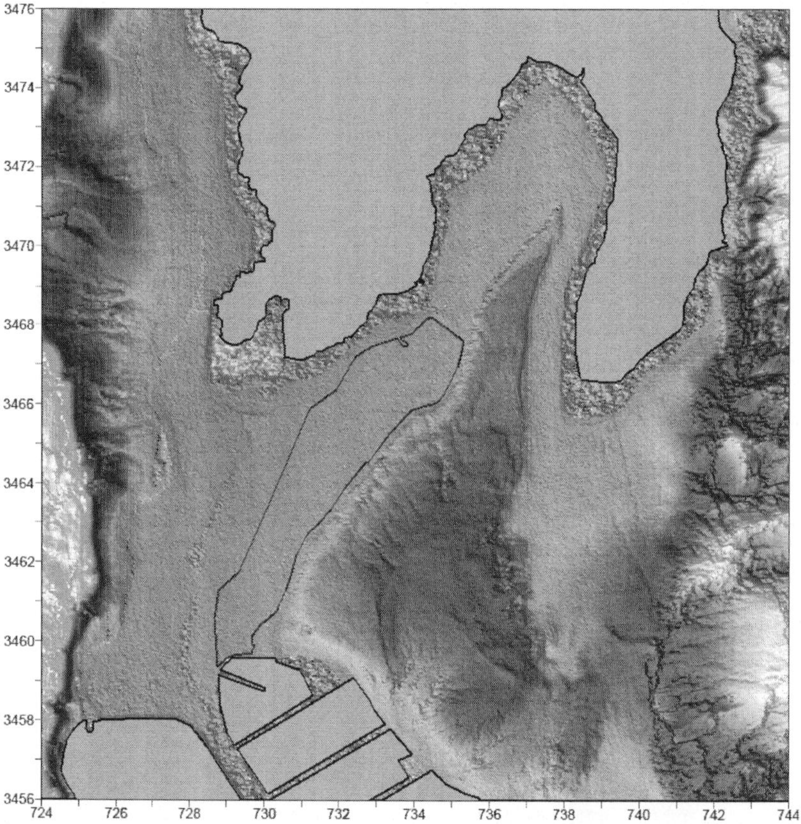

Fig. 7. Geocoded digital surface model of the Lisan area based on the "tandem" pair of ERS images acquired in ascending mode, period: 15-Dec-95 to 16-Dec-95; perpendicular baseline: 334 m; temporal baseline 1 day.

Water surfaces have been masked with data deriving from an acquisition of 2009. As the Dead Sea level is dropping at a rate of about 1 m/yr, the mask does not fit exactly with the water surface of December 1995. The disturbed zone along the shore corresponds to the surface that appeared between 1995 and 2009. West of the Lisan, the external limit of a major saltpan is represented by a black line.

Spatial Analyst tools dedicated to surface and hydrological available in ESRI ArcGIS have been used to extract features such as contours, curvature, slopes, orientations, drainage pattern, watersheds… They allowed a better description and understanding of the geomorphology. The model presented on Figure 7 shows 4 different zones:

- A relatively high land area corresponding to the former peninsula which is an undulated to flat plain with isolated hills (dark to light brown - coord. 734-3462);
- The badlands characterized by steep slopes and deep canyons occurs mainly in the southern part and along the periphery of the high land zone (yellow - 734-3459);
- The low lands occupy the eastern part of the Lisan (yellow-light green - 738-3464);
- The wave-cut platform surrounding the former peninsula in the west, north and northeastern sides (green - 734-3466).

The regional drainage pattern is radial, locally it can be dendritic or parallel such as in the southern and western parts. In all cases the talwegs network reflects the local weakness zones related either to the salt rising or the strike-slip fault displacement. Visual extraction of linear features had been carried out by using a large panel of color palettes emphasing specific topographic contrasts.

The accuracy of the digital surface model is given by the image of the interferometric phase coherence (measure the similarity between the two scenes backscattering) associated with the curve of the theoretical standard deviations of the elevations. Figure 8 presents a standard deviation map based on the coherence image. A quarter of the full scene is represented. Because the estimator of the standard deviations is only valid for the higher coherence values, the computation had been done for the coherence values upper than 0.5. The histogram of the standard deviations shows that the theoretical average accuracy over the scene is about one meter.

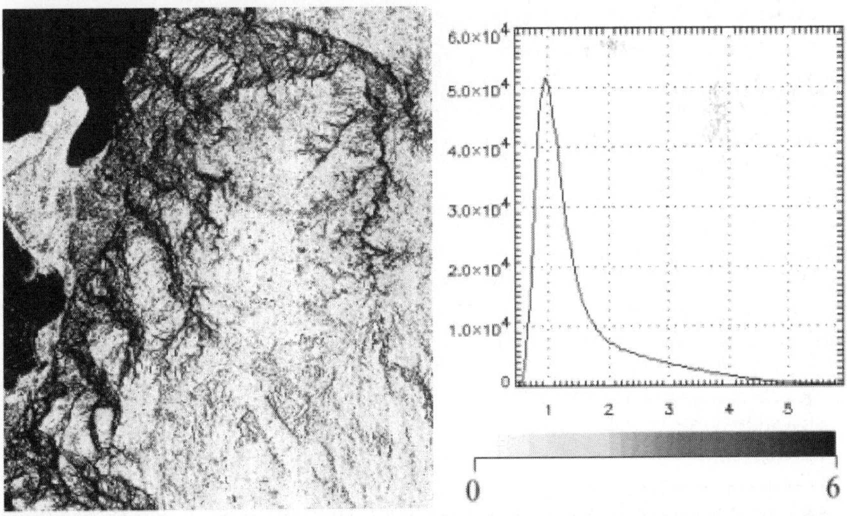

Fig. 8. Standard deviation map based on the coherence image (Derauw, 1999; Closson 2005b).

4.2 Ground displacements

It is impossible to present the whole interferograms that have been computed in this research. As an example, Table 3 shows 15 possible combinations but only 10 are able to bring valuable information. Indeed, four cases lead to a total decorrelation in the interferogram. They are the combibations characterized by the spatial baseline of 547 m and the temporal baseline of 2159, 1564 and 1563 days. Among the whole combinations of the dataset presented in Table 2, a selection of four relevant unwrapped and geocoded interferogram is presented (Figures 9-12). Each case presented in Table 4 emphasizes a particular relation between the spatial- temporal baseline and the type of sensors.

			Spatial – Temporal baselines			
			Small - Short	Small - Wide	Large - Short	Large - Wide
ERS	C- band	Asc				Figure 12
ENVISAT		Dsc		Figure 10		
		Asc				
		Dsc			Figure 11	
ALOS	L- band	Asc	Figure 9			
		Dsc				

Table 4. Characteristics of the various interferograms that can be generated from the available dataset. This table is helpful when comparing results.

4.2.1 Deformation fields with small spatial baseline and short temporal baseline – ALOS data

Figure 9 shows an interferogram computed from two ALOS Palsar scenes acquired the 02-Feb-08 and the 19-Mar-08, i.e. 46 days apart. The normal baseline is 110 m leading to an altitude of ambiguity of 582 m. The Lisan peninsula is only covered by 1/6 of topographic fringe. The wavelength is 23.6 cm (L-band), therefore one fringe cycle will represent a displacement in the line of sigh of 118 mm, about four times more than in the case of ERS.

A total decorrelation (dark purple color) occurred over the high land area (e.g. from 734-3460 to 737-3466) and over the farming zones east of the peninsula (from 739-3461 to 741-3466). The decorrelation over the high land is caused by the very low backscattering of the surface. The smooth (relatively to the wavelength) surface acts as specular reflectors and do not reflect much signal back to the radar.

By comparison with Figure 11, practically no decorrelation occurred in the Lynch Strait, at the exception of the surfaces covered by water (727-3457).

Strong subsidences are found in:

- 728/729-3460/3461, very close to the dike of the major saltpan occupying the western wave-cut platform of the Lisan;
- 730-3468, the delta of wadi Araba which is always aggrading due to the lake level lowering;
- and e.g. in 739-3470, the recently emerged shoreline constitute a ribbon of subsidence buffering the northern Lisan.

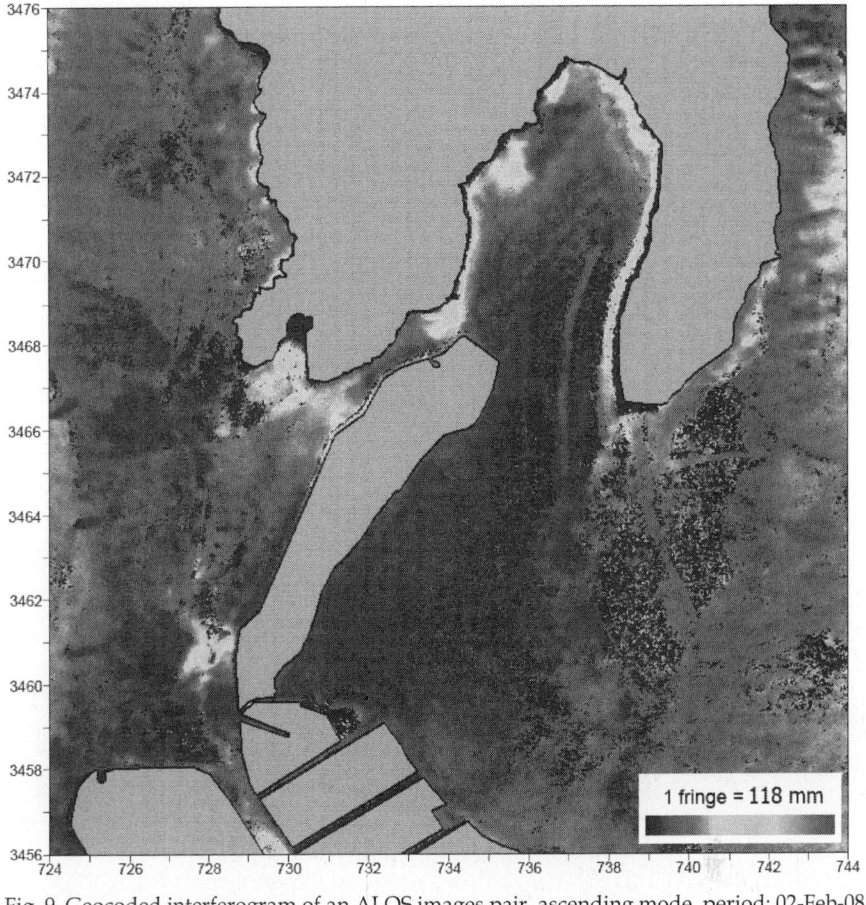

Fig. 9. Geocoded interferogram of an ALOS images pair, ascending mode, period: 02-Feb-08 to 19-Mar-08. Perpendicular baseline: 110 m; temporal baseline 46 days.

4.2.2 Deformation fields with small spatial baseline and wide temporal baseline – ERS data

Figure 10 shows an unwrapped interferogram of an ERS images pair covering the period 29-Jul-95 to 14-Jul-96. The perpendicular baseline is 13 m and the temporal baseline 350 days. The altitude of ambiguity is 770 m. Taking into account that the maximum elevation over the Lisan is -317 m and the minimum -409 m (in Jul-95: -408.7 m; in Jul-96: -409.6; Arab Potash Company gauge station), the Lisan is only covered by 1/8 of topographic fringe (770 m / [-317 m - -409 m]). No atmospheric artifact has been detected by combining the two images with other images of the available dataset (Table 1). The cycle of spectrum color palette corresponding to one fringe indicate that surface displacements occurred all over the peninsula (see Figure 13 for details) during the period of observation. Each cycle represents a ground displacement in the line of sight with amplitude of 28 mm, i.e. half of the ERS-AMI sensor wavelength. The main zones of interest are located below:

- 735-3466, 736-3468, 736-3469, and 737-3470, four uplift areas have affected the northern part of the Lisan;
- 737-3472, 738-3472, 736-3461, 737-346, 735.5-3459, 734.5-3458, 737-3458 correspond to seven kilometric shallow subsidence areas;
- the noisy surfaces (dark blue), mainly east and west of the peninsula, correspond either to agricultural crops in the graben of Mazra'a - Ghor Al Haditha or to soil moisture, especially in the southern Lynch Strait area. It is also suspected that in several other zones, decorrelation is related to ground displacements whose amplitude is too important to be detected in C-band.

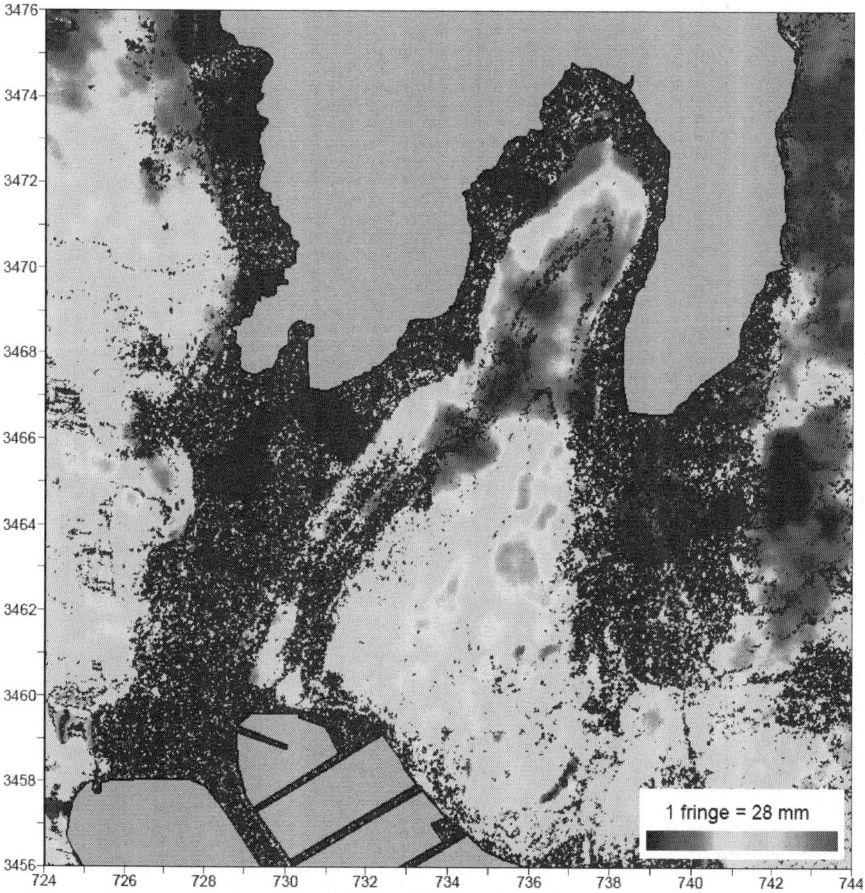

Fig. 10. Geocoded interferogram of an ERS images pair, descending mode, period: 29-Jul-95 to 14-Jul-96. Perpendicular baseline: 13 m; temporal baseline 350 days.

4.2.3 Deformation fields with large spatial baseline and short temporal baseline – ENVISAT data

Figure 11 shows a differential interferogram computed for the period 23-Mar-08 to 27-Apr-08. It extends the period of observation of Figure 9 (02-Feb-08 to 19-Mar-08) but in C-band rather than in L-band. Table 2 indicates that the fringes' rate of Envisat/ERS interferograms are four times greater than the one of ALOS and the general accuracy is 1.6 times less for ALOS too. The normal baseline is 342 m and the altitude of ambiguity 19 m. The topographic phase component had been retrieved from the interferogram with the SRTM data (digital surface model acquired in February 2000). Therefore, the coastal area that appeared between 2000 and 2008 (about 8 m of bathymetric elevation) does not present correct values. As expected, the deformation fields displayed on Figure 11 are in agreement with ones of Figure 9. It is worth noting that the loss of coherence is more important for ALOS than for ENVISAT because of the sensitivity of L-band to the roughness of the ground of the Lisan.

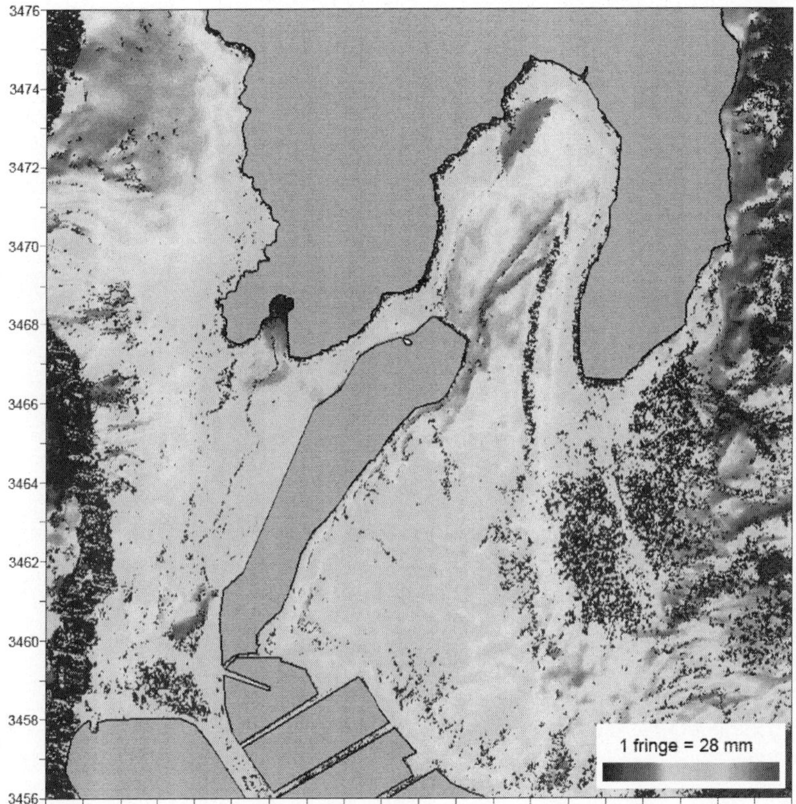

Fig. 11. Geocoded interferogram of an ENVISAT images pair, descending mode, period: 23-Mar-08 to 27-Apr-08. Perpendicular baseline: 342 m; temporal baseline 35 days.

4.2.4 Deformation fields with large spatial baseline and large temporal baseline – ERS data

Figure 12 shows an interferogram spanning over about 3.5 years. The deformations information recorded on Figure 10 are thus included here and fill out with data covering the period not included in Figure 10. The coherence decreased in many places due to the industrial activities over the wave-cut platform. A large part of the northern peninsula had been used as quarries during the period of observation. The zones of interest are located in:

- 732-3460, a large uplift area affected the southwestern part of the Lisan;
- 736/738-3472, kilometric shallow subsidences have affected the northern part of wave-cut platform surrounding the peninsula;
- 736-3461, 734-3459, 737-3458, curvilinear subsidences;

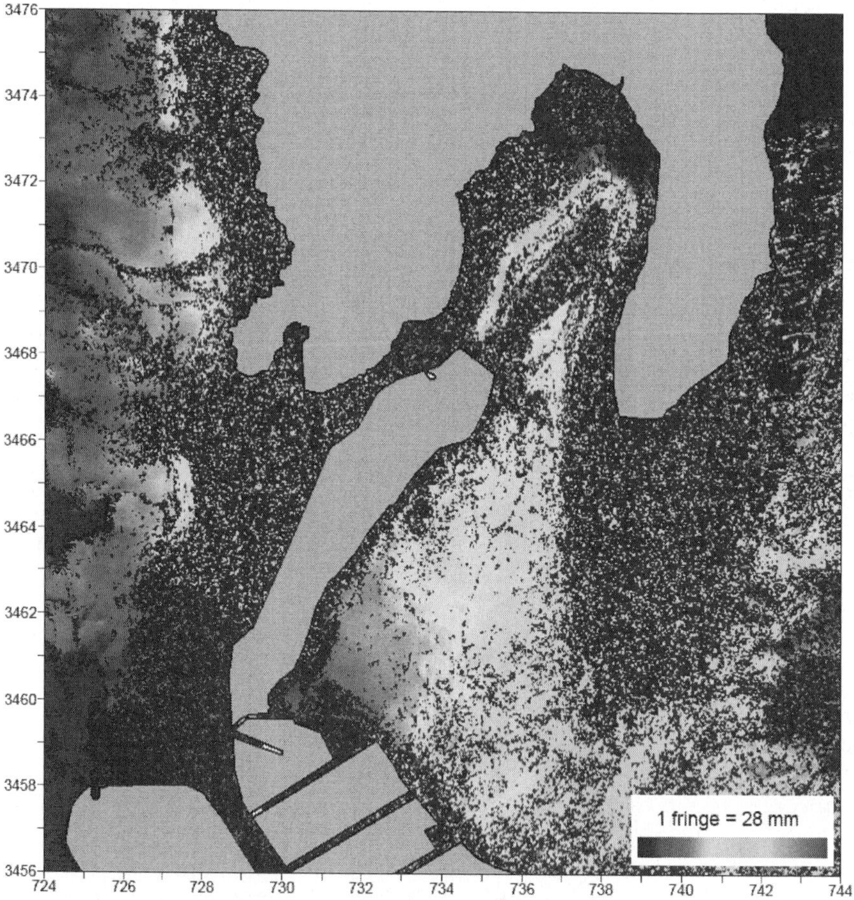

Fig. 12. Geocoded interferogram of an ERS images pair, ascending mode, period: 12-Dec-95 to 29-May-99. Perpendicular baseline: 221 m; temporal baseline 1261 days.

5. Discussion

The knowledge gained from the differential interferogram analysis are the rates of vertical displacements, the location of uplifted and subsidence areas, their spatial extension and evolution through the period of observation. The investigation is done by comparing time series of interferograms and of cross sections such as illustrated in Figure 13. Six cross sections through the interferogram displayed on Figure 10 have been drawn together with their projection over a hillshaded view of the digital surface model displayed in Figure 7.

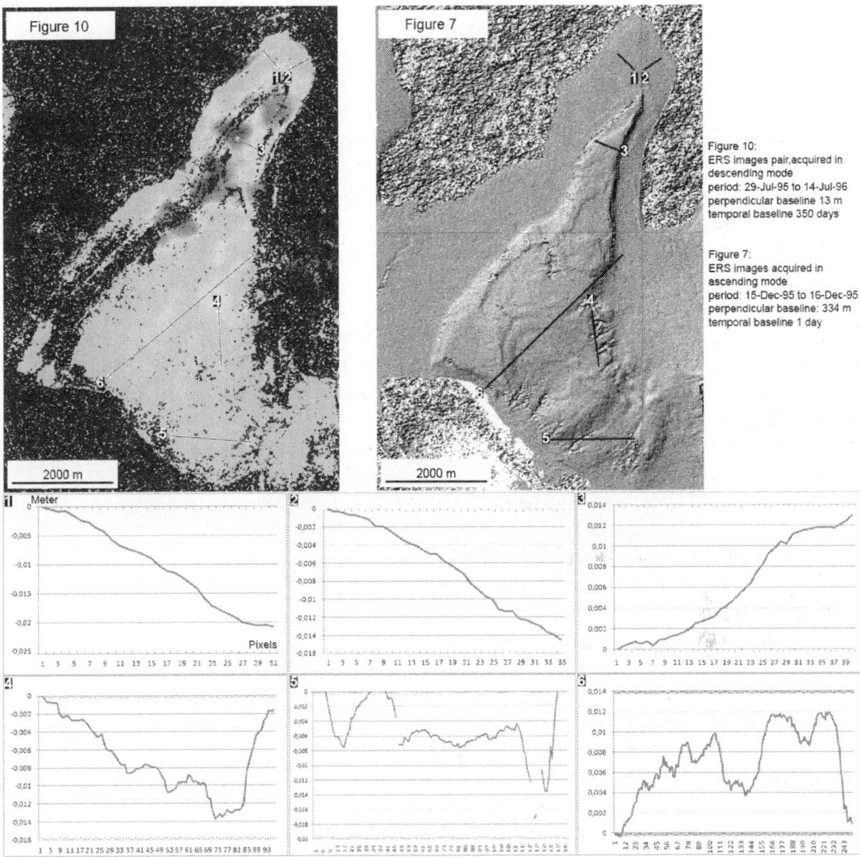

Fig. 13. Evaluation of displacements through cross sections. The ERS Inferferogram is characterized by a perpendicular baseline of 13 m and a temporal baseline of 350 days. The altitude of ambiguity is 770 m and the Lisan is only covered by 1/8 of topographic fringe.

Due to the limited spatial extent of the Lisan area it was not possible to define a reference point far away from the salt diapir which could be considered as a zero phase value for the phase unwrapping process. In consequence, to make comparable the phase variations both in space and time within the Lisan area we have considered arbitrarily the reference sites at each starting point of the cross sections for which the observed deformation is assumed to be minimal over the studied period.

To avoid this problem, one strategy - not realized in this work - consists to create a reference phase value for each starting point by averaging the unwrapped phase values from the surrounding pixels. This minimal deformation phase value is then used as a common reference for the whole time series of unwrapped interferograms. By this process, the resulting field deformation maps give a relative measure of the ground displacement with respect to this fixed reference point.

5.1 Vertical movements affecting the northern Lisan wave cut platform in relation with the Dead Sea level lowering

The interferograms computed with the image dataset (Table 1) have shown two distinct kilometric areas of subsidence affecting the edge of the northern Lisan wave cut platform. The most extended one is sinking northwestward, the second northeastward (see cross sections 1 and 2, Figure 13). A hectometric sill where ground movements are less important separates them. The location of this sill and of lineaments over the wave cut platform suggested that a major fault N-S oriented crosses this zone (Closson et al., 2003; Closson et al., 2005).

Cross sections 1 and 2 bring knowledge of the vertical movements (X-scale bar is expressed in pixels while Y-scale bar is in meter). The pixel size of the interferogram is 20 meters and the cross sections are nearly similar in length. Cross section 1 indicates that pixel n°31 has been lowered by about 20 mm compared to pixel n°1 between 29-Jul-95 and 14-Jul-96. In the same way, cross section 2 informs that pixel n°35 dropped by 14 mm compared to pixel n°1. Subsidence in area 1 was more active than subsidence zone 2 during this period of 350 days.

Other measurements have confirmed that the northern part of the wave-cut platform was in strong subsidence during the 1990s. For example, based on the interferogram displayed on Figure 12, subsidence measurements of 71 mm (area 1) and 70 mm (area 2) were recorded over a period of 1261 days (from 12-Dec-95 to 29-May-99) leading to a rate of 20 mm/year in both areas.

It is worth to be mentioned that during the 1990s, the Arab Potash Company built two major salt evaporation ponds (USD 70M) all along the western Lisan wave cut platform in order to meet its increasing fertilizer production needs. The northernmost unit was destroyed in March 2000 during filling operation. About 1650 m of earthen dike disappeared during the rapid emptying of 55 million cubic meters of Dead Sea brine (Figure 14). The fact is that the place where the dike started to break up is precisely at the location where a maximum of subsidence was recorded (e.g. during the period 29-Jul-95 to 14-Jul-96, see Figure 13 - cross section 1). Fieldwork conducted between 2004 and 2009 in this area have clearly shown that in tens of places the remaining dike is fissuring at an increasing rate. Some flanks of the dikes have also slid, proving the existence of significant ground motions during the present decade since rainfall is a marginal phenomena. In 2010, the southern saltpan still exists. Its external limit had been reproduced on Figure 2 (saltpan, coord. 732-3465). It had been repaired during five years (USD 16M), between 2001 and 2006, and is under continuous surveillance. Fieldwork revealed that sinkholes pierced the earthen dike and the bottom of this saltpan (Closson & Abou Karaki, 2009). In some places, engineers tripled the wide of hectometric dike segments to increase the safety factor... In the area of cross section 2, consolidation and drainage work have been realized at the end of the 1990s until the day of the saltpan destruction. They are still visible in 2010 and attest of the difficulties encountered to stabilize the salt evaporation pond.

Long term and short term measurements carried out with ALOS and ENVISAT data (Table 1) support field observations and above all show a drastic increase in the rate of subsidence by comparison with the rates recorded in the 1990s. For example, Figure 14 presents the result of a cross section in area 1. The rate of subsidence is four times greater than the one observed e.g. by Baer et al. (2002) or by Shimoni et al (2002): 80 mm/year between 18-Dec-2007 to 20-Dec-2008 in place of 20 mm/year between 29-Jul-95 and 14-Jul-96 or between 12-Dec-95 and 29-May-99. By comparison, in area 2, the rate is 25% less than in area 1, i.e. 60 mm/year between 18-Dec-2007 to 20-Dec-2008. In general, the rates of subsidence in area 1 and 2 are quite important but area 1 seems a more active than area 2. This could explain why the dike collapsed in the area 1 in place of area 2.

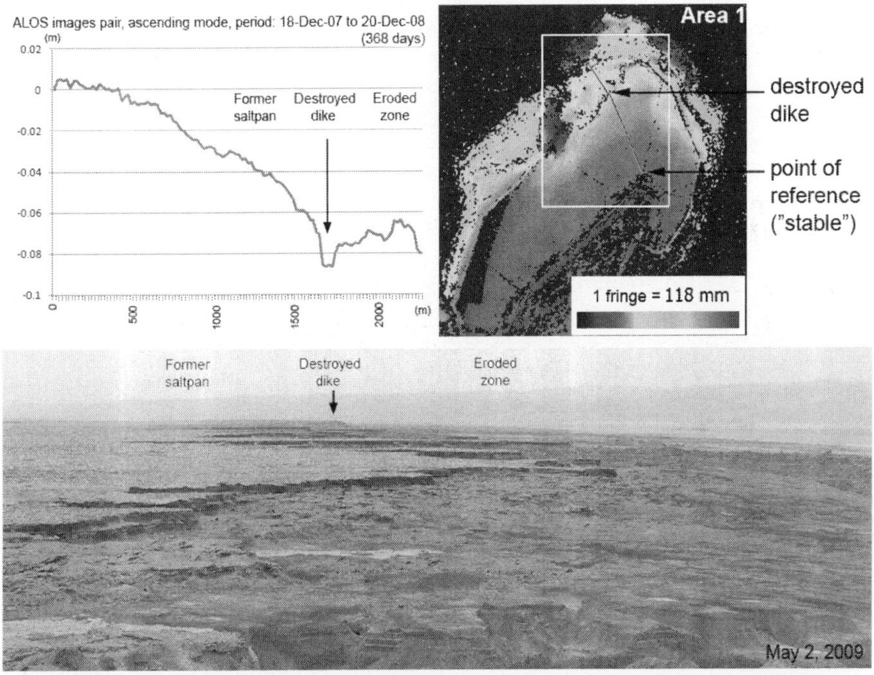

Fig. 14. Area 1 - ground movements detected by ALOS during the period of observation from 18-Dec-2007 to 20-Dec-2008. Northern Lisan, picture of the area where a dike was destroyed in March 2000 (USD 38M). View from the northern remaining dike segment toward SW direction.

The ground displacement observed are localized along coastal zones recently emerged. There, ground-water lowers following the Dead Sea level drop (Closson et al., 2007). As a consequence, salty-marls/argile compaction phenomena occur and trigger land subsidence (Baer et al., 2002). Differential land subsidence produced extensive areas of earth fissures which affect earthen dikes of the Arab Potash Company. Other effects include an increased landslide hazards (Closson et al., 2009).

Shimoni et al. (2002), Closson et al. (2003) and Closson (2005b) have shown that the northern subsidence areas are also at the periphery of rising sub-domes (e.g. cross section 3). This geographical proximity suggested a possible connection between the two phenomena. However this assumption is not supported by the drastic increase in the subsidence rates revealed by ALOS and ENVISAT data in 2008. The increasing rate of subsidence is following a similar trend in the rate of sinkholes proliferation (Closson et al., 2010) which is connected with the Dead Sea level lowering.

5.2 Uplifts and subsidence in relation with the Lisan salt diapir

Cross sections 3-6 (Figure 13) illustrate the surface displacements caused by differential rising rates in the central and southern Lisan. Cross section 3 characterizes a sub-dome rising at a rate of 13 mm/year for the period of observation. It is representative of four sub-domes existing in this area (see locations on Figure 10: 735-3466, 736-3468, 736-3469, and 737-3470). At the opposite, cross section 4 and 5 characterize subsidence linear and curvilinear structural features.

Cross section 4 focuses on a linear zone oriented N-S presenting a subsidence of 14 mm/year at the intersection with another linear structure oriented E-W. The minima areas such as the one discussed here are generally surrounded by several rising sub-domes (e.g. Figure 10: yellow spots surrounding blue/cyan spots in 736/737-3461).

Cross section 5 concerns a wide sub-dome in the southern Lisan (Figure 15, inset).

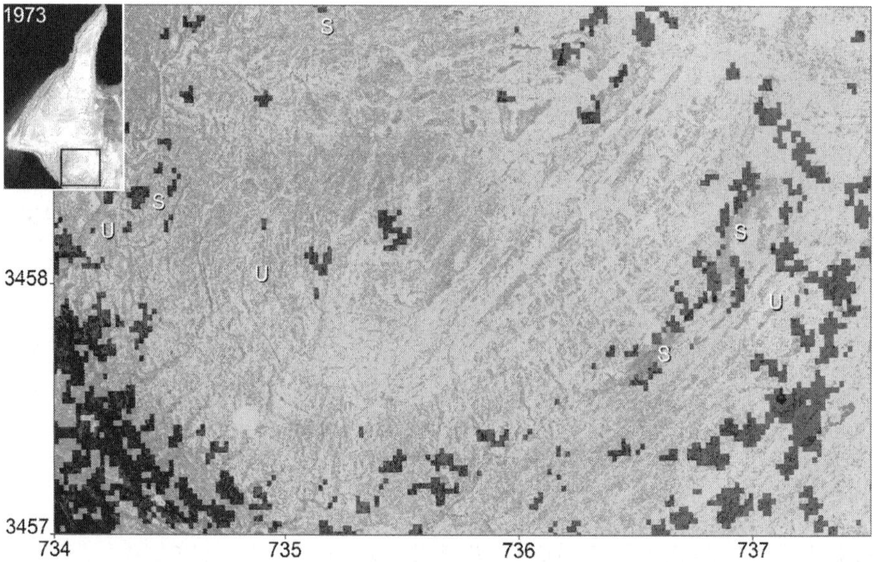

Fig. 15. Uplift and subsidence areas (U/S letters) are superimposed to a Corona image acquired on August 5, 1970. ERS interferogram covers the period 29-Jul-95 to 14-Jul-96. Dark blue pixels correspond to decorrelated areas. Inset: Landsat image of the Lisan acquired in 1973.

Background image is a high resolution Corona photograph acquired in August 5, 1970. A curvilinear structure characterized by blue colors is centered over the place of the maximum vertical thrust (735.5-3458). Yellow and cyan/blue colors derive from the interferogram displayed in Figure 10. They correspond relatively to uplifts (U) and subsidence (S). Subsidence appears at the synclinal area between the flanks of the folds. They attest that this dome is active.

Cross section 6 (Figure 13) brings an overview of the displacement through the Lisan. At least, it illustrates the great complexity of movements affecting the sub-domes of the Lisan diapir structure. In terms of ground displacements, the surface of the salt dome can be compared as a bubbling surface. It is particularly obvious over Figure 10 while Figure 9 shows less evidence of surface displacements. When comparing displacement fields of Figure 11 and 12, one can observe respectively that the northern Lisan is uplifted over a very short period of time (35 days) and the southwestern part records most of the rise during a period of 1261 days. The variations of speeds are important in space and time. It supports the idea that the Lisan diapir uplift is episodic and not monotonous. This observation was already mentioned by Shimoni et al. (2002) and had been confirmed by the processing of ALOS and ENVISAT data for the period 2007-2008. The main ground movement activities are localized in the northern part of the Lisan (uplift) and in the southern Lynch Strait area (subsidence, Closson et al., 2010b). A possible explanation to explain this increase of activity could be that 90M cubic meters of Dead Sea brine have been introduced in 2006 in a saltpan repaired during the period 2001-2005. This load in an area recognized as relatively unstable during the 1990s, could have reactivated numbers of concealed faults and fractures.

6. Summary and conclusions

The problem of the measurement of ground displacements caused by the rising of a salt diapir in the region of the Lisan, Dead Sea, Jordan had been tackled by applying radar interferometry techniques to 27 satellite images acquired in 1992-1999 and 2007-2008.

Baer et al. (2002) and Shimoni et al. (2002) have studied the Lisan by applying the same techniques to 16/17 ERS images (C-band). Their work covered the period 1992-2000. Here, two additional sensors have been used: ALOS (L-band) and ENVISAT (C-band) satellites. The superimposition of radar phase images acquired by ERS, ALOS and ENVISAT satellite radar from different epochs have been compared, resulting in the precise detection of mm scale vertical changes from observed interference fringes.

The processing of ERS data led to the same conclusions than Shimoni et al. (2002). The rate of the deformation over the Lisan is irregular during the 1990s. Cross-sections have shown that the highest deformation occurred in the northern part of the peninsula, at the place where one production unit of the Arab potash Company was destroyed in March 2000. Most of the zones in subsidence are observed in the peninsula periphery while the main uplifted areas are located in the central and southern parts.

The comparison with the measures realized with the dataset of 2007-2008 shows a drastic increase in the rate of subsidence at the periphery of the area of interest. The rising of the northern part of the peninsula is more pronounced than the one of the south and central parts. ALOS data revealed that the most active part during the period 2007-2008 was located in the Lynch Strait area. The use of L-band sensor allows the detection of ground displacements leading to decorrelation in C-band and C-band preserve coherence in places where L-band decorrelate due to specular reflection.

7. References

Abou Karaki, N. (1987). *Synthèse et carte sismotectonique des pays de la bordure de la Méditerranée: sismicité du système de faille du Jourdain - Mer Morte*. Université Louis Pasteur, Strasbourg, France.

Al-Zoubi, A. & ten Brink, U.S. (2001). Salt Diapir in the Dead Sea and their Relationship to Quaternary Extensional Tectonics. *Marine and Petroleum Geology*, 18, 7, August 2001 779-797, ISSN: 0264-8172.

Arieh, E.; Rotstein, Y. & Peled, U. (1982). The Dead Sea Earthquake of April 1979. *Bulletin of the Seismological Society of America*, 72, 5, October 1982, 1627-1634, ISSN: 0037-1106.

Baer, G.; Schattner, U.; Wachs, D.; Sandwell, D.; Wdowinski, S. & Frydman, S. (2002). The Lowest Place on Earth is Subsiding - An InSAR (Interferometric Synthetic Aperture Radar) Perspective. *Geological Society of America Bulletin*, 114, 1, January 2002, 12-23. ISSN: 0016-7606.

Bartov, Y. (1999). *The Geology of the Lisan Formation in Massada Plain and the Lisan Peninsula*. The Hebrew University of Jerusalem, Israel.

Bartov, Y.; Stein, M.; Enzel, Y.; Agnon, A.; Reches, Z. (2002). Lake-levels and sequence stratigraphy of Lake Lisan, the late Pleistocene precursor of the Dead Sea. *Quaternary Research*, 57, 1, January 2002, 9–21, ISSN: 0033-5894.

Bartov, Y. (2004a). Paleolakes at the Dead Sea basin. Chapter 2. In: *Fungal life in the Dead Sea (Biodiversity of cyanoprokaryotes, algae and fungi of Israel)*, Nevo, E., Oren, A., Wasser, S., (Ed). Gantner Verlag, 325 pp, ISBN: 3-906166-10-4.

Bartov, Y. (2004b). *Sedimentary fill analysis of a continental basin – the Late Pleistocene Dead Sea*. Ph.D. thesis, Hebrew University, Jerusalem.

Bartov, Y. & Sagy, A. (2004). Late Pleistocene extension and strike-slip in the Dead Sea. *Geological Magazine*, 141, 5, September 2004, 565–572. ISSN: 0016-7568.

Bartov, Y. ; Enzel, Y. ; Stein, M. ; Agnon, A. (2004). Transgressional erosion - implication on the Holocene uplift rate of the Lisan diapir. *Geological Society of Israel Annual Meeting*, Abstracts, p. 12.

Bartov, Y.; Agnon, A.; Enzel, Y. & Stein, M. (2006). Late Quaternary faulting and subsidence in the central Dead Sea basin. *Israel Journal of Earth Sciences*, 55, 1, 18–31, ISSN: 0021-2164.

Ben-Avraham, Z. (1997). Geophysical Framework of the Dead Sea: Structure and Tectonics, In: *The Dead Sea, the Lake and its Setting*, Niemi T.M., Ben-Avraham Z., Gat J. (Ed), 22-35, Oxford University Press, ISBN-10: 0195087038, ISBN-13: 978-0195087031, USA.

Ben-Avraham, Z. & Lazar, M. (2006). The structure and development of the Dead Sea basin: recent studies. In: *New frontiers in Dead Sea paleoenvironmental research: Geological Society of America Special Paper 401*, Enzel, Y., Stein, M., Agnon, A., (Ed.), 1-13, Geological Society of America Special Papers, ISBN 0-8137-2401-5, Boulder CO.

Bender, F.K. (1974). *Geology of Jordan* (supplementary edition of volune 7). Gebrüder Borntraeger, ISBN: 3-443-11707-4, Berlin-Stuttgart, Germany.

Closson, D.; Abou Karaki, N.; Hansen, H.; Derauw, D.; Barbier, C. & Ozer, A. (2003a). Space-borne Radar Interferometric Mapping of Precursory Deformations of a Dyke Collapse - Dead Sea Area - Jordan. *International Journal of Remote Sensing*, 24, 4, February 2003, 843-849, ISSN 0143-1161.

Closson, D.; Abou Karaki, N.; Hussein, M. J.; Al-Fugha, H.; Ozer, A. & Mubarak, A. (2003b). Subsidence and Sinkholes along the Jordanian Coast of the Dead Sea: Contribution of Gravimetry and Radar Differential Interferometry. *Comptes Rendus Geosciences*, 335, 12, October 2003, 869-879, ISSN 1631-0713.

Closson, D.; Abou Karaki, N.; Klinger, Y. & Hussein, M.J. (2005). Subsidence Hazards Assessment in the Southern Dead Sea Area, Jordan. *Pure and Applied Geophysics*, 162, 2, 221-248, ISSN 0033-4553.

Closson, D. (2005a). Structural Control of Sinkholes and Subsidence Hazards along the Jordanian Dead Sea Coast. *Environmental Geology*, 47, 2, 290-301, ISSN: 0943-0105.

Closson, D. (2005b). *Cooccurrence between the Geo-Hazards Induced by the Dead Sea Level Lowering and the Geological Setting - Lisan Peninsula, Lynch Strait, Ghor Al Haditha - Jordan*. PhD Thesis, University of Liege, Belgium.

Closson, D.; LaMoreaux, P.E.; Abou Karaki, N. & Al-Fugha, H. (2007). Karst system developed in salt layers of the Lisan Peninsula, Dead Sea, Jordan. *Environmental Geology*, 52, March 2007, 155–172, ISSN: 0943-0105.

Closson, D. & Abou Karaki, N. (2009a). Human-induced geological hazards along the Dead sea coast. *Environmental Geology - Special Issue*, 58, 2, July 2009, 371-380, ISSN: 0943-0105.

Closson, D. & Abou Karaki, N. (2009b). Salt karst and tectonics: sinkholes development along tension cracks between parallel strike-slip faults, Dead Sea, Jordan. *Earth Surface Processes and Landforms*, 34, August 2009, 1408-1421. ISSN: 0197-9337.

Closson, D.; Abou Karaki, N. & Hallot, F. (2010a). Landslides along the Jordanian Dead Sea coast triggered by the lake level lowering. *Environmental Earth Sciences*, 94, February 2010, 1417-1430, ISSN: 1866-6280.

Closson, D.; Abou Karaki, N.; Milisavljević, N.; Hallot, F. & Acheroy, M. (2010b). Salt-dissolution-induced subsidence in the Dead Sea area detected by applying interferometric techniques to ALOS Palsar Synthetic Aperture Radar images. *Geodinamica Acta*, 23, 65-78. ISSN: 0985-3111.

Derauw, D. (1999). *Phasimétrie radar à synthèse d'ouverture: théorie et applications*. PhD Thesis, University of Liege, Belgium..

Diabat, A.A. (2005). Sinkholes related to Tectonic Factor at Ghor Al Haditha Area, Dead Sea/Jordan. *Hydrogeologie und Umwelt*, 33, 1-17, ISSN 09309- 3757.

Ferry, M.; Meghraoui, M.; Abou Karaki, N.; Al-Taj, M.; Amoush, H.; Al-Dhaisat, S. & Barjous, M. (2007). A 48-kyr-long slip rate history for the Jordan Valley segment of the Dead Sea Fault. *Earth and Planetary Science Letters*, 260, 3-4, 30 August 2007, 394-406, ISSN 0012-821X.

Freund, R.; Zak, I. & Garfunkel, Z. (1968). Age and Rate of the Sinistral Movement along the Dead Sea Rift. *Nature*, 220, 19 October 1968, 253 – 255, ISSN : 0028-0836.

Freund, R.; Garfunkel, Z.; Zak, I.; Goldberg, M.; Weisbrod, T. & Derin, B. (1970). The Shear along the Dead Sea Rift. *Philosophical Transactions Royal Society London*, 267, 29 October 1970, 107-130, ISSN: 0080-4614.

Gardosh, M.; Reches, Z. & Garfunkel, Z. (1990). Holocene Tectonic Deformation along the Western Margins of the Dead Sea. *Tectonophysics*, 180, 123-137, ISSN: 0040-1951.

Gardosh, M.; Kashai, E.; Salhov, S.; Shulman, H. & Tannenbaum, E. (1997). Hydrocarbon Exploration in the Southern Dead Sea Area, In: *The Dead Sea, The Lake and its Setting*,

Niemi T.M., Ben-Avraham Z., Gat J. (Ed). 57-72, Oxford University Press, ISBN-10: 0195087038, ISBN-13: 978-0195087031, USA.

Garfunkel, Z.; Zak, I. & Freund, R. (1981). Active Faulting in the Dead Sea Rift. *Tectonophysics*, 80, 1-4, 10 December 1981, 1-26, ISSN: 0040-1951.

Garfunkel, Z. & Ben Avraham, Z. (1996). The Structure of the Dead Sea Basin. *Tectonophysics*, 266, 1-4, 15 December 1996, 155-176, ISSN: 0040-1951.

Garfunkel, Z. (1997). The History and Formation of the Dead Sea Basin, In: *The Dead Sea, the Lake and its Setting*, Niemi T.M., Ben-Avraham Z., Gat J. (Ed), 36-55, Oxford University Press, ISBN-10: 0195087038, ISBN-13: 978-0195087031, USA.

Ginat, H.; Enzel, Y. & Avoni, Y. (1998). Translocated Plio-Pleistocene Drainage Systems Along the Arava Fault of the Dead Sea Transform. *Tectonophysics*, 284, 1-2, 15 January 1998, 151-160, ISSN: 0040-1951.

Hall J.K. (1997). Topography and Bathymetry of the Dead Sea Depression, In: *The Dead Sea, the Lake and its Setting*, Niemi T.M., Ben-Avraham Z., Gat J. (Ed), 11-21, Oxford University Press, ISBN-10: 0195087038, ISBN-13: 978-0195087031, USA.

Klinger, Y.; Avouac, J.P.; Abou Karaki, N.; Dorbath, L.; Bourles, D. & Reyss, J.L. (2000). Slip-Rate on the Dead Sea Transform Fault In Northern Araba Valley (Jordan). *Geophysical Journal International* 142, 3, September 2000, 755-768, ISSN 0956-540X .

Neev, D. & Emery, K.O. (1967). The Dead Sea - Depositional Processes and Environments of Evaporates. *Geological Survey of Israel Bulletin*, 41, 1-147.

Niemi, T.; Zhang, H.; Atallah, M. & Harrison, J.B. (2001). Late Pleistocene and Holocene Slip Rate of the Northern Wadi Araba Fault, Dead Sea Transform, Jordan. *Journal of Seismology*, 5, 3, 449-474, ISSN: 1383-4649.

Pe'eri, S.; Wdowinski, S.; Shtibelman, A.; Bechor, N.; Bock, Y.; Nikolaidis, R. & Van Domselaar, M. (2002). Current Plate Motion across the Dead Sea Fault from three Years of Continuous GPS Monitoring. *Geophysical Research Letter*, 29, 14, 26 July 2002, 42/1 - 42/4, ISSN 0094-8276.

Quennell, A. (1958). The Structural and Geomorphic Evolution of the Dead Sea Rift. *Quarterly journal of the Geological Society of London*, 114, 1-4, 1-24, ISSN 0016-7649.

Shimoni, M.; Hanssen, R.F.; Van Der Meer, F.; Kampes, B.M. & Ben Dor, E. (2002). Salt Diapir Movements Using SAR Interferometry in the Lisan Peninsula, Dead Sea Rift. *Proceedings of SPIE* , pp. 151-160, ISBN 0-8194-4268-2, Toulouse, September 17-18, 2001, SPIE, Bellingham WA, USA.

Sunna, B.F. (1986). The Geology of Salt Deposits in the Lisan Peninsula-Dead Sea. *Seminar on salt in the Arab World*, Ministry of Energy and Mineral Resources, Natural Resources Authorities, May 4-6, 1986, Amman, Jordan.

Weinberger, R.; Begin, Z.B.; Waldmann, N.; Gardosh, M.; Baer, G.; Frumkin, A. & Wdowinski, S. (2006). Quaternary rise of the Sedom diapir, Dead Sea basin. In: *New frontiers in Dead Sea paleoenvironmental research: Geological Society of America Special Paper 401*, Enzel, Y., Stein, M., Agnon, A., (Ed.), 33–51, Geological Society of America, , ISBN 0-8137-2401-5, Boulder CO.

Zak, I. (1967). *The Geology of Mount Sedom*. Hebrew University of Jerusalem, Jerusalem, Israel.

Part 7

Tectonics and Petroleum

13

Mantle-like Trace Element Composition of Petroleum – Contributions from Serpentinizing Peridotites

Peter Szatmari[1], Teresa Cristina O. da Fonseca[1] and Norbert F. Miekeley[2]
[1]*Petrobras Research Center (CENPES), Rio de Janeiro, Brazil 21941-915,*
[2]*PUC-Rio, Rio de Janeiro, Brazil*

1. Introduction

The origin of trace metals in petroleum is intimately connected with the origin of petroleum itself, one of the most exciting topics in chemistry and geology for the past 150 years. Because of its fluidity and consequent migration, the origin of petroleum is more difficult to interpret than that of most rocks. Biogenic and abiogenic models developed side by side, each championed by some of the greatest names in chemistry and geology. According to biogenic models, petroleum forms by thermal breakdown, in the presence of water, of complex, ultimately biogenic organic polymers (kerogen) (Tissot & Welte, 1984; Lewan, 1997). Conversely, according to abiogenic models, petroleum forms by reductive polymerization of simple carbon compounds such as CO and CO_2. Experiments and calculations indicate that such polymerization is thermodynamically stable only at high temperatures and pressures (above 1300°C and 30 kbar; Kenney *et al.*, 2002) but it takes place also at low temperatures (<400°C) and low pressures in the presence of metallic hydrogenation catalysts in the Fischer-Tropsch-type synthesis, FTTS (Fischer, 1926; Szatmari, 1989).

Although the biogenic, organic model has been the one generally accepted by the petroleum industry almost since its birth, abiogenic, inorganic models recurrently emerge, proposed by geologists and, more often, chemists.

Major contribution to oil deposits of hydrogen and hydrocarbon liquids formed by FTTS during serpentinization was suggested by Szatmari (1989). The process was demonstrated by the discoveries of the Rainbow (Holm and Charlou, 2001) and Lost City (Kelley *et al.*, 2001; Früh-Green *et al*, 2003; Kelley, 2005; Proskurowski *et al.*, 2008; Bach and Früh-Green, 2010) hydrothermal fields at the Mid-Atlantic ridge, where hydrothermal plumes rising over serpentinizing peridotites contain abundant hydrogen, methane, and aliphatic hydrocarbons in the C_{16}-C_{29} range, formed by FTTS. This finding, together with the discovery of aliphatic and cyclic hydrocarbons in the Orgueuil meteorite and, more recently, in the pristine Tagish Lake meteorite (Pizzarello *et al.*, 2001; Nakamura et al., 2003), as well as new studies on hydrocarbon stability at high P-T conditions (Kenney *et al.*, 2002), gave renewed stimulus to the abiogenic models. The Tagish Lake meteorite is a new type of water and carbon-rich Type 2 carbonaceous chondrite that contains about 5% total carbon, of which the organic carbon reaches 1.3 wt. %; the temperature of the meteorite has not risen above 120°C since the formation of the organics (Brown et al.,2000; Nakamura et al., 2003).

2. Trace metals in oils: Sedimentary-biogenic source

The most abundant trace elements in petroleum are V, Ni, and Fe. Treibs (1934, 1935) was the first to demonstrate the presence of metalloporphyrins of V and Fe in crude oil, bitumens, coals and shales; those of Ni were identified later. The study of trace metals in petroleum was early extended to elements other than V and Ni, whose molecular species were still largely unknown (Filby & Olsen, 1994). Ball *et al.* (1960) analyzed 28 trace elements in 24 U.S. oils. Filby & Shah (1975) determined a wide range of trace metals using neutron activation analysis. Cr, Fe, Co, Cu, Zn, and Hg were found in Tertiary California oil to occur in non-porphyrin form, similar to that of Ni and V, incorporated into the asphaltene sheets through complexing at holes bordered by N, S, or O atoms (Filby, 1975). Jones (1977) compiled data for 29 metals in oils worldwide. Curiale (1987) analyzed ten transition metals in 82 heavy crude oils and 21 solid bitumens from unspecified locations worldwide.

Major works on metals in petroleum include the pioneering contribution of Yen and coworkers (1975), analytical studies by Jones (1977), reviews by Valkovic (1978), Barwise & Whitehead (1983), and extensive studies by Filby and coworkers (1975, 1987, 1994).). Lewan & Maynard (1982) and Lewan (1984) interpreted V-Ni proportionality in petroleum by the varying availability of these metals controlled by varying pH-E_H in the sedimentary environment. Curiale (1987, 1993) pointed to the importance of the biota as a possible source of the metals. Modern techniques of ICP-MS and others resulted in an abundance of data, still incompletely interpreted.

Treibs (1934, 1935) showed that these porphyrins derive from bacterial and plant chlorophylls, and to a lesser extent from hemins, which during sedimentation lose their original chelating metals (respectively Mg and Fe) and become chelated by V and other metals. 2 to 54% of the vanadium and 1 to 47% of the nickel in crude oils was reported to occur as extractable metalloporphyrins (Filby & Van Berkel, 1987), while the rest is present in other, non-extractable, largely unknown organic structures. These may include metalloporphyrins strongly associated or incorporated into the asphaltenes so that they cannot be separated; metal naphthenates; metal species produced by the decomposition of metalloporphyrins during oil maturation; and metals complexed into the asphaltene structure by asphaltene functionalities (Filby, 1994).

The source of V, Ni, and Fe in the oils was also suggested to lie in the fossil organic matter (Treibs 1934, 1935; Yen, 1975), because average transition metal concentrations in crude oil and marine organisms are generally close, within an order of magnitude of one another (Jones 1977, Curiale, 1987), whereas transition metal concentrations in seawater are several (three or more) orders of magnitude lower, less than 1 ppb for V and 0.48 ppb for Ni (Quinby-Hunt & Turekian, 1983).

Lewan & Maynard (1982) and Lewan (1984), however, attributed the prevalence of Ni and V in the oils to selective chelation of the organic matter by sea water, in which most other tetravalent and trivalent ions that could form stable porphyrins (e.g., Si, Al, Ti) are either almost absent or occur in oxygen-bearing hydrolyzed forms unavailable for chelation. Under reducing conditions many divalent metals precipitate as sulfides, becoming equally unavailable, whereas Ni(II) and VO(II) continue to be available, even if at low concentrations (Filby, 1994). Their availability in the depositional environment depends on pH and redox conditions and sulfide activity (Lewan & Maynard, 1982; Lewan, 1984). In strongly reducing, anoxic environments, where the amount of bacterially generated sulfide exceeds the amount of available iron, vanadyl ions are available in the pH 4-8 range whereas nickel is precipitated as aqueous nickel sulfide complexes, so that vanadyl-porphyrins would predominate in the sediments; expelled oils formed by catagenesis of organic matter in these sediments would have high V/Ni ratios and high sulfur contents. Conversely, in less anoxic, H_2S poor environment vanadium occurs in the quinquivalent form, not available for metallation, whereas Ni^{2+} cations are available. This would favor the formation of nickelo-porphyrins; expelled oils thus would have low V/Ni ratio and low sulfur content (Lewan, 1984; Filby, 1994). To compensate for the low concentrations of Ni and V in seawater, Lewan (1984) suggested that when sedimentation is slow, these ions migrate by diffusion from the entire overlying water column into the sediments, which they progressively enrich by becoming selectively retained by the porphyrins.

This model of synsedimentary metallation was somewhat modified when evidence from the Deep Sea Drilling Project (Louda & Baker, 1981; Baker & Louda, 1986) indicated that demetallation of chlorophyll and chelation of porphyrins by V and Ni occur later during diagenesis, at temperatures over 40°C and significantly below the sediment-water interface, when the compacting buried sediments are no more in open contact with the overlying water column. It was therefore suggested that metallation of organic species during diagenesis takes place in ion exchange sites of active clay mineral surfaces that retain high concentrations of metal ions acquired from the depositional environment (Filby & Van Berkel, 1987; Filby, 1994). The USGS Central Region Mineral Resources Team was assessing metal partitioning during hydrocarbon generation from black shale, a process of which both the mechanisms and controls are little known.

2.1 Metals in black shales from seafloor springs

Black shales rich in organic matter are often also rich in chalcophilic metals, gold, and platinum group elements (PGE). Even in average black shales, metal levels often exceed crustal levels by 10^5-10^6 times. Studies of Proterozoic, Ordovician, Devonian, Permian, Cretaceous-Tertiary and other black shales in the United States, Canada, the Baltic Shield, Siberia, and China (e.g., Granch & Huyck, 1989; Wignall, 1994; Fedikow *et al.*, 1998) suggest metal sources in the depositional environment: in anoxic or euxinic depositional environments chalcophilic (Cu, Co, Zn, Pb, Cd, Hg, Ag, As, Ni) and precious metals (Au and PGE) may be incorporated in bottom sediments through precipitation of sulfides, whereas oxo-cations (Mo, Cr, U, V, Se) may be reduced and the elements scavenged from seawater by the fine particulate organic matter falling through the water column. Mass balance considerations indicate, however, that seawater alone does not explain element abundances even for the black shales of the Cretaceous-Tertiary boundary event (Brumsack, 2003), requiring other (fluvial and/or hydrothermal) inputs. For shale-hosted Ni-Zn-Mo-PGE deposits several genetic models have been suggested, reflecting the limited data available and the unusual presence of PGE without ultramafic rocks. Siliceous venting tubes and chert beds in the underlying beds in the Yukon suggest a hydrothermal source for metals (Lefebure & Coveney, 1995); syngenetic deposition from seafloor springs with deposition of metals on or just below the seafloor is the most favored model.

3. Methodology

If there is significant contribution from hydrating mantle peridotites to petroleum formation, it should be reflected in the trace element composition of petroleum. In the present study, we analyzed 24 trace elements by internally coupled plasma-mass spectrometry (ICP-MS) in 68 oils sampled in all seven producing sedimentary basins of Brazil and, for comparison, in 9 oils from major oil-producing areas outside Brazil. All analyses were made by one of us (Fonseca, 2000). We examined correlations between the individual trace elements, differences in trace element compositions among basins of different tectonic and sedimentary settings, and the effect of thermal cracking related to igneous activity. Studies of this kind, covering a wide range of trace metals in petroleum and their regional distribution over a large area, are still rare in the published literature.

We also studied correlations of trace element compositions of the oils with major geochemical earth reservoirs such as chondrite, fertile mantle, primitive mantle, continental crust, and seawater. Brazil is particularly favorable for testing the relationship between the oils and mantle peridotites because most of its oil deposits occur along the Atlantic margin, where early Cretaceous rifting between South America and Africa thinned the continental crust and partially unroofed the mantle lithosphere (Zalan et al., 2010).

4. Geologic setting

Petroleum is produced in Brazil in several basins along the South Atlantic continental margin and inland (Fig. 1).

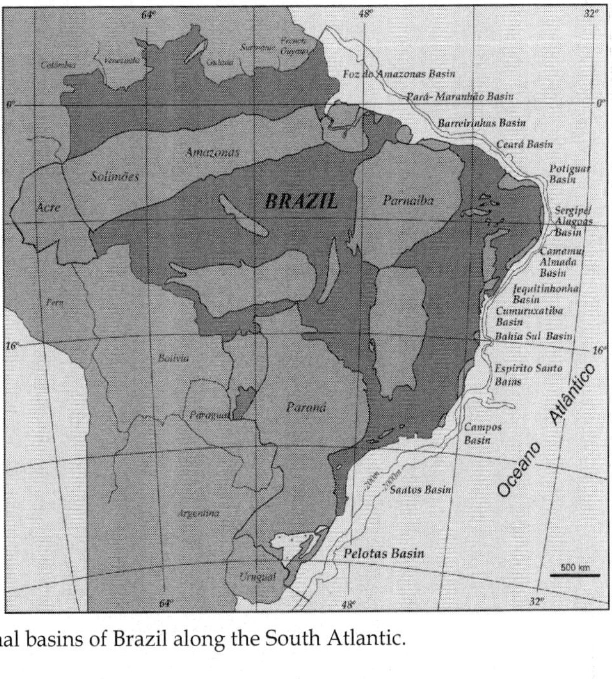

Fig. 1. Marginal basins of Brazil along the South Atlantic.

Except for the Paleozoic Amazonas-Solimões Basin far inland, the oil-producing basins lie over rifted continental crust along or near the continental margin and formed during early Cretaceous continental breakup between South America and Africa. Production in 2009 exceeded 2 million barrels per day. Most of the oils are low in sulfur and vanadium; their biomarkers mostly derive from early Cretaceous syn-rift lacustrine (freshwater and saline) source rocks and from largely post-rift Aptian siliciclastic source rocks deposited in a marine hypersaline environment (Katz & Mello, 2000). The geologic setting and organic geochemistry of Brazilian oils have been amply discussed in the published literature (e.g., Katz & Mello, 2000; Schiefelbein *et al.*, 2000; Milani & Zalan, 1998; Cainelli & Mohriak, 1998; Szatmari, 2000); here we provide only a short summary. The basins sampled (Fig. 1) include the Paleozoic Upper Amazon (Solimões) basin in Amazonas state far inland and six Cretaceous to Tertiary basins along the rifted South Atlantic margin: the Potiguar basin in Rio Grande do Norte state, the Sergipe-Alagoas basin in the homonymous states, the aborted Recôncavo rift in Bahia state, the Espírito Santo basin in Espírito Santo state, the Campos basin in Rio de Janeiro state, and the southern Santos basin. These basins are respectively referred to, in both the text and figures, by the abbreviations *am, rn, se, ba, es, rj,* and *bs*. The pre-salt accumulations recently discovered in the Santos Basin are not discussed in this study. In addition, nine oil samples from outside Brazil, marked *for* (foreign), were taken from rich petroleum producing areas worldwide, with mostly marine source rocks.
Brazil's oldest oil producer is the mostly onshore Recôncavo basin in Bahia state, NE Brazil, an aborted continental rift separated from the continental margin by the wide Salvador-Jacuipe horst. The sedimentary sequence of the basin consists mostly of late Jurassic to early Cretaceous fluvial to lacustrine siliciclastic strata that contain both the reservoirs and the organic-rich source rocks. Igneous rocks, evaporites and marine sediments are absent. The oils are viscous but mostly free of biodegradation; their high viscosity is due to long-chain paraffins, attributed to lacustrine plant sources.

Also in NE Brazil, the oil of the Potiguar basin is sourced in part from the passive margin and in part from an aborted lacustrine rift inland. The Sergipe-Alagoas basin lies along the passive margin and contains thick marine evaporites. Reservoir rocks in both basins are mostly of Cretaceous age.

In SE Brazil, the partly offshore Espírito Santo and the wholly offshore Campos and Santos basins along the continental margin contain thick Aptian evaporites, underlain by thick sediments and early Cretaceous basalts related to the large continental flood basalts of the Paraná Basin. The reservoirs are mostly of Cretaceous and Tertiary age; biodegradation is often intense.

Light oils and condensates are produced in the offshore Cretaceous southern Santos and the Paleozoic inland Amazon-Solimões basins. In the latter basin the oil is thought to be sourced from Devonian shales and produced from Carboniferous sediments overlain by Carboniferous evaporites and intruded by Juro-Triassic diabase sills and dikes. Thermal cracking has been demonstrated by diamondoid studies (Dahl *et al.*, 1999); asphaltene content is low or absent.

5. Analytical method

We have taken 68 oil samples from all oil-producing sedimentary basins in Brazil, over an area more than 3,000 km across (nearly 2,000 miles; Fig. 1). Nine additional samples were taken from major oil producing areas outside Brazil. Each sample was analyzed for 24 trace

elements: Ti, V, Cr, Mn, Fe, Co, Ni, Cu, Zn, Ga, As, Rb, Sr, Y, Mo, Ag, Ba, La, Ce, Pr, Nd, W, Pb, and U, using a Perkin-Elmer-Sciex, Model Elan 5000, internally coupled plasma-mass spectrometer (ICP-MS). Of the 24 elements analyzed, 13 (Ti, V, Cr, Mn, Fe, Co, Ni, Cu, Zn, Y, Mo, Ag, W) were transition metals; 4 (La, Ce, Pr, Nd) were rare earth elements. Pr, W, and U, which were below the sensitivity limit of the method in nearly two thirds of the samples, were only partially evaluated. Sr was contributed in some samples in part by the residual brine retained in the oil samples, especially in viscous *ba* oils.
Microwave digestion was used to prevent escape of volatile components; the isotopes to be measured were selected so as to minimize interference by isobaric ions and plasma gas ions of coincident mass (Filby & Olsen, 1994).

6. Results and discussion

6.1 Correlations between trace metals in the oils

For each trace metal and metal ratio, frequency distributions, means and medians were calculated for individual basins and for the total dataset. Frequency distributions are close to lognormal. The geometric means are close to the medians (oilmed77 for the 77 oils) and to the mode whereas the arithmetic means are higher and more influenced by outliers (Fig 2).

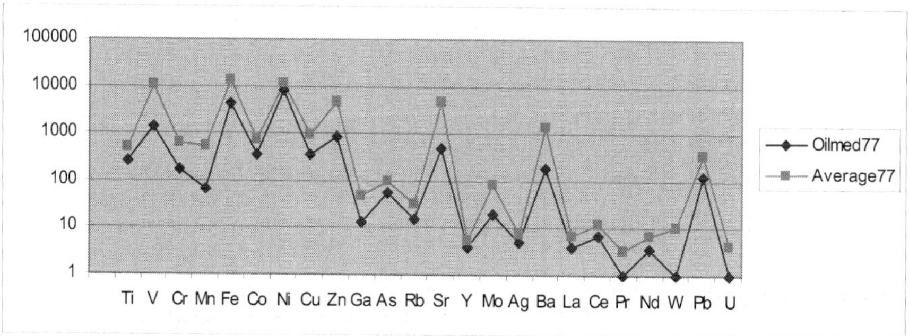

Fig. 2. Comparison of medians and arithmetic averages (means) of the 77 oils analyzed. The medians are better representations, less influenced by outliers, and close to the means of logarithms.

The medians of the oils, calculated for the individual basins and for the whole dataset, nearly coincide with the means of the logarithms of elemental abundances for which standard deviations were calculated. Median trace metal compositions for the 68 Brazilian and the 9 foreign oils are similar (Fig. 3) but the foreign oils tended to be higher in V and lower in Co and As.
Frequency distributions are closest to lognormal for As, Fe, Ce and, less clearly (perhaps because of their low concentrations) for La, Nd, Y, Mo, and Ga. Frequencies are skewed from lognormal toward smaller values for Sr, Ba, Rb, Mn, Cu, and Pb and to a lesser extent also for Cr and Ti. Zn and Ag are more irregular; their frequency distributions are flatter. Relative to lognormal, the distributions are skewed toward higher values for Ni and Co, and less regularly for V, elements most characteristic of asphaltenes.

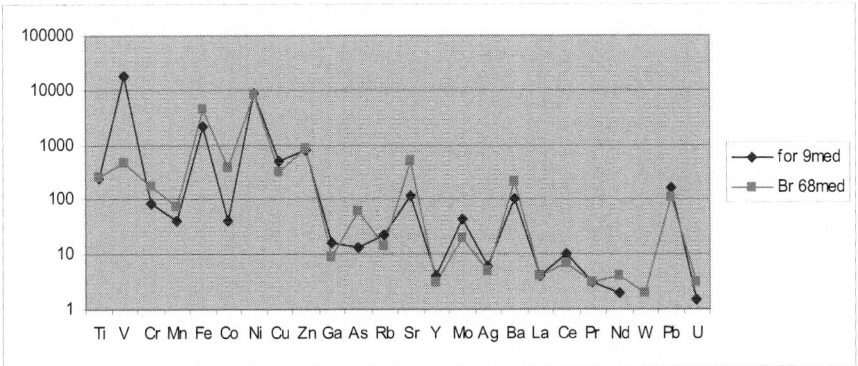

Fig. 3. Comparison of median trace element distributions (ppb) of 68 Brazilian and of 9 foreign oils. Note great similarity, with higher V and lower Co and As in the foreign oils.

Standard statistical methods including cluster analysis were used to measure correlations between elements. Logarithms were preferred for correlation because of the lognormal distribution of the elements. Log-log correlations in the 68 Brazilian oils are best ($r^2>0.85$) for the La-Ce and Pr-Nd couples; very good ($r^2>0.7$) for the Co-Ni, Ce-Nd, and Sr-Ba couples; and less good ($r^2>0.55$) for the V-Ni, Ga-Mo, and Mn-Fe couples.

6.2 Trace element groups

We compared trace element abundances and ratios in the oils with major geochemical reservoirs: CI carbonaceous chondrite (Anders & Grevesse, 1989), serpentinized fertile upper mantle peridotite from the Vosges Mts, France (UB-N geostandard, Govindaraju, 1982, 1995; Meisel et al., 2003), primitive mantle (McDonough & Sun, 1995), oceanic ("8% MgO" MORB) and continental crust (Rudnick & Fountain, 1995; Wedepohl, 1995), and seawater (Broeker & Peng, 1982; Quinby-Hunt & Turekian, 1983), extensively using the home page of the Geochemical Earth Reference Model (GERM).

Below we shall briefly review the regional distribution of each trace element group, using their median values in each basin. The basins are indicated by abbreviations: *rn* (mainly Potiguar Basin in Rio Grande do Norte state), *se* (Sergipe), *ba* (Recôncavo Basin in Bahia state), *es* (Espírito Santo) , *rj* (Campos Basin in offshore Rio de Janeiro), *bs* (southern, post-salt part of the Santos Basin), and *am* (the inland Amazonas-Solimões basin). *For* designates the 9 foreign oils we analyzed for comparison, from Saudi Arabia, Iran and Yemen in the Middle East; Peru, Venezuela, Ecuador and Mexico in the Americas, and one from the Niger Delta.

We distinguished four groups of trace elements, without rigid limits among the groups:

Mantle-derived elements, Group 1: Co, Ni, Ga, (Cu, Zn), La, Ce, Nd

Mantle-derived elements, Group 2: Cr, Mn, Fe, (Ti)

V, Mo

Elements enriched in hydrothermal brines: (Cu, Zn, Mo), As, Ag, Pb

Mantle-derived elements Group I : Ni, Co, Ga, (Cu, Zn), La, Ce, Nd

The oils contain about 10,000 ppb for Ni in most of the Brazilian and foreign oils we analyzed. In the thermally cracked light *am* oils of the Amazon-Solimões basin, affected by

Juro-Triasssic subvolcanic and hydrothermal activity, Ni is reduced to below 1000 and often below 100 ppb. Co is 1000 to 100 ppb in most oils; but only 1-5 ppm in most thermally cracked, light *am* oils.

Ga/Ni vs. Ce/Ni ratios (Fig.4) and the Ga/Ni vs. Co/Ni ratios (Fig. 5) plot close to mantle (M) and chondrite (CH) values; the distributions of La and Nd are similar to Ce. Co/Ni, Ga/Ni, and Ce/Ni ratios show little differentiation between the oils of the various basins. The *ba* oils have the lowest Ga/Ni ratios. The high Ce/Ni and Ga/Ni ratios of the *am* oils reflect the lowered Ni content of these thermally cracked low-asphaltene oils.

Cu/Ni and Zn/Ni also plot close to mantle and chondrite, but they are modified by hydrothermal processes, with a partial shift toward crustal and seawater values (Fig. 6).

Mantle- derived elements Group II : Cr, Mn, Fe, (Ti)

The abundances of these elements in the various basins are similar; irregularities of the curves show relatively minor local variations. Fe is a few 1000 ppb, Ti and Cr are a few 100 ppb; Mn is mostly below 100 ppb.

Compared to chondritic or mantle values, Mn, Fe and Cr are very low in the oils, but their ratios are close to mantle (M) and chondrite (CH) values, except for the *ba* oils and the thermally cracked *am* oils (Fig. 7).

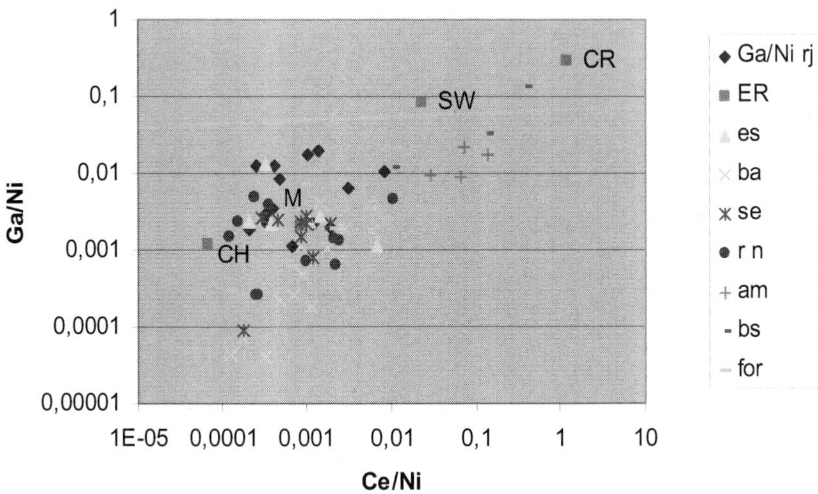

Fig. 4. Ga/Ni versus Ce/Ni ratios in 68 Brazilian and 9 foreign oils. Note clustering around mantle and chondrite values. ER - reservoir data from the Geochemical Earth Reference Model (GERM): CH – CI chondrite; M – mantle; CR – continental crust; SW – seawater. Brazilian basins: rj – Campos (Rio de Janeiro); es – Espírito Santo; ba – Recôncavo (Bahia); se – Sergipe-Alagoas; rn – Potiguar (Rio Grande do Norte); am – Amazonas-Solimões; bs – southern Santos (post-salt); for – foreign samples.

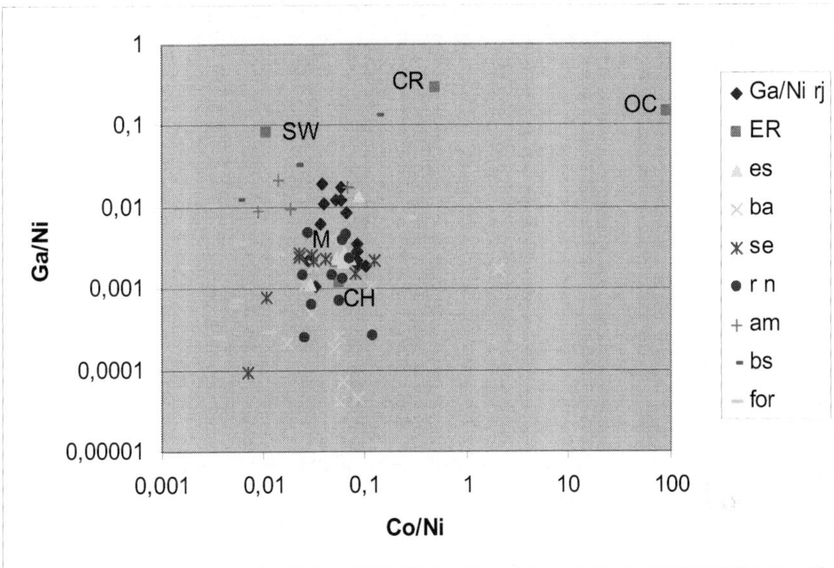

Fig. 5. Ga/Ni versus Co/Ni ratios in 68 Brazilian and 9 foreign oils. Note clustering around mantle and chondrite values. Abbreviations as above; OC - oceanic crust added.

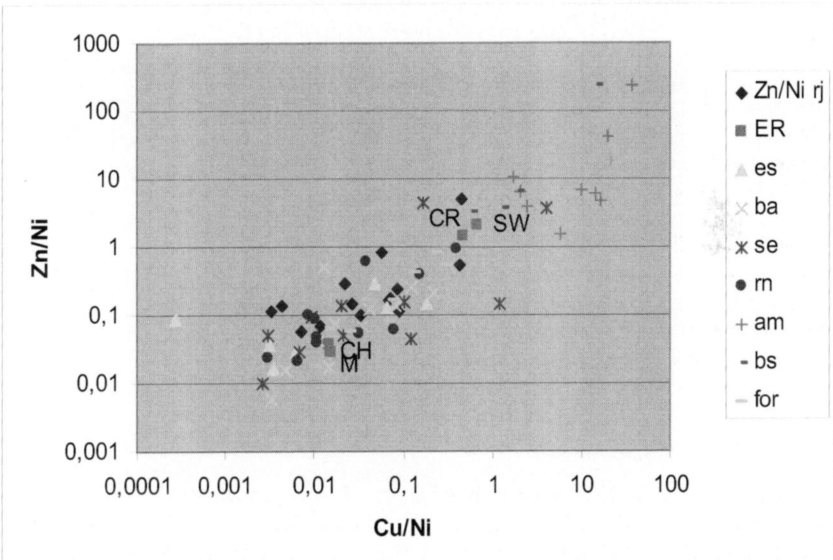

Fig. 6. Zn/Ni vs. Cu/Ni ratios plot dominantly close to the nearly identical chondrite and mantle values, but a large part of them plots close to continental crust and seawater (even beyond in the thermally cracked *am* oils) reflecting their mobility in hydrothermal brines.

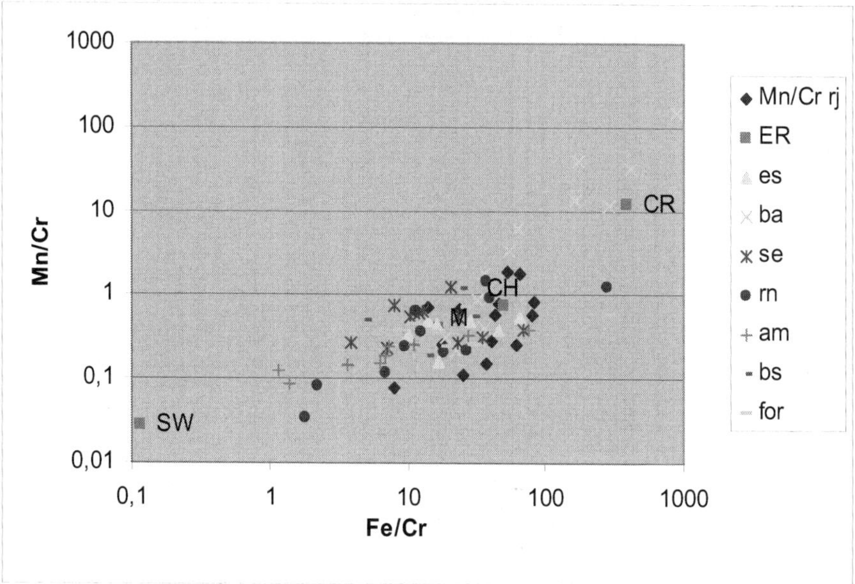

Fig. 7. Mn/Cr versus Fe/Cr in 68 Brazilian and 9 foreign oils. Note clustering around mantle and chondrite values.

V, Mo

V varies more widely in the Brazilian oils than any of the other elements, it shows the greatest variation between basins and sometimes even between individual samples. Mo also varies widely, ranging from a few ppb to nearly 1000 ppb. The dispersion of V in our dataset, defined as the ratio of the third and first quartiles, is more than 100, compared with 27 for Co and 14 or less for all other elements, including Ni. V is above 10,000 ppb in the *rj* and above 100,000 ppb in the *for* oils, but below 1000 ppb in the *ba* oils and mostly below 25 ppb in the thermally cracked *am* oils in which asphaltenes are nearly absent. As a result, V/Co vs. V/Ni ratios plot along the whole range from chondritic to seawater values (although this last one only in foreign oils) (Fig. 8).

V/Ni ratios (Fig. 8) are very low, chondritic to mantle-like (close to 0,01), in the *ba* oils of the aborted Recôncavo rift and only slightly higher (0.04) in some of the *rn* oils from the aborted rift of the Potiguar basin. They rise somewhat above mantle values (0.1-1) in the *se, es* and the rest of the *rn* oils of the Sergipe-Alagoas, Espírito Santo and Potiguar basins, and reach levels close to crustal values (1-2) in most of the *rj* oils in SE Brazil. Still higher levels (2-10), ranging from crustal to seawater values, were measured in various marine-sourced *for* oils we analyzed from Venezuela, Ecuador, Peru, Mexico, Iran, and Yemen.

Similarly, V/Co ratios are very low, chondritic to mantle-like (0.1 to 1), in the V-poor oils of the fresh-water rift sequences of the aborted rifts of the Recôncavo (*ba*) and Potiguar basins, and vary from mantle to crustal values (1-10) in most oils along the continental margin, reaching higher levels (10 to 100) in the *rj* oils of SE Brazil. Even higher ratios (100 to 2000), close to seawater values,were found in various marine-sourced *for* oils from outside Brazil

(Venezuela, Ecuador, Peru, Mexico, Iran, and Yemen). V/Co ratios vs. V/Ni ratios lie along a chondrite-seawater mixing line (Fig. 8). The highest V/Co ratios reflect the sharp drop of Co abundances and Co/Ni ratios in high-V (>100,000 ppb) non-Brazilian oils, such as the *for* oils of our dataset and Curiale's (1987) heavy oils.

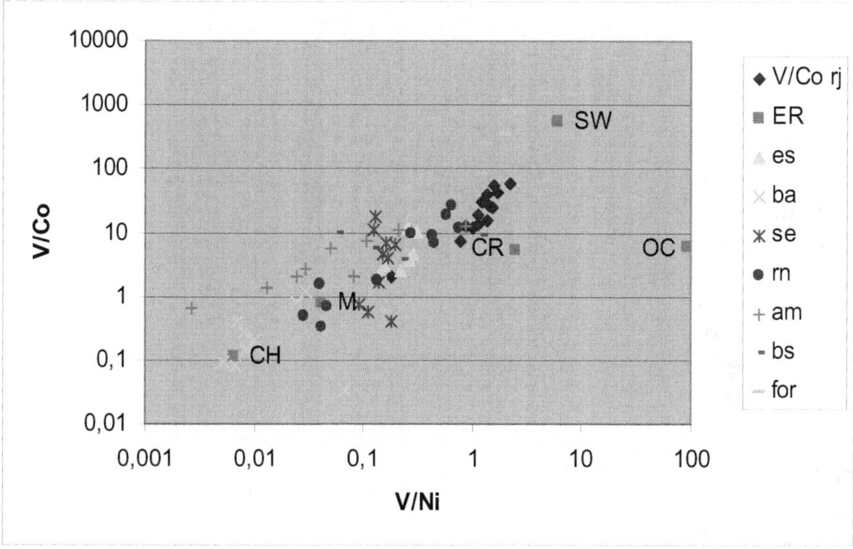

Fig. 8. V/Co versus V/Ni ratios in 68 Brazilian and 9 foreign oils. Note the wide range of distribution between the medians of the various basins, ranging from chondritic to seawater values. Abbreviations as above.

Differently from the wide range of distributions of V/Co vs. V/Ni (Fig. 8), V/Mo vs. Co/Ni values (Fig. 9) are close to the chondrite to mantle range (except for the cracked *am* oils), suggesting that V and Mo both derived from the mantle and were modified by similar processes after leaving their mantle source.

Elements enriched in hydrothermal brines: (Cu, Zn, Mo), As, Ag, Pb

As shown above, most of the Cu/Ni vs. Zn/Ni ratios plot near mantle and chondrite, but a long tail to crustal and seawater values testifies their hydrothermal mobility (Fig. 6).

In the thermally cracked light *am* oils of the Amazon-Solimões basin, affected by Juro-Triasssic subvolcanic and hydrothermal activity, Cu rises five times, Zn and Pb two times above the median of Brazilian oils. Zn is highest, close to 1000 ppb, Cu about 300 ppb, Pb about 100 ppb, and Ag below 10 ppb. Variations between the other basins are small, less than an order of magnitude.

In the *am* oils, Ni-normalized ratios of these elements are higher than crustal values, mainly owing to the loss of Ni from the thermally cracked oils, but also because of the higher levels of Cu, Zn, and Pb in this basin intruded by thick diabase sills.

Reflecting the greater hydrothermal mobility of Mo, Mo/Ni ratios move away from the mantle-chondrite area. Mo/Ni is lower in the V-poor *ba* oils of the Recôncavo rift (Fig.10).

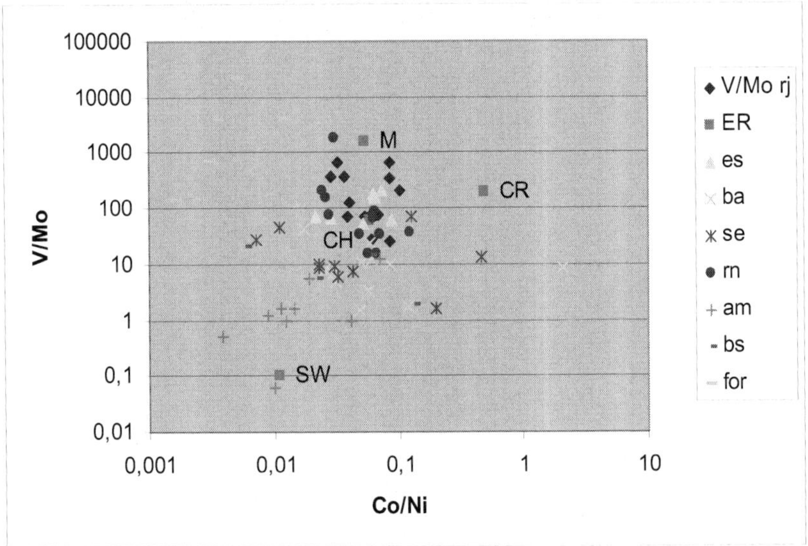

Fig. 9. V/Mo versus Co/Ni ratios in 68 Brazilian and 9 foreign oils. Note clustering around mantle and chondrite values. Abbreviations as above.

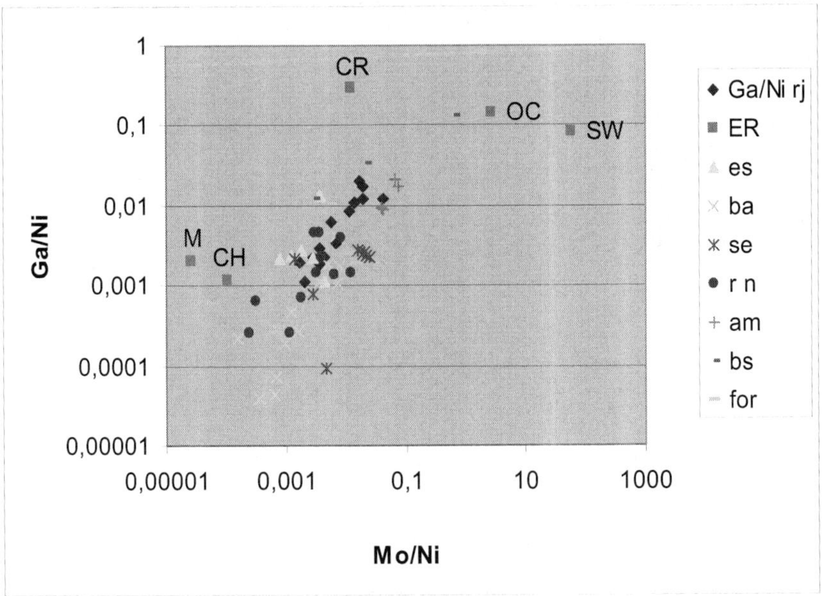

Fig. 10. Ga/Ni versus Mo/Ni ratios in 68 Brazilian and 9 foreign oils. Note clustering away from mantle and chondrite values, reflecting the hydrothermal mobility of Mo. Abbreviations as above.

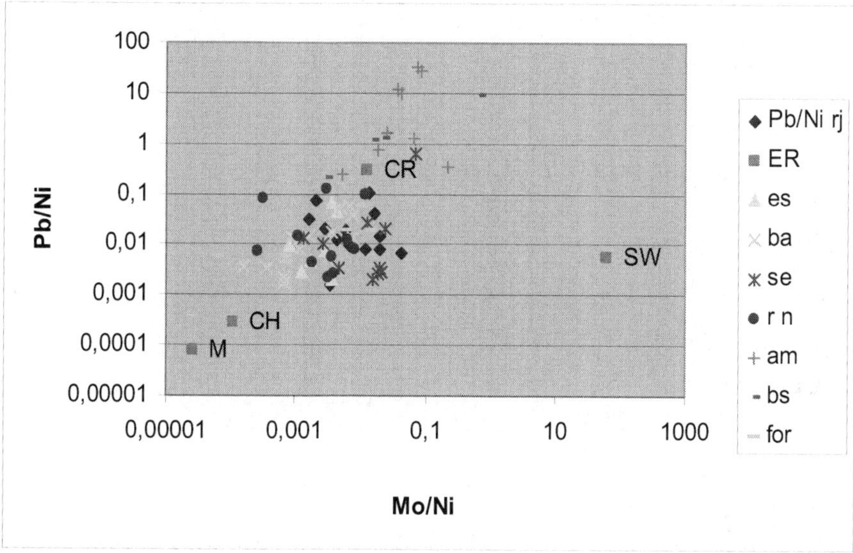

Fig. 11. Pb/Ni versus Mo/Ni ratios in 68 Brazilian and 9 foreign oils. Note clustering away from mantle and chondrite values, close to crustal ones, reflecting the hydrothermal mobility of both Pb and Mo. Abbreviations as above.

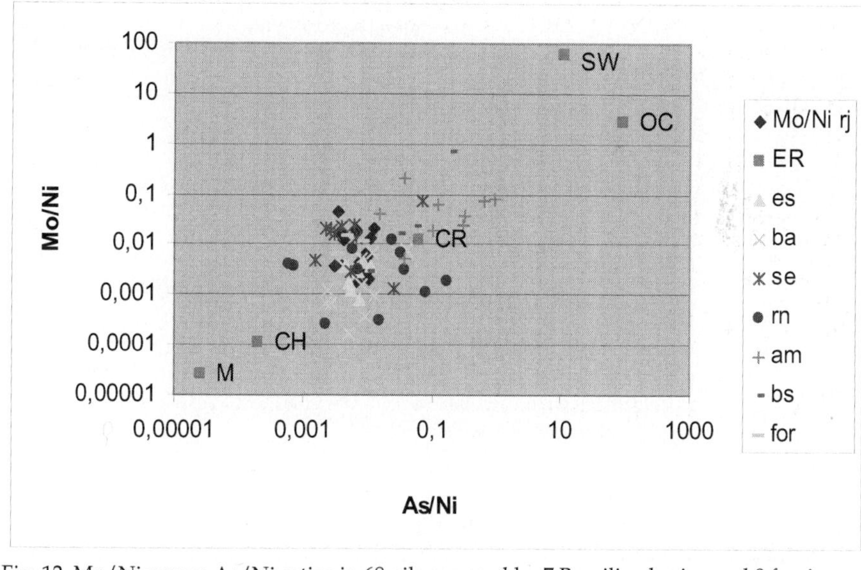

Fig. 12. Mo/Ni versus As/Ni ratios in 68 oils grouped by 7 Brazilian basins and 9 foreign oils. Note clustering away from mantle and chondrite, moving close to crustal values, reflecting the hydrothermal mobility of both As and Mo. Abbreviations as above.

The shift to near-crustal values is conspicuous also for Pb/Ni vs. Mo/Ni ratios (Fig.11) and for Mo/Ni vs. As/Ni ratios (Fig. 12). As/Ni and Mo/Ni ratios are correlated and range from chondritic-serpentinitic to crustal values (and even higher in the *am* oils that have low Ni content). As/Ni ratios in the various basins are similar.

7. Correlations with geochemical Earth reservoirs

We found that log-log correlations of the median composition of the 24 trace elements in the Brazilian oils (Sr is contained in residual brines in some of the oils) are very good with CI chondrite (r^2=0.80) and serpentinized fertile mantle (r^2=0.79); good with the primitive mantle (r^2=0.61); worse with oceanic (r^2=0.41) and continental crust (r^2=0.36); and none with seawater (r^2=0.02). Log-log correlations of the median of the nine non-Brazilian oils (taken from major oil producing areas in the Middle East, Mexico, Venezuela, Ecuador, Peru, Angola, and the Niger Delta) are good with chondrite (r^2=0.62), serpentinized fertile mantle (r^2=0.63), and the primitive mantle (r^2=0.51); worse with oceanic (r^2=0.39) and continental crust (r^2=0.33); and none with seawater (r^2=0.04). The somewhat lower correlations for the non-Brazilian oils result from their much higher V and lower Co content. All these correlations are robust and change little when considering smaller sample groups or individual oils instead of the median of the entire dataset, nor are they significantly altered by removing any single element.

Table 1 compares some trace element ratios in the oils with the primitive mantle and the continental crust, showing their similarity to the former and dissimilarity to the latter.

Metal Ratios	Primitive mantle	Oil median	Oil geometric mean	Cont. crust	Oil 1 st quart.	Oil 3rd quart.
Fe/Cr	23.8	23.9	23.20	388.2	10.8	45.7
Mn/Cr	0.40	0.43	0.55	11,76	0.24	0.80
Co/Ni	0.054	0.049	0.044	0.49	0.026	0.070
Ga/Ni	0.0020	0.0021	0.0021	0.29	0.0011	0.055
Cu/Ni	0.015	0.052	0.077	0.47	0.011	0.40
Cu/Ni*	0.015	0.027	0.028	0.47	0.0079	0.088
V/Ni	0.042	0.158	0.155	2.51	0.0417	0.591
Ni/Ce	1170.1	994.9	613.3	0.85	282.9	2710.8

Table 1. Comparison of some trace element ratios in the 68 Brazilian oil samples with the primitive mantle (McDonough & Sun, 1995), and the continental crust (Rudnick & Fountain, 1995; Wedepohl, 1995). Note excellent correlation with the primitive mantle. (Cu/Ni*: omitting basins in which the oil was thermally cracked).

Thus the median composition of trace elements analyzed in 68 Brazilian oils correlates poorly with continental or oceanic crust and not at all with seawater (Fig. 13).

In contrast, correlations with CI chondrite (Fig. 14) and with the mantle are good. The mantle that is not serpentinized (spinel peridotite, primitive and depleted mantles) differs from the oil in the absence of hydrothermally enriched elements As, Mo, Ag, Pb, whereas chondrite and serpentinized fertile mantle UB-N, like the oils, are enriched in these elements (Fig. 15-17).

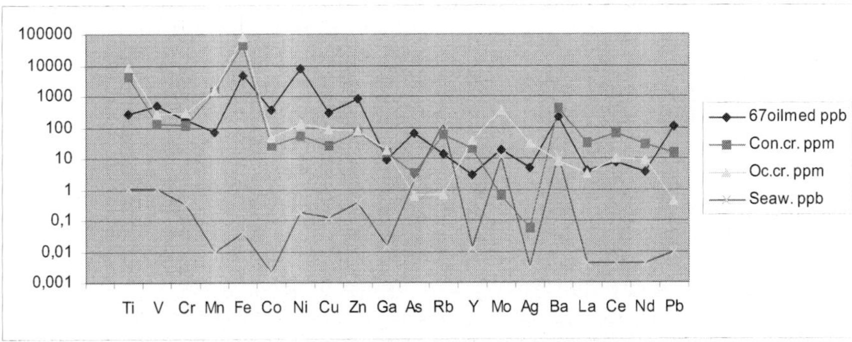

Fig. 13. Comparison of the median of 67 Brazilian oils with continental and oceanic crust and seawater. Correlations are poor with the crusts; there is no correlation with seawater.

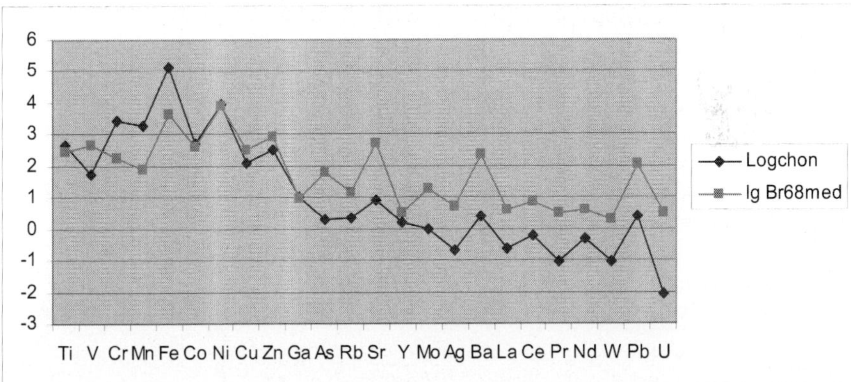

Fig. 14. Median composition of 68 Brazilian oils compared to chondrite.

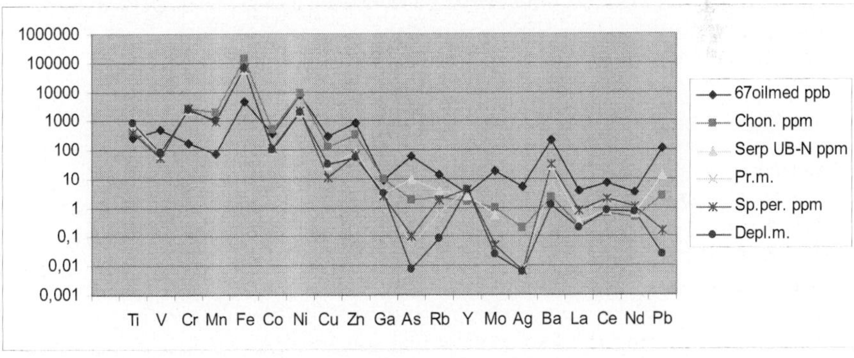

Fig. 15. Comparison of trace element medians of 67 Brazilian oils (ppb) with chondrite, serpentinized mantle UB-N, primitive mantle, spinel peridotite mantle and depleted mantle (ppm). Correlations are good, best with serpentinized mantle.

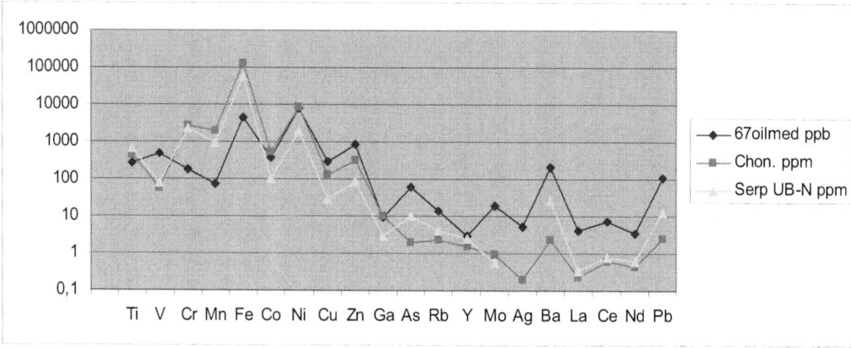

Fig. 16. Comparison of median trace element composition of the 67 Brazilian oils with chondrite and with serpentinized fertile mantle UN-B. Note the excellent correlations; the serpentinized mantle parallels more closely the hydrothermal enrichment of As, and Pb in the oils.

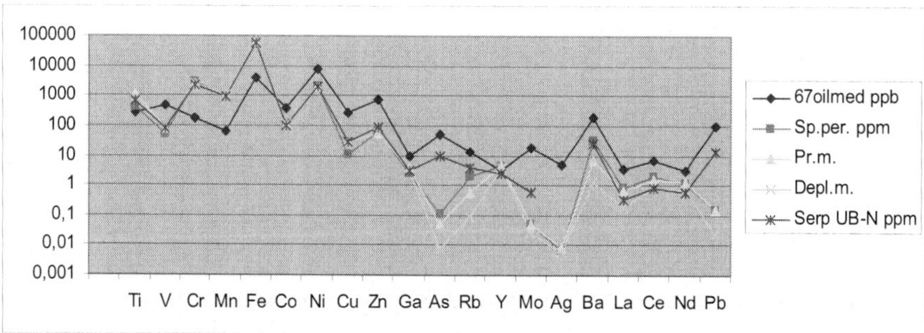

Fig. 17. Good correlation of the medians of 67 Brazilian oils (ppb) with four types of mantle (ppm): spinel peridotite mantle, primitive mantle, depleted mantle and serpentinized fertile spinel-peridotite mantle (UB-N). High relative abundances of hydrothermally enriched elements As, Mo, Pb are enriched only in the oils and in the serpentinized mantle.

7.1 Correlations of trace element compositions with mantle peridotite

Normalizing the medians of the Brazilian oils to chondrite and to spinel peridotite mantle, element abundances form indistinct plateaus (Fig. 18).

When we normalize the oil medians of the individual basins to spinel peridotite mantle/1000, these plateaus become more distinct (Fig. 19). Three groups of elements can be distinguished:

1. **Mantle-derived elements Group 1 and V:** Co, Ni, Ga, (Cu, Zn), La, Ce, Nd are mostly close to 10. Cu and Zn are somewhat higher, reflecting hydrothermal enrichment; V varies highly (from 0.2 to 40) between the individual basins.
2. M**antle-derived elements Group 2:** Fe, Cr, and Mn are two orders of magnitude lower, (0.02-0.07) but their ratios are mantle-like reflecting their retention in secondary magnetite and related spinels formed during serpentinization. Fe/Cr and Mn/Cr ratios

are the same as in the mantle, Ti/Cr is somewhat higher. Fe/Cr and Mn/Cr ratios in our oils plot about chondritic and mantle values (Table 1), as do to the average Fe/Cr and Mn/Cr ratios for 88 Alberta crude oils analyzed by Hitchon *et al.* (1975). Some freshwater *ba* oils and a Libyan oil analyzed by Filby & Shah (1975) showed, however, higher Fe/Cr and Mn/Cr ratios.

3. Finally, mantle-normalized abundances of **hydrothermally enriched elements** As, Mo, Ag, and Pb in the oils are orders of magnitude higher than the elements of the first group (about 1000), reflecting their enrichment by hydrothermal processes.

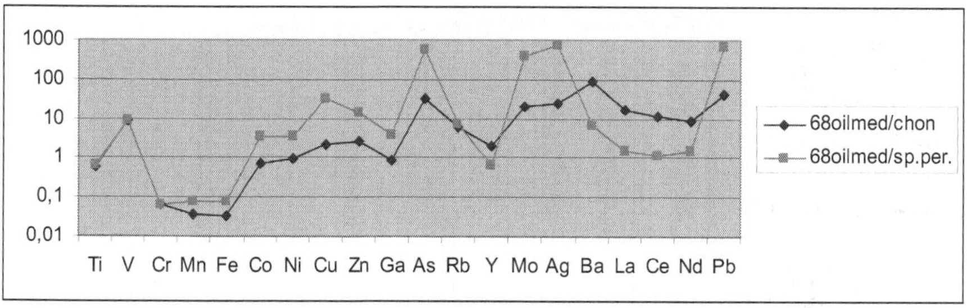

Fig. 18. Medians of 68 Brazilian oils (ppb) normalized to chondrite and to spinel peridotite mantle (ppm).

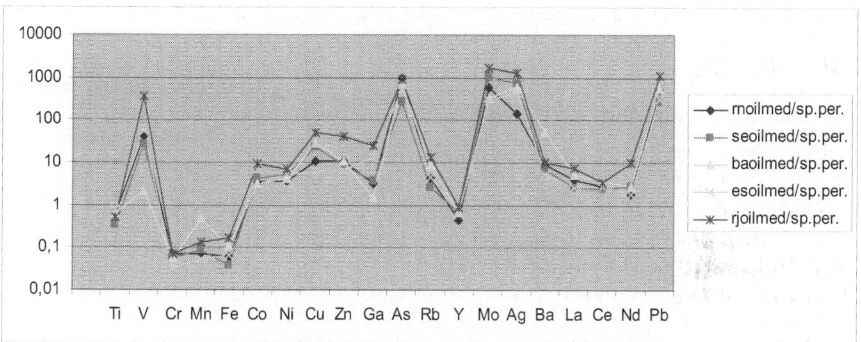

Fig. 19. Median trace metal abundances of oils (ppb) from five Brazilian basins normalized to spinel peridotite mantle (ppm). The plateau of mantle-derived elements Group I is about 10 ppb/ppm, that of Group II (Cr, Mn, Fe) about 0.1 ppb/ppm, and that of the hydrothermally enriched elements (As, Mo, Ag, Pb) about 1000 ppb/ppm.

7.2 Correlations of trace element compositions with serpentinized fertile peridotite

Serpentinized mantle peridotite shows even better correlation with the oils (Fig. 20) than the non-hydrated mantle shown above (Fig. 19). We used the geologic reference material UB-N, a serpentinized fertile peridotite from Col de Bagnelle in the Vosges Mountains, France, that was repeatedly analyzed in many laboratories for major and trace element concentrations (Govindaraju, 1982, 1995). Their Re-Os isotope systematics are now well understood, attesting to a serpentinized fertile upper mantle garnet-spinel peridotite (Meisel et al., 2003).

Textural and mineral chemistry data suggest that the UB-N serpetinite originally formed in the garnet peridotite stability field at high pressure and subsequently re-equilibrated in the spinel peridotite stability field (Meisel et al., 2003). The serpentinizing fluid was a seawater-derived brine as indicated by the high (600 ppm) Cl content of the serpentinite. As and Pb, presumably introduced by the seawater-derived serpentinizing brine, are enriched by two orders of magnitude, Mo by one order of magnitude relative to fertile peridotites and the primitive mantle (McDonough et al., 1995).

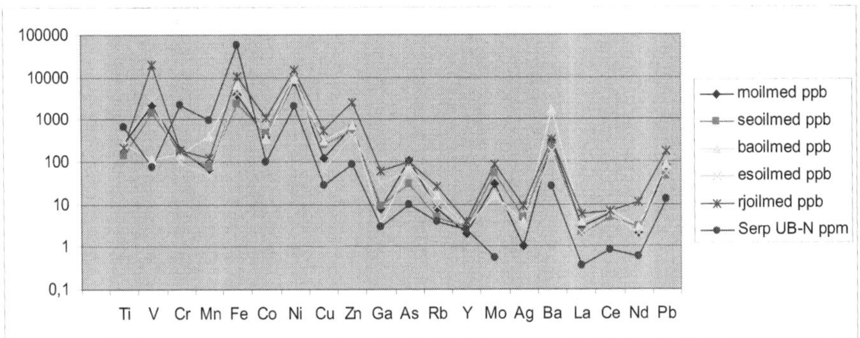

Fig. 20. Median trace elements distributions (ppb) in oils from five Brazilian basins, compared to serpentinized mantle UB-N. Note close parallelism.

When the trace element compositions of our oil samples are normalized to this serpentinized peridotite (Fig. 21, pink), the medians of the hydrothermally enriched elements join the mantle derived elements Group I on the same plateau. Thus, normalized to UB-N serpentinite/1000, the median abundances of the elements we analyzed in the Brazilian oils form only two plateaux: one near 10 for V, Co, Ni, Cu, Zn, Ga, As, Mo, Ba, La, Ce, Nd, Pb, and another near 0.1 for Fe, Cr, and Mn (Ti lies between the two levels).

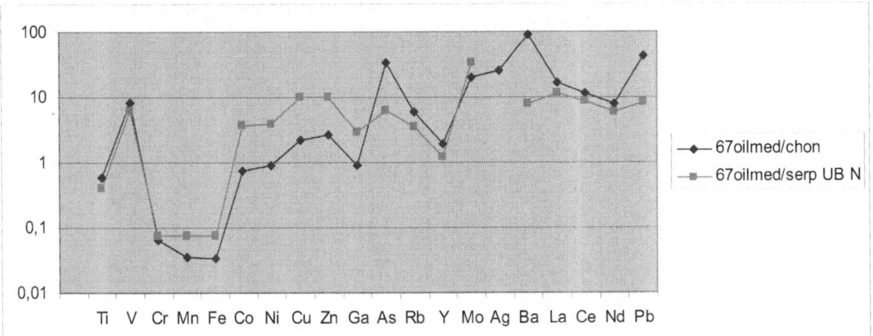

Fig. 21. Median trace element abundances of 67 Brazilian oils (ppb) normalized to chondrite/1000 and to serpentinized fertile mantle peridotite UB-N/1000. Values are in ppb for the oils, in ppm for the earth reservoirs. Note rising values when normalized to chondrite and a plateau near 10 for most elements when normalized to serpentinized mantle UB-N.

8. Abiogenic sources of trace metals and hydrocarbons

8.1 Carbonaceous chondrites

The ten-years-old Tagish Lake meteorite, recovered in pristine condition over lake ice in Canada, is a new type of water and carbon-rich Type 2 carbonaceous chondrite that contains about 5% total carbon, of which the organic carbon reaches 1.3 wt. %. The temperature of the meteorite has not risen above 120°C since the formation of the organics (Brown et al.,2000; Nakamura et al., 2003). It contains aliphatic hydrocarbons including both saturated and cyclic and/or unsaturated species, with peak abundance of normal alkanes at C_{23} and $\delta^{12}C$ in the terrestrial range at -18 to –29 per mil. Carboxylic acids and n-alkanes in the meteorite display a distinct linear chain preference suggesting catalytic surface processes accompanied by aqueous processing of minerals on the parent body forming serpentine and saponite (Pizzarello *et al.*, 2001).

8.2 Serpentinizing mantle peridotites

Rising hydrogen was observed in shallow pools over outcropping peridotites in Oman (Neal & Stanger, 1983). Using seawater in experiments to serpentinize oceanic peridotites at temperatures of 200°C and 300°C, at a pressure of 500 bar, and in the absence of CO_2, Janecky & Seyfried (1986) observed magnetite precipitation and significant hydrogen production as water was reduced by the oxidation of Fe^{2+} ions. In nature, serpentinization in the absence of CO_2 generates hydrogen and creates one of the most reducing geologic environments on the Earth's surface, in which oxygen fugacities drop to the IW (iron-wüstite) buffer. Most of Fe, Cr, and Mn (and part of Ti) in serpentinizing peridotite are immobilized in magnetite and other spinels that form during the serpentinization process, whereas most of the other trace elements pass into serpentine minerals (antigorite, lizardite) or into aqueous solutions. Part of Ni and Fe become reduced to metallic awaruite (Ni_3Fe) and other Ni-Fe alloys that have been found extensively in outcropping serpentinite bodies (Alt & Shanks, 1998). These metals are widely used in the industry as hydrogenation catalysts (Pajonk & Teichner, 1986); their presence during serpentinization is essential for the production of methane-rich hydrothermal fluids from dissolved CO_2 (McCollom & Seewald, 2001).

Serpentinization results from the hydration of peridotites at temperatures below 600°C; it is favored by tectonic deformation as in collision zones and in active oceanic and continental rifts (O'Hanley, 1996). The circulation of water is controlled mostly by thermal convection: colder water (seawater or fresh water) sinks into the partially unroofed peridotites, heats up, and rises back along faults. Oceanic serpentinization results in considerable water increase, local CaO decrease, and uptake of trace amounts of Sr by the peridotites while the rare earth elements remain immobile (Scambelluri et al., 2001). The Ocean Drilling Project demonstrated the presence of an extensive ridge of tectonically exhumed and serpentinized mantle peridotites that formed during Mesozoic breakup along the continental margin of Iberia, where the sedimentary sequence is anomalously thin (Boillot *et al.*, 1989; Whitmarsh *et al.*, 1998, 2001; Beard et al., 2002, Manatschal, 2004) and where the role of shear zones channeling fluid flow during serpentinization was demonstrated by oxygen isotope studies (Skelton & Valley, 2000). Skelton *et al.* (2003) suggest that syntectonic serpentinization also aided the transition from pure shear to simple shear rifting, the latter suggested also along the South Atlantic rifted margins. If dissolved CO_2 is present during serpentinization, it reacts with the molecular hydrogen generated, forming formate and methane (Berndt *et al.*, 1996; Horita & Berndt, 1999, McCollom & Seewald, 2001). Heavier hydrocarbons may also form.

Szatmari (1989) suggested that serpentinization of the mantle provided major contributions of hydrogen, hydrocarbon liquids, and metals to petroleum formation by FTTS near plate boundaries and proposed that Ni, a metal rare in the continental crust but abundant in petroleum, may derive from serpentinizing mantle peridotites. The evolution of the abiogenic model based on serpentinization of the mantle was presented by Szatmari *et al.* at successive meetings of the Geological Society of America (2004), the American Association of Petroleum Geologists Hedberg Research Conference on Origin of Petroleum -- Biogenic and/or Abiogenic and Its Significance in Hydrocarbon Exploration and Production (2005; Katz *et al.*, 2008), the American Geophysical Union (2003, 2010), the International Meeting of Organic Geochemistry (2007), the International Geological Congress (2008) and Goldschmidt Conferences (2009, 2010).

FTTS over oceanic serpentinites was demonstrated in the Rainbow (Holm and Charlou, 2001) and Lost City hydrothermal fields at the Mid-Atlantic ridge. In the ultramafic, reduced, low-temperature environment of the off-axis Lost City Hydrothermal Field (LCHF), where exothermic serpentinization reactions are believed to drive hydrothermal activity, rich microbial activity developed in the hydrogen and methane-rich environment. Hydrothermal plumes rising over serpentinizing peridotites contain abundant hydrogen, methane, and aliphatic hydrocarbons in the C_{16}-C_{29} range, formed by FTTS (Holm & Charlou, 2001, Kelley *et al.*, 2001; Shrenk *et al.*, 2002; Früh-Green *et al*, 2003; Kelley, 2005; Proskurowski *et al.*, 2008; Bach and Früh-Green, 2010). In a high-temperature (364°C) water plume rich in hydrogen and methane, in the Rainbow hydrothermal field where the Mid-Atlantic Ridge is intersected by an active fault zone that exposes serpentinizing peridotite rocks, Holm & Charlou (2001) reported normal paraffins with chain lengths of 16 to 29 carbon atoms formed by FTTS. This finding, together with the discovery of aliphatic and cyclic hydrocarbons in the Orgueil meteorite and in the more recent pristine Tagish Lake meteorite (Pizzarello *et al.*, 2001; Nakamura et al., 2003), as well as new studies on hydrocarbon stability at high P-T conditions (Kenney *et al.*, 2002), gave renewed stimulus to the abiogenic models.

8.3 Serpentinizing basalts

Alteration of basaltic rocks by groundwater or seawater may also contribute to hydrogen generation and base metal mobility (Seewald & Seyfried, 1990; Sacoccia *et al.*, 1994; Stevens & McKinley, 1995). In the outcropping Columbia River basalt, hydrogen generation may be impeded by the inflow of surface waters in equilibrium with the atmosphere. Experiments found small amounts of H_2 produced at pH 6 but not at pH 8, whereas the pH of groundwater in basalt aquifers is buffered about 8 (Anderson *et al.*, 1998). This is because the breakdown of minerals unstable at low temperatures consumes H^+ and releases Ca and HCO_3 precipitating $CaCO_3$. Geochemical modeling by Wallendahl & Treiman (1999) also showed that if the basalts are open to ground water inflow carrying CO_2 and O_2, as in the case of the Columbia River basalt, then abundant calcium carbonate will form, buffering pH at 8, so that production of molecular hydrogen by reduction of water and oxidation of Fe and Fe-Mg silicates will be minimal. If, however, the system is closed, serpentine and clay minerals will predominate as alteration products and abundant H_2 will be produced, so that hydrocarbons may form by FTTS from the CO_2 already present in the reservoirs. Chapelle *et al.* (2002) describe a unique hydrogen-consuming, methane-producing subsurface microbial community from Idaho, sustained by hydrogen rising with hydrothermal waters circulating in deeply buried igneous rocks.

8.4 A model of rising fluids carrying hydrogen, hydrocarbon, and metals

It has not been established when metallation occurs in the diagenetic conversion of chlorophyll to metalloporphyrin, or the mechanism by which the reaction takes place (Filby, 1994). During the formation of some metalliferous black shales there are indications for an additional, hydrothermal source of the chalcophilic metals and platinum group elements, hence it is conceivable that metals in kerogen-rich shales and petroleum also derive, in part, from non-sedimentary sources. These may include hydrogen- and hydrocarbon-bearing hydrothermal fluids formed as Fe(II)-bearing mantle minerals react with water and carbon compounds, either deep in the mantle (Kenney et al., 2002) or during serpentinization. Metals in the rising fluids may be carried as chelated complexes or transported as ionic solutions. Below we briefly outline a model that derives a significant portion of both trace metals and hydrocarbons in petroleum and kerogen-rich shales from hydrothermal serpentinization of peridotites.

During strong deformation of the mantle lithosphere, as by rifting or initial subduction, cold water descends along faults, shear zones and breccias into partially unroofed, deforming mantle peridotites, causing serpentinization. Fe(II) in the peridotites and associated basalts becomes oxidized by the infiltrating water while hydrogen is generated creating a strongly reducing environment. In this hydrogen-rich environment, reduced carbon species including hydrocarbon gases and liquids of varying chain lengths may form at 200-300°C from carbonaceous and CO_2 inclusions in the peridotites by Fischer-Tropsch-type synthesis, catalyzed by Fe-Ni alloys such as awaruite, often present in serpentinized peridotites. Heteroatomic C, S, and N compounds introduced by the infiltrating water or forming *in situ* may be chelated by metals from the peridotites in approximately mantle-like proportions; other metals may be transported in ionic solution. Where the infiltrating fluid is seawater rich in sulfates, or basalts are abundant, sulfide activity and with it V rises while Co falls relative to Ni , raising V/Ni (and to a lesser extent Mo/Ni and Ga/Ni) ratios from mantle-like toward crustal and seawater values.

Rapid rise of the fluid at high pressures, from deeply buried, serpentinizing peridotites, may permit hydrocarbons formed by FTTS to reach the sedimentary sequence uncracked but undergoing cyclicization and aromatization, forming new oil deposits or adding to existing ones. The rising fluid may interact with kerogen-rich shales that adsorb heavier hydrocarbons and metal-bearing polar and heteroatomic compounds from it while enriching it in biomarkers and biogenic hydrocarbons desorbed from the shales, creating the impression that these are the only source of all oil and metals in the oils. Adsorption and asphaltene precipitation increase, desorption decreases upward with decreasing temperatures and pressures. Hydrogen also decreases upward in the sediment column, increasing oxidation and polymerization of the transported organic matter that is being added to the fossil kerogen. A bacterial biota may feed on these fluids at depth, as shown for the Lost City vent field (Kelley et al., 2005). Part of the hydrogen and hydrocarbon-bearing fluid may seep through the sediments into the aquifer leading, in isolated basins, to the deposition of new organic and metal-rich shales.

9. Conclusions

Inorganic chemical evidence indicates a relationship between mantle and petroleum. 24 trace elements were analyzed by ICP-MS in 68 oils (one of them heavy), sampled from oil-producing sedimentary basins over the subcontinent of Brazil that produces more than 2 million barrels of oil per day. The analyses showed good correlation of the oils with CI chondrite and mantle peridotites, worse correlation with oceanic and continental crust, and none with seawater.

Mantle-normalized abundances of Co, Ni, Cu, Zn, Ga, La, Ce, Nd are similar to each other, indicating a common, mantle origin; V is variable. Normalizing the oils to serpentinized mantle also includes the hydrothermally mobile elements As, Ag, Mo, Pb with this group. Mantle-normalized Cr, Mn, and Fe are also similar to each other but their abundances are two orders of magnitudes lower than those of other mantle-derived elements, reflecting their lesser availability from secondary magnetite and other minerals that formed during serpentinization.

In the thermally cracked light oils of the Amazon-Solimões basin, affected by Juro-Triasssic subvolcanic and hydrothermal activity, the asphaltene-related elements V and Co abundances are more than a hundred times, Ni forty times, Ga, Mo, and As two to four times less than the median of the Brazilian oils, whereas Cu, Zn, and Pb abundances are two to four times above that median.

The dispersion of V values in our dataset, defined as the ratio of the third and first quartiles, is one to two orders of magnitudes higher than that of any other trace element including Ni. This leads to wide variation in the V/Ni ratios which are chondritic to mantle-like in most Brazilian oils but rise to crustal levels in the *rj* oils and in most of the major foreign oils we analyzed. Although Ni reaches high levels of about 10,000 ppb in oils of both the aborted Recôncavo rift and the passive margin, the corresponding V contents differ by orders of magnitude. Such wide variation may reflect differences in pH, redox conditions and sulfide activity in the depositional environment of source shales, as proposed by the biogenic-synsedimentary model (Lewan & Maynard, 1982; Lewan, 1984), or high activity of V-bearing organisms in certain marine environments (Curiale, 1987). Alternatively, the variation may be due to differences in the sulfate content of the water infiltrating through the rifted crust and causing serpentinization in shallow mantle peridotites. In the fresh-water Reconcavo rift, where the water available for infiltration into the lithosphere is sulfate-poor and basaltic rocks are absent, the oils are poor in V (about 100 ppb). Conversely, along the passive margin and especially near the southern barrier of the South Atlantic Aptian evaporite basin, where the infiltrating water is sulfate-rich seawater and basalts are abundant, the V content of the oils reaches and exceeds their Ni content, raising V/Ni ratios to crustal and sea water values. Sulfate reduction during serpentinization would decrease the availability of Ni (and Co), but not of V (and Ga, Mo) that do not form sulfides, in the same way as suggested for the synsedimentary availability of these metals by Lewan & Maynard (1982) and Lewan (1984). There appears to exist a worldwide correlation between the V/Ni ratio and the size of petroleum reserves, with the richest oil provinces generally having the highest V/Ni ratio (Persian Gulf, Venezuela, Campos basin in Brazil).

Hydrogen and hydrocarbons form when bivalent Fe in mantle peridotites is oxidized by water in the presence of CO_2. This process may take place both at high (Gold, 1999; Kenney *et al.*, 2002) and low P-T conditions. At high pressures and temperatures water, carbonates, and organic carbon are introduced by subducted slabs into the Earth's interior, whereas at low P-T conditions infiltrating water carrying dissolved organic and inorganic carbon compounds serpentinizes faulted and partially unroofed subcontinental mantle peridotites. In this paper we explored this second process. Hydrothermal plumes bearing aliphatic hydrocarbons and rising over serpentinizing peridotites in the Rainbow field at the Mid-Atlantic ridge (Holm & Charlou, 2001) and aliphatic and cyclic hydrocarbons contained in the serpentinized material of the Tagish Lake meteorite (Pizzarello *et al.*, 2001) may serve as possible analogs.

Heteroatomic organic compounds introduced with the infiltrating water into or forming in the hydrogen-rich environment of serpentinization may take up transition metals and rare earth elements from peridotites in approximately mantle-like proportions (Szatmari *et al.*, 2000, 2002). Metals in organic compounds and in ionic solution rise with the hydrogen- and hydrocarbon-bearing hydrothermal fluids along faults from the serpentinizing peridotites into the sedimentary sequence, where they may enter petroleum reservoirs, become partially adsorbed by clay-rich shales during diagenesis, or seep to the surface. The details and partition coefficients of such metal transport are still poorly known.

Thus an internally consistent model can be conceived, deriving a major part of trace metals in the oils from mantle peridotites reacting with water, either at lower temperatures during serpentinization, or in the deeper mantle. The internal coherence of the model needs to be weighed carefully against the enormous database of the standard model which favors a sedimentary origin for both metals and hydrocarbons in petroleum. Systematic analysis of metals in algal-bacterial organic matter (including bacteria living over serpentinizing peridotites), hydrocarbons from the Tagish Lake meteorite (Pizzarello *et al.*, 2001), and hydrocarbon-bearing plumes rising over serpentinizing peridotites near the Mid-Atlantic Ridge (Holm & Charlou, 2001) will help to distinguish contributions from the various sources.

10. Acknowledgements

We thank Petrobras for support and permission to publish.

11. References

Alt, J. C., & Shanks, W. C. III. (1998). Serpentinization processes and microbial sulfate reduction. *Journal of Geophysical Research*, 103, 9917.

Anders, E., & Grevesse N., (1989). Abundances of the elements: Meteoritic and solar. *Geochimica Cosmochimica Acta*, 53, 197-214.

Anderson, R.T., Chapelle, F.H. & Lovley, D.R. (1998). Evidence against hydrogen-based microbial ecosystems in basalt aquifers. *Science* 281, 976-977.

Bach, W. & Früh-Green, G.L. (2010) Alteration of the oceanic lithosphere and implications for seafloor processes: *Elements* v. 6(3), 173-178.

Baker, E.W. & Louda, J.W., (1986). Porphyrins in the geological record. In: Johns, R.B. (ed.) *Biological Markers in the Sedimentary Record*, Amsterdam: Elsevier, 125-225.

Ball, J.S., Wenger, W.J., Hyden, H.J., Horr, C.A. & Myers, A.T. (1960). Metal content of twenty-four petroleums: *Journal of Chemical and Engineering Data* 5, 553-557.

Barwise, A.J.G. & Whitehead, E.V. (1983). Fosssil fuel metals. In: Augustithis, S.S. (ed.) *The significance of trace elements in solving petrogenic problems and controversies*. Athens: Theophrastus Publications, 599-643.

Beard, J.S., Fullagar, P.D. & Sinha, A.K. (2002) Gabbroic pegmatite intrusions, Iberia abyssal plain, ODP Leg 173, Site 1070: Magmatism during a transition from non-volcanic rifting to sea-floor spreading. *Journal of Petro*logy 43, 885-905.

Berndt, M. E., D. E. Allen & Seyfried, W. E., Jr. (1996). Reduction of CO_2 during serpentinization of olivine at 300 KC and 500 bar. *Geology* 24, 351-354.

Berthelot, M., (1866). Sur l'origine des carbures et des combustibles minéraux. *Ann. de Chimie et des Physiques 4. ser.* 9, 481-483.

Boillot, G., G., Féraud, M. Recq & Girardeau, J. (1989). Undercrusting by serpentinite beneath rifted margins: *Nature* 341, 523-525.

Broeker, W. S., & Peng, T.-H. (1982). *Tracers in the Sea.* Eldigio Press.

Brown, P. G.,Hildebrand, A. R., Zolensky, M. E., Grady, M., et al., (2000) The Fall, Recovery, Orbit, and Composition of the Tagish Lake Meteorite: A New Type of Carbonaceous Chondrite. *Science,* 290, 320-325.

Brumsack, H,.-J. (2003). Major and minor element geochemistry of black shales from the Cenomanian Turonian Boundary Event (CTBE). *Paleoenvironmental constraints. Ocean Drilling Project, Contributions,* University of Oldenburg, Institute of Microbiochemistry http://eagle.icbm.uni-oldenburg.de

Cainelli, C. & Mohriak, W.U.(1998). Geology of Atlantic Eastern Brazilian Basins (Short Course, Coordinator M. Carminatti): *AAPG International Conference & Exhi*bition, Rio de Janeiro, Brazil, 67 p.

Chapelle, F.H., O'Neill, K., Bradley P.M., Methé, B.A., Clufo, S.A., Knobel, L.L. &Lovley, D.R. (2002). A hydrogen-based subsurface microbial community dominated by methanogens. *Nature* 415, 312-315.

Charlou, J. L., Fouquet, Bougalt, Y. H., Donval, J. P., Etoubleau, J., Ph.Jean-Baptiste, Ph., Dapoigny, A., Appriou, P. & Rona, P. S. (1998). Intense CH_4 plumes generated by serpentinization of ultramafic rocks at the intersection of the 15°20'N fracture zone and the Mid-Atlantic Ridge: *Geochimica Cosmochimica* Acta 62, 2323-2333.

Curiale, J.A., (1987). Distribution and occurrence of metals in heavy crude oils and solid bitumens implications for petroleum exploration. In: Meyer, R.F. (ed.) *Exploration for heavy crude oil and natural bitumen. AAPG Studies in Geology* 25, 207-219.

Curiale, J.A. (1993). Occurrence and significance of metals in solid bitumens: an organic geochemical approach. In: Parnell, J., Kucha, H., & Landais, P. (eds.) *Bitumens in ore deposits.* New York: Springer-Verlag, 461- 474.

Dahl, J. E., Moldowan, J. M., Peters K. E. et al., (1999). Diamondoid hydrocarbons as indicators of natural oil cracking. *Nature* 399, 54-56.

Fedikow, M.A.F., Bezys, R.K., Bamburak, J.D., Conley, G.G. & Garrett, R.G. (1998). Geochemical database for Phanerozoic black shales in Manitoba. *Manitoba Energy and Mines, Geological Services, Open File Report OF98-2,* 124 p. and one diskette.

Filby, R. H., (1975). The nature of metals in petroleum. In: Yen, T. F. (ed.) *The Role of Trace Metals in Petroleum,* Ann Arbor: Ann Arbor Science, 31-58

Filby, R. H., (1994). Origin and nature of trace element species in crude oils, bitumens and kerogens: implications for correlation and other geochemical studies. In: Parnell, J. (ed.) *Geofluids: Origin, Migration and Evolution of Fluids in Sedimentary Basins.* Geological Society of London Special Publications 78, 203-219.

Filby, R. H., & Van Berkel, G. J. (1987). Geochemistry of metal complexes in petroleum, source rocks and coals: an overview. In: Filby R. H. & Branthaver, J. F. (eds.) *Metal Complexes in Fossil Fuels: Geochemistry, Characterization and Processing,* American Chemical Society, 2-39.

Filby, R. H., & S. D. J. Olsen (1994). A comparison of instrumental neutron activation analysis and inductively coupled plasma-mass spectrometry for trace element determination in petroleum geochemistry. *Radioanaytical and Nuclear Chemistry*, 80, 285-294.

Filby, R.H. & Shah, K.R., (1975). Neutron activation methods for trace elements in crude oils. In: Yen, T.F. (ed.) *The Role of Trace Metals in Petroleum.* Ann Arbor: Ann Arbor Science Publishers, 89-110.

Fischer, Franz (1925). *The conversion of coal into oils,* Ernest Benn, London.

Fonseca, T. C. Oliveira da (2000). Caracterização inorgânica dos petróleos brasilieros pela técnica de ICP-MS para fins de exploração geoquímica, *Ph.D. thesis, P.U.C.*, Rio de Janeiro, 135 p.

Foustoukos, D.I. & William E. Seyfried, W.E.,Jr.(2004). Hydrocarbons in hydrothermal vent fluids: The role of chromium-bearing catalysts. *Science* 304, 1002-1005.

Früh-Green, G.L., Kelley, D.S., Bernasconi, M., Karson, J.A., Ludwig, K.A., Butterfield, D.A., Boschi, C., Proskurowski, G. (2003). 30,000 years of hydrothermal activity at the Lost City vent field. *Science* 301, 495-498.

Gold, T., (1999). *The deep hot biosphere.* New York: Springer.

Govindaraju, K. (1982) Report (1967–1981) on four ANRT rock reference samples: diorite DR-N, serpentinite UB-N, bauxite BX-N and disthene DT-N. *Geostandard Newsletter* **6,** 91–159.

Govindaraju, K. (1995). 1994 compilation of working values and sample descriptions for 383 geostandards. *Geostandards Newsletter Special Issue* 19, 1-32.

Granch, R.I. & Huyck, H.L.O. (eds.) (1989). *Metalliferous Black Shales and Related Ore Deposits.* United States Geological Survey, Circular 1058.

Hitchon, B., Filby, R.H., & Shah, K. (1975).Geochemistry of trace elements in crude oils, Alberta, Canada. In: Yen, T.F. (ed.) *The Role of Trace Metals in Petroleum.* Ann Arbor: Ann Arbor Science Publishers, 111-121.

Holm, N.G., & Charlou, J. (2001). Initial indications of abiotic formation of hydrocarbons in the Rainbow ultramafic hydrothermal system, Mid-Atlantic Ridge. *Earth and Planetary Science Letters* 191, 1-8.

Horita, J., & M.E. Berndt (1999). Abiogenic methane formation and isotopic fractionation under hydrothermal conditions. *Science* 285, 1055-1057.

Janecky, D. R., & W. E. Seyfried (1986). Hydrothermal serpentinization of peridotite within the continental crust: Experimental investigations of mineralogy and major element chemistry. *Geochimica Cosmochimica Acta* 50, 1357-1378.

Jones, P., (1977) The presence of trace elements in crude oils and allied substances. *Forum on Oil and Ore in Sediments, Proceedings,* London: Imperial College 40-53.

Katz, B. J., & M. R. Mello (2000). Petroleum systems of South Atlantic marginal basins—an overview. In: M. R. Mello & B. Katz (eds.) *Petroleum systems of South Atlantic margins, AAPG Memoir,* 73, 1-14.

Kelley, D.S. et al., (2005).A serpentinite-hosted ecosystem: The Lost City hydrothermal field, *Science* 307, 1428-1444.

Kenney, J.F., Kutcherov, V.A., Bendeliani, N.A., & Alekseev, V.A. (2002). The evolution of multicomponent systems at high pressures: VI. The thermodynamic stability of the hydrogen-carbon system, the genesis of hydrocarbons, and the origin of petroleum. *Proceedings of the National Academy of Science*, PNAS 2002 99: 10976-10981; published online before print as 10.1073/pnas.172376899

Kevin K. Roe, Geoff T. Lebon, Pete Rivizzigno & the AT3-60 Shipboard Party (2001). An off-axis hydrothermal vent field near the Mid-Atlantic Ridge at 30° N. *Nature* 412, 145-149.

Lancet, M., & E. Anders (1970). Carbon isotope fractionation in the Fischer-Tropsch synthesis and in meteorites: *Science*, 170, 980-982.

Lefebure, D.V. & Coveney, R..M. Jr.(1995). Shale-hosted Ni-Zn-Mo-PGE. In: Selected British Columbia Mineral Deposit Profiles, Volume 1 - Metallics and Coal,

Lefebure, D.V. & Ray, G.E. (eds.) *British Columbia Ministry of Energy of Employment and Investment*, Open File 1995-20, 45-48.

Lewan, M.D. (1997) Experiments on the role of water in petroleum formation. Geochimica Cosmochimica Acta 61, 3691-3723.

Lewan, M. D. (1984). Factors controlling the proportionality of vanadium to nickel; ratios in crude oils. *Geochimica Cosmochimica Acta*, 48, 2231-2238.

Lewan, M. D., & Maynard, J. B. (1982). Factors controlling enrichment of vanadium and nickel in the bitumen of organic sedimentary rocks: *Geochimica Cosmochimica Acta*, 46, 2547-2560.

Louda, J.W. & Baker, E.W. (1981). Geochemistry of tetrapyrrole, carotenoid and perylene pigments in sediments from the San Miguel Gap (Site 467) and Baja California Borderland (Site 471). In: *Initial Reports of the Deep Sea Drilling Project*, 63, 785- 818.

Manatschal G. (2004). New models for evolution of magma-poor rifted margins based on a review of data and concepts from West Iberia and the Alps. *Int. J. Earth Sci.*, 93, 432-466.

McCollom, T. M., & Seewald, J. S. (2001). A reassessment of the potential for reduction of dissolved CO_2 to hydrocarbons during serpentinization of olivine: *Geochimica Cosmochimica Acta* 65, 3769-3778.

McDonough, W.F., & Sun, S.-S. (1995). The composition of the Earth. *Chemical Geology* 120, 223-253.

Meisel, T. Reisberg, L., Moser, J., Carignan, J., Melcher, F. & Brügman, G. (2003). Re–Os systematics of UB-N, a serpentinized peridotite reference material, *Chemical Geology*, 201, 161-179.

Milani, E.J. & Zalan, P.V. (1998).The geology of Paleozoic cratonic basins and Mesozoic interior rifts of Brazil. (Short Course, Coordinator M. Carminatti). *AAPG International Conference & Exhibition*, Rio de Janeiro, Brazil, 184 p.

Nakamura et al. (2003) *International Astrobiology* 1, 179.

Neal, C., & Stanger, G. (1983). Hydrogen generation from mantle source rocks in Oman: *Earth Planetary Science Letters* 66, 315-320.

O'Hanley, D., (1996) *Serpentinites: Records of Tectonic and Petrologic History*: Oxford University Press.

Pajonk, G. M., & Teichner, S. J. (1986). Carrier effect on hydrogenation properties of Metals. In: L. Cerveny (ed.) *Catalytic Hydrogenation*. Amsterdam: Elsevier 277-304.

Pizzarello, S., Huang, Y., Becker, L., Poreda, R. J., Nieman, R. A., Coopeer, G. & Williams, M. (2001). *Science* 293, 2236-2239.

Proskurowski, G., Marvin D. Lilley,M.D., Seewald, J.S., Früh-Green, G.L. Olson, E.J., Lupton, J.E., S.P. and Kelley, D.S. (2008) Abiogenic hydrocarbon production at Lost City hydrothermal field: *Science* 319, no. 5863, pp. 604 - 607.

Quinby-Hunt, M. S., & Turekian, K. K. (1983). Distribution of elements in sea water: *EOS*, p. 130-132.

Robinson, R., 1963, Duplex origin of petroleum: *Nature*, 199, 113-114.

Rudnick, R. L., & Fountain, D. M. (1995). Nature and composition of the continental crust - a lower crustal perspective: *Review of Geophysics*, 33, 267-309.

Saccocia, P.J., Ding K., Berndt M.E., Seewald J.S. & Seyfried W.E. Jr. (1994). Experimental and theoretical perspectives on crustal alteration at mid-ocean ridges. In: Lentz, D.R. (ed.) *Alteration Processes Associated with Ore-forming Systems.* Geological Association of Canada, Short Course Notes 11, 403-431.

Scambelluri, M., Rampone, E. & Piccardo, G.B. (2001). Fluid and element cycling in subducted serpentinite: a trace-element study of the Erro-Tobbio high-pressure ultramafites (western Alps, NW Italy). *Journal of Petrology* 42, 55-67.

Schiefelbein, C.F., Zumberge, Cameron, N.C., & Brown, S.W. (2000). Geochemical comparison of crude oil along South Atlantic margins. In: M. R. Mello & B. Katz (eds.) *Petroleum systems of South Atlantic margins, AAPG Memoir*, 73, 1-14.

Seewald, J.S. & Seyfried W.E. Jr. (1990). The effect of temperature on heavy and base metal mobility in subseafloor hydrothermal systems: constraints from basalt alteration experiments and field studies. *Earth Planetary Science Letters* 101, 388-403.

Shrenk, M.O., Cimino, P., Kelley, D.S., & Baross, J.A. (2002). Microbial diversity of carbonate chimneys at the Lost City Hydrothermal Field: Implications for life-sustaining systems in peridotite seafloor environments. *American Geophysical Union, Fall Meeting* Abstract #B71B-0742.

Skelton, A., Koyi, H., Whitmarsh, R., & Arghe, F. (2003). Extent, rate, timing and implications of serpentinization at the zone of exhumed continental mantle, west of Iberia. *EGS - AGU - EUG Joint Assembly*, Nice, France, Abstract #5038.

Skelton A.D.L. & Valley, J.W. (2000). *Earth Planetary Science Letters* 178, 327-338.

Speight, J.G. & Long, R.G. (1996). The concept of asphaltenes revisited. Fuel Sci. & Tech. Int. 14, 1-12.

Stevens, T.O. & McKinley, J.P. (1995). Lithoautotrophic microbial ecosystems in deep basalt aquifers. *Science* 270, 450-454

Strong, D. & R.H. Filby, (1987). Vanadylporphyrin distribution in the Alberta oil-sand bitumens. In: R.H. Filby & J.F. Branthaver (eds) *Metal Complexes in Fossil Fuels: Geochemistry, Characterization and Processing.* American Chemical Society, 154-172.

Szatmari, P. (1989). Petroleum formation by Fischer-Tropsch synthesis in plate tectonics. *AAPG Bulletin* 73, 989-998.

Szatmari, P. (2000). Habitat of petroleum along South Atlantic margins. In: Mello, M.R. & Katz B. (eds.) *Petroleum systems of South Atlantic margins.* AAPG Memoir 73, 69- 76.

Szatmari, P., T. C. Oliveira da Fonseca, N. F. Miekeley (2000). Trace metals in petroleum derived from serpentinizing peridotites. *Geological Society of America Annual Meeting,* Reno, Nevada, 9-18 Nov 2000.

Szatmari, P., T. C. Oliveira da Fonseca, N. F. Miekeley (2002). Petroleum formation during serpentinization: the evidence of trace elements: *American Geophysical Union Spring Meeting,* Washington, 13-15 May, 2002.

Szatmari, P., T.C. Oliveira da Fonseca, N.F. Miekeley (2005) Trace element evidence for major contribution to commercial oils by serpentinizing mantle peridotites. AAPG Research Conference "Origin of Petroleum—Biogenic and/or abiogenic and its significance in hydrocarbon exploration and production", Calgary, Canada.

Takahiro, H., Zolensky, M. E., Pieters, C. M. (2000) The Tagish Lake meteorite: A possible sample from a D-type asteroid. *Science* 293, 2234-2236.

Tissot B.P. & Welte, D.H. (1984) *Petroleum formation and occurrence - second revised and enlarged edition.* Berlin: Springer Verlag, 699 p.

Valkovic, V. (1978) *Trace elements in petroleum.* Tulsa: Petroleum Publishing, 269 p.

Treibs, A., 1934, 1935, Chlorophyll- und Häminderivate in bituminosen Gesteinen, Erdöl, Erdwachsen und Asphalten, Annalen der Chemie 509, 103-114 (1934); 510, 42- 62 (1934); 517, 172-196 (1935).

United States Geological Survey Central Region Mineral Resources Team (2003) URL: http://minerals.cr.usgs.gov/projects/brines/task2.html

Wallendahl, A. & Treiman, A.H. (1999).Geochemical models of low-temperature alteration of Martian rocks. [Abstract #1268] Lunar and Planetary Science Conference *XXX*, LPI Contribution No. 964, Lunar and Planetary Institute, Houston (CD-ROM).

Wedepohl, H., (1995). The composition of the continental crust: *Geochimica Cosmochimica Acta* 59, 1217-1239.

Whitmarsh, R. B., M. O. Beslier, P. J. Wallace *et al.* (1998). *Proceedings of the Ocean Drilling Program, Initial Reports* 173. College Station, Texas.

Whitmarsh, R.B., Manatschal G. & Minshull, T.B. (2001). *Nature* 413, 150-154.

Wignall, P.B. (1994). *Black shales.* Geology and Geophysics Monographs 30. Oxford: Oxford University Press, 130 p.

Yen, T.F. (1975). Chemical aspects of metals in native petroleum. In: Yen, T.F. (ed.) *The Role of Trace Metals in Petroleum.* Ann Arbor: Ann Arbor Science Publishers, 1-30.

Zalán, P. et al. (2009) Stretching and Thinning of the Upper Lithosphere and Continental-Oceanic Transition in Southeastern Brazil, AAPG ICE Rio de Janeiro, Brazil.